FROM HABITABILITY TO LIFE ON MARS

FROM HABITABILITY TO LIFE ON MARS

Edited by

NATHALIE A. CABROL

EDMOND A. GRIN

Elsevier
Radarweg 29, PO Box 211, 1000 AE Amsterdam, Netherlands
The Boulevard, Langford Lane, Kidlington, Oxford OX5 1GB, United Kingdom
50 Hampshire Street, 5th Floor, Cambridge, MA 02139, United States

Notices
Knowledge and best practice in this field are constantly changing. As new research and experience
broaden our understanding, changes in research methods, professional practices, or medical treatment may
become necessary.

Practitioners and researchers must always rely on their own experience and knowledge in evaluating and
using any information, methods, compounds, or experiments described herein. In using such information
or methods they should be mindful of their own safety and the safety of others, including parties for
whom they have a professional responsibility.

To the fullest extent of the law, neither the Publisher nor the authors, contributors, or editors, assume
any liability for any injury and/or damage to persons or property as a matter of products liability, negligence
or otherwise, or from any use or operation of any methods, products, instructions, or ideas contained in
the material herein.

Library of Congress Cataloging-in-Publication Data
A catalog record for this book is available from the Library of Congress

British Library Cataloguing-in-Publication Data
A catalogue record for this book is available from the British Library

ISBN: 978-0-12-809935-3

For information on all Elsevier publications
visit our website at https://www.elsevier.com/books-and-journals

Working together
to grow libraries in
developing countries

www.elsevier.com • www.bookaid.org

Publisher: Candice Janco
Acquisition Editor: Marisa LaFleur
Editorial Project Manager: Katerina Zaliva
Production Project Manager: Prem Kumar Kaliamoorthi
Cover Designer: Mark Rogers

Typeset by SPi Global, India

CONTENTS

7. Siliceous Hot Spring Deposits: Why They Remain Key Astrobiological Targets   179

Sherry L. Cady, John R. Skok, Virginia G. Gulick, Jeff A. Berger, Nancy W. Hinman

8. Habitability and Biomarker Preservation in the Martian Near-Surface Radiation Environment   211

Luis Teodoro, Alfonso Davila, Richard C. Elphic, David Hamilton,
Christopher McKay, Richard Quinn

9. UV and Life Adaptation Potential on Early Mars: Lessons From Extreme Terrestrial Analogs   233

Donat-Peter Häder, Nathalie A. Cabrol

LIST OF CONTRIBUTORS

Abigail C. Allwood
Jet Propulsion Laboratory, California Institute of Technology, Pasadena, CA, United States

Raymond E. Arvidson
Department of Earth and Planetary Sciences, McDonnell Center for the Space Sciences, Washington University in Saint Louis, Saint Louis, MO, United States

Pietro Baglioni
ESA/ESTEC; ExoMars Team, Noordwijk, The Netherlands

David Beaty
Jet Propulsion Laboratory, California Institute of Technology, Pasadena, CA, United States

Luther W. Beegle
Jet Propulsion Laboratory, California Institute of Technology, Pasadena, CA, United States

Jeff A. Berger
Department of Physics, University of Guelph, Guelph, ON, Canada

Rohit Bhartia
Jet Propulsion Laboratory, California Institute of Technology, Pasadena, CA, United States

Jean-Pierre Bibring
Institut d'Astrophysique Spatiale, Orsay, France

Janice L. Bishop
SETI Institute, Carl Sagan Center; SETI Institute NAI Team, Mountain View, CA, United States

André Brack
Centre de biophysique moléculaire, CNRS, Rue Charles Sadron, Orléans, France

William Brinckerhoff
NASA Goddard Space Flight Center, Greenbelt, MD, United States

Adrian J. Brown
NASA Headquarters, Washington, DC, United States

Nathalie A. Cabrol
SETI Institute Carl Sagan Center; SETI Institute NAI Team, Mountain View, CA, United States

Sherry L. Cady
Environmental Molecular Sciences Laboratory, Pacific Northwest National Lab, Richland; SETI Institute NAI Team, Mountain View, CA, United States

Jeffrey G. Catalano
Department of Earth and Planetary Sciences, McDonnell Center for the Space Sciences, Washington University in Saint Louis, Saint Louis, MO, United States

Valérie Ciarletti
LATMOS/IPSL, UVSQ Université Paris-Saclay, UPMC Université Paris 06, CNRS, Guyancourt, France

Andrew J. Coates
Mullard Space Science Laboratory, University College London, United Kingdom

Alfonso Davila
NASA Ames, Mountain View, CA, United States

M. Cristina De Sanctis
Istituto di Astrofisica e Planetologia Spaziali INAF, Roma, Italy

Richard C. Elphic
NASA Ames, Mountain View, CA, United States

Kenneth A. Farley
Division of Geologic and Planetary Sciences, California Institute of Technology, Pasadena, CA, United States

Jack D. Farmer
School of Earth and Space Exploration, Arizona State University, Tempe, AZ, United States

David T. Flannery
Planetary Science Section, NASA Jet Propulsion Laboratory, Pasadena, CA, United States

Fred Goesmann
Max-Planck-Institut für Sonnensystemforschung, Göttingen, Germany

Edmond A. Grin
SETI Institute Carl Sagan Center, Mountain View, CA, United States; SETI Institute NAI Team, Mountain View, CA, United States

Virginia G. Gulick
SETI Institute Carl Sagan Center, Mountain View, CA, United States. NASA Ames Research Center and SETI Institute NAI Team

Donat-Peter Häder
Department of Biology, Emeritus from Friedrich-Alexander University, Möhrendorf, Germany

David Hamilton
University of Glasgow, Glasgow, United Kingdom

Svein-Erik Hamran
Faculty of Mathematics and Natural Sciences, University of Oslo, Oslo, Norway

Michael H. Hecht
Massachusetts Institute of Technology Haystack Observatory, Westford, MA, United States

Nancy W. Hinman
College of Humanity and Sciences, University of Montana, Missoula, MT; SETI Institute NAI Team, Mountain View, CA, United States

Joel A. Hurowitz
Department of Geosciences, Stony Brook University, Stony Brook, NY, United States

Ralf Jaumann
DLR Institut für Planetenforschung, Berlin, Germany

Jean-Luc Josset
SPACE-X, Space Exploration Institute, Neuchâtel, Switzerland

Manuel de la Torre Juarez
Jet Propulsion Laboratory, California Institute of Technology, Pasadena, CA, United States

Gerhard Kminek
ESA/ESTEC, Noordwijk, The Netherlands

Oleg Korablev
Space Research Institute of the Russian Academy of Sciences (IKI), Moscow, Russia

Sylvestre Maurice
Institute de Recherche en Astrophysique et Planétologie, Toulouse, France

Alfred S. McEwen
Lunar and Planetary Lab, University of Arizona, Tucson, AZ, United States

Christopher McKay
NASA Ames, Mountain View, CA, United States

Sarah Milkovich
Jet Propulsion Laboratory, California Institute of Technology, Pasadena, CA, United States

Igor Mitrofanov
Space Research Institute of the Russian Academy of Sciences (IKI), Moscow, Russia

Jeffrey Moersch
University of Tennessee, Department of Earth and Planetary Sciences, Knoxville, TN; SETI Institute NAI Team, Mountain View, CA, United States

Nora Noffke
Old Dominion University, Ocean, Earth & Atmospheric Sciences, Norfolk, VA; SETI Institute NAI Team, Mountain View, CA, United States

Cynthia Phillips
Jet Propulsion Lab, Caltech, CA, United States; SETI Institute NAI Team, Mountain View, CA, United States

Richard Quinn
NASA Ames, Mountain View, CA, United States

François Raulin
Laboratoire Interuniversitaire des Systèmes Atmosphériques (LISA), Université Paris-Est Créteil, Paris, France

Daniel Rodionov
Space Research Institute of the Russian Academy of Sciences (IKI), Moscow, Russia

Jose A. Rodriguez-Manfredi
Department of Instrumentation, Centro de Astrobiología (INTA-CSIC), Madrid, Spain

Fernando Rull
Unidad Asociada UVA-CSIC, Universidad de Valladolid, Spain

Elliot Sefton-Nash
ESA/ESTEC, Noordwijk, The Netherlands

John R. Skok
SETI Institute Carl Sagan Center, Mountain View, CA, United States

Pablo Sobron
SETI Institute Carl Sagan Center; SETI Institute NAI Team, Mountain View, CA, United States

Kathryn M. Stack
Jet Propulsion Laboratory, California Institute of Technology, Pasadena, CA, United States

David Summers
SETI Institute Carl Sagan Center; SETI Institute NAI Team, Mountain View, CA, United States

Roger E. Summons
Department of Earth, Atmospheric, and Planetary Sciences, Massachusetts Institute of Technology, Cambridge, MA, United States

Håkan Svedhem
ESA/ESTEC, Noordwijk, The Netherlands

Luis Teodoro
BAERI, NASA Ames, Mountain View, CA, United States

Jorge L. Vago
ESA/ESTEC, Noordwijk, The Netherlands

Malcolm R. Walter
School of Biological, Earth and Environmental Sciences, University of New South Wales, Sydney, Australia

Kimberley Warren-Rhodes
SETI Institute Carl Sagan Center, SETI Institute NAI Team, Mountain View, CA, United States

Frances Westall
CNRS-OSUC-, Centre, de Biophysique Moléculaire, Orléans, France

David S. Wettergreen
The Robotics Institute, Carnegie Mellon University, Pittsburgh, PA; SETI Institute NAI Team, Mountain View, CA, United States

Roger C. Wiens
Los Alamos National Laboratory, Los Alamos, NM, United States

Kenneth H. Williford
Jet Propulsion Laboratory, California Institute of Technology, Pasadena, CA, United States

Diane Winter
Algal Analysis, LLC, Missoula, MT, United States

Pierre Zippi
Biostratigraphy.com, LLC, Garland; Department of Earth Science, Southern Methodist University, Dallas, TX, United States

FOREWORD

My life and the US space program changed on August 7, 1996. That was the day a group of scientists in the NASA Headquarters auditorium in Washington, District of Columbia, announced they had found evidence of life in a meteorite from Mars that landed on Earth roughly 13,000 years earlier. That meteorite was the first recovered in a 1984 collecting expedition to Antarctica's Allan Hills, hence its designation, ALH84001. On that summer day in 1996, I had been working at NASA Headquarters for eight years. I watched the press event from a windowless conference room because the auditorium was filled with television cameras and folks higher than me on the NASA food chain. As of that day I had never taken a biology course—not even in high school. But I knew that working on the questions raised by the announcement, particularly *Has there ever been life on Mars and even beyond?*, was going to be too exciting not to be a part of. So I embarked on a path of learning and practice that led to the Directorship of the NASA Astrobiology Institute (NAI) and two decades of immersion in the quest to understand the potential of the universe to harbor life beyond Earth.

The changes to the space program that began that day have led to countless outcomes, this book among them. For the Space Summit that President Clinton called for that afternoon as he left the White House for California resulted in NASA's Origins Program with its doubling of funding for Mars exploration and its creation of the NAI. And after his arrival in California, speaking to students and educators at the John Muir Middle School in San Jose, he linked the announcement to education.

> *…what it says is that…if we can nurture scientific interest and capacity in our young people…they will be able to do work and discover things that we have not imagined yet.*

And thus began a journey of exploration and inspiration—personal, national, and international—that continues to this day. It really doesn't matter that the scientific community reached a consensus (with some notable hold-outs) that there was no evidence of life in ALH84001. What matters is that it raised the consciousness of scientists and politicians and the public alike to the potential for life beyond Earth and the tools within our grasp to evaluate that potential.

As we progress on this journey, our thinking evolves. The titular transition of this volume—*From Habitability to Life*—represents such an evolution. We have thought of a planet's potential for life in terms of its habitability. But as one of the editors of this book has pointed out,[1] habitability is a generality. Life exists in *habitats*. And so we must think in terms of habitats and ecosystems as we explore for life beyond Earth.

---

[1] Cabrol, N.A., 2018. The coevolution of life and environment on mars: an ecosystem perspective on the robotic exploration of biosignatures. Astrobiology 18, 1–27.

From Habitability to Life can also represent the evolution of a planet. But perhaps those planetary states are closer than we have thought. Life may be a planetary-scale phenomenon—something that happens not *on* a planet, but rather *to* a planet.[2] From this perspective, we are asking the question *Was Mars ever alive?* The answer may depend on whether the chemical disequilibrium conditions of early Mars led not only to abiotic processes of free energy dissipation, but to biotic ones as well.[3]

Should we find that biology was involved, it will be one of the most important and profound discoveries in human history. For if we find Martian life that shares a common origin with life on Earth, we will add branches to the tree of life that will illuminate evolution in a way never before possible. And if we find Martian life that originated separately from life on Earth, we will have found the holy grail of astrobiology—a second example of life ($N=2$)—with all its implications for distinguishing what is necessary from what is contingent in life. And if, as we explore Mars, we begin to conclude it has never been alive despite at least transient conditions of habitability, it may affect how we regard the significance of life on our own planet.

So our exploration of Mars and the quest of astrobiology are contributing to society in many ways, with the potential for even greater contributions to come. Readers of and contributors to this volume are joined in a journey of discovery addressing fundamental questions of our existence: *What is the nature of life? How does it arise?* Mars has a role to play in this journey not so much as a destination, but as an intellectual waypoint. A place to stop and consider a particular set of circumstances that may or may not have led to life. Beyond Mars are Europa, Enceladus, Titan, Triton, and countless worlds and moons around other stars. Each presents different circumstances and possibilities. The potential of the universe is vast. The potential for life seems vast as well, but it is only through the journey on which we have embarked that we will come to understand how, and whether, that potential has been realized.

Carl B. Pilcher
Blue Marble Space Institute of Science

[2] Credit for this formulation goes independently to David Grinspoon, Sara Walker, and Eric Smith.
[3] For much more on this point, see Smith, E., Morowitz, H.J., 2016. The Origin and Nature of Life on Earth, Cambridge University Press.

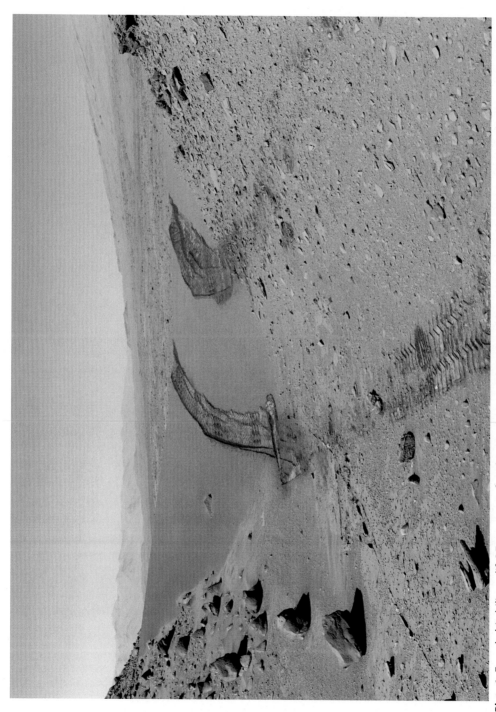

Plate 1 From habitability to life? Forty years of exploration has provided converging evidence that early Mars was habitable for life as we know it. Whether life ever took hold is a question that will be addressed by the next generation of missions, starting with Mars 2020 and ExoMars. Credit image: NASA/JPL–Caltech/MSSS. Dingo Gap captured by the Curiosity rover at Gale crater, Mars.

Plate 2 Elements of Martian habitability. *Top left*—Mastcam image of a section of the Murray Buttes at Gale crater captured by the Curiosity rover on 8 September 2016. The buttes and mesas are eroded remnants of ancient sandstones. Credits: NASA/JPL-Caltech/MSSS. *Top right*—Lava deposits and sequences of sedimentary layered deposits in the background captured in the foothill of Mount Sharp, Gale crater, Mars. Credits: NASA/JPL-Caltech/MSSS. *Middle panel*—The Comanche outcrop holds key mineralogical evidence for an ancient lake in Gusev crater. Image by NASA/JPL-Caltech/ Cornell University: Mars Exploration Rover mission and Spirit rover. *Bottom left*—Opaline silica deposits in a series of depressions in Noctis Labyrinthus as viewed from orbit by HiRISE onboard MRO. The opal could have formed by chemical weathering of basaltic lava flows or ash in the presence water. Credit image: NASA/JPL/University of Arizona. Image ID: ESP_023359_1710. *Bottom right*—Mastcam image showing a pattern typical of a lake-floor sedimentary deposit not far from where flowing water entered a lake. Credits: NASA/JPL-Caltech/MSSS: NASA's Curiosity Mars rover.

CHAPTER 1

Habitability as a Tool in Astrobiological Exploration

Jack D. Farmer

Contents

1.1 OVERVIEW

The exploration strategy currently guiding the search for life in the solar system begins with a short list of the fundamental resources required by terrestrial life. The most important of these requirements is liquid water. Second in importance are sources of biologically essential elements, the so-called CHNOPS elements (required for all living systems), plus a dozen or so transition metals that fulfill important roles in coenzyme functions in cells. Third, life requires sources of energy, obtained through chemical redox reactions. This simple, three-pronged strategy is usually embodied in the phrase "follow the water." This approach to habitability has served astrobiology well, with recent missions reporting evidence for past and present water on Mars and in the subsurface of tidally heated icy moons (e.g., Enceladus and Europa). While the "follow the water" strategy has proved successful in discovering potentially habitable zones (HZs) of liquid water in these and potentially other extraterrestrial environments, understanding the potential for life to actually develop and persist in these places remains unclear. On Earth, habitability depends on the coexistence of three things: liquid water, energy sources and chemical building blocks. However, it is also clear that, on Earth, life exists within a complex web of ecological interactions that, through evolution, have continually reshaped the origin, nature, and distribution of species. Stated differently, life on Earth is a powerful ecological force that itself shapes habitability.

From Habitability to Life on Mars
https://doi.org/10.1016/B978-0-12-809935-3.00002-5

1

As a testimonial to the power of eco-evolutionary forces to shape the history of the biosphere, discoveries of extremophiles have revealed that terrestrial life occupies a much broader range of environmental extremes than once thought possible. On planetary surfaces, organisms have evolved metabolic strategies that extract energy from sunlight, over an impressive range of environmental conditions (e.g., temperature, pH, and water activity). Particularly impactful, however, are life forms that do not require sunlight but rather subsist on chemical energy from their surrounding environment. These "chemo-trophic" microorganisms are part of an extensive subsurface biosphere that resides in deep subsurface habitats on the Earth. Of particular interest for astrobiological exploration are "chemolithoautotrophic" microorganisms that obtain energy from the chemical by-products of aqueous weathering of mafic crustal rocks. These microbes require no connection to the surface and can exist unseen from surface exploration.

While "follow the water" has been an extraordinarily effective strategy in the search for habitable environments in the solar system, astrobiologists will require a more refined approach to exploration during the next phase of exploration when we will begin to target specific habitable sites for deploying a new generation of in situ life detection experiments. Identifying the best sites for in situ experiments or for selecting samples for return to Earth will likely require a more refined knowledge of past and/or present aqueous environments, with an ability to detect HZs at the microscale. Success in detecting extraterrestrial life may require a spatially integrated sampling strategy that includes the ability to measure multiple microenvironmental factors at each study site. Such microscale, multidimensional approaches have been employed in the past by microbial ecologists to refine and quantify biological concepts like the niche. Such approaches may prove useful in more effectively conveying the concept of habitability.

1.2 INTRODUCTION

The concept of "habitability" has emerged as a core principle in NASA's exploration strategy to search for signs of extraterrestrial life in the solar system and beyond. Goal 1 of NASA's Astrobiology Roadmap seeks to "understand the nature and distribution of habitable environments in the universe" (Des Marais et al., 2008). The general concept of habitability has been used widely to identify potential sites for future astrobiology missions. However, definitions of the term are typically very generalized and in some instances downright confusing. Looking beyond simple strategies based on the presence of liquid water, elemental building blocks, and energy sources, at a deeper level, there are environmental factors that, in combination, clearly challenge our assessments of potential habitability. The purpose of this chapter is to (1) broadly review the concept of habitability, particularly as it relates to astrobiology and exploration strategies for extraterrestrial life, and (2) illustrate how the concept of the ecological niche, originally developed by Hutchinson (1957), may provide a multivariate statistical approaches with the potential to contribute more broadly based discussions of habitability.

1.3 DEFINING HABITABILITY

Habitability has been defined as the capability of an environment to support life. Often, the term is used to target specific locations ("habitable zones") within the solar system, on other planets, or moons where extraterrestrial life may be present. The potential for habitability is presently based on comparisons with the known requirements for life on Earth.

1.3.1 Follow the Water

Arguments for the habitability of other planetary bodies in the solar system have focused on the inferred presence of liquid water at the surface, or within the subsurface of a planet or moon. Because all living organisms require liquid water for survival and growth, it is often referred to as a *universal requirement of life*. Given the close relationship between liquid water and life, it is not surprising that the search for water, in all its forms (liquid, ice, or vapor), has provided a consistent strategic focus in NASA's search for past or present HZs in the solar system. This is reflected in the widely embraced mantra "follow the water." Indeed, the effectiveness of this strategy is evident in recent discoveries of water (past and present) by NASA's missions to Mars (e.g., Squyres et al., 2004; Ruff et al., 2011; Grotzinger et al., 2013) and the outer solar system icy moons, Europa (e.g., Roth et al., 2013; Lowell and DuBosse, 2005) and Enceladus (Meyer and Wisdom, 2007; McKay et al., 2014; Glein et al., 2015).

But why water? Water is often identified as a "universal solvent." In large part, this traces to the basic dipolar structure of the water molecule and its ability to form hydrogen bonds with other water molecules and/or cations/anions in solution. In addition to water's extraordinary solvent properties, it easily outcompetes other potential biological solvents in its combination of biologically favorable physical and chemical properties (Table 1.1; Plaxco and Gross, 2011). Water remains liquid over a broad temperature range and has a high molar density, heat capacity, and dielectric constant, all of which can favor habitability. Given its properties, it is perhaps not surprising that liquid water is the required medium for carrying out all of the basic cellular functions of organisms (e.g., energy transduction, reproduction, and locomotion).

However, the recent discovery of hydrocarbon lakes on Titan (Mitri et al., 2007) highlights the importance of considering the potential for alternative solvents for life. A comparison of some of the basic physical and chemical properties of water and other solvents (see Table 1.1) places this in perspective. While a strong case has been made for water as the solvent for life, it is not the only possibility. Some forms of life may have followed different evolutionary pathways involving other hydrocarbon-based solvent systems (e.g., HF or NH_3). It seems prudent to keep this in mind as we think about habitability.

Another property that is important to consider with regard to habitability is the chemical activity of water (A_w). This provides a measure of water's availability to carry out

Table 1.1 Alternative solvents for life based on their common physical and chemical properties
Physical properties of potential biological solvents

Solvent	Formula	Liquid range (°C at 1 atm)	Molar density (mol/L)	Heat capacity (cal/g °C)	Dielectric constant
Water	H_2O	From 0 to +100	55.5	1.0	80
Hydrogen fluoride	HF	From −83 to +19	48.0	0.8	84
Ammonia	NH_3	From −78 to −34	40.0	1.1	25
Hydrogen sulfide	H_2S	From −85 to −6	26.8	0.5	9
Methane	CH_3	From −182 to −161	26.4	0.7	25
Hydrogen	H_2	From −259 to −253	35.0	0.002	1

From Plaxco, K.W., Michael, G., 2011. Astrobiology: A Brief Introduction, second ed. Johns Hopkins University Press, Baltimore. ISBN 978-1-4214-0096-9.

chemical reactions in aqueous solutions. A_w provides a thermodynamic measure of salinity, another variable of common interest in planetary exploration. A_w values range from 0.0 to 1.0, where 1.0 (the value for pure water) indicates that water is 100% available to host reactions. An A_w value of 0.0 indicates that water is unavailable. With minor exceptions, A_w values for life on Earth fall between 1.0 and 0.6 (Beuchat, 1983). A_w has the potential to provide a more sensitive indicator of habitability than just "follow the water" by addressing the specific composition of the water (e.g., hypersaline brine vs fresh water). This was shown in geochemical modeling of Martian water activities by Tosca et al. (2008). This study revealed how the salinity of the water at Meridiani Planum may have exceeded the limit for terrestrial life based on the range of A_w values obtained using models that were constrained by independent in situ measurements of Martian mineralogy.

Astrobiologists are also interested in the search for liquid water in far-flung places beyond the solar system. Astrophysicist Su-Shu Huang (1960) was the first to suggest the presence of an orbital zone around sun-like, main-sequence stars where planetary surface environments are likely to meet two fundamental requirements for the long-term habitability of a planet. These requirements include a sustained energy source (the central star), coexisting with liquid water on the surface of the planet. This orbital region where liquid water is stable at the surface of a planet is referred to as the HZ (Kasting et al., 1993). It is notable that the HZ has also been applied in searching for habitable extrasolar planets in orbit around other stars in the nearby galaxy (Lammer et al., 2016). Near the end of 2013, the Kepler mission team announced that there could be as many as 40 billion

Earth-sized planets orbiting within HZs of sun-like stars and red dwarfs within the Milky Way Galaxy.

The discovery of subsurface zones of liquid water within the interiors of some outer solar system icy moons, maintained by internal frictional heating from tidal flexing (Peale, 2003), has further broadened our view of the HZ to include icy moons of gas giants in the outer solar system (Williams et al., 1997). And with the discovery of a deep, hot biosphere on Earth (Gold, 1999; Farmer, 2000), populated by subsurface thermophilic chemotrophs, an even broader shift in thinking has been required by astrobiologists. This has opened up potentially large subsurface zones of liquid water within the interiors of terrestrial planets and moons. More recently, the interior HZ concept has been expanded further to include subsurface hydrothermal habitats populated by subsurface chemolithoautotrophic organisms that obtain energy by fixing carbon dioxide (CO_2) released by the aqueous weathering of mafic crustal rocks. These organisms convert the CO_2 derived from aqueous weathering to energy-storing glucose (Kuenen, 2009). The impact of these discoveries on potential habitability is noteworthy (Chapelle et al., 2002) and indicates that such ecosystems may survive independently of surface energy sources, with the potential to thrive within extensive, stable subsurface environments that are basically undetected by orbiting spacecraft.

1.3.2 Follow the Bioessential Elements

We now know of many potential destinations in our solar system where zones of liquid water likely exist to support life. And while liquid water is usually regarded as the most crucial requirement of living systems, as noted above, life also requires sources of elements for the synthesis of complex macromolecules that comprise the basic building blocks of living systems. Almost 99% of the human body consists of six elements: oxygen, carbon, hydrogen, nitrogen, calcium, and phosphorus (i.e., the so-called "CHNOPS" elements). In the human body, <1% is accounted for by just five elements: potassium, sulfur, sodium, chlorine, and magnesium. The remaining elements are present in trace amounts. These trace elements are mainly transition metals that include V, Cr, Mo, Mn, Fe, Co, Ni, Cu, Zn, and Se. Despite their low abundance, these trace elements are considered to be micronutrients that fill essential cellular functions, mainly as components of coenzymes.

While they are essential for life, the CHNOPS elements also fulfill many nonbiological roles. Thus, they may provide fewer constraints for inferring habitability. It can be argued that the bioessential trace elements may actually be more sensitive indicators of habitability because they are micronutrients that may be limiting and may require biological concentration mechanisms for their enrichment in organisms. In other words, the trace metals that fulfill vital roles in living systems, while far less abundant, may be more useful as biological indicators than the CHNOPS elements.

1.3.3 Follow the Energy Sources

Life also requires sources of energy to support basic cellular functions. Energy for metabolism is produced through electron transfers during oxidation-reduction reactions. Redox-based energy systems have been discussed by Nealson and Conrad (1999) who concluded that the extreme metabolic plasticity observed among prokaryotes allows them to exploit essentially every redox couple available on Earth.

Metabolic plasticity in energy production is well illustrated by the bacterium *Shewanella putrefaciens* (Nealson and Saffarini, 1994). In the absence of oxygen, this species is able to use a number of alternative electron sources for respiration. That terrestrial organisms obtain energy using transduction systems that share many similarities in common suggests that these pathways have been widely conserved in biological evolution (Hoehler et al., 2007; Thauer et al., 1977).

As noted, metabolic flexibility is a general feature of prokaryotes (see Anderson et al., 2014) and is particularly widespread in all low-oxygen environments. Nealson and Conrad (1999) noted the importance of this "metabolic extremophile" in shaping our approaches to the exploration for extraterrestrial life. Certainly, such examples impact our view of potential habitability by opening up alternative environments (e.g., low oxygen) that, until recently, were not believed possible. This illustrates how the discovery of alternative biochemistries on Earth can provide an important reality check when considering the potential of an extraterrestrial environment to support extreme forms of "weird life" (National Research Council, 2007).

1.4 EXPLORING THE EXTREMES OF LIFE

On Earth, life occupies a broad range of environmental extremes (Table 1.2; see also National Research Council, 2009; Rothschild and Mancinelli, 2001; Plaxco and Gross, 2011). The upper temperature limit for microbial growth is presently reported as $\sim 121^{\circ}$C (Kashefi and Lovley, 2003), with a survival temperature of perhaps 130°C. At $>150^{\circ}$C, complex biomolecules are degraded to their basic components. In between, specialized survival mechanisms, such as heat shock proteins, work to stabilize molecular structures (e.g., Trent et al., 1994).

In moderate-temperature surface environments, where water is lost through high rates of evaporation (e.g., warm playa lakes), the main challenge for life is high salinity and alkalinity. But salinity can also be an important constraint on the habitability of low-temperature environments where interstitial brines form by freeze exclusion. Some groups of psychrophilic halophiles have been shown to grow and reproduce down to -18°C, well below the freezing point of fresh water. They do this by being halophilic and by living within brine films around grains of rock and soil found within permafrost, or in brine-filled microfractures or fluid inclusions in glacial ice (Mazur, 1984).

Table 1.2 Environmental limits for microbial species

Parameter	Classification	Definition	Examples
Temperature	Hyperthermophile Thermophile Psychrophile	>80°C Growth from 60°C to 80°C Growth <15°C Active at −18°C	Archaeal strain 121; 121°C *Pyrolobus fumarii*; ~116°C *Synechococcus lividis*; ~73°C *Psychrobacter* *Himalayan midge*
pH	Acidophile Alkaliphile	Low pH (<5) High pH (>9)	*Ferroplasma acidarmanus* pH 0 *Alkaliphilus transvaalensis*, pH 12.5 *Natronobacterium*; pH 10.5
Salinity	Halophile	2–5 Molar NaCl	Halobacteriaceae
Oxygen tension	Aerobe Microaerophile Anaerobe	Requires O_2 Tolerates some O_2 Not tolerant of O_2	Bacteria, archaea Neutral pH Fe^{2+}-oxidizing bacteria Methanogens, SO_4^{2-} reducers
Dessication	Xerophile	Anhydrobiotic	Lichens, cyanobacteria; arid deserts
Radiation	Radiophile	Ionizing radiation to 15 kGy	*Deinococcus radiodurans*
Pressure	Piezophile	Pressure-loving	Obligate strain MT41
Chemical extremes	Gases Metals	Metalotolerant	*Cyanidium caldarium* *Ferroplasma*

Modified from NAS, 2009.

Psychrophilic organisms have been shown to stabilize bimolecular structures with heat shock proteins that are similar to those employed by hyperthermophiles (Trent et al., 1994). The main problem faced by microorganisms living near the freezing point of such brines is the potential for cell lysis due to the formation of ice crystals (Mazur, 1984).

Even where the minimum requirements for sustaining metabolism, growth, and reproduction are absent, many organisms survive combined extremes of temperature, pH, radiation, and desiccation (including freezing) as endospores, or other resistant structures. In some cases, survival has been documented over extremely long periods of time. Navarro-González et al. (2003) confirmed a lower limit for habitability in the Atacama Desert of Chile where water activities fell below 0.6. Indeed, below this value, living organisms were absent, as well as endospores, suggesting the environmental limits for both cells and endospores had been exceeded.

Life has also been shown to occupy the full range of pH, with the fungus, *Ferroplasma acidarmanus*, growing at a pH of 0.77, in waters that originate from an acid mine drainage (Bond et al., 2000). At the other end of the pH scale, hypersaline alkaline lakes support microbial alkaliphiles at pH values >10. Adaptations for extremes in radiation are seen in terrestrial microbes that have evolved extremely rapid DNA repair systems, such as *Deinococcus radiodurans* (Battista, 1997). This organism lives in association with natural radioactive mineral deposits and on fuel rods in nuclear power plants (Rothschild and Mancinelli, 2001). Rapid repair of DNA basically allows *Deinococcus* to maintain a working genome.

The observations presented support the following conclusions: terrestrial life is extremely robust and has evolved to occupy a broad range of (as yet unknown) environmental and metabolic limits, seemingly constrained by only the presence of liquid water, sources of biogenic elements, and metabolic energy. This perspective has had a profound impact on our perception of habitability, greatly expanding the scope of environmental possibilities for life on Mars and other planets or moons in our solar system. However, as researchers debate the habitability of the ocean beneath the icy crust of Europa, or in the deep subsurface of Mars where liquid, likely hypersaline water, circulates through sediments and cracks in permeable igneous rocks, the first-order question is whether life is permitted to be present at all. To answer, this question requires more specific studies of habitability, in particular, those that can lead to quantification.

1.5 NICHE-BASED MULTIVARIATE APPROACH TO HABITABILITY

The concept of the ecological niche is rooted in a complex and interesting history (see MacArthur, 1968), with early definitions ranging from the niche as the role played by a species in a community, to the type of environment a species occupies. In 1957, immanent ecologist G.E. Hutchinson formulated a new definition using set theory to quantify the concept of the niche (Hutchinson, 1978). In this new definition, the niche was represented as an *n-dimensional hypervolume* in multidimensional biospace, occupied by the species being investigated, including the complete range of environmental factors (dimensions) required for successful reproduction of the population (Green, 1971). In this niche model, the n-dimensional hypervolume occupied by the species, including all of its required resources, was deemed the *fundamental niche* of the species. This approach has achieved popularity among ecologists, primarily because there are so many biologically relevant environmental factors in nature that can be easily measured and quantified (Chase and Leibold, 2003; Austin, 2006).

To further illustrate Hutchinson's niche concept, Fig. 1.1 shows three environmental variables (conditions) displayed in one-, two-, and three-dimensional "biospace." Each dimension of the fundamental niche represents an *independent* environmental variable that is required for the growth and survival of the species. The realized niche is the portion of

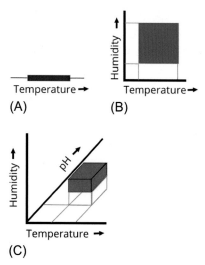

Fig. 1.1 Three environmental variables ("conditions") mapped in (A) one-, (B) two-, and (C) three-dimensional "biospace."

the fundamental niche that is actually occupied by the species in question, whereas the unoccupied part of the hypervolume represents the unrealized portion of the niche (essentially the potential biospace not yet occupied).

As discussed by Guisan et al. (2006), there have been a number of modifications to the Hutchinsonian niche concept, including the development of multivariate statistical approaches (e.g., multiple discriminate analysis; see Green, 1971). Another example uses habitat suitability models (HSMs; Hirzel and Le Lay, 2008) that relate a set of environmental variables to the likelihood of occurrence of a particular species. Results from HSMs are usually displayed as 2D maps in *environmental space* that can be quickly compared with the 2D spatial distribution of a species in *geographic space*. Such comparisons of spatial (geographic) distributions and correlated environmental data can lead to testable hypotheses regarding the nature of species-environment interactions.

Returning to the simplified, unscaled *3D representation* of Hutchinson's niche (Fig. 1.1C), we can see how the hypervolume/biospace approach might be repurposed to represent "habitability" space. For example, we could redefine the three axes (conditions) to represent environmental variables, such as temperature, salinity and water activity, and/or spatial dimensions. To expand the analysis to *n*-dimensional habitability space, we could add the environmental limits for each of the extremophilic species shown in Table 1.1. By including the full range of environmental data for all known microbial species, we could begin to define (qualitatively) the "shape" of *n*-dimensional habitat space for terrestrial extremophiles. We could then add additional dimensions to the distribution (e.g., radiation tolerance, redox, and hydrostatic pressure). The uncolored part of the boxed region in Fig. 1.1A–C could be redefined as "prospective" habitability space. A multidimensional data set for extremophiles may provide new insights into the nature of habitability and how certain groups (e.g., polyextremophiles) have evolved to occupy certain region's habitability space and how the competition for resources might have occurred, driving evolution. Factors selected for analysis could focus more specifically on key environmental dimensions for extremophiles that hold high scientific interest for understanding the limits of habitability on Earth. This information could be used in strategic mission planning.

1.6 CONCLUSIONS

While we acknowledge the recent successes of solar system exploration and extrasolar planet research based on the state and distribution of water, we still lack a system-level understanding of how other factors required by life influence habitability. In articulating a more complete conceptual framework for habitability and one that extends beyond simply "follow the water, elements, and energy," we need to also consider the importance of biological factors that directly contribute to habitability. It seems clear that astrobiological exploration of the solar system has benefitted greatly from basic discoveries in extremophile research, which have significantly expanded the environmental limits of life on Earth. This work has been foundational for astrobiology in expanding our understanding of where extraterrestrial life could exist and how we might be able to detect it. But next steps are needed to further refine our understanding of habitability. On Earth, species and populations are organized into ecosystems made up of interactive networks of individuals that manage energy flow through various ecological interactions. The success of future life detection experiments could ultimately depend on recognizing spatially integrated ecological interactions between species at the microscale, with the capability to quantify associated processes in situ. Multidimensional ecological models like those developed by Hutchison for the niche could provide useful direction to lead us to more focused strategies for astrobiological exploration, including science-driven technology developments that will allow us to identify and measure dimensions of the environment most critical for life.

REFERENCES

Anderson, R.E., Sogin, M., Baross, J.A., 2014. Biogeography and ecology of the rare and abundant microbial lineages in deep-sea hydrothermal vents. FEMS Microbiol. Ecol. 91 (1), 1–11. https://doi.org/10.1093/femsec/fiu01.

Austin, M., 2006. Species distribution models and ecological theory: a critical assessment and some possible new approaches. Ecol. Model. 200, 1–19.

Battista, J.R., 1997. Against all odds: the survival strategies of *Deinococcus radiodurans*. Annu. Rev. Microbiol. 51, 203–224.

Beuchat, L.R., 1983. Influence of water activity on growth, metabolic activities, and survival of yeasts and molds. J. Food Prot. 46, 135–141.

Bond, P.L., Smriga, S.P., Banfield, J.F., 2000. Phylogeny of microorganisms populating a thick, subaerial, predominantly lithotrophic biofilm at an extreme acid mine drainage site. Appl. Environ. Microbiol. 66 (9), 3842–3849.

Chapelle, F.H., O'Neill, K., Bradley, P.M., Methé, B.A., Ciufo, S.A., Knobel, L.R.L., Lovley, D.R., 2002. A hydrogen-based subsurface microbial community dominated by methanogens. Nature 415, 312–315. https://doi.org/10.1038/415312.

Chase, J.M., Leibold, M.A., 2003. Ecological Niches: Linking Classical and Contemporary Approaches. Univ. Chicago Press, Chicago.

Des Marais, D.J., Nuth III, J.A., Allamandola, L.J., Boss, A.P., Farmer, J.D., Hoehler, T.M., Jakosky, B.M., Meadows, V.S., Pohorille, A., Runnegar, B., Spormann, A.M., 2008. The NASA astrobiology roadmap. Astrobiology 8 (4), 715–730. https://doi.org/10.1089/ast.2008.0819.

Farmer, J.D., 2000. Hydrothermal systems: doorways to early biosphere evolution. GSA Today 10 (7), 1–9.

Glein, C.R., Baross, J.A., et al., 2015. The pH of Enceladus' ocean. Geochim. Cosmochim. Acta 162, 202–219.

Gold, T., 1999. The Deep Hot Biosphere. Springer, New York. ISBN 0-387-98546-8.

Green, R.H., 1971. A multivariate statistical approach to the Hutchinsonian niche: bivalve molluscs of central Canada. Ecology 52 (4), 543–566.

Grotzinger, J.P., Sumner, D.Y., Kah, L.C., Stack, K., Gupta, S., Edgar, L., Rubin, D., Lewis, K., Schieber, J., Mangold, N., Milliken, R., Conrad, P.G., Des Marais, D., Farmer, J., et al., 2013. A habitable fluvio-lacustrine environment at Yellowknife Bay, Gale crater. Mars. Sci. 342. https://doi.org/10.1126/science.1242777.

Guisan, A., Broennimann, O., Engler, R., Vust, M., Yoccoz, N.G., Lehmann, A., Zimmermann, N.E., 2006. Using niche-based models to improve the sampling of rare species. Conserv. Biol. 20 (2), 501–511.

Hirzel, A.H., Le Lay, G., 2008. Habitat suitability modeling and niche theory. J. Appl. Ecol. 45, 1372–1381. https://doi.org/10.1111/j.1365-2664.2008.01524.x.

Hoehler, T.M., Amend, J.P., Shock, E.L., 2007. A "follow the energy" approach for astrobiology. Astrobiology 7 (6), 819–823. https://doi.org/10.1089/ast.2007.0207.

Huang, S.-S., 1960. Life-supporting regions in the vicinity of binary systems. Publ. Astron. Soc. Pac. 72 (425), 106–114. https://doi.org/10.1086/127489.

Hutchinson, G.E., 1957. Concluding remarks. Cold Spring Harb. Symp. Quant. Biol. 22, 415–427.

Hutchinson, G.E., 1978. An Introduction to Population Ecology. Yale Univ Press, New Haven, CT.

Kashefi, K., Lovley, D.R., 2003. Extending the upper temperature limit for life. Science 301 (5635), 934. https://doi.org/10.1126/science.1086823.PMID.12920290.

Kasting, J.F., Whitmire, D.P., Reynolds, R.T., 1993. Habitable zones around main sequence stars. Icarus 101 (1), 108–118. https://doi.org/10.1006/icar.1993.1010.

Kuenen, G.W., 2009. Oxidation of inorganic compounds by chemolithotrophs. In: Lengeler, J., Drews, G., Schlegel, H. (Eds.), Biology of the Prokaryotes. John Wiley & Sons, U.S.A. Indianapolis, IN, ISBN 978-0-632-05357, p. 242.

Lammer, H., Bredehöft, J.H., Coustenis, A., Khodachenko, M.L., et al., 2016. What makes a planet habitable? Astron. Astrophys. Rev. 17, 181–249. https://doi.org/10.1007/s00159-009-0019-z.

Lowell, R.P., DuBosse, M., 2005. Hydrothermal systems on Europa. Geophys. Res. Lett. 32(5). https://doi.org/10.1029/2005GL022375.

MacArthur, R.H., 1968. The theory of the niche. In: Lewontin, R.C. (Ed.), Population Biology and Evolution. Syracuse Univ. Press, Syracuse, NY, pp. 159–176.

Mazur, P., 1984. Freezing of living cells: mechanisms and implications. Am. J. Phys. 247 (3 Pt 1), C125–C142.

McKay, C.P., Anbar, A.D., Porco, C., Tsou, P., 2014. Follow the plume: the habitability of Enceladus. Astrobiology 14 (4), 352–355.

Meyer, J., Wisdom, J., 2007. Tidal heating in Enceladus. Icarus 188 (2), 535–539. https://doi.org/10.1016/j.icarus.2007.03.001.

Mitri, G., Showman, A.P., Lunine, J.I., Lorenz, R.D., 2007. Hydrocarbon lakes on titan. Icarus 186 (2), 385–394. https://doi.org/10.1016/j.icarus.2006.09.004.

National Research Council, 2007. The Limits of Organic Life in Planetary Systems. The National Academies Press, Washington, DC. https://doi.org/10.17226/11919.

National Research Council, 2009. Assessment of Planetary Protection Requirements for Mars Sample Return Missions. Space Studies Board, Division on Engineering and Physical Sciences, National Research Council of the National Academies, National Academy Press, Washington, DC, 80 pp.

Navarro-González, R., Rainey, F.A., Molina, P., Bagaley, D.R., Hollen, B.J., de la Rosa, J., Small, A.M., Quinn, R.C., Grunthaner, F.J., Cáceres, L., Gomez-Silva, B., McKay, C.P., 2003. Mars-like soils in the Atacama Desert, Chile, and the dry limit of microbial life. Science 302 (5647), 1018–1021.

Nealson, K.H., Conrad, P.G., 1999. Life: past, present and future. Philos. Trans. R. Soc. Lond. B 354, 1923–1939.

Nealson, K.H., Saffarini, D., 1994. Iron and manganese in anaerobic respiration environmental significance, physiology and regulation. Ann. Rev. Microbiol. 48, 311–343.

Peale, S.J., 2003. Tidally induced volcanism. Celest. Mech. Dyn. Astron. 87, 129–155.

Plaxco, K.W., Gross, M., 2011. Astrobiology: A Brief Introduction, second ed. Johns Hopkins University Press, Baltimore. ISBN 978-1-4214-0096-9.

Roth, L., Saur, J., Retherford, K.D., Strobel, D.F., Feldman, P.D., McGrath, M.A., Nimmo, F., 2013. Transient water vapor at Europa's south pole. Science 343 (6167), 171–174.

Rothschild, L.J., Mancinelli, R.L., 2001. Life in extreme environments. Nature 409, 1092–1101. https://doi.org/10.1038/35059215.

Ruff, S.W., Farmer, J.D., Calvin, W.M., Herkenhoff, K.E., Johnson, J.R., Morris, R.V., Rice, M.S., Arvidson, R.E., Bell III, J.F., Christensen, P.R., Squyres, S.W., 2011. Characteristics, distribution, origin, and significance of opaline silica observed by the Spirit rover in Gusev crater, Mars. J. Geophys. Res. 116, E00F23. https://doi.org/10.1029/2010JE003767.

Squyres, S.W., Arvidson, R.E., Bell III, J.F., Bruckner, J., Cabrol, N.A., Calvin, W., Carr, M.H., Christensen, P.R., Clark, B.C., Crumpler, L., Des Marais, D.J., d'Huston, C., Economou, T., Farmer, J., et al., 2004. The opportunity Rover's Athena science investigation at Meridiani Planum, Mars. Science 306, 1698–1703.

Thauer, R.K., Jungermann, K., Decker, K., 1977. Energy conservation in chemotrophic anaerobic bacteria. Bacteriol. Rev. 41, 100–180.

Tosca, N.J., Knoll, A.H., McLennan, S.M., 2008. Water activity and the challenge for life on early Mars. Science 320 (5880), 1204–1207. https://doi.org/10.1126/science.1155432.

Trent, J.D., Gabrielsen, M., Jensen, B., Neuhard, J., Olsen, J., 1994. Acquired thermotolerance and heat shock proteins in thermophiles from the three phylogenetic domains. J. Bacteriol. 176 (19), 6148–6152.

Williams, D.M., Kasting, J.F., Wade, R.A., 1997. Habitable moons around extrasolar giant planets. Nature 385 (6613), 234–236. https://doi.org/10.1038/385234a0.

FURTHER READING

Stoker, C.R., Zent, A., Catling, D.C., Douglas, S., Marshall, J.R., Archer, D., Clark, B., Kounaves, S.P., Lemmon, M.T., Quinn, R., Renno, N., Smith, P.H., Young, S.M.M., 2010. Habitability of the phoenix landing site. J. Geophys. Res. Planets 115, 1–24.

Westall, F., Loizeau, D., Foucher, F., Bost, N., Betrand, M., Vago, J., Kminek, G., 2013. Habitability on Mars from a microbial point of view. Astrobiology 13, 887–897.

CHAPTER 2

An Origin of Life on Mars?

André Brack

Contents

2.1 OVERVIEW

Mapping of Mars by orbiters revealed significant geologic evidence of past water activity, including enormous outflow channels carved by floods, ancient river valley networks, and lake beds. Further clues of water activity have come from the identification of aqueously altered rocks at the surface and from Martian meteorites. Three possible sources for organics are considered: a primitive atmosphere, hydrothermal systems, and space delivery of meteorites and micrometeorites. So far, the inventory of organic molecules found on Mars is rather poor, mostly in Martian meteorites. The early histories of Mars and Earth clearly show similarities, having in common liquid water and organic material. Chemistry being reproducible, processing identical ingredients should have produced the same effects. There are, however, some limitations: the complexity of the origin of terrestrial life and the three specific features of the Earth (generalized plate tectonics, global magnetic field, and a large moon generating oceanic tides). The discovery of a second independent genesis of life would demonstrate that life is not a magic one-shot process, but probably a rather common phenomenon.

2.2 INTRODUCTION

The upcoming Mars Science Laboratory (Grotzinger et al., 2012; Mahaffy et al., 2015a, b) and ExoMars (Vago et al., 2017) missions are largely motivated by the potential that Mars

From Habitability to Life on Mars
https://doi.org/10.1016/B978-0-12-809935-3.00001-3

harbored life in the past or even at the present. This chapter aims to present the chemical arguments supporting a possible emergence of life on the red planet. When listing the requirements for life to originate on Mars, it is generally assumed that life on Mars could have originated as carbon-based organized molecular systems in water, as did life on Earth. Apart from an anthropomorphically skewed perspective, there are specific reasons to believe that searching for a carbon-based life is the most appropriate.

2.3 LIQUID WATER

As parts of an open system, the constituents of a living system must be able to diffuse at a reasonable rate. Solid-state life is generally discarded, the constituents being unable to migrate and to be easily exchanged. A gaseous phase would allow fast diffusion of the parts, but the limited inventory of stable volatile organic molecules would constitute a severe restriction. A liquid phase offers the most efficient environment for the diffusion and the exchange of dissolved organic molecules.

Liquid water is a fleeting substance that can persist only above 0°C and under a pressure higher than 6 mbar. Salts dissolved in water (brines) depress the freezing point. For instance, the 5.5% (by weight) salinity of the Dead Sea depresses the freezing point of seawater by 2.97°C. Large freezing-point depressions are observed for 15% (wt%) LiCl (23.4°C) and for 22% (wt%) NaCl (19.2°C). Monovalent and divalent salts are essential for terrestrial life because they are required as cocatalysts in many enzymatic activities. Usually, the tolerated salt concentrations are quite low (<0.5%) because high salt concentrations disturb the networks of ionic interactions that shape biopolymers and hold them together. However, extreme halophiles tolerate a wide range of salt concentrations (1%–20%), and some bacteria have even managed to thrive in hypersaline biotopes (salines and salt lakes) up to 25%–30% sodium chloride (Ollivier et al., 1994).

According to its molecular weight, water should be a gas under standard terrestrial conditions by comparison with CO_2, SO_2, and H_2S, for example. Its liquid state is due to the ability of the water molecules to exchange hydrogen bonds. The polymeric network of water molecules via H-bonds is so tight that the boiling point of water is raised from 40°C, a temperature inferred from the boiling point of the smallest alcohols, to 100°C.

In addition to the H-bonding capability, water exhibits a large dipole moment of 1.85 D. This large dipole moment supports the dissociation of ionizable groups such as $-NH_2$ and $-COOH$ leading to ionic groups, which can form additional H-bonds with water molecules, thus improving their solubility. Water is also an outstanding dielectric ($\varepsilon = 80$). When oppositely charged organic groups are formed, their recombination is disadvantaged, the attraction force of recombination being proportional to $1/\varepsilon$. This is also true for mineral ions, which have probably been associated with organic molecules since the beginning of life's history.

Water facilitates the formation of clay minerals via aqueous alteration of silicate minerals. As soon as liquid water appears on the surface of a rocky planet, clay minerals can accumulate and become suspended in putative oceans or lakes. Bernal (1949) listed the advantageous features of clays: (i) their ordered arrangement, (ii) their large adsorption capacity, (iii) their shielding capacity against sunlight, (iv) their ability to concentrate organic chemicals, and (v) their ability to serve as polymerization templates. Since the seminal hypothesis of Bernal, many prebiotic scenarios involving clays have been written (e.g., Cairns-Smith and Hartman, 1986), and numerous prebiotic experiments have used clays. The most impressive results were obtained by Jim Ferris's group (Huang and Ferris, 2007). Oligomers up to 50 monomers long formed for both nucleotides and amino acids in the presence of montmorillonite for nucleotides and of illite for amino acids.

Water is also a good heat dissipator. Hydrothermal vents are often disqualified as efficient reactors for the synthesis of prebiotic molecules because of the high temperature. However, the products that are synthesized in hot vents are rapidly quenched in the surrounding cold water thanks to the good heat conductivity of the milieu (see below the production of hexaglycine under simulated hydrothermal conditions).

Water can also act as a driver for chemistry. Adding the activating agent N,N'-carbonyldiimidazole (CDI) to free amino acids in organic solvent led to aminoacylimidazolides, a carboxylic acid derivative, very sensitive to hydrolysis and therefore useless for making peptides in aqueous phase. Adding the activating agent to amino acid directly in water formed N-imidazoyl-(1)-carbonyl amino acid, an amino activation that generated pure oligo-L-leucines in 70% yield with a molecular weight of eight. Oligomers up to the 11-mer were identified with glutamic acid. A mixture of amino acids containing both protein and nonprotein amino acids, close to that found in the Murchison meteorite, was treated with CDI in water, and the condensate was enriched in protein amino acids (Brack, 1987).

2.4 CARBON CHEMISTRY

Life is autocatalytic in essence and must be able to evolve, that is, to increase its diversity. The molecules bearing the hereditary memory must be able to be extended and diversified by combinatorial dispersive reactions. This can be best achieved with a scaffolding of polyvalent atoms. From a chemical viewpoint, carbon chemistry is, by far, the most productive in this respect. Another clue in favor of carbon is provided by radio astronomers: About 110 carbon-containing molecules have been identified in the interstellar medium, while only 9 silicon-based molecules have been detected. Looking for a carbon-based life appears therefore as a highly plausible prerequisite when searching for life elsewhere.

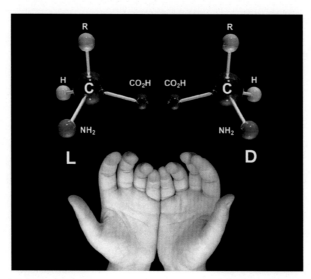

Fig. 2.1 L-form and D-form, mirror image enantiomers of a generic amino acid that is chiral.

Carbon chemistry generates another remarkable feature, namely, one-handedness, also called homochirality (from the Greek kheiros, the hand). Pasteur was probably the first to recognize that homochirality could best distinguish between inanimate matter and life. The carbon atom occupies the center of a tetrahedron. When the four substituents at the apexes are different, the carbon atom becomes asymmetrical and shows two mirror images, a left-handed form and a right-handed form (Fig. 2.1).

Theoretical models show that autocatalytic systems fed with both left- and right-handed molecules must become one-handed in order to perpetuate. The use of one-handed molecules sharpens the sequence information needed to replicate a chain. For a polymer made of n units, the number of sequence combinations will be divided by 2^n when the system uses only one-handed monomers. Considering macromolecules made up of hundreds of monomers, the tremendous gain in simplicity offered by the use of homochiral monomers is self-evident.

2.5 WATER ON MARS

The size of Mars and its distance to the Sun allowed the young planet to be located at the edge of the habitable zone, that is, a zone where water can exist in the liquid phase (Kasting et al., 1993). Determining whether liquid water existed on Mars is central to understanding its potential for the emergence of life. The present-day Martian climate is too cold to be conducive to abundant liquid water, with an average temperature of $-53°C$, and the 6 mb CO_2 atmosphere is too thin for water to be stable at the surface (e.g., Carr and Head, 2015). Mars does nevertheless have water but mostly as vapor in the atmosphere, as polar ice deposits, and as water ice in the subsurface as observed by the SHAllow RADar (SHARAD) instrument on Mars Reconnaissance Orbiter (MRO) (Holt et al., 2008). Epithermal neutron data from the Mars Odyssey Neutron Spectrometer showed hydrogen-rich mineralogy far from the poles, including about 10 wt% water equivalent hydrogen on the flanks of the Tharsis Montes and >40 wt% at the Medusae Fossae Formation, suggesting the presence of bulk water ice (Wilson et al., 2018).

Mars mapping by orbiters (Mariner 9, Viking 1 and 2, Mars Global Surveyor, Mars Odyssey, Mars Express, and Mars Reconnaissance Orbiter) revealed much geologic evidence of past water including outflow channels carved by floods (Craddock and Howard, 2002; Fairén et al., 2003; Rodriguez et al., 2015a), ancient river valley networks as on Fig. 2.2 (Hynek et al., 2010), deltas (di Achille and Hynek, 2010; Salese et al., 2016), and lake beds (Cabrol et al., 1998; Cabrol and Grin, 1999, 2010; Fassett and Head, 2008; Grotzinger et al., 2014; Cardenas et al., 2017). Long-term subsurface water circulation has also been suggested (Ehlmann et al., 2011; Michalski et al., 2013).

Further clues of water activity come from the identification of aqueously altered minerals and rocks at the surface. Hematite has been identified on the surface using thermal emission spectroscopic measurements (Christensen et al., 2000). Hematite is an iron oxide weathering product inferred to have been precipitated from water flowing on the surface or through the crust. The identification of clay minerals with Observatoire pour la Minéralogie, l'Eau, les Glaces et l'Activité (OMEGA) on Mars Express (Poulet et al., 2005) and Compact Reconnaissance Imaging Spectrometer for Mars (CRISM) currently flying on MRO (Murchie et al., 2009) implies that liquid water was present on Mars (Carter et al., 2013). Curiosity's Sample Analysis at Mars (SAM) experiment measured the D/H ratio of Hesperian water within a Yellowknife Bay

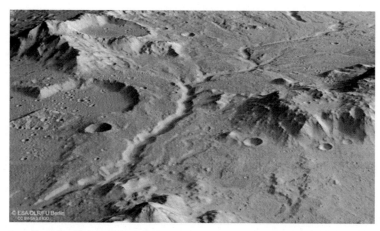

Fig. 2.2 Ancient fluvial valleys in Libya Montes, Mars. Credit: ESA/Mars Express HRSC/DLR/FU Berlin.

mudstone, whose clay minerals were formed in an active lacustrine environment. These measurements helped constrain the volume of water lost through escape processes over the past 3 billion years (Mahaffy et al., 2015a, b).

Hydrated minerals are present in several of the Martian meteorites as well. Prior to the Viking missions, it was shown that SNC meteorites (after their type specimens Shergotty, Nakhla, and Chassigny) had comparatively young crystallization ages of 1.3 Ga or less (Jagoutz and Wänke, 1986). EETA 79001 was found on Antarctica in 1979. The meteorite had gas trapped within glass pockets, which both compositionally and isotopically matched, in all respects, the makeup of the Martian atmosphere as measured by the mass spectrometer used to assess the level of organic compounds present in the soil (Bogard and Johnson, 1983; Becker and Pepin, 1984; Carr et al., 1985). The data provide a very strong argument that at least that particular SNC meteorite came for Mars. There are now 96 Martian meteorites known, and the number continues to increase. It can be demonstrated that they are all related by comparing their oxygen isotopic compositions. Only these specimens (out of a total of >20,000 meteorites) define a correlation line of slope 0.5 on a plot of $\delta^{18}O$ versus $\delta^{17}O$ with $\Delta^{17}O = 0.321 \pm 0.013‰$, that is, displaced from the Earth reference line. Hydrated minerals of clay-type structure have been found in Martian meteorites (Hutchison, 1975; Gooding and Muenow, 1986; Gooding et al., 1988, 1991; Treiman et al., 1993) as well as evidence for subsurface hydrothermal alteration of clay (Chatzitheodoridis et al., 2014).

It remains difficult to estimate the total amount of water that may have existed on the surface of Mars. The estimated depth ranges from a few meters to several hundred meters. Observations at powerful ground-based observatories, Very Large Telescope (VLT), Keck, and InfraRed Telescope Facility (IRTF), of atmospheric water and its deuterated form (HDO) across the Martian globe showed strong isotopic anomalies, and a significant deuterium/hydrogen (D/H) enrichment indicative of great water loss and also that of early Mars (4.5 billion years ago) had a global equivalent layer (GEL) at least 137 m deep (Villanueva et al., 2015). Geomorphological records on Mars do indicate a wetter past, up to >2000 m GEL. Yet, as summarized by Carr and Head (2003) and confirmed by subsurface observations with the Mars Advanced Radar for Subsurface and Ionosphere Sounding (Mouginot et al., 2012), the best estimate is provided by the Vastitas Borealis Formation, which implies a 156 m GEL, in relatively good agreement with the 137 m GEL inferred from the D/H enrichment. The density profiles of water and water ions in the ionosphere/thermosphere have been computed. They are in fairly good agreement with the measured values by the Neutral Gas and Ion Mass Spectrometer instrument on the Mars Atmosphere and Volatile EvolutioN (MAVEN) spacecraft (Fox et al., 2015).

2.6 THE TIMING OF AQUATIC HABITATS

Several features suggest that aquatic habitats existed during the late Noachian about 3.8 billion years ago (Grotzinger et al., 2014), at the end of the Hesperian about 3 billion years ago (McKay and Davis, 1991), and during the Early Amazonian (Fairén et al., 2009; Rodriguez et al., 2015b), although the presence of liquid water on the surface of Mars at any time is not universally agreed upon (e.g., Hynek, 2016; Wordsworth, 2016). At that time, most of the surface of Mars was emerged land, even when liquid water was abundant (Carr and Head, 2015).

Even some of the youngest features on Mars appear to show evidence for liquid water. Gullies have been identified on the walls of canyons, channels, and impact craters (Malin and Edgett, 2000; Cabrol et al., 2001). These gullies are 10–30 m wide and appear to emanate from a region typically 100–300 m below the rims of the walls. Debris eroded from the gullies has accumulated as fans at the bottom of the walls. The sharp, unweathered appearance of such small features and the fact that their debris overlie features that are themselves thought to be very young suggest that the gullies were formed very recently and possibly within only the last 1–2 million years. Their origin from below the rims of cliffs is consistent with a formation by seepage of some fluid from within the crust, liquid water being the most likely (Kolb et al., 2010). This can occur at shallow depths in places where the surface materials have low thermal conductivity, which means that temperatures rise very quickly with depth (Mellon and Phillips, 2001). As an alternative mechanism, liquid CO_2 has also been suggested as an agent for carving the gullies, based on the fact that the pressure within the crust at the depth from which many of the seeps occur is very close to the pressure at which CO_2 liquefies (Musselwhite et al., 2001). There are two issues that appear to rule out this explanation: First, pressurized liquid CO_2 in the crust that is released to the surface would catastrophically vaporize upon expansion, a process that would not produce the observed gullies (Stewart and Nimmo, 2002). Second, no plausible physical mechanism has been suggested by which the crust could be filled with liquid CO_2.

Recurring slope lineae (RSL), which are narrow streaks of lower reflectance compared with the surrounding terrain, appear and grow incrementally in the downslope direction during warm seasons when temperatures reach about 250–300 K, a pattern consistent with the transient flow of a volatile species (e.g., McEwen et al., 2011, 2013; Ojha et al., 2014). Spectral data from the Compact Reconnaissance Imaging Spectrometer for Mars (CRISM) instrument on board MRO (Ojha et al., 2015) strongly support the hypothesis that RSL form as a result of contemporary water activity on Mars (see also Chapter 10).

2.7 POSSIBLE SOURCES OF ORGANIC MOLECULES ON MARS

There is abundant evidence for liquid water on early Mars, thus attesting the presence of an atmosphere thicker than the current one (for a review, see Lammer et al., 2013). Could this atmosphere dominated by carbon dioxide have generated organic molecules? When Stanley Miller replaced methane by carbon dioxide in his historic sparking experiments, only small yields of amino acids were obtained (Schlesinger and Miller, 1983).

More recent studies show that the low yields previously reported could be the outcome of oxidation of the organic compounds during hydrolytic reprocessing by nitrite and nitrate produced in the reactions. The yield of amino acids is greatly increased when oxidation inhibitors, such as ferrous iron, are added prior to hydrolysis, suggesting that synthesis from neutral primitive atmospheres may be more important than previously thought (Cleaves et al., 2008). Nevertheless, the production of large amounts of organic molecules by reducing carbon dioxide directly in the atmosphere was probably limited. More recent studies suggest the presence of methane and H_2 components in the early Martian atmosphere (e.g., Ramirez, 2017; Kite et al., 2017; Wordsworth et al., 2017).

The early Martian atmosphere was also capable of decelerating meteorites and micrometeorites. The carbonaceous chondrites delivered organic materials to the Earth (Pizzarello et al., 2001; Glavin et al., 2006; Pizzarello and Shock, 2010, 2017). They contain from 1.5 to 4% of carbon, for the most part as organic materials. The Murchison meteorite, a CM2-type carbonaceous chondrite that fell in Australia in 1969, has been extensively analyzed (Pizzarello, 2007; Pizzarello and Shock, 2010; Cooper and Riosa, 2016). For instance, the amino acid diversity has been analyzed in detail. The total number of amino acids detected in meteorites is about 100. All the possible α-amino alkylamino acids up to seven carbons were identified, as well as large abundances of N-substituted, cyclic, β-, γ-, δ-, and ε-amino acids (Cronin et al., 1988). Eight biological amino acids (glycine, alanine, proline, leucine, isoleucine, valine, aspartic acid, and glutamic acid) have been found. Nucleic acid bases, purines and pyrimidines, have also been found in the Murchison meteorite (Stoks and Schwartz, 1982; Callahan et al., 2011). No ribose (the sugar linking together the nucleic acid building blocks) was detected in meteorites. Droplet-forming fatty acids have been extracted from different carbonaceous meteorites (Deamer, 1985, 1998). A combination of high-resolution analytic methods, including organic structural spectroscopy applied to the organic fraction of Murchison extracted under mild conditions, has extended its indigenous chemical diversity to tens of thousands of different molecular compositions and likely millions of diverse structures (Schmitt-Kopplin et al., 2010; Hertkorn et al., 2015).

Micrometeorite collections in the Greenland and Antarctica ice sheets (Maurette, 1998, 2006) show that the Earth captures interplanetary dust as micrometeorites at a rate of about 20,000 t/year. About 99% of this mass is carried by micrometeorites in the 50–500 μm size ranges (Fig. 2.3). This value is about 2000 times higher than the most reliable estimate of the meteorite flux, about 10 t/year (Bland et al., 1996). At least approximately 20 wt% of these micrometeorites survives unmelted upon atmospheric entry. As their kerogen-like fraction represents about 2.5 wt% of carbon, this amounts to a total mass of $\sim 2.5 \times 10^{22}$ g of kerogen-like material on the early Earth surface (Maurette and Brack, 2006).

One amino acid, α-amino isobutyric acid, has been identified in Antarctic micrometeorites (Brinton et al., 1998). These grains also contain a high proportion of metallic

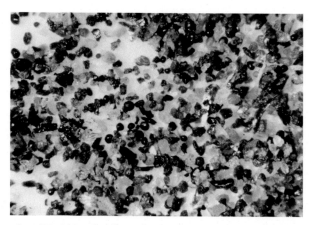

Fig. 2.3 Micrometeorites (50–100 μm) collected in Antarctica ice. *(Courtesy of M. Maurette.)*

sulfides, oxides, and clay minerals, a rich variety of inorganic catalysts that could have promoted the reactions of the carbonaceous material leading to the origin of life. Many similarities can be found between Antarctic micrometeorites and samples from Comet Wild 2, in terms of chemical, mineralogical, and isotopic compositions, and in the structure and composition of their carbonaceous matter (Dobrica et al., 2013). The cometary origin has been confirmed by a zodiacal cloud model based on the orbital properties and lifetimes of comets and asteroids and on the dynamic evolution of dust after ejection. The model is not only quantitatively constrained by Infrared Astronomical Satellite observations of thermal emission but also qualitatively consistent with other zodiacal cloud observations, meteor observations, spacecraft impact experiments, and the properties of recovered micrometeorites (Nesvorny et al., 2010).

The reducing conditions in hydrothermal systems, which are due to serpentinization reactions (reviewed in Holm et al., 2015), may have been an important source of biomolecules (Baross and Hoffman, 1985; Holm, 1992; Holm and Andersson, 1998, 2005). These reducing environments result from the flow of substances dissolved in seawater passing through very hot crustal material. The reduced compounds flow from the hydrothermal system, and the inorganic sulfides formed will precipitate when they mix with the cold (4°C) ocean water. For example, hydrocarbons and oxidized organic compounds have been detected in hydrothermal fluids from the Rainbow and Lost City ultramafic-hosted vents (Konn et al., 2009). There is much evidence for ancient hydrothermal systems on Mars, such as the hydrothermal seafloor deposits in Eridania basin (Michalski et al., 2017) and hydrothermal clay systems at Nili Fossae (e.g., Ehlmann et al., 2011; Mustard and Tarnas, 2017), observed by MRO's Compact Reconnaissance Spectrometer for Mars, which would have provided favorable environments for these processes to take place.

2.8 WHERE ARE THE MARTIAN ORGANIC MOLECULES?

The early histories of Mars and Earth clearly show similarities. Mars should have therefore inherited an abundance of micrometeoritic organic material. Yet, so far, the inventory of such a material is very poor. The Viking landers could not find any organic carbon in the Martian soil by gas chromatography-mass spectrometry (Biemann et al., 1976; Biemann, 2007). The lack of GC-MS evidence for nonterrestrial organic carbon at either Viking landing site led to speculation that chemical oxidation processes (Biemann et al., 1977; Chun et al., 1978; Encrenaz et al., 2004; Oyama et al., 1977; Yen et al., 2000) and/or ultraviolet and ionizing radiation (Moores and Schuerger, 2012; Oró and Holzer, 1979; Stalport et al., 2009; ten Kate et al., 2005) could have either destroyed or transformed the organic material into forms that are not readily detectable. It was concluded that the most plausible explanation for these results was the presence, at the surface, of highly reactive oxidants like H_2O_2 that would have been photochemically produced in the atmosphere (Hartman and McKay, 1995). The Viking landers could not sample soils below 6 cm, and therefore, the depth of this apparently organic-free and oxidizing layer is unknown. Bullock et al. (1994) have calculated that the depth of diffusion for H_2O_2 is <3 m. Direct photolytic processes can also be responsible for the dearth of organics at the Martian surface (Stoker and Bullock, 1997). However, nonvolatile salts of benzenecarboxylic acids and perhaps oxalic and acetic acid should be metastable intermediates of meteoritic organics under oxidizing conditions. Salts of these organic acids would have been largely invisible to gas chromatography-mass spectrometry (Benner et al., 2000).

The SAM instrument on board the Mars Science Laboratory Curiosity rover detected chlorobenzene and C2–C4 dichloroalkanes with the SAM gas chromatography-mass spectrometry and detection of chlorobenzene in the direct evolved gas analysis mode, in multiple portions of the fines from the Cumberland drill hole in the Sheepbed mudstone at Yellowknife Bay (Freissinet et al., 2015).

At least three Martian meteorites (EETA 79001, ALH84001, and Nakhla) that have carbonates with elevated $\delta^{13}C$ have coexisting organic matter (Wright et al., 1989; Grady et al., 1994; Becker et al., 1999). EETA 79001 is the best studied, including a variety of lithologies investigated. Without exception, specimens without carbonate have a uniformly low organic matter content, as estimated by the ignition temperature established during stepped combustion experiments. In contrast, for the carbonate-rich fractionations, organic contents are significant. The amounts of carbon are a factor of five higher than any other bulk Martian meteorite. The implication is that the organic matter associated with carbonates came to Earth with the meteorite and is cogenetic. None of the organic matter in EETA 79001 or for that matter Nakhla and ALH 84001, with the exception of minute amounts of PAHs in the latter (MacKay et al., 1996), has been fully characterized in terms of compound class or individual structures.

2.9 SEVERAL POSSIBLE WAYS TO START LIFE ON MARS

Two types of originating living species are generally considered depending on their use of organic molecules. Autotrophs produce their own organics from CO_2, while heterotrophs use already made organics. In a "metabolism-first approach," the proponents of an autotrophic life call for the spontaneous formation of simple molecules from carbon dioxide and water to rapidly generate life (Wächtershäuser, 1988). In the second hypothesis, the "primeval soup scenario," rather complex organic molecules accumulated in a warm little pond before leading to life (Haldane, 1929; Oparin, 1924).

Carbon dioxide was abundant in the primitive Martian atmosphere. The energy source required to reduce the gas might have been provided by the oxidative formation of pyrite from iron sulfide and hydrogen sulfide. Pyrite has positive surface charges and bonds the products of carbon dioxide reduction, giving rise to a two-dimensional reaction system, a "surface metabolism" (Wächtershäuser, 1994, 2007). Laboratory work has provided some support for this hypothesis. An early laboratory simulation of hydrothermal synthetic reactions is the reduction of carbon dioxide to organic sulfides in the presence of FeS and H_2S provided mainly methyl and ethyl thiol along with smaller amounts of other thiols containing up to five carbon atoms. The CO_2 was also converted to CS_2 and COS (Heinen and Lauwers, 1996). The direct reduction of CO_2 to acetic acid, acetaldehyde, ethanol, and smaller amounts of carbon compounds containing up to six carbon atoms was observed to take place at 350°C and high pressure in the presence of magnetite (Fe_3O_4) and small amounts of water (Chen and Bahnemann, 2000). Along with the scenario proposed by Michael Russel (Martin et al., 2008), a laboratory setup simulating conditions prevailing in alkaline hydrothermal vents generated low yields of simple organics (Herschy et al., 2014). So far, the proponents of a metabolism-first approach have not been able to produce large enough precursor prebiotic molecules to create simple primitive life in a test tube.

The primeval soup approach, also called "replication first," supposes complex organic molecules accumulating in a warm little pond, "à la" Darwin. By analogy with contemporary living systems, it is tempting to speculate that life emerges as a cell-like system. Such a system requires, at least, boundary molecules able to isolate a system from the aqueous environment (membrane), catalytic molecules able to conduct the basic chemical work of the cell (like enzymes), and information molecules able to store and to transfer the information needed for reproduction (like nucleic acid polymers). Great efforts have been deployed in laboratories to produce these three prerequisites, albeit separately.

Fatty acids are known to form vesicles when their hydrocarbon chains contain >10 carbon atoms. Such vesicle-forming fatty acids have been identified in the Murchison meteorite, as already mentioned. However, the membranes obtained with these simple amphiphiles (amphiphilic molecules bearing both water-soluble and water-insoluble portions) are not stable over a broad range of environmental conditions (Deamer, 1998, 2017; Pohorille and Deamer, 2009). How different prebiotically available building blocks could have become precursors of phosphorus-containing lipids that form vesicles has been reviewed (Fiore and Strazewski, 2016).

Chemical reactions able to selectively condense protein amino acids, at the expense of the nonprotein ones in water, have been identified (Brack, 1987). Helical and sheet structures can be modeled with the aid of only two different amino acids, one hydrophobic, the other hydrophilic. Polypeptides with alternating hydrophobic and hydrophilic residues adopt water-soluble layered β-sheet structures (β-sheets) due to hydrophobic side-chain clustering (Fig. 2.4).

Owing to the formation of a β-sheet, alternating sequences gain good resistance toward chemical degradation. Aggregation of alternating sequences into β-sheets is possible only with all-L or all-D polypeptides. Short peptides have also been shown to exhibit catalytic properties (Brack, 2007). Matsuno and colleagues (Imai et al., 1999) reported peptide formation in a flow reactor that mimicked the conditions in a hydrothermal system. The team was able to polymerize glycine monomers up to six units in the presence of Cu ions. With this setup, the authors were able to demonstrate polymerization at temperatures of 200–250°C, contrary to the popular belief that organic molecules are unstable under high temperatures. Rodriguez-Garcia et al. (2015) mimicked this hydrothermal system and developed an automated method to expose glycine monomers to prolonged dehydration-hydration cycles, and interestingly, chain lengths of 20 amino acids were observed.

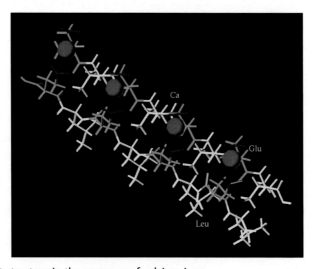

Fig. 2.4 A β-sheet structure in the presence of calcium ions.

In a cell-like system, the hereditary memory is stored in nucleic acids, long chains built from nucleotides. Each nucleotide is composed of a base (purine or pyrimidine), a sugar (ribose for RNA and deoxyribose for DNA), and a phosphate group. The synthesis of nucleotides is a complex issue due mainly to the instability of sugars and the poor yield of sugar formation in the formose reaction (Decker et al., 1982). Ribose has not been detected in meteorites; however, irradiation by UV photons of interstellar ice analogues made of water, methanol, and ammonia at 78 K in a high-vacuum chamber generated substantial quantities of ribose and a diversity of structurally related sugar molecules such as arabinose, xylose, and lyxose (de Marcellus et al., 2015; Meinert et al., 2016). Binding phosphorus to sugars is another complex issue. Studies using the mineral schreibersite, $(Fe,Ni)_3P$, conducted independently by Pasek (2017) and Kee et al. (2013), provide a possible solution to this problem. Diamidophosphate (DAP), a plausible prebiotic agent produced from trimetaphosphate, efficiently phosphorylates a wide variety of prebiotically relevant sugar molecules (Krishnamurthy et al., 2000). Phosphorylation of various sugars can also proceed spontaneously in aqueous microdroplets containing a simple mixture of sugars and phosphoric acid (Nam et al., 2017).

Instead of trying to assemble the three subunits of a ribonucleotide in several steps, some laboratories turn to a single-pot approach. Sutherland's team produced simultaneously precursors of nucleic acid, amino acids, and lipids starting with hydrogen cyanide, hydrogen sulfide, and UV light (Patel et al., 2015), a step toward the congruence hypothesis advocated by de Duve (2003). DAP, already mentioned, efficiently phosphorylates a wide variety of potential building blocks—nucleosides/nucleotides, amino acids, and lipid precursors—under aqueous conditions. Significantly, higher-order structures—oligonucleotides, peptides, and liposomes—are formed under the same phosphorylation reaction conditions. This plausible prebiotic phosphorylation process running under similar reaction conditions could enable the chemistry of the three classes of prebiotic molecules and their oligomers, in a single-pot aqueous environment (Gibard et al., 2017).

The discovery of the ribozyme, a class of RNA with catalytic properties (Zaug and Cech, 1986), opened the RNA world hypothesis. For example, they increase the rate of hydrolysis of oligoribonucleotides and can also act as polymerization templates. Since their primary discovery, the catalytic spectrum of these ribozymes has been considerably enlarged by directed test-tube molecular evolution experiments (Prywes et al., 2016; Horning and Joyce, 2016). Since RNA was shown to be able to act simultaneously as an information molecule and as a catalytic molecule, RNA has been considered as a possible primordial living entity. A ribozyme-based "RNA world" has been modeled in some detail and reviewed (Higgs and Lehman, 2015). The route to a possible RNA world has been paved by the impressive work of Jim Ferris who polymerized RNA-like monomers on clays (Ferris 2005, 2006). One should, however, remember that the complete synthesis of RNA under "prebiotic conditions" remains an unsolved challenge. It seems therefore unlikely that life could start with RNA molecules; they are not simple enough, yet too difficult to assemble.

For some scientists, the first step toward the origin of life could be the spontaneous condensation of amphiphilic molecules to form vesicles (Morowitz, 1992). Examples of autocatalytic micelle growth have been described (Bachmann et al., 1992). However, these simple autocatalytic systems do not store hereditary information and cannot therefore evolve by natural selection. Szostak and colleagues (Hanczyc et al., 2007; Schrum et al., 2010) found that the presence of naturally occurring clay minerals, such as montmorillonite, can bring RNA into the interior of the vesicles, thus providing information to the vesicular system (for a comprehensive review, see Meierhenrich et al., 2010).

Chemists are also tempted to consider that primitive replicating systems can use simpler information-retaining molecules rather than biological nucleic acids or their analogues and looked for simple self-sustaining chemical systems capable of self-replication, mutation, and selection. It has been shown that simple molecules unrelated to nucleotides can actually provide exponentially replicating autocatalytic models. Autocatalysis is observed when the product of the reaction catalyzes its own formation. Von Kiedrowski (Terfort and von Kiedrowski, 1992) tested different templates, while Burmeister (1998) reported information transfer templates in complex systems. In most cases, the rate of the autocatalytic growth did not vary in a linear sense, in contrast to most autocatalytic reactions known so far. Two preformed fragments of a peptide have been demonstrated to be autocatalytically ligated by the whole peptide acting as a template (Lee et al., 1996; Saghatelian et al., 2001). However, the replicated molecules possess a very low level of information.

Interplanetary transfer of life represents another way to start life on Mars. The different steps for such a process include (1) the escape process (i.e., the removal to space of biological material that has survived being lifted off from the surface of a parent body to high altitudes), (2) the travel conditions in space (i.e., the survival of the biological material over timescales comparable with an interplanetary journey and exposure to extreme conditions such as temperature, pressure, and UV), and (3) the entry process (i.e., the likelihood for the nondestructive deposition of the biological material on Mars). Following the identification of meteorites of lunar and Martian origin, the escape of material ranging from small particles up to boulder-size particles from a planet after the impact of a large asteroid is a feasible process. Bacterial spores have been shown to survive shockwaves produced by a simulated meteorite impact (Horneck et al., 2001) and huge accelerations (Roten et al., 1998). In order to study step (2), the survival of resistant microbial forms in the upper atmosphere and free space, *Bacillus subtilis* spores, bacteria, bacteria-infecting virus, tobacco mosaic virus, microbes adapted to high salt concentrations (osmophilic), cyanobacteria, and lichens have been exposed aboard balloons, rockets, spacecraft, and space stations—such as Gemini, Apollo, Spacelab, Long Duration Exposure Facility, Foton, Eureca, and ISS (Fig. 2.5)—and their responses

Fig. 2.5 EXPOSE facility on board the International Space Station (credit ESA).

investigated after recovery (Horneck et al., 2010; Raggio et al., 2011; Panitz et al., 2015; Mancinelli, 2015).

Laboratory experiments under simulated interstellar medium conditions point to a remarkably less damaging effect of UV radiation at low temperatures. Treating *B. subtilis* spores with three simulated factors simultaneously (UV, vacuum, and temperature of 10 K) produces an unexpectedly high survival rate, even at very high UV fluxes. It has been estimated that under average conditions in space spores may survive for hundreds of years (Weber and Greenberg, 1985). Based on the mean sizes and numbers of meteorites ejected (e.g., from Mars) and percentages falling on Earth, models for galactic cosmic rays, and laboratory responses to accelerated heavy ions of *B. subtilis* spores and *Deinococcus radiodurans* cells, it has been calculated that viable transfer of microbes from Mars to Earth via impact ejecta is possible due to the high number of meteorites and the impressive resistance of microorganisms to the dangers of space (Mileikowsky et al., 2000).

2.10 THE ODDS FOR AN ORIGIN OF LIFE

In addition to liquid water and carbon molecules, specific environmental components and conditions were needed for the origin of life, that is, reactive rocks and minerals bathed by warm to hot hydrothermal fluids (Westall et al., 2015). Such minerals that existed on Earth and probably on Mars as well (Bibring et al., 2006; Christensen et al., 2004; Yen et al., 2005) could have supported the concentration of organic molecules and contributed to their conformation and stabilization and to the complexification of larger, stable molecules. Like a pastry recipe, chemistry is reproducible, and processing identical ingredients leads necessarily to the same products. There are however some limitations. First, the complexity of the emergence of terrestrial life is still unknown. By demonstrating in 1953 that it was possible to form amino acids—the building blocks of proteins—from a gas mixture, Stanley Miller generated the ambitious hope that chemists will be able to recreate life in a test tube. They have a good knowledge of the environmental landscape and of some of the chemical actors, but no working scenario for a self-replicating and evolving molecular system has been released so far (Dass et al., 2016). The simplicity of such a process (or chemists' skillfulness) is therefore questionable. Second, compared with Mars, the Earth has some specific features, that is, global plate tectonics, global magnetic field, and a large moon generating oceanic tides. For instance, tides could have boosted prebiotic chemistry by allowing wet-dry cycling chemical reactions (Mamajanov et al., 2014; Higgs, 2016). The magnetic field generated a magnetosphere, which prevented the loss of Earth's atmosphere via stripping by the solar wind, unlike what happened on Mars (Jakosky et al., 2015). On the other hand, if despite these differences life had originated on Mars, then it would clearly demonstrate that the existence of these terrestrial features is not mandatory for worlds where liquid water and organics are present.

2.11 CONCLUSION

In addition to the societal impact, the discovery of a second genesis of life would bring great advances to the scientific study of the origin of life (Brack, 1997; Brack and Pillinger, 1998). The discovery of a second independent genesis of life on Mars, a body presenting environmental conditions similar to those that prevailed on the primitive Earth, would strongly support the idea of a rather simple genesis of terrestrial life. It would also demonstrate that life is not a random one-shot process, but likely a rather common occurrence. However, Martian life would have to be sufficiently different from terrestrial life to rule out the possibility of planetary transfer of microorganisms to Mars. Conversely, although Mars is generally considered as the twin planet of the Earth, the odds for the emergence of a carbon-based life on the red planet have several limitations. The fact that, despite tremendous efforts, chemists have not been able to recreate life in a test tube under plausible prebiotic conditions seems to indicate that the origin of life is not really simple. The possible role of Earth's specific features—plate tectonics, global magnetic field, and a large moon generating oceanic tides—is still unknown. Consequently, the astrobiology community must be prepared to explain to the public that searching for life on Mars is a real challenge and that failure is a real possibility. Finding an overabundance of organics on the surface of Mars compared with the composition of the mantle would itself be a great discovery. Any extraterrestrial explorer searching for life on Earth would probably be struck by the fact that carbon, which constitutes 17.9% of the biomass, holds only 0.094% of the mantle. Surface missions at Mars are investigating the type and abundance of organics and biogenic elements in order to provide clues about the possibility of life on Mars. One of the preferred pieces of advice from the late Stanley Miller was "never demand too much."

REFERENCES

Bachmann, P.A., et al., 1992. Autocatalytic self-replicating micelles as models for prebiotic structures. Nature 357, 57–59.

Baross, J.A., Hoffman, S.E., 1985. Submarine hydrothermal vents and associated gradient environment as sites for the origin and evolution of life. Orig. Life Evol. Biosph. 15, 327–345.

Becker, L., et al., 1999. The origin of organic matter in the Martian meteorite ALH84001. Earth Planet. Sci. Lett. 167, 71–79.

Becker, R.H., Pepin, R.O., 1984. The case for a Martian origin of the shergottites: nitrogen and noble gases in EETA 79001. Earth Planet. Sci. Lett. 69, 225–242.

Benner, S.A., et al., 2000. The missing organic molecules on Mars. Proc. Natl. Acad. Sci. U. S. A. 97, 2425–2430.

Bernal, J.D., 1949. The physical basis of life. Proc. Phys. Soc. A 62, 537–558.

Bibring, J.-P., et al., 2006. Global mineralogical and aqueous Mars history derived from OMEGA/Mars express data. Science 312, 400–404. https://doi.org/10.1126/science.1122659.

Biemann, K., et al., 1976. Search for organic and volatile inorganic compounds in two surface samples from the chryse planitia region of Mars. Science 194, 72–76.

Biemann, K., et al., 1977. The search for organic substances and inorganic volatile compounds in the surface of Mars. J. Geophys. Res. 82, 4641–4658.

Biemann, K., 2007. On the ability of the Viking gas chromatograph-mass spectrometer to detect organic matter. Proc. Natl. Acad. Sci. U. S. A. 104, 10310–10313.

Bland, P.A., et al., 1996. The flux of meteorites to the earth over the last 50000 years. Mon. Not. R. Astron. Soc. 233, 551–565.

Bogard, D.D., Johnson, P., 1983. Martian gases in an Antarctic meteorite? Science 221, 651–654.

Brack, A., Pillinger, C.T., 1998. Life on Mars: chemical arguments and clues from Martian meteorites. Extremophiles 2, 313–319.

Brack, A., 2007. From interstellar amino acids to prebiotic catalytic peptides. Chem. Biodivers. 4, 665–679.

Brack, A., 1997. Life on Mars: a clue to life on Earth? Chem. Biol. 4, 9–12.

Brack, A., 1987. Selective emergence and survival of early polypeptides in water. Orig. Life Evol. Biosph. 17, 367–379.

Brinton, K.L.F., et al., 1998. A search for extraterrestrial amino acids in carbonaceous Antarctic micrometeorites. Orig. Life Evol. Biosph. 28, 413–424.

Bullock, M.A., et al., 1994. A coupled soil-atmosphere model of H_2O_2 on Mars. Icarus 107, 142–154.

Burmeister, J., 1998. Self-replication and autocatalysis. In: Brack, A. (Ed.), The Molecular Origins of Life: Assembling Pieces of the Puzzle. Cambridge University Press, Cambridge, UK, pp. 295–312.

Cabrol, N., Grin, E.A., 2010. Lakes on Mars. Elsevier.

Cabrol, N.A., Grin, E.A., 1999. Distribution, classification, and ages of Martian impact crater lakes. Icarus 142, 160–172.

Cabrol, N.A., et al., 2001. Recent aqueous environments in martian impact craters: an astrobiological perspective. Icarus 154, 98–112. https://doi.org/10.1006/icar.2001.6661.

Cabrol, N.A., et al., 1998. Duration of the Ma'Adim Vallis/Gusev Crater hydrogeologic system, Mars. Icarus 133, 98–108.

Cairns-Smith, A.G., Hartman, H., 1986. Clay Minerals and the Origin of Life. Cambridge University Press, Cambridge, UK.

Callahan, M.P., et al., 2011. Carbonaceous meteorites contain a wide range of extraterrestrial nucleobases. Proc. Natl. Acad. Sci. U. S. A. 108, 13995–13998.

Cardenas, B.T., et al., 2017. Fluvial stratigraphy of valley fills at Aeolis dorsa, Mars: evidence for base-level fluctuations controlled by a downstream water body. Geol. Soc. Am. Bull. 129. https://doi.org/10.1130/B31567.1.

Carr, M.H., Head, J.W., 2015. Martian surface/near-surface water inventory: sources, sinks, and changes with time. Geophys. Res. Lett. 42, 726–732.

Carr, M.H., Head, J.W., 2003. Oceans on Mars: an assessment of the observational evidence and possible fate. J. Geophys. Res. 108 (E5), 5042. https://doi.org/10.1029/2002JE001963.

Carr, M.H., et al., 1985. Martian atmospheric carbon dioxide and weathering products in SNC meteorites. Nature 314, 245–250.

Carter, J., et al., 2013. Hydrous minerals on Mars as seen by the CRISM and OMEGA imaging spectrometers: updated global view. J. Geophys. Res. 118, 831–858.

Chatzitheodoridis, E., et al., 2014. A conspicuous clay ovoid in Nakhla: evidence for subsurface hydrothermal alteration on Mars with implications for astrobiology. Astrobiology 14, 651–693.

Chen, Q.W., Bahnemann, D.W., 2000. Reduction of carbon dioxide by magnetite: implications for the primordial synthesis of organic molecules. J. Am. Chem. Soc. 122, 970–971.

Christensen, P.R., et al., 2000. Detection of crystalline hematite mineralization on Mars by the thermal emission spectrometer: evidence for near-surface water. J. Geophys. Res. 105, 9623–9642.

Christensen, P.R., et al., 2004. Mineralogy at Meridiani Planum from the mini-TES experiment on the opportunity rover. Science 306, 1733–1739.

Chun, S.F.S., et al., 1978. Photocatalytic oxidation of organic-compounds on Mars. Nature 274, 875–876.

Cleaves, H.J., et al., 2008. A reassessment of prebiotic organic synthesis in neutral planetary atmospheres. Orig. Life Evol. Biosph. 38, 105–115.

Cooper, G., Riosa, A.C., 2016. Enantiomer excesses of rare and common sugar derivatives in carbonaceous meteorites. Proc. Natl. Acad. Sci. U. S. A. 113, E3322–E3331.

Craddock, R.A., Howard, A.D., 2002. The case for rainfall on a warm, wet early Mars. J. Geophys. Res. 107 (E11), 5111. https://doi.org/10.1029/2001JE001505.

Cronin, J.R., et al., 1988. Organic matter in carbonaceous chondrites, planetary satellites, asteroids, and comets. In: Kerridge, J.F., Matthews, M.S. (Eds.), Meteorites and the Early Solar System. University of Arizona Press, Tucson, pp. 819–857.

Dass, A.V., et al., 2016. Stochastic prebiotic chemistry within realistic geological systems. ChemistrySelect 1, 4906–4926.

de Duve, C., 2003. A research proposal on the origin of life. Orig. Life Evol. Biosph. 33, 559–574.

de Marcellus, P., et al., 2015. Aldehydes and sugars from evolved precometary ice analogs: importance of ices in astrochemical and prebiotic evolution. Proc. Natl. Acad. Sci. U. S. A. 112, 965–970.

Deamer, D.W., 1985. Boundary structures are formed by organic components of the Murchison carbonaceous chondrite. Nature 317, 792–794.

Deamer, D.W., 1998. Membrane compartments in prebiotic evolution. In: Brack, A. (Ed.), The Molecular Origins of Life: Assembling Pieces of the Puzzle. Cambridge University Press, Cambridge, UK, pp. 189–205.

Deamer, D.W., 2017. The role of lipid membranes in life's origin. Life 7, 5. https://doi.org/10.3390/life7010005.

Decker, P., et al., 1982. Bioids. X. Identification of formose sugars, presumable prebiotic metabolites, using capillary gas chromatography/gas chromatography-mass spectrometry of *n*-butoxime trifluoroacetates on OV-225. J. Chromatogr. A 244, 281–291.

di Achille, G., Hynek, B.M., 2010. Ancient ocean on Mars supported by global distribution of deltas and valleys. Nat. Geosci. 3, 459–463.

Dobrica, E., et al., 2013. Connection between micrometeorites and wild 2 particles: from Antarctic snow to cometary ices. Meteorit. Planet. Sci. 44, 1643–1661.

Ehlmann, B.L., et al., 2011. Subsurface water and clay mineral formation during the early history of Mars. Nature 479, 53–60.

Encrenaz, T., et al., 2004. Hydrogen peroxide on Mars: evidence for spatial and seasonal variations. Icarus 170, 424–429.

Fairén, A.G., et al., 2003. Episodic flood inundations of the northern plains of Mars. Icarus 165, 53–67.

Fairén, A.G., et al., 2009. Evidence for Amazonian acidic liquid water on Mars—a reinterpretation of MER mission results. Planet Space Sci. 57, 276–287.

Fassett, C.I., Head, J.W., 2008. Valley network-fed, open-basin lakes on Mars: distribution and implications for Noachian surface and subsurface hydrology. Icarus 198, 37–56.

Ferris, J.P., 2005. Mineral catalysis and prebiotic synthesis: montmorillonite-catalyzed formation of RNA. Elements 1, 145–149.

Ferris, J.P., 2006. Montmorillonite-catalysed formation of RNA oligomers: the possible role of catalysis in the origins of life. Phil. Trans. R. Soc. B 361, 1777–1786.

Fiore, M., Strazewski, P., 2016. Prebiotic lipidic amphiphiles and condensing agents on the early Earth. Life 6, 17. https://doi.org/10.3390/life6020017.

Fox, J.L., et al., 2015. Water and water ions in the Martian thermosphere/ionosphere. Geophys. Res. Lett. 42, 8977–8985. https://doi.org/10.1002/2015GL065465.

Freissinet, C., et al., 2015. Organic molecules in the Sheepbed mudstone, Gale Crater, Mars. J. Geophys. Res. Planets 120, 495–514. https://doi.org/10.1002/2014JE004737.

Gibard, C., et al., 2017. Phosphorylation, oligomerization and self-assembly in water under potential prebiotic conditions. Nat. Chem. https://doi.org/10.1038/nchem.2878.

Glavin, D.P., et al., 2006. Amino acid analyses of Antarctic CM2 meteorites using liquid chromatography time of flight-mass spectrometry. Meteorit. Planet. Sci. 41, 889–902.

Gooding, J.L., Muenow, D.W., 1986. Martian volatiles in shergottite EETA 79001: new evidence from oxidized sulfur and sulfur-rich aluminosilicates. Geochim. Cosmochim. Acta 50, 1049–1059.

Gooding, J.L., et al., 1988. Calcium carbonate and sulfate of possible extraterrestrial origins in the EETA79001 meteorite. Geochim. Cosmochim. Acta 52, 909–915.

Gooding, J.L., et al., 1991. Aqueous alteration of the Nakhla meteorite. Meteoritics 26, 135–143.

Grady, M.M., et al., 1994. Carbon and nitrogen in ALH 84001. Meteoritics 29, 469.

Grotzinger, J.P., et al., 2014. A habitable fluvio-lacustrine environment at Yellowknife Bay, Gale Crater, Mars. Science. 343. https://doi.org/10.1126/science.1242777.

Grotzinger, J.P., et al., 2012. Mars Science Laboratory mission and science investigation. Space Sci. Rev. 170, 5–56. https://doi.org/10.1007/s11214-012-9892-2.

Haldane, J.B.S., 1929. The origin of life. Ration. Annu. 148, 3–10.

Hanczyc, M.M., et al., 2007. Mineral surface directed membrane assembly. Orig. Life Evol. Biosph. 37, 67–82.

Hartman, H., McKay, C.P., 1995. Oxygenic photosynthesis and the oxidation state of Mars. Planet Space Sci. 43, 123–128.

Heinen, W., Lauwers, A.M., 1996. Sulfur compounds resulting from the interaction of iron sulfide, hydrogen sulfide and carbon dioxide in an anaerobic aqueous environment. Orig. Life Evol. Biosph. 26, 131–150.

Herschy, B., et al., 2014. An origin-of-life reactor to simulate alkaline hydrothermal vents. J. Mol. Evol. 79, 213–227.

Hertkorn, N., et al., 2015. Nontarget analysis of Murchison soluble organic matter by high-field NMR spectroscopy and FTICR mass spectrometry. Magn. Reson. Chem. 53, 754–768.

Higgs, P.G., 2016. The effect of limited diffusion and wet-dry cycling on reversible polymerization reactions: implications for prebiotic synthesis of nucleic acids. Life 6, 24–40. https://doi.org/10.3390/life6020024.

Higgs, P.G., Lehman, N., 2015. The RNA world: molecular cooperation at the origins of life. Nat. Rev. Genet. 16, 7–17.

Holm, N.G., Andersson, E.M., 2005. Hydrothermal simulation experiments as a tool for studies of the origin of life on earth and other terrestrial planets: a review. Astrobiology 5, 444–460.

Holm, N.G., Andersson, E.M., 1998. Organic molecules on the early Earth: hydrothermal systems. In: Brack, A. (Ed.), The Molecular Origins of Life: Assembling Pieces of the Puzzle. Cambridge University Press, Cambridge, UK, pp. 86–99.

Holm, N.G., et al., 2015. Serpentinization and the formation of H and CH on celestial bodies (planets, moons, comets). Astrobiology 15, 587–600.

Holm, N.G., 1992. Why are hydrothermal systems proposed as plausible environments for the origin of life? Orig. Life Evol. Biosph. 22, 5–14.

Holt, J.W., et al., 2008. Radar sounding evidence for ice within Lobate Debris Aprons near Hellas Basin, Mid-Southern Latitudes of Mars. Lunar Planet Sci. XXXIX, 2441.

Horneck, G., et al., 2001. Bacterial spores survive simulated meteorite impact. Icarus 149, 285–290.

Horneck, G., et al., 2010. Space microbiology. Microbiol. Mol. Biol. Rev. 74, 121–156.

Horning, D.P., Joyce, G.F., 2016. Amplification of RNA by an RNA polymerase ribozyme. Proc. Natl. Acad. Sci. U. S. A. 113, 9786–9791.

Huang, W., Ferris, J.P., 2007. One-step, regioselective synthesis of up to 50-mers of RNA oligomers by montmorillonite catalysis. J. Am. Chem. Soc. 128, 8914–8919.

Hutchison, R., 1975. Water in non-carbonaceous stony meteorites. Nature 256, 714–715.

Hynek, B.M., et al., 2010. Updated global map of martian valley networks and implications for climate and hydrologic processes. J. Geophys. Res. 115. https://doi.org/10.1029/2009JE003548.

Hynek, B.M., 2016. RESEARCH FOCUS: the great climate paradox of ancient Mars. Geology 44, 879–880.

Imai, E.-I., et al., 1999. Elongation of oligopeptides in a simulated submarine hydrothermal system. Science 283, 831–833.

Jagoutz, E., Wänke, H., 1986. Sr and Nd isotopic sematics of Shergotty meteorite. Geochim. Cosmochim. Acta 50, 939–953.

Jakosky, B.M., et al., 2015. MAVEN observations of the response of Mars to an interplanetary coronal mass ejection. Science. 350(6261). https://doi.org/10.1126/science.aad0210.

Kasting, J.F., Whitmire, D.P., Reynolds, R.T., 1993. Habitable zones around main sequence stars. Icarus 10 (1), 108–128.

Kee, T.P., et al., 2013. Phosphate activation via reduced oxidation state phosphorus (P): mild routes to condensed-P energy currency molecules. Life 3, 386–402.

Kite, E.S., et al., 2017. Methane bursts as a trigger for intermittent lake-forming climates on post-Noachian Mars. Nat. Geosci. 10, 737–740.

Kolb, K.J., et al., 2010. Modeling the formation of bright slope deposits associated with gullies in Hale Crater, Mars: implications for recent liquid water. Icarus 205, 113–137.

Konn, C., et al., 2009. Hydrocarbons and oxidized organic compounds in hydrothermal fluids from Rainbow and Lost City ultramafic-hosted vents. Chem. Geol. 258, 299–314.

Krishnamurthy, R., et al., 2000. Regioselective α-phosphorylation of aldoses in aqueous solution. Angew. Chem. Int. Ed. 39, 2281–2285.

Lammer, H., et al., 2013. Outgassing history and escape of the Martian atmosphere and water inventory. Space Sci. Rev. 174, 113–154. https://doi.org/10.1007/s11214-012-9943-8.

Lee, D.H., et al., 1996. A self-replicating peptide. Nature 382, 525–528.

MacKay, D.S., et al., 1996. Search for past life on Mars: possible relic biogenic activity in martian meteorite ALH 84001. Science 273, 924–930.

Mahaffy, P.R., et al., 2015a. The imprint of atmospheric evolution in the D/H of Hesperian clay minerals on Mars. Science 347, 412–414.

Mahaffy, P.R., et al., 2015b. Volatile and isotopic imprints of ancient Mars. Elements 11, 51–56. https://doi.org/10.2113/gselements.11.1.51.

Malin, M.C., Edgett, K.S., 2000. Evidence for recent ground water seepage and surface runoff on Mars. Science 288, 2330–2335.

Mamajanov, I., et al., 2014. Ester formation and hydrolysis during wet-dry cycles: generation of far-from-equilibrium polymers in a model prebiotic reaction. Macromolecules 47, 1334–1343.

Mancinelli, R.L., 2015. The affect of the space environment on the survival of Halorubrum chaoviator and Synechococcus (Nägeli): data from the space experiment OSMO on EXPOSE-R. Int. J. Astrobiol. 14, 123–128.

Martin, W., et al., 2008. Hydrothermal vents and the origin of life. Nat. Rev. Microbiol. 6, 805–814.

Maurette, M., Brack, A., 2006. Cometary petroleum in Hadean time? Meteorit. Planet. Sci. 41, 5247.

Maurette, M., 1998. Carbonaceous micrometeorites and the origin of life. Orig. Life Evol. Biosph. 28, 385–412.

Maurette, M., 2006. Micrometeorites and the Mysteries of Our Origins. Springer Berlin, Heidelberg, Germany.

McEwen, A.S., et al., 2011. Seasonal flows on warm Martian slopes. Science 333, 740–743.

McEwen, A.S., et al., 2013. Recurring slope lineae in equatorial regions of Mars. Nat. Geosci. 7, 53–58.

McKay, C.P., Davis, W.L., 1991. Duration of liquid water habitats on early Mars. Icarus 90, 214–221.

Meierhenrich, U.J., et al., 2010. On the origin of primitive cells: from nutrient intake to elongation of encapsulated nucleotides. Angew. Chem. Int. Ed. Eng. 49, 3738–3750.

Meinert, C., et al., 2016. Ribose and related sugars from ultraviolet irradiation of interstellar ice analogs. Science 352, 208–212.

Mellon, M.T., Phillips, R.J., 2001. Recent gullies on Mars and the source of liquid water. J. Geophys. Res. 106, 23165–23180. https://doi.org/10.1029/2000JE001424.

Michalski, J.R., et al., 2017. Ancient hydrothermal seafloor deposits in Eridania basin on Mars. Nat. Commun. 8, 15978. https://doi.org/10.1038/ncomms15978.

Michalski, J.R., et al., 2013. Groundwater activity on Mars and implications for a deep biosphere. Nat. Geosci. 6, 133–138.

Mileikowsky, C., et al., 2000. Natural transfer of viable microbes in space. Part 1. From Mars to Earth and Earth to Mars. Icarus 145, 391–427.

Moores, J.E., Schuerger, A.C., 2012. UV degradation of accreted organics on Mars: IDP longevity, surface reservoir of organics, and relevance to the detection of methane in the atmosphere. J. Geophys. Res. 117, E08008. https://doi.org/10.1029/2012JE004060.

Morowitz, H., 1992. Beginnings of Cellular Life. Yale University Press, New Haven, CT.

Mouginot, J.A., et al., 2012. Dielectric map of the Martian northern hemisphere and the nature of plain filling materials. Geophys. Res. Lett. 39, L02202. https://doi.org/10.1029/2011GL050286.

Murchie, S.L., et al., 2009. A synthesis of Martian aqueous mineralogy after 1 Mars year of observations from the Mars Reconnaissance Orbiter. J. Geophys. Res. 114(E2). https://doi.org/10.1029/2009JE003342.

Musselwhite, D.S., et al., 2001. Liquid CO_2 breakout and the formation of recent gullies on Mars. Geophys. Res. Lett. 28, 1283–1285.

Mustard, J.F., Tarnas, J.D., 2017. Hydrogen production from the upper 15 km of Martian Crust via serpentinization: implications for habitability. Lunar Planet Sci. Conf. XLVIII, Abstract #2384.

Nam, I., et al., 2017. Abiotic production of sugar phosphates and uridine ribonucleoside in aqueous microdroplets. Proc. Natl. Acad. Sci. U. S. A. https://doi.org/10.1073/pnas.1714896114.

Nesvorny, D., et al., 2010. Cometary origin of the zodiacal cloud and carbonaceous micrometeorites: implications for hot debris disks. Astrophys. J. 713, 816–836.

Ojha, L., et al., 2015. Spectral evidence for hydrated salts in recurring slope lineae on Mars. Nat. Geosci. 8, 829–833.

Ojha, L., et al., 2014. HiRISE observations of Recurring Slope Lineae (RSL) during southern summer on Mars. Icarus 231, 365–376.

Ollivier, B., et al., 1994. Anaerobic bacteria from hypersaline environments. Microbiol. Rev. 58, 27–38.

Oparin, A.I., 1924. Proikhozndenie Zhizni. Izvestia Moskowski Rabochi, Moscow, Russia.

Oró, J., Holzer, G., 1979. The photolytic degradation and oxidation of organic compounds under simulated Martian conditions. J. Mol. Evol. 14, 153–160.

Oyama, V.I., et al., 1977. Preliminary findings of Viking gas-exchange experiment and a model for Martian surface-chemistry. Nature 265, 110–114.

Panitz, C., et al., 2015. The SPORES experiment of the EXPOSE-R mission: *Bacillus subtilis* spores in artificial meteorites. Int. J. Astrobiol. 14, 105–114.

Pasek, M.A., 2017. Schreibersite on the early Earth: scenarios for prebiotic phosphorylation. Geosci. Front. 8, 329–335.

Patel, B.H., et al., 2015. Common origins of RNA, protein and lipid precursors in a cyano sulfidic proto metabolism. Nat. Chem. 7, 301–307.

Pizzarello, S., Shock, E., 2017. Carbonaceous Chondrite Meteorites: the Chronicle of a Potential Evolutionary Path between stars and life. Orig. Life Evol. Biosph. 47, 249–260.

Pizzarello, S., Shock, E., 2010. The organic composition of carbonaceous meteorites: the evolutionary story ahead of biochemistry. Cold Spring Harbor Persp. Biol. 2, a002105.

Pizzarello, S., et al., 2001. The organic content of the Tagish Lake meteorite. Science 293, 2236–2239.

Pizzarello, S., 2007. The chemistry that preceded life's origin: a study guide from meteorites. Chem. Biodivers. 4, 680–693.

Pohorille, A., Deamer, D.W., 2009. Self-assembly and function of primitive cell membranes. Res. Microbiol. 160, 449456.

Poulet, F., et al., 2005. Phyllosilicates on Mars and implications for early Martian climate. Nature 438, 623–627.

Prywes et al., 2016. Nonenzymatic copying of RNA templates containing all four letters is catalyzed by activated oligonucleotides. Life. 517756.

Raggio, J., et al., 2011. Whole lichen thalli survive exposure to space conditions: results of Lithopanspermia experiment with *Aspicilia fruticulosa*. Astrobiology 11, 281–292.

Ramirez, R.M., 2017. A warmer and wetter solution for early Mars and the challenges with transient warming. Icarus 297, 71–82.

Rodriguez, J.A.P., et al., 2015. Martian outflow channels: how did their source aquifers form, and why did they drain so rapidly? Sci. Rep. 513404. https://doi.org/10.1038/srep13404.

Rodríguez, J.A.P., et al., 2015. Did the martian outflow channels mostly form during the Amazonian period? Icarus 257, 387–395.

Rodriguez-Garcia, M., et al., 2015. Formation of oligopeptides in high yield under simple programmable conditions. Nat. Commun. 6, 1–7.

Roten, C.-A.H., Gallusser, A., Borruat, G.D., Udry, S.D., Karamata, D., 1998. Impact resistance of bacteria entrapped in small meteorites. Bull. Soc. Vaud. Sci. Nat. 86, 1–17.

Saghatelian, A., et al., 2001. A chiroselective peptide replicator. Nature 409, 797–801.

Salese, F., et al., 2016. Hydrological and sedimentary analyses of well-preserved paleofluvial-paleolacustrine systems at Moa Valles, Mars. J. Geophys. Res. Planets 121, 194–232.

Schlesinger, G., Miller, S.L., 1983. Prebiotic syntheses in atmospheres containing CH_4, CO and CO_2. 1. Amino acids. J. Mol. Evol. 19, 376–382.

Schmitt-Kopplin, P., et al., 2010. High molecular diversity of extraterrestrial organic matter in Murchison meteorite revealed 40 years after its fall. Proc. Natl. Acad. Sci. U. S. A. 107, 2763–2768.

Schrum, J.P., et al., 2010. The origins of cellular life. Cold Spring Harbor Persp. Biol. 2, a002212.

Stalport, F., et al., 2009. Investigating the photostability of carboxylic acids exposed to Mars surface ultraviolet radiation conditions. Astrobiology 9, 543–549.

Stewart, S.T., Nimmo, F., 2002. Surface runoff features on Mars: testing the carbon dioxide formation hypothesis. J. Geophys. Res. 107, 5069–5080. https://doi.org/10.1019/2000JE001465.

Stoker, C.R., Bullock, M.A., 1997. Organic degradation under simulated Martian conditions. J. Geophys. Res. 102, 10881–10888.

Stoks, P.G., Schwartz, A.W., 1982. Basic nitrogen-heterocyclic compounds in the Murchison meteorite. Geochim. Cosmochim. Acta 46, 309–315.

ten Kate, I.L., et al., 2005. Amino acid photostability on the Martian surface. Meteorit. Planet. Sci. 40, 1185–1193.

Terfort, A., von Kiedrowski, G., 1992. Self-replication by condensation of 3-aminobenzamidines and 2-formyl-phenoxyacetic acids. Angew. Chem. Int. Ed. Engl. 31, 654–656.

Treiman, A.H., et al., 1993. Preterrestrial aqueous alteration of the Lafayette (SNC) meteorite. Meteoritics 28, 86–97.

Vago, J.L., et al., 2017. Habitability on early Mars and the search for biosignatures with the ExoMars Rover. Astrobiology 17, 471–510.

Villanueva, G.L., et al., 2015. Strong water isotopic anomalies in the Martian atmosphere: probing current and ancient reservoirs. Science 348, 218–221.

Wächtershäuser, G., 1988. Before enzymes and templates: theory of surface metabolism. Microbiol. Mol. Biol. Rev. 52, 452–484.

Wächtershäuser, G., 1994. Life in a ligand sphere. Proc. Natl. Acad. Sci. U. S. A. 91, 4283–4287.

Wächtershäuser, G., 2007. On the chemistry and evolution of the pioneer organism. Chem. Biodivers. 4, 584–602.

Weber, P., Greenberg, J.M., 1985. Can spores survive in interstellar space? Nature 316, 403–407.

Westall, F., et al., 2015. Biosignatures on Mars: what, where, and how? Implications for the search for Martian life. Astrobiology 15, 998–1029.

Wilson, J.T., et al., 2018. Equatorial locations of water on Mars: improved resolution maps based on Mars Odyssey Neutron Spectrometer data. Icarus 299, 148–160.

Wordsworth, R.D., 2016. The climate of early Mars. Annu. Rev. Earth Planet. Sci. 44, 381–408.

Wordsworth, R.D., et al., 2017. Transient reducing greenhouse warming on early Mars. Geophys. Res. Lett. 44, 665–671.

Wright, I.P., et al., 1989. Organic materials in a Martian meteorite. Nature 340, 220–222.

Yen, A.S., et al., 2005. An integrated view of the chemistry and mineralogy of Martian soils. Nature 435, 49–54.

Yen, A.S., et al., 2000. Evidence that the reactivity of the Martian soil is due to superoxide ions. Science 289, 1909–1912.

Zaug, A.J., Cech, T.R., 1986. The intervening sequence RNA of Tetrahymena is an enzyme. Science 231, 470–475.

CHAPTER 3

Remote Detection of Phyllosilicates on Mars and Implications for Climate and Habitability

Janice L. Bishop

Contents

3.1 OVERVIEW

Identification of phyllosilicates and short-range ordered (SRO) materials on Mars has led to new insights about the early martian climate and habitability. Phyllosilicates were first conclusively identified on Mars in 2005 using visible/near-infrared (VNIR) remote sensing by the Observatoire pour la Minéralogie, l'Eau, les Glaces et l'Activité (OMEGA) instrument. Following these detections, phyllosilicates and poorly crystalline aluminosilicates were documented in numerous large and small outcrops using the Compact Reconnaissance Imaging Spectrometer for Mars (CRISM) VNIR instrument. Analyses of the Thermal Emission Spectrometer (TES) imagery also support the presence of phyllosilicates and SRO materials in several locations on Mars. These orbital detections are reinforced by measurements taken by the Chemistry and Mineralogy (CheMin) instrument on the Mars Science Laboratory (MSL) rover at Gale Crater. Characterization of Mars analog materials containing phyllosilicates such as smectites, kaolin-serpentine

From Habitability to Life on Mars
https://doi.org/10.1016/B978-0-12-809935-3.00003-7
37

group clay minerals, chlorites, mica, talc, prehnite, and SRO phases such as hydrated silica, allophane, imogolite, ferrihydrite, schwertmannite, and akaganéite has enabled constraints to be placed on the geochemical conditions of formation for alteration materials on Mars. Association of phyllosilicates with related minerals such as carbonates, sulfates, iron oxides/hydroxides, and Cl salts on Mars provides further constraints on the temperature, pH, salinity, and water/rock ratio of potentially habitable environments on Mars. These observations together support warm and wet conditions on early Mars where liquid water was stable on the surface during parts of the Noachian, followed by periods during the Hesperian and Amazonian where surface water was only transient. Subsurface alteration and formation of clays also occurred on Mars during its early history and may have enabled the formation of clay minerals more recently when surface conditions no longer could support liquid water. Both surface and subsurface aqueous environments could have provided important niches for astrobiology on Mars. The phyllosilicates and associated minerals formed in subaqueous and subaerial surface environments on Mars provide constraints on the climate. Smectite clays that formed in surface environments mark a time when liquid water was stable on the surface with temperatures likely 20°C or warmer. Transitions to SRO materials rather than phyllosilicates on Mars indicate a change in climate to colder conditions where water was only transiently present on the surface.

3.2 PRESENCE OF PHYLLOSILICATES AND POORLY CRYSTALLINE ALUMINOSILICATES ON MARS

The current understanding of Mars as a planet containing abundant phyllosilicates on the surface (Murchie et al., 2009a; Ehlmann and Edwards, 2014; Carter et al., 2015a) was formed over the past decade. The martian surface chemistry measured by the Viking landers first led scientists to believe clay minerals were likely present on Mars. Modeling of the major element data from instruments on Viking was found to be consistent with 60–80 wt% smectite (Baird et al., 1977; Toulmin III et al., 1977). However, telescopic spectra of Mars collected over regions spanning hundreds of kilometers in the 1980s and 1990s did not provide evidence of any clay minerals (McCord et al., 1982; Singer, 1985; Bibring et al., 1990; Bell III et al., 1994). As the signal-to-noise ratio of these data improved and the spot size became smaller and still no clay minerals were found (Christensen et al., 2001), skepticism about clay minerals on Mars arose. Phyllosilicates were definitively identified on Mars in 2005 using VNIR spectra collected by the OMEGA instrument at 1–3 km spot sizes (Poulet et al., 2005).

Soon after OMEGA's confirmation of clays on Mars, the CRISM instrument was launched on the Mars Reconnaissance Orbiter and began collecting targeted high-resolution (18 m/pixel) spectral images in 2006. Analyses of these VNIR CRISM images are largely responsible for characterization of the clay-rich outcrops on Mars (e.g., Mustard et al., 2008; Murchie et al., 2009a). These phyllosilicate-rich regions include a wide range of clay minerals and are frequently mixed with other aqueous alteration products including opal, sulfates, carbonates, iron oxides/oxyhydroxides, and poorly crystalline aluminosilicates (e.g., Bishop et al., 2008b, 2013a; Bishop and Rampe, 2016; Ehlmann et al., 2008b, 2009, 2010; Weitz et al., 2011, 2014b; Milliken et al., 2010; McKeown et al., 2009a). Phyllosilicates have different spectral properties depending on their crystal structures (Clark et al., 1990; Bishop et al., 2008a). The formation conditions (water/rock ratio, temperature, pH, and water chemistry) vary for each type of phyllosilicate (e.g., Chamley, 1989; Velde, 1995), and for this reason, phyllosilicates can be used to provide information about the aqueous geochemical environment and climate at the time of clay formation or subsequent alteration. In fact, it is the abundance of phyllosilicates in ancient martian rocks that shaped the view that early Mars was warm and wet (Bibring et al., 2006), although this argument was presented previously to explain geologic features such as valley networks, dendritic channels, and deltas that indicate frequent running water and fluvial erosion on early Mars (Ansan et al., 2008; Fassett and Head, 2011; Craddock and Howard, 2002). Phyllosilicates likely formed in a variety of environments on Mars but primarily during the Noachian period (Fig. 3.1).

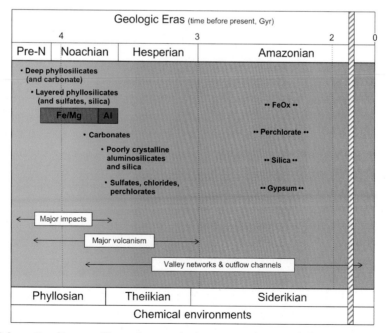

Fig. 3.1 Schematic diagram illustrating geologic eras and mineral formation on Mars. This approximate timeline of the formation of phyllosilicates and associated minerals was derived based on this study and previous studies (Bibring et al., 2006; Murchie et al., 2009a, b, 2018; Ehlmann et al., 2011a, b).

Phyllosilicates and associated alteration phases form in aqueous environments on the surface of Mars and also in subsurface aqueous environments. Subsurface environments were proposed for the formation of "crustal" clays on Mars that require elevated temperatures and different conditions from those found on the surface (Ehlmann et al., 2011b, 2013). The NE Syrtis region neighboring Isidis basin hosts multiple outcrops of phyllosilicates including mixtures of Fe/Mg smectite, chlorite, prehnite, serpentine, and possibly talc (Mustard et al., 2007, 2009; Ehlmann et al., 2009; Brown et al., 2010; Viviano et al., 2013). Many of these Mg-rich phyllosilicate assemblages in the Nili Fossae region occur as varying outcrops with different morphologies in neighboring environments, which is also consistent with multiple, distinct subsurface regions controlled by different chemical environments. These crustal clays formed at elevated temperatures near 150–300°C where chlorite or serpentine is present (Velde, 1995) and in some cases up to 400°C where prehnite is present (Velde, 1995; Ehlmann et al., 2011a). These subsurface conditions are not associated with the surface environment or climate on Mars.

In contrast, phyllosilicate outcrops dominated by smectites on Earth are typically formed in subaqueous or subaerial surface environments (Chamley, 1989). These phyllosilicates are a common alteration product of volcanic ash and tephra (e.g., Grim, 1968; Chamley, 1989; Cuadros et al., 1999). Smectites are swelling clays with expandable interlayer regions. They occur as trioctahedral (three cations per formula unit in octahedral sites, Mg^{2+}- or Fe^{2+}-rich) and dioctahedral (two cations per formula unit in octahedral sites, Al- or Fe^{3+}-rich) phyllosilicates. Smectites typically form in temperate to warm climates with alternating wet (>50 cm/year) and dry seasons. Mawrth Vallis is a location on Mars at the border of the southern highlands and northern plains in Arabia Terra that exhibits wide expanses of Fe/Mg smectite with vertical profiles of phyllosilicates, iron oxides/oxyhydroxides, and sulfates (Bishop et al., 2013a) consistent with formation in surface environments. These outcrops are similar to the multiple, colored, and clay-bearing horizons observed at the Grand Canyon and Painted Desert in Arizona (e.g., McKeown et al., 2009b). A common stratigraphy containing phyllosilicates, hydrated silica, and sulfates is found across hundreds to thousands of kilometers in this region (Noe Dobrea et al., 2010; Bishop et al., 2013a). A 150–200 m-thick unit containing abundant Fe/Mg smectite (e.g., Mg-rich nontronite) is the dominant aluminosilicate material observed here (Loizeau et al., 2007, 2010). It is covered by sulfates in some areas (Wray et al., 2010; Farrand et al., 2009; Bishop et al., 2016) and then a 50 m-thick Al-phyllosilicate/opal unit (Loizeau et al., 2010; Bishop et al., 2013a). Such dioctahedral smectite-rich outcrops and laterally extensive vertical profiles of Fe/Mg smectites, sulfates, and Al-rich clay assemblages are more consistent with formation in surface environments (Bishop et al., 2018).

The importance of poorly crystalline aluminosilicates including allophane and imogolite as soil components and protoclays on Earth was recognized long ago

(Wada et al., 1972; Farmer et al., 1979, 1983; Parfitt et al., 1980). More recently, the importance of these for remote sensing on Mars has been recognized (Rampe et al., 2012; Bishop and Rampe, 2016). Poorly crystalline aluminosilicates such as Fe-rich opal and allophane were observed at Coprates Chasma (Weitz et al., 2014a), the nanophase iron hydroxide akaganéite was observed in the region surrounding Gale Crater (Carter et al., 2015b), and allophane- and imogolite-type materials were observed at Mawrth Vallis (Bishop and Rampe, 2016). The investigation at Mawrth Vallis found that the poorly crystalline aluminosilicates are present at the top of the clay profile and that these were linked to a change in climate from the warm and wet conditions necessary for smectite formation to a colder environment where water activity was limited (Bishop and Rampe, 2016).

3.3 REMOTE DETECTION OF PHYLLOSILICATES AND RELATED MATERIALS AT MARS

Identification and characterization of phyllosilicates and poorly crystalline aluminosilicates and other SRO materials on Mars is mainly conducted using orbital VNIR reflectance spectra from $\sim$0.4–4 μm acquired by CRISM on the Mars Reconnaissance Orbiter (Mustard et al., 2008; Murchie et al., 2009a) and from $\sim$0.4 to 5 μm using OMEGA on Mars Express (Bibring et al., 2005). Thermal infrared (TIR) spectra collected in the mid-IR (MIR) region from $\sim$200 to 2000 cm^{-1} ($\sim$5–50 μm) by TES on Mars Global Surveyor (Christensen et al., 2001) also provide orbital information about mineralogy including phyllosilicates and poorly crystalline aluminosilicates. The MSL rover uses an X-ray diffraction (XRD) instrument called CheMin to identify minerals on the surface at the landing site inside Gale Crater (Blake et al., 2013). CheMin results have been coordinated with APXS data to infer the composition of phyllosilicates and SRO materials (e.g., Vaniman et al., 2014).

Remote sensing of phyllosilicates and SRO materials is performed at Mars using reflectance spectra at VNIR wavelengths with the CRISM and OMEGA instruments and using thermal emission spectra at mid-infrared (MIR) wavelengths with the TES instrument. Remote detection of phyllosilicates and SRO materials in the MIR region with TIR spectra uses Si(Al)-O stretching and bending vibrations and M-OH bending vibrations (where M is a metal cation such as Al, Fe, or Mg), while NIR detections of phyllosilicates and related materials employ primarily overtone and combination bands for H_2O and OH vibrations. Fe excitation features are used in the VNIR region as well to characterize Fe-bearing materials. Coordinated investigations using both VNIR and MIR regions enable optimal detection of rocks and minerals on Mars. XRD identifies phyllosilicates using their mineral structures. The instrument must be in contact with the sample, so XRD can be used on surface missions but not in orbit.

3.3.1 Detection of Phyllosilicates and SRO Materials on Mars Using VNIR Spectra

Detection of phyllosilicates and SRO materials in the VNIR region varies with mineral structure and cation type and has been summarized recently (Bishop et al., 2017). Remote detection of phyllosilicates and associated phases on Mars has been enabled through decades of lab investigations of pure clay minerals (King and Clark, 1989; Clark et al., 1990; Post and Noble, 1993; Bishop et al., 1994, 2008a; Petit et al., 1999, 2004; Decarreau et al., 2008), SRO materials (e.g., Anderson and Wickersheim, 1964; Bishop et al., 2013b, 2015; and Bishop and Murad, 2002), phyllosilicate mixtures (e.g., Madejová et al., 2002; McKeown et al., 2011; Cuadros et al., 2015; Michalski et al., 2015), and clay-bearing Mars analogs (e.g., Bishop et al., 2002; Hamilton et al., 2008; Ehlmann et al., 2012). VNIR spectra of selected phyllosilicates and SRO materials are shown in Fig. 3.2 in order to illustrate how these minerals/phases can be detected on Mars.

Smectites are the most common phyllosilicates detected on Mars and include Al-rich montmorillonite and beidellite, Fe-rich nontronite, and Mg-rich hectorite and saponite, as well as mixed cation smectites. All smectite spectra exhibit an H_2O stretching overtone at 1.41 μm and a H_2O combination band centered at 1.91 μm (Fig. 3.2, Bishop et al., 1994). VNIR smectite spectra include MOH combination bands as well that depend on the presence of Al, Fe, or Mg cations in octahedral sites in the mineral structure and are often the most diagnostic for smectites and other phyllosilicates on Mars.

Montmorillonites exhibit an Al_2OH stretching plus bending combination band near 2.20–2.21 μm, depending on if there is some Fe or Mg substitution for Al in the octahedral sites (Bishop et al., 2002). The Al_2OH stretching overtone for montmorillonite lies at 1.41 μm, at the same wavelength as the overtone of the H_2O stretching vibration, which was a source of confusion for decades in making band assignments. Beidellite spectra exhibit Al_2OH bands at 1.40 and 2.18 μm (Bishop et al., 2011), which occur at shorter wavelengths than the related bands observed in montmorillonite spectra due to increased tetrahedral Al substitution for Si in beidellite. Nontronite spectra (Bishop et al., 2002) include an $Fe_2^{3+}OH$ overtone at 1.43 μm and an $Fe_2^{3+}OH$ combination band at 2.29 μm, while saponite spectra (Post, 1984) include an Mg_3OH stretching overtone at 1.39 μm and an Mg_3OH combination band at 2.31 μm. VNIR smectite spectra include an additional OH band that is diagnostic of the mineral chemistry at 2.38 μm for saponite, 2.39 μm for hectorite, 2.41 μm for nontronite, 2.44 μm for beidellite, and 2.45 μm for montmorillonite (e.g., Bishop et al., 2008b).

Several additional phyllosilicates have been found on Mars as well, but they are less common than smectites. Clays from the kaolinite-serpentine group are also frequently detected on Mars. Except for halloysite, phyllosilicates in this group do not have H_2O bands. Kaolinite spectra exhibit an Al_2OH stretching overtone doublet near 1.4 μm and a doublet at 2.17 and 2.21 μm for the stretching plus bending combination bands. Serpentine spectra include an Mg_3OH stretching overtone at 1.39 μm and an Mg_3OH

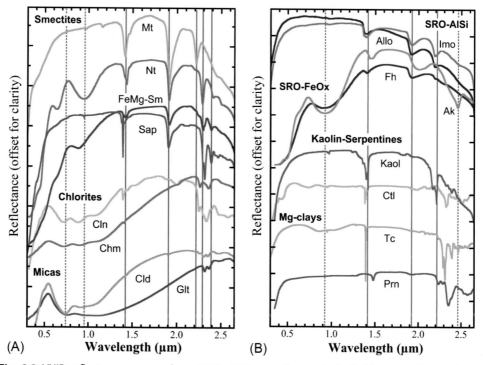

Fig. 3.2 VNIR reflectance spectra from 0.3 to 2.65 μm of several phyllosilicates and SRO phases observed on Mars. The spectra are offset for clarity and grouped by type. (A) Smectites: montmorillonite (Mt), nontronite (Nt), Fe/Mg smectite (Fe/Mg-Sm), and saponite (Sap); chlorites: clinochlore (Cln) and chamosite (Chm); micas: celadonite (Cld) and glauconite (Glt). (B) SRO aluminosilicates (AlSi): allophane (Allo) and imogolite (Imo); SRO iron oxides/oxyhydroxides (FeOx): ferrihydrite (Fh) and akaganéite (Ak); kaolinite-serpentines: kaolinite (Kaol) and chrysotile (Ctl); other Mg-rich clays: talc (Tal) and prehnite (Prn); gray lines mark features near 1.4, 1.9, 2.2, 2.3, 2.4, and 2.5 μm in order to facilitate the comparison of the spectral features among different phyllosilicates. Dotted lines mark features due to Fe: the Fe^{2+} electronic excitation near 0.75 μm in chlorites and micas, the Fe^{3+} electronic excitation near 0.95 μm in nontronite, and an FeOH combination band at 2.46 μm in akaganéite. These spectra were measured as particulate samples at Brown University's RELAB for previous studies (Bishop et al., 2008a, 2013b) or at the USGS Spectroscopy Lab (Clark et al., 2007).

combination band near 2.33 μm. Many serpentine spectra also include bands near 2.52–2.57 μm.

Chlorites or mixed-layer smectite-chlorites (corrensites) have also been observed in VNIR spectra of Mars and include similar OH bands to serpentines near 1.39 and 2.33–2.35 μm. These band positions vary depending on the relative abundance of Fe and Mg in the clay structure. Chlorites can be distinguished from serpentines by the presence of an additional AlMgOH (or $AlFe^{2+}OH$) combination band near 2.25 μm. Additional phyllosilicates such as mica, illite, palygorskite, sepiolite, and talc have spectral features similar to the others, where water bands occur near 1.4 and 1.9 μm and OH combination bands occur near 2.20–2.22 μm for Al-rich clays and near 2.30–2.36 μm for Mg-rich clays.

SRO materials such as nanophase or amorphous aluminosilicates, opal, and ferrihydrite can be difficult to characterize using XRD data but may be readily detectable using spectroscopy. The VNIR spectral features for opal include H_2O and OH stretching overtones near 1.39–1.41 μm depending on the degree of H-bonding, an H_2O combination band near 1.91 μm, and a Si-OH combination band centered near 2.21–2.23 μm with a shoulder toward longer wavelengths depending on the degree of H-bonding. The nanophase (np) aluminosilicates allophane and imogolite have VNIR spectral features similar to those of Al-rich smectites and opal. The primary difference is the broadened spectral features for the np-aluminosilicates due to a distribution of AlOH and Si-OH sites in the structure. Another difference is that the H_2O combination band is centered at 1.92 μm instead of 1.91 μm. The corresponding MOH bands are also shifted. The Al_2OH stretching overtone occurs as a doublet at 1.38 and 1.40 μm for allophane and at 1.37 and 1.39 μm for imogolite, while the Al_2OH combination band occurs near 2.19 μm for both of these materials. The H_2O bands for ferrihydrite are typically shifted toward even longer wavelengths toward 1.42 and 1.93 μm, and FeOH combination bands occur near 2.3 μm in some cases.

3.3.2 Detection of Phyllosilicates and SRO Materials on Mars in TIR Spectra

TES spectra of Mars have provided global information about rock composition across the planet (Christensen et al., 2001) through modeling TES spectra (Rogers and Christensen, 2007) with lab TIR spectra of numerous minerals (Christensen et al., 2000b) and rocks (Hamilton and Christensen, 2000). Detailed lab studies of clay minerals (Michalski et al., 2005, 2006) and related materials such as zeolites (Ruff, 2004), glass (Minitti et al., 2002; Byrnes et al., 2007), and amorphous phases (Kraft et al., 2003; Rampe et al., 2012) were performed to support detection of these phases by TES. Phyllosilicates are typically pressed into powders in order to improve the band strength in emission spectra (Michalski et al., 2005). Pressing finely grained materials into powders decreases pore space and reduces internal scattering, which strengthens the reststrahlen bands (reflectance maximum or emission minimum). Most importantly, clay minerals occurring naturally within rocks behave spectroscopically similar to coarse materials or pressed pellets, whereas clays in fine-grained regolith or dust behave more like fine particulates. TIR spectral studies of clay-bearing rocks were also performed (e.g., Wyatt et al., 2001; Michalski et al., 2004; Hamilton et al., 2008; Ehlmann et al., 2012; Rampe et al., 2013) in order to support the identification of clays on planetary surfaces.

3.3.2.1 Si-O Stretching Vibrations

The strongest bands for the identification of clay minerals in thermal remote sensing typically result from vibrations of the tetrahedral sheet $Si(Al,Fe)O_4$ (Michalski et al.,

2006). However, related $Si(Al,Fe)O_4$ vibrations are observed in spectra of other silicates as well. Salisbury (1993) describes the Christiansen feature (CF) as a reflectance minimum or emission maximum located at the short-wavelength edge of the Si-O reststrahlen feature. The CF occurs where the imaginary component of the index of refraction approaches 0 and the real component approaches 1. Salisbury (1993) noted shifts in the wavelength of this feature with mineral composition. Typically, the CF occurs at longer wavelengths for clay minerals compared with mafic silicates (Salisbury, 1993). Michalski et al. (2005) observed a shift in the SiO_4 stretching vibrations in spectra of phyllosilicates, aluminosilicate glasses, and altered phases, such that the band near $1000 \ cm^{-1}$ transitions toward lower wavenumbers (longer wavelengths) with increasing abundance of tetrahedral Al and Fe and decreasing abundance of Si. This is consistent with the trends observed by Salisbury et al. (1991) for a large collection of silicate minerals and with observed transmittance spectra of clay minerals (Farmer, 1974). For example, Michalski et al. (2005) noted an emissivity minimum at $8.8 \ \mu m$ ($1135 \ cm^{-1}$) for montmorillonite with a Si/O ratio of 0.399 compared with an emissivity minimum at $9.5 \ \mu m$ ($1056 \ cm^{-1}$) for nontronite with a Si/O ratio of 0.349.

The $Si(Al,Fe)O_4$ vibrations observed in emission spectra of several phyllosilicates are shown in Fig. 3.3. The Si-O stretching band center exhibits a trend of increasing wavelength across the smectites as the octahedral cation changes from Al to Fe to Mg. The Si-O stretching band position for Al-rich smectites is similar to that observed for kaolinite and aluminosilicate gel. The Si-O stretching band occurs at longer wavelengths for the poorly crystalline allophane and imogolite and also for phyllosilicates rich in Fe and Mg (Fig. 3.3).

3.3.2.2 Si-O Bending Vibrations

The bending vibrations for tetrahedral SiO_4 groups and lattice deformation modes such as $Si-O-M_{oct}$ absorptions (where M_{oct} is an octahedral cation such as Al, Fe, and Mg) occur near $400-550 \ cm^{-1}$ ($\sim18-25 \ \mu m$) in emission spectra when little or no tetrahedral substitution occurs (Michalski et al., 2005; Salisbury et al., 1991). Spectra of serpentines, chlorites, and biotite contain a strong SiO_4 bending vibration centered near $470-485 \ cm^{-1}$ ($\sim20.6-21.3 \ \mu m$). In tetrahedral-octahedral-tetrahedral (TOT) configuration clays, bands associated with $Si-O-M_{oct}$ absorptions partially overlap with the Si-O bending absorption. In trioctahedral clays, the overlap between the Si-O-Si and $Si-O-Mg_{oct}$ bending vibrations is significant, resulting in a single, strong band located near $480 \ cm^{-1}$ ($\sim20.8 \ \mu m$), which requires modeling to resolve the individual vibrational components. However, for dioctahedral clays, the $Si-O-Al_{oct}$ and $Si-O-Fe^{3+}_{oct}$ bending bands shift to shorter wavelengths ($510-540 \ cm^{-1}$ or $18.6-19.6 \ \mu m$), separating them from the Si-O-Si bending vibrations. This difference is important because the resulting single or double band can be used to distinguish dioctahedral and trioctahedral clay compositions using long-wavelength thermal emission remote sensing data

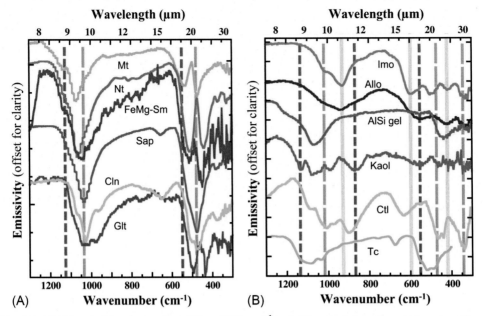

Fig. 3.3 Mid-IR emissivity spectra from 300 to 1300 cm^{-1} ($\sim$8–30 μm) of clay minerals found on Mars. The spectra are offset for clarity and grouped by type. (A) Smectites: montmorillonite (Mt), nontronite (Nt), Fe/Mg smectite (FeMg-Sm), and saponite (Sap); chlorite: clinochlore (Cln); mica: glauconite (Glt). (B) Poorly crystalline aluminosilicates: imogolite (Imo), allophane (Allo), and aluminosilicate gel (Al/Si gel); kaolinite-serpentines: kaolinite (Kaol) and chrysotile (Ctl); other Mg-clay: talc (Tc). In panel A, medium gray dashed lines mark the Si-O stretching band of saponite near 1035 cm^{-1} and the bending band near 480 cm^{-1}, while dark gray dashed lines mark bands near 1125 and 545 cm^{-1} found for montmorillonite. In panel B, light gray lines mark features near 930, 595, and 415 cm^{-1} found in imogolite; medium gray dashed lines mark features near 1015, 470, and 340 cm^{-1} found in serpentine; and dark gray dashed lines mark features near 1130, 865, and 550 cm^{-1} found in kaolinite. These spectra were measured as pressed powders or particulate samples at the Mars Space Flight Facility at Arizona State University for previous studies (Michalski et al., 2005; Bishop et al., 2008a, 2013b; Rampe et al., 2012).

(Michalski et al., 2010a). Coordinated bending also occurs along the Si-O-M$_{oct}$ bonds from the tetrahedral Si-O into octahedral O-M bonds and is responsible for the weaker bands that often occur near 550–670 cm^{-1} ($\sim$15–18 μm) for Al and Fe cations and near 400–450 cm^{-1} ($\sim$22–24 μm) for Mg cations. Si-O-Mg$_{oct}$ bending vibrations are present in the emission spectra of the Mg-rich clays saponite, chrysotile, talc, and clinochlore (Fig. 3.3).

3.3.3 Detection of Phyllosilicates and SRO Materials on Mars by Rovers

The CheMin instrument has identified phyllosilicates, SRO materials, and numerous minerals at several sites inside Gale Crater (Blake et al., 2013; Bish et al., 2013;

Vaniman et al., 2014). CheMin results indicate that mobile surface dust is primarily composed of basalt without evidence of aqueous alteration (Bish et al., 2013), while mudstones investigated at a region called Yellowknife Bay within Gale Crater contain trioctahedral smectite clays and amorphous components that are thought to have formed in situ by aqueous alteration of olivine (Bristow et al., 2015; Vaniman et al., 2014). SRO phases and phyllosilicates were also identified by CheMin in sandstone rocks from the Windjana drill site in the Kimberley area (Treiman et al., 2016). The presence of smectite clays with little alteration to chlorite as observed by CheMin is consistent with orbital observations of Fe/Mg smectite using CRISM imagery (Milliken et al., 2010; Wray, 2013). Mudstones collected from the Murray formation at Gale Crater contain ~25 wt% phyllosilicate, including both dioctahedral Al-rich smectite and trioctahedral Mg-rich smectite, and ~30 wt% SRO phases (Bristow et al., 2017).

The Yellowknife Bay sediments contain ~20 wt% trioctahedral smectite identified through the $CoK\alpha$ patterns between 8–11°2θ (or 9.4–12 Å) for the 001 reflection and 22–23°2θ for the *021* reflection with some variability between the John Klein and Cumberland samples (Bristow et al., 2015). The position of the broad *021* reflection near 22.5°2θ is characteristic of trioctahedral smectites (Moore and Reynolds, 1997), while the shape and position of the 001 and *021* peaks are most consistent with synthetic Fe^{2+}-bearing saponite (Chemtob et al., 2015) or Fe^{3+}-bearing saponite from Griffith Park (Treiman et al., 2014). The Windjana sediments contain ~25 wt% SRO materials and phyllosilicates where the bulk of this is X-ray amorphous material similar to allophane and ferrihydrite with some Fe/Mg smectite and kaolinite (Treiman et al., 2016). If the clays are poorly crystalline, they could be contributing to the X-ray amorphous phase. The poorly crystalline Fe^{3+} oxyhydroxide mineral akaganéite, found in Yellowknife Bay sediments (Ming et al., 2014) and the Windjana sample (Treiman et al., 2016), has been identified from orbit as well near Gale Crater (Carter et al., 2015b) using spectral features near 2.46 μm (Bishop et al., 2015).

The clay-bearing rocks at Yellowknife Bay are Hesperian (~3.4–3.7 Ga) and were derived from a low-temperature, neutral environment under low water/rock ratio conditions with limited chemical alteration (Grotzinger et al., 2014; McLennan et al., 2014). Diagenetic features such as Ca-sulfate veins, raised ridges, fractures, and concretions in the Yellowknife Bay sediments were analyzed together with geochemistry in order to deduce that alteration under diverse depositional and diagenetic sedimentary environments was responsible for the formation of the phyllosilicates there (McLennan et al., 2014). Degradation of the rim of Gale Crater is thought to have allowed water to flow into the system bringing sediments and forming a lake basin with a depth of 75 m or more (Grotzinger et al., 2015). This lake may have existed intermittently for thousands to millions of years producing thick sequences of sedimentary rocks (Grotzinger et al., 2015) and enabling the formation of the phyllosilicates, amorphous materials, sulfates, Cl salts, and iron oxide-bearing minerals (Bristow et al., 2015).

Characterizing clays and SRO materials with the Mars Exploration Rover (MER) at Gusev crater and Meridiani Planum was more difficult because neither a VNIR spectrometer nor an XRD instrument were included on these rovers. However, results from the Mössbauer instrument, mini-TES, and chemistry from the alpha particle X-ray spectrometer (APXS) were coordinated to identify silica, carbonate, and sulfates at Gusev crater (e.g., Morris et al., 2004; Ming et al., 2006; Parente et al., 2009; Rice et al., 2010; McSween et al., 2008; Ruff and Farmer, 2016). Fe/Mg-carbonate was detected from orbit at the Home Plate site using VNIR spectral data from CRISM (Carter and Poulet, 2012) that coordinated well with the surface detection of carbonate using Mössbauer, mini-TES, and chemistry data (Morris et al., 2008).

The Meridiani Planum region contains friable sandstone bedrock that includes sulfates, FeMg smectite, and hematite concretions (e.g., Christensen et al., 2000a; Squyres et al., 2006; Wray et al., 2009; Wiseman et al., 2010; Noe Dobrea et al., 2012). The sulfates include jarosite (Klingelhöfer et al., 2004), gypsum (Grotzinger et al., 2005), and Mg-sulfates (Wiseman et al., 2010) identified through analyses of the Mössbauer, APXS, and mini-TES data from Opportunity and VNIR CRISM spectra from orbit. Later in the mission, Opportunity drove to Endeavour crater and found fine-grained, layered rocks with Ca-sulfate veins (Arvidson et al., 2014). Based on chemistry measured by the rover and phyllosilicate signatures from orbit, both Al-smectites (Arvidson et al., 2014) and Fe-rich smectites appear to be present in the layered rocks (Fox et al., 2016).

3.3.4 Detection of Phyllosilicates and SRO Materials in Martian Meteorites

Fe-rich phyllosilicates have been detected together with iron oxides/hydroxides and carbonates in several nakhlites, shergottites, and chassignites (Gooding, 1992; Treiman et al., 1993). The orthopyroxenite ALH 84001 also contains small amounts of an Fe-rich poorly crystalline material as part of the aqueous alteration phase (McKay et al., 1996). Unfortunately, the phyllosilicates present in meteorites tend to occur in extremely tiny aliquots, on the scale of nanometer or smaller, and are poorly crystalline, thus challenging researchers to identify and characterize them. The nakhlite meteorite group contains the largest quantity of phyllosilicates, and for this reason, they have been characterized in more detail. The most abundant phyllosilicates present in nakhlites include poorly crystalline trioctahedral Fe-bearing saponite and serpentine (Hicks et al., 2014; Hallis et al., 2014). Modeling of alteration pathways for the formation of the observed phyllosilicates and associated phases in martian meteorites suggests evaporation from low-temperature brines in contact with parent igneous rocks (Bridges et al., 2001). More recent modeling efforts on Fe-rich smectites in nakhlites indicate temperatures of 150–200°C with pH ~6–8 and a water/rock ratio ≤300 (Bridges and Schwenzer, 2012).

Identification of SRO aluminosilicates and iron oxides/hydroxides in martian meteorites has been increasing as technology is enabling characterization of smaller regions

(McKay et al., 1996; Hicks et al., 2014). There may not be much relationship between the phyllosilicates observed in martian meteorites and the clays observed on the surface of Mars (Bishop and Velbel, 2017), because of differences in the ages and compositions of these materials. The mixed clays and amorphous materials observed in martian meteorites are typically Amazonian (Lee et al., 2015, 2017), while nearly all occurrences of phyllosilicates on the martian surface are Noachian. The one martian meteorite containing Noachian alteration materials, ALH 84001, contains carbonate and magnetite, but not phyllosilicates (McKay et al., 1996; Treiman, 1998). One possible explanation for this discrepancy is that the clays in rocks on the martian surface were destroyed on impact or ejection at Mars or on entry to Earth.

3.4 CHARACTERIZATION OF PHYLLOSILICATES AND SRO MATERIALS ON MARS

Clay minerals have been primarily detected on Mars from orbit through analysis of VNIR spectral images acquired by the OMEGA (Bibring et al., 2005) and CRISM (Murchie et al., 2009a) imaging spectrometers. Most of the planet has been mapped at VNIR wavelengths at kilometers to ~200 m resolution using OMEGA and CRISM multispectral mode, while CRISM targeted observations at ~18 m resolution have been collected across the planet at sites of mineralogical interest. TES orbital imaging provides additional mineralogical context with near global coverage at 1–5 km resolution (Christensen et al., 2001). Clay minerals frequently occur on Mars in small outcrops that are below the spatial resolution of TES. However, modeling developed by the TES team enables estimations of major and minor surface components including clays and related altered aluminosilicate-bearing materials (Ramsey and Christensen, 1998; Rogers and Christensen, 2007; Rampe et al., 2012). Band parameters have also been developed for detection of Fe/Mg-phyllosilicates using thermal emission imaging system (THEMIS) imagery (Viviano and Moersch, 2012), which trades better spatial resolution for fewer channels (Christensen et al., 2003). Despite detection challenges, analyses of TES spectra have identified regions containing altered basalt, glass, zeolite, high-Si phases, and poorly crystalline aluminosilicates (Michalski et al., 2013; Michalski and Fergason, 2009; Rogers and Christensen, 2007; Rampe et al., 2012; Ruff, 2004). Recently, coordinated studies of CRISM and TES analyses have found that including poorly crystalline aluminosilicates in the model enables detection of clay minerals as well in TES data at Mawrth Vallis, one of the most phyllosilicate-rich sites of Mars (Bishop and Rampe, 2016).

3.4.1 Global Observations of Phyllosilicates, SRO Phases and Aqueous Alteration on Mars

The global distribution of clay-bearing outcrops on Mars is not homogenous (Fig. 3.4), with the majority of detections in the southern, Noachian-aged (>3.8 Gyrs old)

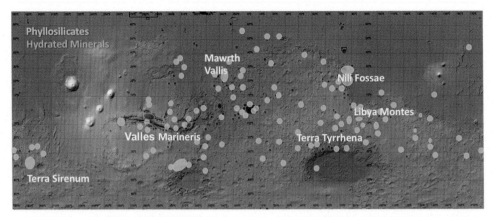

Fig. 3.4 Global map of phyllosilicates and hydrated minerals on Mars. Colored regions indicate detections from CRISM and OMEGA images, where phyllosilicates are displayed in green and associated hydrated materials are shown in blue (data from multiple sources summarized in Carter et al., 2013; Ehlmann et al., 2013).

highlands (Carter et al., 2013; Ehlmann et al., 2013). Terrains younger than the Late Hesperian (~3.5–3.8 Gyrs) are nearly devoid of phyllosilicate signatures, indicating that the bulk of the clay-forming environments occurred before that time. Phyllosilicate exposures are often small in size (from a few hundred meters to a few kilometers across), appearing in crater walls or ejecta and eroded surfaces where the ancient surface material is visible. Recent reviews of orbital remote sensing at Mars describe the current understanding of the martian surface composition including phyllosilicates and SRO materials (Bishop et al., 2017; Hamilton et al., 2019; Murchie et al., 2019). An early interpretation of Mars surface mineralogy and aqueous processes based on the analyses of OMEGA data found that clay minerals largely formed during the ancient Noachian period (Poulet et al., 2005; Bibring et al., 2006), which continues to hold for most of the planet (Ehlmann and Edwards, 2014; Murchie et al., 2019). Younger clay-rich materials have been observed in isolated regions such as Noctis Labyrinthus (e.g., Weitz et al., 2011), Coprates Chasma (e.g., Weitz et al., 2014a), and some impact craters (Carter et al., 2010). Early investigations of clay-bearing outcrops at Mawrth Vallis revealed that Al-rich phyllosilicates always occurred in strata above Fe/Mg-rich smectites (Bishop et al., 2008b). A more recent study found that this trend is common across most of the planet where clay exposures are observed (Carter et al., 2015a).

TES spectra provide a global perspective at a coarse spatial resolution (hundreds of square kilometers) on the occurrence and abundance of minerals including clays on Mars. Martian dark regions are spectrally dominated by basaltic materials in the thermal IR (Christensen et al., 2000c). Linear spectral unmixing of TES data suggests that dark volcanic surfaces are composed of approximately 25%–40% plagioclase feldspar, 25%–40% pyroxene, 15%–35% clay minerals or high-silica phases, and 5%–10% olivine

(Michalski et al., 2006; Rogers and Christensen, 2007). Spectral unmixing depends heavily on the spectral shape in the 900–1300 cm^{-1} region, where Si-O absorptions occur in all silicate minerals and amorphous phases. In this spectral range, phyllosilicates exhibit somewhat broad unremarkable features related to Si-O vibrations that are shared by other poorly crystalline materials with similar Si/O molar ratios (Michalski et al., 2005). Early results from TES showing 10%–15% of phyllosilicates or similar materials in the dark, low-latitude regions were difficult to interpret in the pre-OMEGA era but are now accepted as an average composition of Noachian martian crust. Such detections likely also include SRO aluminosilicates and iron oxide-bearing materials, possibly associated with weathered volcanic ash in the ancient crust (Rampe et al., 2012; Bishop and Rampe, 2016).

VNIR phyllosilicate exposures are observed in the peaks, walls, and ejecta of many impact craters over Noachian-aged terrains (e.g., Carter et al., 2013; Ehlmann et al., 2013). This indicates that clays are often buried up to a few kilometers. Impact craters over Hesperian- and Amazonian-aged terrains rarely excavate phyllosilicates, with the exception of larger craters (typically >20–200 km), which are thought to have penetrated the deeply buried and altered Noachian crust. Fewer phyllosilicates are detected in the northern lowlands of Mars, likely because Noachian-aged rocks are seldom exposed at the surface, where the few phyllosilicate signatures are restricted to remnant Noachian units and the largest craters.

Phyllosilicate exposures are numerous in nonimpact sites as well and include several expansive deposits spread across the southern highlands (e.g., Poulet et al., 2005; Bishop et al., 2008a, b, 2013c; Mustard et al., 2008; Murchie et al., 2009a, b; Ehlmann et al., 2009; Mckeown et al., 2009a; Milliken et al., 2010; Loizeau et al., 2012a; Carter et al., 2015a, b; Weitz et al., 2012, 2015; Weitz and Bishop, 2016). A variety of clay minerals have been observed in these outcrops (Ehlmann and Edwards, 2014; Murchie et al., 2019): smectites including nontronite, saponite, beidellite and montmorillonite, vermiculite, chlorite, mica, halloysite/kaolinite, serpentine, and prehnite, as well as mixed-layer clays, zeolite, opal, allophane, and imogolite. The most common phyllosilicate globally on Mars is an Fe-rich smectite with OH bands near 2.29–2.30 and 2.38–2.39 μm. Nearly all of these Fe-rich smectite occurrences differ spectrally from either nontronite or saponite and were termed Fe/Mg smectite because the spectral bands are consistent with Mg-bearing nontronite or smectite containing both Fe and Mg in octahedral sites (Bishop et al., 2008b). Variations also occur in the band centers of these spectra that are consistent with variations in the Fe and Mg abundance in octahedral sites.

Correlations between the clays and related mineralogy and their morphological context across the planet have been interpreted as originating from a number of surface and subsurface environments (Ehlmann et al., 2011b). These include detrital and authigenic lacustrine clays, climate-mediated surface weathering (involving rain or snowfall), evaporitic playa environments, and hydrothermal systems (deep crustal, epithermal, and possibly spring deposits). Geochemical conditions proposed for these environments range from highly acidic (pH ~2) to mildly alkaline, near freezing to >400°C, water-dominated to rock-dominated, mostly oxidizing, and with varying ion activities (Ehlmann et al., 2013; Carter et al., 2015a; Murchie et al., 2019).

The globally common Fe/Mg smectite (Carter et al., 2015a) is frequently mixed with chlorite or other high-temperature clays and termed simply Fe/Mg-phyllosilicate (Ehlmann et al., 2011b; Carter et al., 2013; Michalski et al., 2015). Pure chlorites are observed in ~20% of phyllosilicate detections, while Al-rich phyllosilicates (montmoril-lonite, beidellite, halloysite, and kaolinite) are found in about one-third of phyllosilicate detections. Hydrated SRO phases such as opal, hydrated silica, allophane, and imogolite are also observed in numerous aqueous outcrops. Hydrated sulfates, carbonates, and zeolites are also found in altered materials and are frequently associated with phyllosilicates (e.g., Murchie et al., 2019).

3.4.2 Regional Characterization of Phyllosilicates and Aqueous Alteration on Mars

Phyllosilicates have been investigated in detail at numerous sites on Mars. The most abundant clay-bearing units are present at Mawrth Vallis (Bishop et al., 2008b) and Nili Fossae (Ehlmann et al., 2009). However, phyllosilicates are also present across a broad region of Syrtis to NE Syrtis (Skok et al., 2010a, b; Ehlmann and Mustard, 2012; Bramble et al., 2017; Brown et al., 2010; Michalski et al., 2010a; Viviano et al., 2013; Mustard et al., 2009) and throughout the Libya Montes/Terra Tyrrhena region (Loizeau et al., 2012a; Bishop et al., 2013c), surrounding Argyre (Buczkowski et al., 2010) and at Tharsis (Viviano-Beck et al., 2017). Phyllosilicates have been detected at many of the rover landing sites (e.g., Wray et al., 2009; Milliken et al., 2010; Arvidson et al., 2014; Fox et al., 2016; Seelos et al., 2014; Poulet et al., 2014; Quantin et al., 2016), and they have been observed together with sulfates and other hydrated phases in Noctis Labyrinthus (Weitz et al., 2011; Thollot et al., 2012) and along chasma and plateaus near Valles Marineris (e.g., Milliken et al., 2008; Murchie et al., 2009b; Weitz et al., 2010, 2012, 2014b, 2015; Roach et al., 2010; Flahaut et al., 2015; Weitz and Bishop, 2016; Quantin et al., 2012; Le Deit et al., 2012). Variations in the clay mineralogy can sometimes be linked to specific geologic units with distinct morphologies. The main limitation in characterizing these clay-bearing units is probably linked to the availability of high-resolution CRISM images. Additional limitations come from masking of the surface by dust or caprock and the lack of sufficiently high spatial resolution to detect small outcrops. Another challenge is the lack of a direct analog on Earth for some of the alteration processes and geologic contexts observed on Mars. Three sites are described in detail as examples of the types of phyllosilicate-rich alteration observed on Mars.

3.4.2.1 The Phyllosilicate-Rich Mawrth Vallis Region in Eastern Chryse Planitia

The Chryse Planitia basin provides a diversity of terrains with ages ranging from Noachian to Amazonian (Tanaka et al., 2005). This basin has long been infilled by mass wasting, catastrophic flood deposits, and aeolian and volcanic resurfacing processes. The Chryse Planitia region exhibits the signatures of several aqueous environments present during the Noachian to early Hesperian time period. The Mawrth Vallis plateau in the eastern part of Chryse Planitia at the dichotomy of the southern highlands and northern lowlands contains a wealth of phyllosilicates and associated hydrated materials (Fig. 3.5). This region has been recognized as an example of surface pedogenesis

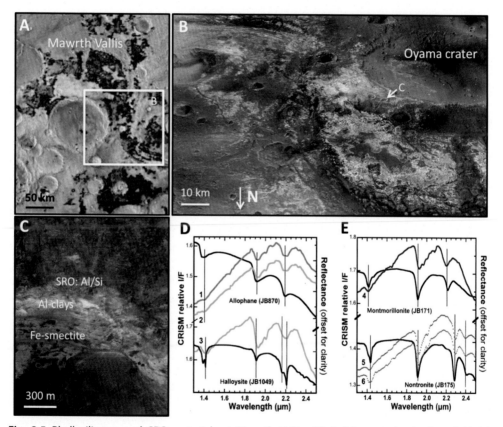

Fig. 3.5 Phyllosilicates and SRO materials at Mawrth Vallis. (A) Fe/Mg smectite (red) and Al-rich phyllosilicate (blue) detections (Carter et al., 2015a) are mapped over THEMIS IR imagery color coded (green-brown tones) with altimetry from the Mars Orbiter Laser Altimeter (MOLA) instrument (Smith et al., 2003). (B) An oblique view from the high-resolution stereo camera (HRSC) imager (Neukum et al., 2004) of light-toned phyllosilicate-rich material and fluvial surface features at Mawrth Vallis (5× vertical). (C) HiRISE view of layered phyllosilicates in a crater wall over an HRSC DTM with mineralogy from CRISM (Fe-rich smectite in red, Al-phyllosilicates in blue, and allophane in green). (D and E) Sample CRISM spectra illustrating the materials in (C) compared with lab spectra from the Bishop collection.

weathering on Mars based on its stratigraphy and composition (Michalski et al., 2010b; Bishop et al., 2013a). The lower part of the clay profile is a unit dominated by layered Fe/ Mg smectite ~100–200 m thick that likely formed by the deposition of altered volcanic sediments (e.g., Michalski and Noe Dobrea, 2007; Wray et al., 2008; Bishop et al., 2008b; McKeown et al., 2009a; Loizeau et al., 2010, 2012b; Bishop and Rampe, 2016). Upper, highly leached strata consist of mixtures containing Al-rich phyllosilicates, hydrated silica and opal (McKeown et al., 2011; Bishop et al., 2013a), sulfates including bassanite (Wray et al., 2010), jarosite (Farrand et al., 2009), and hydrated Fe sulfates (Farrand et al., 2014), and SRO phases such as allophane and imogolite (Bishop and Rampe, 2016). Changes in climate conditions from near neutral to mildly acidic are proposed to have caused this chemical gradient in the Mawrth Vallis region (Bishop et al., 2016).

Layered outcrops in the Mawrth Vallis region of Mars contain a great diversity of aqueous alteration materials that provide an opportunity to infer past aqueous environments. Numerous orbital investigations have documented aluminous and siliceous clay-bearing units overlying the thick Fe/Mg smectite unit (e.g., Bishop et al., 2008b; Wray et al., 2008; McKeown et al., 2009a; Michalski et al., 2010b; Noe Dobrea et al., 2010; Loizeau et al., 2012b). Multiple secondary minerals were identified in the upper units (e.g., Bishop et al., 2013a), but the presence of poorly crystalline phases was reported only recently (Bishop and Rampe, 2016). Further, allophane and imogolite comprise a significant portion of the uppermost stratum of the clay profiles, covering the Al-phyllosilicate-rich and opal-rich unit (Fig. 3.5). These results signify a change in climate on Mars from a warm and wet environment to one where water was sporadic and likely depleted rapidly (Bishop and Rampe, 2016). Further, Bishop et al. (2016) describe an ancient wet and warm geologic record that formed the thick Fe/Mg smectite unit, a period of wet/dry cycling to create acid alteration and sulfates in some areas, followed by leaching or pedogenesis to result in aluminous clay minerals, and finally a drier, colder climate that left the altered ash in the form of SRO aluminosilicates, rather than crystalline clay minerals.

3.4.2.2 The Clay-Bearing Region West and South of Isidis Planitia

After the Chryse Planitia basin, the regions surrounding the Isidis basin have the next greatest abundance of phyllosilicate-rich material (Figs. 3.6 and 3.7). The NE Syrtis region northwest of Isidis hosts pristine olivine and pyroxene outcrops (Mustard et al., 2009) and abundant and varied clay-bearing units in several provinces with multiple, distinct mineral assemblages (Mustard et al., 2008; Ehlmann et al., 2009; Bramble et al., 2017). Fe/Mg smectite is found throughout the Nili Fossae region in the ancient Noachian rocks. This Fe/Mg smectite unit occurs in eroded terrains in eastern Nili Fossae where it is stratigraphically covered by either kaolinite or magnesite (Ehlmann et al., 2009). In other outcrops, the Fe/Mg smectite is mixed with mica, illite, or chlorite. The presence of zeolites

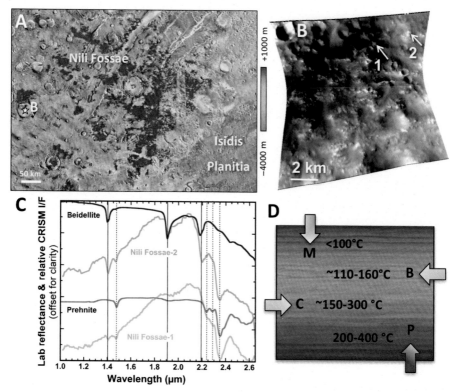

Fig. 3.6 Views of clay-bearing outcrops at NE Syrtis. (A) a portion of the NE Syrtis region near Nili Fossae with Mg-rich phyllosilicates mapped in red and Al-rich phyllosilicate mapped in blue (data from Carter et al., 2015a) with a white star showing the location of B. (B) View of CRISM image FRT0000454E as an illustration of the high-temperature prehnite-bearing exposures with sites marked where the spectra shown in C were collected. (C) Ratio of CRISM spectra of prehnite-bearing units compared with lab spectra of prehnite from the USGS spectral library (Clark et al., 2007) and beidellite (Bishop et al., 2011). (D) Diagram illustrating differences in formation temperature for phyllosilicates observed on Mars, where M refers to montmorillonite, B to beidellite, C to chlorite, and P to prehnite, and the temperatures indicate the approximate ranges for their formation.

and prehnite together with Fe/Mg smectite in and around impact craters southwest of Nili Fossae is an indicator of low-grade metamorphic processes or subsurface hydrothermal alteration (Ehlmann et al., 2009; Marzo et al., 2010; Fairén et al., 2010). Sulfates are observed together with Fe/Mg smectite, magnesite, and other clay minerals along the southwestern part of Nili Fossae between Jezero crater and the Syrtis volcanic region (Ehlmann and Mustard, 2012). Jarosite ridges are observed at this site on top of layered material composed of polyhydrated sulfates and Fe/Mg smectite. The diversity in clay minerals observed across the NE Syrtis and Nili Fossae region and the geologic context of the many clay-bearing outcrops indicate that several episodes of aqueous alteration took

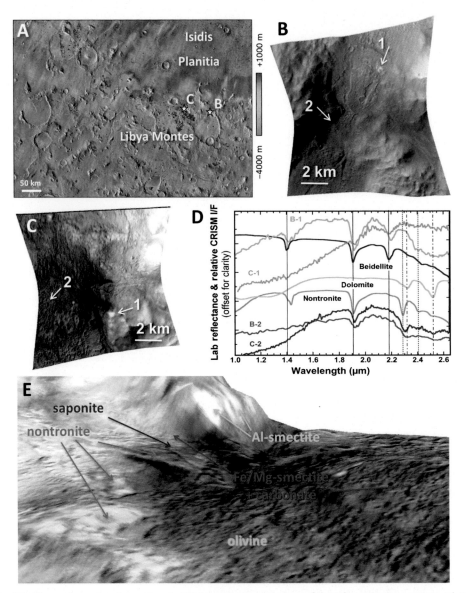

Fig. 3.7 Views of clay-bearing outcrops at Libya Montes. (A) View of the Libya Montes region with Fe/Mg smectites mapped in red and Al-phyllosilicates mapped in blue (data from Carter et al., 2015a) with white stars indicating the locations of the CRISM images shown in B and C. (B) CRISM image FRT0000B0CB with the locations marked where spectra were collected. (C) CRISM image FRT0000A819 with the locations marked where spectra were collected. (D) Example spectra collected from outcrops containing beidellite (B-1 and C-1) and Fe/Mg smectite mixed with dolomite. (E) A 3-D view of CRISM image FRT0000A819 illustrating the locations of phyllosilicates and carbonate in relation to the olivine-bearing unit (adapted from Bishop et al., 2013c).

place (Ehlmann et al., 2009). The delta region at Jezero crater (Ehlmann et al., 2008a; Brown et al., 2016; Schon et al., 2012; Goudge et al., 2017) and the aqueous outcrops at NE Syrtis (Mustard et al., 2009; Ehlmann et al., 2009; Brown et al., 2010; Ehlmann and Mustard, 2012; Bramble et al., 2017; Viviano et al., 2013; Michalski et al., 2010a) represent geochemically exciting areas that may become landing sites for future martian rovers.

Phyllosilicate formation temperatures from terrestrial investigations can be used as geothermometers in remote sensing. This is illustrated in Fig. 3.6D. The Al-smectites montmorillonite and beidellite form at different temperatures in related outcrops (e.g., Guisseau et al., 2007), or beidellite can form from montmorillonite through diagenesis (e.g., Chamley, 1989; Velde, 1995). Studies of geothermal fields at Guadeloupe, Lesser Antilles (Guisseau et al., 2007), and Chipilapa, El Salvador (Papapanagiotou et al., 2003), found that montmorillonite formed in regions where low-temperature waters were present ($<100°C$) and beidellite formed nearby from warmer fluids ($\sim110–160°C$). Additional montmorillonite-to-beidellite transitions were observed with increasing depth and temperature of the fluids in other active geothermal systems (Inoue et al., 2004; Yang et al., 2001). Alternatively, burial diagenesis can occur to transform montmorillonite to beidellite during the initial phase of illitization over the temperature range 100–200°C (Saito et al., 2003; Beaufort et al., 2001). Chlorite formation temperatures are typically in the range 150–300°C (De Caritat et al., 1993), while prehnite generally forms through low-grade metamorphism, hydrothermal processes, or diagenesis in the range 200–400°C (e.g., Chamley, 1989; Velde, 1995). The presence of both chlorite and prehnite in an outcrop supports formation conditions of $\sim200–300°C$ (Ehlmann et al., 2011a). Subsurface diagenesis or hydrothermal processes from impacts could be responsible for the formation of chlorite-prehnite assemblages from Fe/Mg smectite in the NE Syrtis region (Ehlmann et al., 2009; Marzo et al., 2010; Fairén et al., 2010). Exposures of serpentine and magnesite (Ehlmann et al., 2008b, 2010) or mixtures of saponite, chlorite, talc, and magnesite (Brown et al., 2010; Viviano et al., 2013) could also represent hydrothermal alteration at elevated temperatures in this region.

The Libya Montes region lies at the southern rim of the Isidis impact basin (Fig. 3.7) and contains Fe/Mg smectite, dolomite, and beidellite in aqueous outcrops exposed around olivine-bearing and pyroxene-bearing basalt (Tornabene et al., 2008; Mustard et al., 2009; Bishop et al., 2013c). Evidence for fluvial, lacustrine, aeolian, volcanic, and hydrothermal processes has been documented, resulting in a variety of landforms and ample evidence for chemical alteration of the local rocks (Crumpler and Tanaka, 2003; Tornabene et al., 2008; Jaumann et al., 2010; Erkeling et al., 2012). The ancient Noachian basaltic crustal materials experienced extensive aqueous alteration at the time of the Isidis impact, during which the montes were also formed, followed by emplacement of a rough olivine-rich lava or melt and finally the smooth pyroxene-bearing caprock unit (Bishop et al., 2013c). The chemistry of the smectites in this region varies from nontronite with an OH combination (stretching plus bending) band at 2.29 µm to Fe/Mg smectite with a band at 2.30 μm to saponite with a band at 2.31 μm (Fig. 3.7), and the intermediate Fe/Mg smectite is the most common here, as observed elsewhere on the planet. Beidellite is also found in smaller exposures in the Libya Montes region with an Al_2OH band at 2.19 μm rather than montmorillonite, which would have a band at 2.21 μm. Beidellite in these sites likely formed via low-temperature hydrothermal alteration from the Isidis impact. Abundant fluvial features dating from the Noachian to Amazonian time periods traverse the Libya Montes region; however, clay minerals are primarily observed in the ancient Noachian rocks (Bishop et al., 2013c).

3.5 DISCUSSION OF PHYLLOSILICATES AND CLIMATE ON MARS

The nature and stratigraphy of clay-bearing outcrops on Mars have been evaluated recently with the goal of contributing to our understanding of the climate on early Mars (Bishop et al., 2018). There are a variety of phyllosilicates, and they form in many different alteration environments (Chamley, 1989). Phyllosilicate-rich materials on Earth dominated by dioctahedral smectites (e.g., montmorillonite and nontronite), Al-rich micas, and kaolin group clays generally form in low-temperature (~25–50°C) subaqueous or subaerial surface environments (Chamley, 1989). Such temperate to warm climates with alternating wet (>50 cm/year) and dry seasons support soil formation with high smectite contents (up to 90% of phyllosilicates). Nontronite-bearing outcrops on Mars are consistent with formation in surface environments similar to these terrestrial subaqueous or subaerial surface environments. On the other hand, Mg-rich clays such as serpentine and prehnite form at elevated temperatures up to 400°C (e.g., Chamley, 1989; Velde, 1995) and are more likely to have formed on Mars in subsurface "crustal" environments or due to hydrothermal conditions (Ehlmann et al., 2011b; Fairén et al., 2010; Marzo et al., 2010).

Evaluating formation of different clays in terrestrial environments is helpful for understanding and constraining formation environments of clays on Mars. Terrestrial smectite-bearing hydrothermally altered seafloor sediments were used to classify mixed-layer smectite/chlorite/talc assemblages on Mars (Michalski et al., 2015). Alteration of Columbia River Basalt (CRB) also provides clues to formation environments for nontronite and associated minerals. Zeolite is also an indicator of CRB alteration temperature, where nontronite forms in all cases, but zeolite only forms at depth under elevated temperatures (Benson and Teague, 1982). Bishop et al. (2018) proposed recently that occurrences of Mg-rich mixed clays including trioctahedral smectite, chlorite, talc, or serpentine with lateral variations likely formed in subsurface hydrothermal environments on Mars, while dioctahedral (Al or Fe^{3+}-rich) smectite and widespread vertical stratigraphies of Fe/Mg smectites (nontronite and saponite), clay assemblages, and sulfates formed in aqueous surface environments.

The formation environments of poorly crystalline and SRO aluminosilicates on Earth also dictate likely formation environments for these materials on Mars that are increasingly being discovered across the planet. Outcrops on Mars containing these SRO aluminosilicates such as allophane and imogolite are indicators of environments with limited liquid water on the surface or cold environments (Bishop and Rampe, 2016). Studies of allophane and imogolite formation from volcanic glass show that these SRO phases are favored over smectite clay formation in specific environments (Parfitt, 2009; Rasmussen et al., 2010; Chamley, 1989). These include either rainy and temperate, well-drained environments without standing liquid water or cold climates with temperatures near freezing that experienced standing liquid water from melting snow or ice. Examples of cold environments where SRO aluminosilicates form rather than phyllosilicates when liquid water is present include sediments in the Antarctic Dry Valleys (Bishop et al., 2014), glacial deposits in Oregon (Scudder et al., 2017), cold streams in Iceland (Thorpe et al., 2017), and high-elevation sites of the Cascade mountains of California (Rasmussen et al., 2010). Thus, the presence of abundant SRO aluminosilicates without phyllosilicates on Mars could mark the end of the warm and wet surface conditions supporting smectite formation (Bishop and Rampe, 2016).

Mars has long been known to harbor features due to liquid water on the surface (e.g., Carr, 1979), and numerous geologic features are present that represent past flowing water and fluvial erosion (e.g., Craddock and Howard, 2002; Ansan et al., 2008; Fassett and Head, 2011). Many of these surface features are associated with terrains that exhibit spectral signatures of clays and aqueous minerals, but others do not (Murchie et al., 2009a; Carter et al., 2015a; Ehlmann et al., 2011b). One example of a region containing abundant alluvial fans, valley networks, and dendritic channels occurs across the Libya Montes region south of Isidis basin. Several of these features formed by liquid water there vary in age from Noachian to Amazonian, where phyllosilicates are not observed (Bishop et al., 2013c). These features were interpreted to result from short-term liquid water events that did not persist sufficiently long for clay formation (Bishop et al., 2013c). However, these features could have also been shaped by cold water (Fairén, 2010). This is supported by the many recent studies that have found SRO phases instead of crystalline clays in cold environments (e.g., Rasmussen et al., 2010; Bishop et al., 2014; Scudder et al., 2017; Thorpe et al., 2017). This could indicate that cold and wet conditions may have been responsible for the regions bearing fluvial erosion features without the formation of phyllosilicates, while warm and wet conditions were responsible for the regions having both fluvial erosion features and phyllosilicates (Bishop et al., 2018). Bishop et al. (2018) further proposed that short-term (from thousands to hundreds of thousands of years) warm and wet conditions during the Noachian could have been responsible for the formation of surface smectite outcrops if summer high temperatures reached 25–50°C. This model would be consistent with generally cooler Noachian conditions outside of these punctuated warming trends.

3.6 DISCUSSION OF PHYLLOSILICATES AND HABITABILITY ON MARS

The martian geologic timeline includes changes in water availability on the surface. The diagram in Fig. 3.8 depicts a scenario where liquid water was present on the surface of Mars during its early history and phyllosilicates formed in surface environments (Bishop et al., 2013a). Thermodynamics predicts that these clay-rich units would remain after the liquid water is no longer present (Gooding, 1978) and they would likely be buried over time by lava or other materials. Fe/Mg smectite is the most common phyllosilicate observed on Mars globally (e.g., Murchie et al., 2009a, and Carter et al., 2013). Smectite clays are common in regions with high water/rock ratio environments dominated by wet/dry cycling of the climate (e.g., Chamley, 1989). Formation of a planet-wide Fe/Mg smectite unit may have occurred early in the planets' history that was then eroded, buried, or altered depending on local geologic activity. This possibly widespread Fe/Mg smectite unit is observed currently in smaller exposures on the surface (Fig. 3.8) where impact craters or erosion has revealed the Noachian rocks below.

This scenario (Fig. 3.8) describes episodes of transient liquid water on the surface during the Hesperian but rarely sufficient water for clay formation in surface environments (Bishop et al., 2013a). However, clays likely formed in subsurface environments during

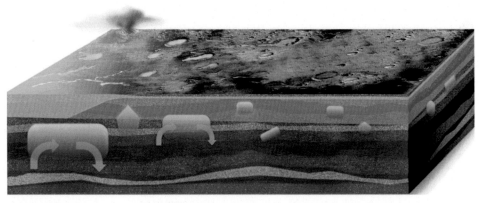

Early Mars (Noachian)
~4.5-3.8 Ga
Liquid water likely on surface

Intermediate Mars (Hesperian)
~3.8-3 Ga
Transient liquid water possible

Recent Mars (Amazonian)
<3 Ga
Liquid water not stable on surface

Fig. 3.8 Martian phyllosilicate timeline. This diagram represents how phyllosilicates may have formed in warm and wet surface environments (blue) on early Mars (~4 billion years ago) and then become buried by other materials. These phyllosilicate-bearing units (green) could exist across much of the martian surface below caprock and dust. Phyllosilicates are currently only visible where the caprock has been eroded away, although they may still be present below the surface. Additionally, formation of phyllosilicates occurred in subsurface environments at elevated temperatures through burial diagenesis or hydrothermal processes related to impacts or volcanism. The timeline for subsurface phyllosilicate formation is less constrained, although much of this activity likely occurred during the Noachian, with some occurring during the Hesperian and possibly Amazonian.

the Noachian and Hesperian in aqueous environments heated through geothermal processes or impacts (Ehlmann et al., 2011b). The timing and extent of these warm subsurface environments are difficult to estimate, but they are likely responsible for the formation of clay minerals such as chlorite, serpentine, and prehnite that all require elevated temperatures. Warm environments supporting serpentine formation could even have provided a viable ecosystem for microbes using H_2 for energy (Mustard and Tarnas, 2008). Over the past 3 billion years, water has not been stable on the surface of Mars according to most models, and surface environments supporting phyllosilicates or life are not viable.

Diagenesis can also provide clues about the availability of subsurface water on Mars because conversion of smectite to illite or chlorite is dependent on temperature, burial history, and the availability of water (e.g., Chamley, 1989; Velde, 1995). Tosca et al. (2008) attempted to constrain smectite alteration through modeling the effects of time and temperature on sediment diagenesis. They concluded that smectites buried for ∼3.5 billion years to a depth of at least 300–400 m should have converted in the presence of water to chlorite (and/or illite if K^+ is available). Because thick Fe/Mg smectite units have persisted at Mawrth Vallis, Gale Crater, and elsewhere on Mars, liquid water has likely been limited since their deposition, or they have remained relatively close to the surface.

Phyllosilicate-bearing rocks on Mars could have supplied reaction templates for prebiotic chemistry and possibly the development of life as well (Bishop et al., 2013a). Comets and asteroids provided delivery of organic molecules to early Earth and other planetary bodies including Mars (e.g., Anders, 1989; Chyba et al., 1990; Delsemme, 1997; Pierazzo and Chyba, 1999). Smectite clays are efficient catalysis agents for a variety of chemical reactions due to their layered configuration and the acidity of their internal surfaces that facilitates bringing molecules together (Pinnavaia, 1983). Organic reactions on montmorillonite surfaces showed that this clay can catalyze the formation of RNA and other precursor molecules required for life (e.g., Ferris and Hagan Jr., 1986; Ferris et al., 1989; Ferris, 2005, 2006; Franchi et al., 2003). In order to sustain life, an environment must provide essential nutrients, biologically accessible energy, and liquid water (Nealson, 1997). Studies of terrestrial marine basalts with chemical compositions consistent with Mars (McSween et al., 2009) have shown that these rocks contain sufficient requirements (e.g., nutrients, water, and radiation protection) to sustain life (Fisk et al., 1998). UV exposure experiments showed that photosynthetic microbes living in an aqueous matrix including smectite clay and ferrihydrite were much better protected from damaging UV rays than microbes without the benefit of these minerals (Bishop et al., 2006). Experiments with soil bacteria and viruses have shown that they can survive in a variety of soil and clay environments including temperature and moisture extremes replicating martian conditions (Hawrylewicz et al., 1962; Foster et al., 1978; Moll and Vestal, 1992). This could have enabled microbial life to form in smectite-rich environments on early Mars and then continue to exist despite limited water availability.

Preservation of biosignatures is favored in rapid burial conditions in fine-grained clay-rich systems (Farmer and Des Marais, 1999), although long-term preservation of biosignatures is a challenge. Biosignature preservation is most successful in host rocks composed of stable minerals that are resistant to weathering and provide an impermeable barrier limiting exposure to external fluids that would alter and/or remove the biosignatures. Mineral precipitates such as phyllosilicates and silica provide an excellent matrix for microbial fossilization (Farmer and Des Marais, 1999). Phyllosilicates have a longer crustal residence time than many other minerals such as carbonates and sulfates, thus improving the chance of preservation of potential biosignatures (e.g., Butterfield, 1990; Summons et al., 2011). The preservation potential is highest for sediments formed in low-permeability environments where temperatures remained low over time (Summons et al., 2011) and diagenesis did not occur. For these reasons, regions with expansive and thick smectite outcrops such as Mawrth Vallis (Bishop et al., 2013a) and Gale Crater (Bristow et al., 2017) would be ideal locations to look for biosignatures on Mars. The abundance of smectite clays at Mawrth Vallis is one reason this area has been proposed as a landing site for martian rovers seeking evidence of life (Gross et al., 2017; Loizeau et al., 2017).

3.7 SUMMARY OF PHYLLOSILICATES AND SRO MATERIALS ON MARS

Spectral remote sensing has enabled detection and characterization of multiple clay minerals on Mars. Fe/Mg smectite is the most abundant clay mineral identified on the surface of Mars and occurs in light-toned, layered outcrops 100–200 m thick in some areas where liquid water was present on the surface of early Mars. Other regions of the planet contain mixtures of trioctahedral Mg-rich clays including smectite, chlorite, serpentine, talc, or prehnite that formed in subsurface hydrothermal, metamorphic, or diagenetic environments that are not related to the martian climate. Both of these surface and subsurface clay-forming environments on Mars could have supported habitats for microbes. Low-temperature, surface environments would provide suitable niches for photosynthetic microbes, while warm subsurface environments could have supported methanogens or chemoautotrophs. Phyllosilicates are associated with carbonates or gypsum in regions where neutral waters were likely to have occurred, while phyllosilicates together with sulfates such as jarosite indicate lower pH systems. Regions of Mars containing SRO materials such as opal, allophane, or imogolite rather than crystalline clays represent a colder climate or insufficient standing water for clay formation and would be less likely to have supported life.

ACKNOWLEDGMENTS

Thanks are due to C. Weitz for helpful editorial comments. Support from the NASA Astrobiology Institute grant NNX15BB01 enabled the preparation and completion of this chapter.

REFERENCES

Anders, E., 1989. Pre-biotic organic matter from comets and asteroids. Nature 342, 255–257.

Anderson, J.H., Wickersheim, K.A., 1964. Near infrared characterization of water and hydroxyl groups on silica surfaces. Surf. Sci. 2, 252–260.

Ansan, V., Mangold, N., Masson, P., Gailhardis, E., Neukum, G., 2008. Topography of valley networks on Mars from Mars Express High Resolution Stereo Camera digital elevation models. J. Geophys. Res. 113. https://doi.org/10.1029/2007JE002986.

Arvidson, R.E., Squyres, S.W., Bell, J.F., Catalano, J.G., Clark, B.C., Crumpler, L.S., De Souza, P.A., Fairén, A.G., Farrand, W.H., Fox, V.K., Gellert, R., Ghosh, A., Golombek, M.P., Grotzinger, J.P., Guinness, E.A., Herkenhoff, K.E., Jolliff, B.L., Knoll, A.H., Li, R., McLennan, S.M., Ming, D.W., Mittlefehldt, D.W., Moore, J.M., Morris, R.V., Murchie, S.L., Parker, T.J., Paulsen, G., Rice, J.W., Ruff, S.W., Smith, M.D., Wolff, M.J., 2014. Ancient aqueous environments at endeavour crater, mars. Science. 343. https://doi.org/10.1126/science.1248097.

Baird, A.K., Castro, A.J., Clark, B.C., Toulmin III, P., Rose Jr., H., Keil, K., Gooding, J.L., 1977. The Viking X ray fluorescence experiment: sampling strategies and laboratory simulations. J. Geophys. Res. 82, 4595–4624.

Beaufort, D., Berger, G., Lacharpagne, J.C., Meunier, A., 2001. An experimental alteration of montmorillonite to a di + trioctahedral smectite assemblage at 100 and 200°C. Clay Miner. 36, 211–225.

Bell III, J.F., Pollack, J.B., Geballe, T.R., Cruikshank, D.P., Freedman, R., 1994. Spectroscopy of Mars from 2.04 to 2.44 μm during the 1993 opposition: absolute calibration and atmospheric vs. mineralogic origin of narrow absorption features. Icarus 111, 106–123.

Benson, L.V., Teague, L.S., 1982. Diagenesis of Basalts from the Pasco Basin, Washington—I. Distribution and composition of secondary mineral phases. J. Sediment. Res. 52, 595–613.

Bibring, J.-P., Combes, M., Langevin, Y., Cara, C., Drossart, P., Encrenaz, T., Erard, S., Forni, O., Gondet, B., Ksanfomality, L., Lellouch, E., Masson, P., Moroz, V., Rocard, F., Rosenqvist, J., Sotin, C., Soufflot, A., 1990. ISM observations of Mars and Phobos: first results. Proc. Lunar Planet. Sci. Conf. XX, 461–471.

Bibring, J.-P., Langevin, Y., Gendrin, A., Gondet, B., Poulet, F., Berthé, M., Soufflot, A., Arvidson, R., Mangold, N., Mustard, J., Drossart, P., 2005. Mars surface diversity as revealed by the OMEGA/Mars Express observations. Science 307, 1576–1581.

Bibring, J.-P., Langevin, Y., Mustard, J.F., Poulet, F., Arvidson, R., Gendrin, A., Gondet, B., Mangold, N., Pinet, P., Forget, F., 2006. Global mineralogical and aqueous Mars history derived from OMEGA/Mars Express data. Science 312, 400–404.

Bish, D.L., Blake, D.F., Vaniman, D.T., Chipera, S.J., Morris, R.V., Ming, D.W., Treiman, A.H., Sarrazin, P., Morrison, S.M., Downs, R.T., Achilles, C.N., Yen, A.S., Bristow, T.F., Crisp, J.A., Morookian, J.M., Farmer, J.D., Rampe, E.B., Stolper, E.M., Spanovich, N., Team, M.S., 2013. X-ray diffraction results from Mars Science Laboratory: mineralogy of Rocknest at Gale Crater. Science. 341. https://doi.org/10.1126/science.123893.

Bishop, J.L., Englert, P.A.J., Patel, S., Tirsch, D., Roy, A.J., Koeberl, C., böttger, U., Hanke, F., Jaumann, R., 2014. Mineralogical analyses of surface sediments in the Antarctic Dry Valleys: coordinated analyses of Raman spectra, reflectance spectra and elemental abundances. Phil. Trans. R. Soc. A. 372. 20140198.

Bishop, J.L., Fairén, A.G., Michalski, J.R., Gago-Duport, L., Baker, L.L., Velbel, M.A., Gross, C., Rampe, E.B., 2018. Surface clay formation during short-term warmer and wetter conditions on a largely cold ancient Mars. Nat. Astron 2, 206–213. https://doi.org/10.1038/s41550-017-0377-9.

Bishop, J.L., Gates, W.P., Makarewicz, H.D., McKeown, N.K., Hiroi, T., 2011. Reflectance spectroscopy of beidellites and their importance for Mars. Clay Clay Miner. 59, 376–397.

Bishop, J.L., Gross, C., Rampe, E.B., Wray, J.J., Parente, M., Horgan, B., Loizeau, D., Viviano-Beck, C.E., Clark, R.N., Seelos, F.P., Ehlmann, B.L., Murchie, S.L., 2016. In: Mineralogy of layered outcrops at Mawrth Vallis and implications for early aqueous geochemistry on Mars.Lunar Planet. Sci. Conf. XLVII, The Woodlands, TX. Abstract #1332.

Bishop, J.L., Lane, M.D., Dyar, M.D., Brown, A.J., 2008a. Reflectance and emission spectroscopy study of four groups of phyllosilicates: smectites, kaolinite-serpentines, chlorites and micas. Clay Miner. 43, 35–54.

Bishop, J.L., Loizeau, D., McKeown, N.K., Saper, L., Dyar, M.D., Des Marais, D., Parente, M., Murchie, S.L., 2013a. What the ancient phyllosilicates at Mawrth Vallis can tell us about possible habitability on early Mars. Planet. Space Sci. 86, 130–149.

Bishop, J.L., Louris, S.K., Rogoff, D.A., Rothschild, L.J., 2006. Nanophase iron oxides as a key ultraviolet sunscreen for ancient photosynthetic microbes. Int. J. Astrobiol. 5, 1–12.

Bishop, J.L., Madeová, J., Komadel, P., Fröschl, H., 2002. The influence of structural Fe, Al and Mg on the infrared OH bands in spectra of dioctahedral smectites. Clay Miner. 37, 607–616.

Bishop, J.L., Michalski, J.R., Carter, J., 2017. Chapter 14: remote detection of clay minerals. In: Gates, W.P., Kloprogge, J.T., Madejová, J., Bergaya, F. (Eds.), Infrared and Raman Spectroscopies of Clay Minerals. Elsevier, The Netherlands, pp. 482–514.

Bishop, J.L., Murad, E., 2002. Spectroscopic and geochemical analyses of ferrihydrite from hydrothermal springs in iceland and applications to mars. In: Smellie, J.L., Chapman, M.G. (Eds.), Volcano-Ice Interactions on Earth and Mars. Geological Society, London. Special Publication No. 202, 357–370.

Bishop, J.L., Murad, E., Dyar, M.D., 2015. Akaganéite and schwertmannite: spectral properties, structural models and geochemical implications of their possible presence on Mars. Am. Mineral. 100, 738–746.

Bishop, J.L., Noe Dobrea, E.Z., McKeown, N.K., Parente, M., Ehlmann, B.L., Michalski, J.R., Milliken, R.E., Poulet, F., Swayze, G.A., Mustard, J.F., Murchie, S.L., Bibring, J.-P., 2008b. Phyllosilicate diversity and past aqueous activity revealed at Mawrth Vallis, Mars. Science 321, 830–833. https://doi.org/10.1126/science.1159699.

Bishop, J.L., Pieters, C.M., Edwards, J.O., 1994. Infrared spectroscopic analyses on the nature of water in montmorillonite. Clay Clay Miner. 42, 702–716.

Bishop, J.L., Rampe, E.B., 2016. Evidence for a changing Martian climate from the mineralogy at Mawrth Vallis. Earth Planet. Sci. Lett. 448, 42–48.

Bishop, J.L., Rampe, E.B., Bish, D.L., Baker, L.L., Abidin, Z., Matsue, N., Henmi, T., 2013b. Spectral and hydration properties of allophane and imogolite. Clay Clay Miner. 61, 57–74.

Bishop, J.L., Tirsch, D., Tornabene, L.L., Jaumann, R., McEwen, A.S., McGuire, P.C., Ody, A., Poulet, F., Clark, R.N., Parente, M., Voigt, J., Aydin, Z., Bamberg, M., Petau, A., McKeown, N.K., Mustard, J.F., Hash, C., Murchie, S.L., Swayze, G., Neukum, G., Seelos, F., 2013c. Mineralogy and morphology of geologic units at Libya Montes, Mars: ancient aqueous outcrops, mafic flows, fluvial features and impacts. J. Geophys. Res. 118, 487–513.

Bishop, J.L., Velbel, M.A., 2017. Comparison of phyllosilicates observed on the surface of Mars with those found in martian meteorites. In: 80th Annual Meeting of the Meteoritical Society, Santa Fe, NM. Abstract #6115.

Blake, D.F., Morris, R.V., Kocurek, G., Morrison, S.M., Downs, R.T., Bish, D., Ming, D.W., Edgett, K.S., Rubin, D., Goetz, W., Madsen, M.B., Sullivan, R., Gellert, R., Campbell, I., Treiman, A.H., McLennan, S.M., Yen, A.S., Grotzinger, J., Vaniman, D.T., Chipera, S.J., Achilles, C.N., Rampe, E.B., Sumner, D., Meslin, P.-Y., Maurice, S., Forni, O., Gasnault, O., Fisk, M., Schmidt, M., Mahaffy, P., Leshin, L.A., Glavin, D., Steele, A., Freissinet, C., Navarro-González, R., Yingst, R.A., Kah, L.C., Bridges, N., Lewis, K.W., Bristow, T.F., Farmer, J.D., Crisp, J.A., Stolper, E.M., Des Marais, D.J., Sarrazin, P., Team, M.S., 2013. Curiosity at gale crater, mars: characterization and analysis of the Rocknest sand shadow. Science. 341. https://doi.org/10.1126/science.1239505.

Bramble, M.S., Mustard, J.F., Salvatore, M.R., 2017. The geological history of Northeast Syrtis Major, Mars. Icarus 293, 66–93.

Bridges, J.C., Catling, D.C., Saxton, J.M., Swindle, T.D., Lyon, I.C., Grady, M.M., 2001. Alteration assemblages in Martian Meteorites: implications for near-surface processes. Space Sci. Rev. 96, 365–392.

Bridges, J.C., Schwenzer, S.P., 2012. The nakhlite hydrothermal brine on Mars. Earth Planet. Sci. Lett. 359–360, 117–123.

Bristow, T.F., Bish, D.L., Vaniman, D.T., Morris, R.V., Blake, D.F., Grotzinger, J.P., Rampe, E.B., Crisp, J.A., Achilles, C.N., Ming, D.W., Ehlmann, B.L., King, P.L., Bridges, J.C., Eigenbrode, J.L., Sumner, D.Y., Chipera, S.J., Moorokian, J.M., Treiman, A.H., Morrison, S.M., Downs, R.T., Farmer, J.D., Marais, D.D., Sarrazin, P., Floyd, M.M., Mischna, M.A., Mcadam, A.C., 2015. The origin and implications of clay minerals from Yellowknife Bay, Gale crater, Mars. Am. Mineral. 100, 824–836.

Bristow, T.F., Blake, D.F., Vaniman, D.T., Chipera, S.J., Rampe, E.B., Grotzinger, J.P., Mcadam, A.C., Ming, D.W., Morrison, S.M., Yen, A.S., Morris, R.V., Downs, R.T., Treiman, A.H., Achilles, C.N.,

Des Marais, D.J., Morookian, J.M., Crisp, J.A., Hazen, R.M., Farmer, J.D., 2017. In: Surveying clay mineral diversity in the Murray Formation, Gale Crater, Mars.Lunar Planet. Sci. Conf. XLVIII, The Woodlands, TX. Abstract #2462.

Brown, A.J., Hook, S.J., Baldridge, A.M., Crowley, J.F., Bridges, N.T., Thomson, B.J., Marion, G.M., Filho, C.R.D.S., Bishop, J.L., 2010. Hydrothermal formation of clay-carbonate alteration assemblages in the Nili Fossae region of Mars. Earth Planet. Sci. Lett. 297, 174–182. https://doi.org/10.1016/j.epsl.2010.06.018.

Brown, A.J., Viviano-Beck, C.E., Bishop, J.L., Cabrol, N.A., Andersen, D., Sobron, P., Moersch, J., Templeton, A.S., Russell, M.J., 2016. In: A serpentinization origin for Jezero Crater carbonates.Lunar Planet. Sci. Conf. XLVII, The Woodlands, TX. Abstract #2165.

Buczkowski, D.L., Murchie, S.L., Clark, R.N., Seelos, K.D., Seelos, F.P., Malaret, E., Hash, C., 2010. Investigation of an Argyre Basin ring structure using MRO/CRISM. J. Geophys. Res. 115. https://doi.org/10.1029/2009JE003508.

Butterfield, N.J., 1990. Organic preservation of non-mineralizing organisms and the taphonomy of the Burgess Shale. Paleobiology 16, 272–286.

Byrnes, J.M., Ramsey, M.S., King, P.L., Lee, R.J., 2007. Thermal infrared reflectance and emission spectroscopy of quartzofeldspathic glasses. Geophys. Res. Lett. 34.

Carr, M.H., 1979. Formation of martian flood features by release of water from confined aquifers. J. Geophys. Res. 84, 2995–3007.

Carter, J., Loizeau, D., Mangold, N., Poulet, F., Bibring, J.-P., 2015a. Widespread surface weathering on early Mars: a case for a warmer and wetter climate. Icarus 248, 373–382.

Carter, J., Poulet, F., 2012. Orbital identification of clays and carbonates in Gusev crater. Icarus 219, 250–253.

Carter, J., Poulet, F., Bibring, J.-P., Murchie, S., 2010. Detection of hydrated silicates in crustal outcrops in the Northern Plains of Mars. Science 328, 1682–1686.

Carter, J., Poulet, F., Bibring, J.P., Mangold, N., Murchie, S., 2013. Hydrous minerals on Mars as seen by the CRISM and OMEGA imaging spectrometers: updated global view. J. Geophys. Res. Planets 118, 831–858.

Carter, J., Viviano-Beck, C., Le Deit, L., Bishop, J.L., Loizeau, D., 2015b. Orbital detection and implications of akaganéite on Mars. Icarus 253, 296–310.

Chamley, H., 1989. Clay Sedimentology. Springer-Verlag, New York.

Chemtob, S.M., Nickerson, R.D., Morris, R.V., Agresti, D.G., Catalano, J.G., 2015. Synthesis and structural characterization of ferrous trioctahedral smectites: implications for clay mineral genesis and detectability on Mars. J. Geophys. Res. Planets 120, 1119–1140.

Christensen, P.R., Bandfield, J.L., Clark, R.N., Edgett, K.S., Hamilton, V.E., Hoefen, T., Kieffer, H.H., Kuzmin, R.O., Lane, M.D., Malin, M.C., Morris, R.V., Pearl, J.C., Pearson, R., Roush, T.L., Ruff, S.W., Smith, M.D., 2000a. Detection of crystalline hematite mineralization on Mars by the Thermal Emission Spectrometer: evidence for near-surface water. J. Geophys. Res. 105, 9623–9642.

Christensen, P.R., Bandfield, J.L., Hamilton, V.E., Lane, M.D., Piatek, J.L., Ruff, S.W., Stefanov, W.L., 2000b. A thermal emission spectral library of rock-forming minerals. J. Geophys. Res. 105, 9735–9739.

Christensen, P.R., Bandfield, J.L., Hamilton, V.E., Ruff, S.W., Kieffer, H.H., Titus, T.N., Malin, M.C., Morris, R.V., Lane, M.D., Clark, R.L., Jakosky, B.M., Mellon, M.T., Pearl, J.C., Conrath, B.J., Smith, M.D., Clancy, R.T., Kuzmin, R.O., Roush, T., Mehall, G.L., Gorelick, N., Bender, K., Murray, K., Dason, S., Greene, E., Silverman, S., Greenfield, M., 2001. Mars global surveyor thermal emission spectrometer experiment: investigation description and surface science results. J. Geophys. Res. 106, 23823–23871.

Christensen, P.R., Bandfield, J.L., Smith, M.D., Hamilton, V.E., Clark, R.N., 2000c. Identification of basaltic component on the Martian surface from Thermal Emission Spectrometer Data. J. Geophys. Res. 105, 9609–9621.

Christensen, P.R., Bandfield, J.L., Bell III, J.F., Gorelick, N., Hamilton, V.E., Ivanov, A., Jakosky, B.M., Kieffer, H.H., Lane, M.D., Malin, M.C., McConnochie, T., McEwen, A.S., McSween Jr., H.Y., Mehall, G.L., Moersch, J.E., Nealson, K.H., Rice Jr., J.W., Richardson, M.I., Ruff, S.W., Smith, M.D., Titus, T.N., Wyatt, M.B., 2003. Morphology and composition of the surface of Mars: Mars Odyssey THEMIS results. Science 300, 2056–2061.

Chyba, C.F., Thomas, P.J., Brookshaw, L., Sagan, C., 1990. Cometary delivery of organic molecules to the early Earth. Science 249, 366–373.

Clark, R.N., King, T.V.V., Klejwa, M., Swayze, G.A., 1990. High spectral resolution reflectance spectroscopy of minerals. J. Geophys. Res. 95, 12653–12680.

Clark, R.N., Swayze, G.A., Wise, R., Livo, E., Hoefen, T., Kokaly, R., Sutley, S.J., 2007. USGS digital spectral library splib06a. U.S. Geological Survey, Digital Data Series, Reston.

Craddock, R.A., Howard, A.D., 2002. The case for rainfall on a warm, wet early Mars. J. Geophys. Res. Planets. 107, 21-1-21-36.

Crumpler, L.S., Tanaka, K.L., 2003. Geology and MER target site characteristics along the southern rim of Isidis Planitia, Mars. J. Geophys. Res. 108. https://doi.org/10.1029/2002JE002040.

Cuadros, J., Caballero, E., Huertas, F.J., Jiménez De Cisneros, C., Huertas, F., Linares, J., 1999. Experimental alteration of volcanic tuff: smectite formation and effect on ^{18}O isotope composition. Clay Clay Miner. 47, 769–776.

Cuadros, J., Michalski, J.R., Dekov, V., Bishop, J.L., 2015. Octahedral chemistry of 2:1 clay minerals and hydroxyl band position in the near-infrared. Application to Mars. Am. Mineral. 101, 554–563.

De Caritat, P., Hutcheon, I., Walshe, J.L., 1993. Chlorite geothermometry: a review. Clay Clay Miner. 41, 219–239.

Decarreau, A., Petit, S., Martin, F., Vieillard, P., Joussein, E., 2008. Hydrothermal synthesis, between 75 and 150C, of high-charge ferric nontronites. Clay Clay Miner. 56, 322–337.

Delsemme, A.H., 1997. The origin of the atmosphere and of the oceans. In: Thomas, P.J., Chyba, C.F., McKay, C.P. (Eds.), Comets and the Origin and Evolution of Life. Springer-Verlag, New York.

Ehlmann, B., Berger, G., Mangold, N., Michalski, J., Catling, D., Ruff, S., Chassefière, E., Niles, P., Chevrier, V., Poulet, F., 2013. Geochemical consequences of widespread clay mineral formation in Mars' ancient crust. Space Sci. Rev. 174, 329–364.

Ehlmann, B.L., Edwards, C.S., 2014. Mineralogy of the martian surface. Annu. Rev. Earth Planet. Sci. 42, 291–315.

Ehlmann, B.L., Mustard, J.F., 2012. An in-situ record of major environmental transitions on early Mars at Northeast Syrtis Major. Geophys. Res. Lett. 39, L11202. https://doi.org/10.1029/2012GL051594.

Ehlmann, B.L., Mustard, J.F., Clark, R.N., Swayze, G.A., Murchie, S.L., 2011a. Evidence for low-grade metamorphism, hydrothermal alteration, and diagenesis on Mars from phyllosilicate mineral assemblages. Clay Clay Miner. 59, 359–377.

Ehlmann, B.L., Mustard, J.F., Fasset, C.I., Schon, S.C., Head, J.W., Des Marais, D.J., Grant, J.A., Murchie, S.L., 2008a. Clay minerals in delta deposits and organic preservation potential on Mars. Nat. Geosci. 1, 355–358. https://doi.org/10.1038/ngeo207.

Ehlmann, B.L., Mustard, J.F., Murchie, S.L., 2010. Geologic setting of serpentine deposits on Mars. Geophys. Res. Lett. 37. https://doi.org/10.1029/2010GL042596.

Ehlmann, B.L., Mustard, J.F., Murchie, S.L., Bibring, J.P., Meunier, A., Fraeman, A.A., Langevin, Y., 2011b. Subsurface water and clay mineral formation during the early history of Mars. Nature 479, 53–60.

Ehlmann, B.L., Mustard, J.F., Murchie, S.L., Poulet, F., Bishop, J.L., Brown, A.J., Calvin, W.M., Clark, R.N., Des Marais, D.J., Milliken, R.E., Roach, L.H., Roush, T.L., Swayze, G.A., Wray, J.J., 2008b. Orbital identification of carbonate-bearing rocks on Mars. Science 322, 1828–1832.

Ehlmann, B.L., Mustard, J.F., Swayze, G.A., Clark, R.N., Bishop, J.L., Poulet, F., Marais, D.J.D., Roach, L.H., Milliken, R.E., Wray, J.J., Barnouin-Jha, O., Murchie, S.L., 2009. Identification of hydrated silicate minerals on Mars using MRO-CRISM: geologic context near Nili Fossae and implications for aqueous alteration. J. Geophys. Res. 114. https://doi.org/10.1029/2009JE003339.

Erkeling, G., Reiss, D., Hiesinger, H., Poulet, F., Carter, J., Ivanov, M.A., Hauber, E., Jaumann, R., 2012. Valleys, paleolakes and possible shorelines at the Libya Montes/Isidis boundary: implications for the hydrologic evolution of Mars. Icarus 219, 393–413.

Fairén, A.G., 2010. A cold and wet Mars. Icarus 208, 165–175.

Fairén, A.G., Chevrier, V., Abramov, O., Marzo, G.A., Gavin, P., Davila, A.F., Tornabene, L.L., Bishop, J.L., Roush, T.L., Gross, C., Kneissl, T., Uceda, E.R., Dohm, J.M., Schulze-Makuch, D., Rodríguez, J.A.P., Amils, R., McKay, C.P., 2010. Phyllosilicates in impact craters on Mars: pre-

or post-impact genesis. Proc. Natl. Acad. Sci. 107, 12095–12100. https://doi.org/10.1073/pnas.1002889107.

Farmer, J.D., Des Marais, D.J., 1999. Exploring for a record of ancient Martian life. J. Geophys. Res. 104, 26977–26995.

Farmer, V.C., 1974. The layer silicates. In: Farmer, V.C. (Ed.), The Infrared Spectra of Minerals. The Mineralogical Society, London.

Farmer, V.C., Adams, M.J., Fraser, A.R., Palmieri, F., 1983. Synthetic imogolite: properties, synthesis, and possible applications. Clay Miner. 18, 459–472.

Farmer, V.C., Fraser, A.R., Tait, J.M., 1979. Characterization of the chemical structures of natural and synthetic aluminosilicate gels and sols by infrared spectroscopy. Geochim. Cosmochim. Acta 43, 1417–1420.

Farrand, W.H., Glotch, T.D., Horgan, B., 2014. Detection of copiapite in the northern Mawrth Vallis region of Mars: evidence of acid sulfate alteration. Icarus 241, 346–357.

Farrand, W.H., Glotch, T.D., Rice Jr., J.W., Hurowitz, J.A., Swayze, G.A., 2009. Discovery of jarosite within the Mawrth Vallis region of Mars: implications for the geologic history of the region. Icarus 204, 478–488.

Fassett, C.I., Head, J.W., 2011. Sequence and timing of conditions on early Mars. Icarus 211, 1204–1214.

Ferris, J.P., 2005. Mineral catalysis and prebiotic synthesis: montmorillonite-catalyzed formation of RNA. Elements 1, 145–149.

Ferris, J.P., 2006. Montmorillonite-catalysed formation of RNA oligomers: the possible role of catalysis in the origins of life. Philos. Trans. R. Soc. Lond. B Biol. Sci. 361, 1777–1786.

Ferris, J.P., Ertem, G., Agarwal, V.K., 1989. The adsorption of nucleotides and polynucleotides of montmorillonite clay. Orig. Life Evol. Biosph. 19, 153–164.

Ferris, J.P., Hagan Jr., W.J., 1986. The adsorption and reaction of adenine nucleotides on montmorillonite. Orig. Life Evol. Biosph. 17, 69–84.

Fisk, M.R., Giovannoni, S.J., Thorseth, I.H., 1998. Alteration of oceanic volcanic glass: textural evidence of microbial activity. Science 281, 978–980.

Flahaut, J., Carter, J., Poulet, F., Bibring, J.P., Van Westrenen, W., Davies, G.R., Murchie, S.L., 2015. Embedded clays and sulfates in Meridiani Planum, Mars. Icarus 248, 269–288.

Foster, T.L., Winans Jr., L., Casey, R.C., Kirschner, L.E., 1978. Response of terrestrial microorganisms to a simulated Martian environment. Appl. Environ. Microbiol. 35, 730–737.

Fox, V.K., Arvidson, R.E., Guinness, E.A., McLennan, S.M., Catalano, J.G., Murchie, S.L., Powell, K.E., 2016. Smectite deposits in Marathon Valley, Endeavour Crater, Mars, identified using CRISM hyperspectral reflectance data. Geophys. Res. Lett. 43, 4885–4892.

Franchi, M., Ferris, J.P., Gallori, E., 2003. Cations as mediators of the adsorption of nucleic acids on clay surfaces in prebiotic environments. Orig. Life Evol. Biosph. 33, 1–16.

Gooding, J.L., 1978. Chemical weathering on Mars. Thermodynamic stabilities of primary minerals (and their alteration products) from mafic igneous rocks. Icarus 33, 483–513.

Gooding, J.L., 1992. Soil mineralogy and chemistry on Mars: possible clues from salts and clays in SNC meteorites. Icarus 99, 28–41.

Goudge, T.A., Milliken, R.E., Head, J.W., Mustard, J.F., Fassett, C.I., 2017. Sedimentological evidence for a deltaic origin of the western fan deposit in Jezero crater, Mars and implications for future exploration. Earth Planet. Sci. Lett. 458, 357–365.

Grim, R.E., 1968. Clay Mineralogy. McGraw-Hill Book Co.

Gross, C., Carter, J., Poulet, F., Loizeau, D., Bishop, J.L., Horgan, B., Michalski, J.R., 2017. In: Mawrth Vallis—an auspicious destination for the ESA and NASA 2020 landers.Lunar Planet. Sci. Conf. XLVIII, The Woodlands, TX. Abstract #2194.

Grotzinger, J.P., Arvidson, R.E., Bell III, J.F., Calvin, W.M., Clark, B.C., Fike, D.A., Golombek, M., Greeley, R., Haldemann, A., Herkenhoff, K.E., Jolliff, B.L., Knoll, A.H., Malin, M.C., McLennan, S.M., Parker, T., Soderblom, L.A., Sohl-Dickstein, J.N., Squyres, S.W., Tosca, N.J., Watters, W.A., 2005. Stratigraphy and sedimentology of a dry to wet eolian depositional system, Burns formation, Meridiani Planum, Mars. Earth Planet. Sci. Lett. 240, 11–72.

Grotzinger, J.P., Gupta, S., Malin, M.C., Rubin, D.M., Schieber, J., Siebach, K., Sumner, D.Y., Stack, K.M., Vasavada, A.R., Arvidson, R.E., Calef, F., Edgar, L., Fischer, W.F., Grant, J.A., Griffes, J., Kah, L.C., Lamb, M.P., Lewis, K.W., Mangold, N., Minitti, M.E., Palucis, M., Rice, M., Williams, R.M.E., Yingst, R.A., Blake, D., Blaney, D., Conrad, P., Crisp, J., Dietrich, W.E., Dromart, G., Edgett, K.S., Ewing, R.C., Gellert, R., Hurowitz, J.A., Kocurek, G., Mahaffy, P., McBride, M.J., McLennan, S.M., Mischna, M., Ming, D., Milliken, R., Newsom, H., Oehler, D., Parker, T.J., Vaniman, D., Wiens, R.C., Wilson, S.A., 2015. Deposition, exhumation, and paleoclimate of an ancient lake deposit, Gale crater, Mars. Science 350.

Grotzinger, J.P., Sumner, D.Y., Kah, L.C., Stack, K., Gupta, S., Edgar, L., Rubin, D., Lewis, K., Schieber, J., Mangold, N., Milliken, R., Conrad, P.G., Desmarais, D., Farmer, J., Siebach, K., Calef, F., Hurowitz, J., McLennan, S.M., Ming, D., Vaniman, D., Crisp, J., Vasavada, A., Edgett, K.S., Malin, M., Blake, D., Gellert, R., Mahaffy, P., Wiens, R.C., Maurice, S., Grant, J.A., Wilson, S., Anderson, R.C., Beegle, L., Arvidson, R., Hallet, B., Sletten, R.S., Rice, M., Bell, J., Griffes, J., Ehlmann, B., Anderson, R.B., Bristow, T.F., Dietrich, W.E., Dromart, G., Eigenbrode, J., Fraeman, A., Hardgrove, C., Herkenhoff, K., Jandura, L., Kocurek, G., Lee, S., Leshin, L.A., Leveille, R., Limonadi, D., Maki, J., McCloskey, S., Meyer, M., Minitti, M., Newsom, H., Oehler, D., Okon, A., Palucis, M., Parker, T., Rowland, S., Schmidt, M., Squyres, S., Steele, A., Stolper, E., Summons, R., Treiman, A., Williams, R., Yingst, A., Team, M.S., Kemppinen, O., Bridges, N., Johnson, J.R., Cremers, D., Godber, A., Wadhwa, M., Wellington, D., Mcewan, I., Newman, C., Richardson, M., Charpentier, A., Peret, L., King, P., Blank, J., Weigle, G., Li, S., Robertson, K., Sun, V., Baker, M., Edwards, C., Farley, K., Miller, H., Newcombe, M., Pilorget, C., Brunet, C., Hipkin, V., Léveillé, R., et al., 2014. A habitable fluvio-lacustrine environment at Yellowknife Bay, Gale Crater, Mars. Science 343. https://doi.org/10.1126/science.aac7575.

Guisseau, D., Mas, P.P., Beaufort, D., Girard, J.P., Inoue, A., Sanjuan, B., Petit, S., Lens, A., Genter, A., 2007. Significance of the depth-related transition montmorillonite-beidellite in the Bouillante geothermal field (Guadeloupe, Lesser Antilles). Am. Mineral. 92, 1800–1813.

Hallis, L.J., Ishii, H.A., Bradley, J.P., Taylor, G.J., 2014. Transmission electron microscope analyses of alteration phases in martian meteorite MIL 090032. Geochim. Cosmochim. Acta 134, 275–288.

Hamilton, V.E., Christensen, P.R., 2000. Determining the modal mineralogy of mafic and ultramafic igneous rocks using thermal emission spectroscopy. J. Geophys. Res. 105, 9701–9716.

Hamilton, V.E., Morris, R.V., Gruener, J.E., Mertzman, S.A., 2008. Visible, near-infrared, and middle infrared spectroscopy of altered basaltic tephras: Spectral signatures of phyllosilicates, sulfates, and other aqueous alteration products with application to the mineralogy of the Columbia Hills of Gusev Crater, Mars. J. Geophys. Res. Planets. 113. https://doi.org/10.1029/2007JE003049.

Hamilton, V.E., Christensen, P.R., Bandfield, J.L., Rogers, A.D., Edwards, C.S., 2019. Thermal infrared spectral analyses of Mars from orbit using TES and Themis. In: Bishop, J.L., Bell III, J.F., Moersch, J.E. (Eds.), Remote Compositional Analysis: Techniques for Understanding Spectroscopy, Mineralogy, and Geochemistry of Planetary Surfaces. Cambridge University Press, Cambridge (Chapter 24).

Hawrylewicz, E., Gowdy, B., Ehrlich, R., 1962. Micro-organisms under a simulated Martian environment. Nature 193, 497.

Hicks, L.J., Bridges, J.C., Gurman, S.J., 2014. Ferric saponite and serpentine in the nakhlite martian meteorites. Geochim. Cosmochim. Acta 136, 194–210.

Inoue, A., Meunier, A., Beufort, D., 2004. Illite-smectite mixed-layer minerals in felsic volcaniclastic rocks from drill cores, Kakkonda, Japan. Clay Clay Miner. 52, 66–84.

Jaumann, R., Nass, A., Tirsch, D., Reiss, D., Neukum, G., 2010. The Western Libya Montes Valley System on Mars: evidence for episodic and multi-genetic erosion events during the Martian history. Earth Planet. Sci. Lett. 294, 272–290.

King, T.V.V., Clark, R.N., 1989. Spectral characteristics of chlorites and Mg-serpentines using high-resolution reflectance spectroscopy. J. Geophys. Res. 94, 13997–14008.

Klingelhöfer, G., Morris, R.V., Bernhardt, B., Schröder, C., Rodionov, D., De Souza, P.A.J., Yen, A.S., Gellert, R., Evlanov, E.N., Zubkov, B., Foh, J., Bonnes, U., Kankeleit, E., Gütlich, P., Ming, D.W., Renz, F., Wdowiak, T.J., Squyres, S.W., Arvidson, R.E., 2004. Jarosite and hematite at Meridiani Planum from Opportunity's Mössbauer spectrometer. Science 306, 1740–1745.

Kraft, M.D., Michalski, J.R., Sharp, T.G., 2003. Effects of pure silica coatings on thermal emission spectra of basaltic rocks: Considerations for Martians surface mineralogy. Geophys. Res. Lett. 30, 1–4.

Le Deit, L., Jessica, F., Quantin, C., Hauber, E., Mège, D., Bourgeois, O., Gurgurewicz, J., Massé, M., Jaumann, R., 2012. Extensive surface pedogenic alteration of the Martian Noachian crust suggested by plateau phyllosilicates around Valles Marineris. J. Geophys. Res. 116E00J05. https://doi.org/10.1029/2011JE003983.

Lee, M.R., Cohen, B.E., Mark, D.F., Boyce, A., 2017. In: Multiphase aqueous alteration of the Nakhlite Northwest Africa 817.80th Annual Meeting of the Meteoritical Society, Santa Fe, NM. Abstract #6186.

Lee, M.R., Maclaren, I., Andersson, S.M.L., Kovács, A., Tomkinson, T., Mark, D.F., Smith, C.L., 2015. Opal-A in the Nakhla meteorite: a tracer of ephemeral liquid water in the Amazonian crust of Mars. Meteorit. Planet. Sci. 50, 1362–1377.

Loizeau, D., Carter, J., Bouley, S., Mangold, N., Poulet, F., Bibring, J.P., Costard, F., Langevin, Y., Gondet, B., Murchie, S.L., 2012a. Characterization of hydrated silicate-bearing outcrops in Tyrrhena Terra, Mars: implications to the alteration history of Mars. Icarus 219, 476–497.

Loizeau, D., Mangold, N., Poulet, F., Ansan, V., Hauber, E., Bibring, J.P., Gondet, B., Langevin, Y., Masson, P., Neukum, G., 2010. Stratigraphy in the Mawrth Vallis region through OMEGA, HRSC color imagery and DTM. Icarus 205, 396–418.

Loizeau, D., Mangold, N., Poulet, F., Bibring, J.-P., Gendrin, A., Ansan, V., Gomez, C., Gondet, B., Langevin, Y., Masson, P., Neukum, G., 2007. Phyllosilicates in the Mawrth Vallis region of Mars. J. Geophys. Res. 112. https://doi.org/10.1029/2006JE002877.

Loizeau, D., Poulet, F., Horgan, B., Bishop, J.L., 2017. In: Mawrth Vallis as a landing site for the NASA Mars2020 mission.Lunar Planet. Sci. Conf. XLVIII, The Woodlands, TX. Abstract #2988.

Loizeau, D., Werner, S.C., Mangold, N., Bibring, J.P., Vago, J.L., 2012b. Chronology of deposition and alteration in the Mawrth Vallis region, Mars. Planet. Space Sci. 72, 31–43.

Madejová, J., Kečkéš, J., Pálková, H., Komadel, P., 2002. Identification of components in smectite/kaolinite mixtures. Clay Miner. 37, 377–388.

Marzo, G.A., Davila, A.F., Tornabene, L.L., Dohm, J.M., Fairén, A.G., Gross, C., Kneissl, T., Bishop, J.L., Roush, T.L., McKay, C.P., 2010. Evidence for Hesperian impact-induced hydrothermalism on Mars. Icarus 208, 667–683.

McCord, T.B., Clark, R.N., Singer, R.B., 1982. Mars: near-infrared spectral reflectance of surface regions and compositional implications. J. Geophys. Res. 87, 3021–3032.

McKay, D.S., Gibson Jr., E.K., Thomas-Keprta, K.L., Vali, H., Romanek, C.S., Clemett, S.J., Chillier, X.D.F., Maechling, C.R., Zare, R.N., 1996. Search for past life on Mars: possible relic biogenic activity in Martian meteorite ALH 84001. Science 273, 924–930.

McKeown, N.K., Bishop, J.L., Cuadros, J., Hillier, S., Amador, E., Makarewicz, H.D., Parente, M., Silver, E., 2011. Interpretation of reflectance spectra of clay mineral-silica mixtures: implications for Martian clay mineralogy at Mawrth Vallis. Clay Clay Miner. 59, 400–415.

McKeown, N.K., Bishop, J.L., Noe Dobrea, E.Z., Ehlmann, B.L., Parente, M., Mustard, J.F., Murchie, S.L., Swayze, G.A., Bibring, J.-P., Silver, E., 2009a. Characterization of phyllosilicates observed in the central Mawrth Vallis region, Mars, their potential formational processes, and implications for past climate. J. Geophys. Res. 114. https://doi.org/10.1029/2008JE003301.

McKeown, N.K., Noe Dobrea, E.Z., Bishop, J.L., Silver, E.A., 2009b. In: Coordinated lab, field, and aerial study of the Painted Desert, AZ, as a potential analog site for phyllosilicates at Mawrth Vallis, Mars.Lunar and Planetary Science Conference XL, Houston, TX. LPS XL, abs. #2509.

McLennan, S.M., Anderson, R.B., Bell, J.F., Bridges, J.C., Calef, F., Campbell, J.L., Clark, B.C., Clegg, S., Conrad, P., Cousin, A., Des Marais, D.J., Dromart, G., Dyar, M.D., Edgar, L.A., Ehlmann, B.L., Fabre, C., Forni, O., Gasnault, O., Gellert, R., Gordon, S., Grant, J.A., Grotzinger, J.P., Gupta, S., Herkenhoff, K.E., Hurowitz, J.A., King, P.L., Le Mouélic, S., Leshin, L.A., Léveillé, R., Lewis, K.W., Mangold, N., Maurice, S., Ming, D.W., Morris, R.V., Nachon, M., Newsom, H.E., Ollila, A.M., Perrett, G.M., Rice, M.S., Schmidt, M.E., Schwenzer, S.P., Stack, K., Stolper, E.M., Sumner, D.Y., Treiman, A.H., Vanbommel, S., Vaniman, D.T., Vasavada, A., Wiens, R.C., Yingst, R.A., Kemppinen, O., Bridges, N., Johnson, J.R., Minitti, M., Cremers, D., Farmer, J., Godber, A., Wadhwa, M., Wellington, D., Mcewan, I., Newman, C., Richardson, M., Charpentier, A., Peret, L., Blank, J., Weigle, G., Li, S., Milliken, R., Robertson, K., Sun, V., Baker, M., Edwards, C., Farley, K., Griffes, J., Miller, H., Newcombe, M., Pilorget, C., Siebach, K., Brunet, C., Hipkin, V., Marchand, G., Sánchez, P.S., Favot, L., Cody, G., Steele, A., Flückiger, L., Lees, D., Nefian, A., Martin, M., Gailhanou, M., Westall, F., Israël, G., Agard, C., Baroukh, J., Donny, C., Gaboriaud, A., Guillemot, P.,

Lafaille, V., Lorigny, E., Paillet, A., et al., 2014. Elemental geochemistry of sedimentary rocks at Yellowknife Bay, Gale Crater, Mars. Science. 343. https://doi.org/10.1126/science.1244734.

McSween, H.Y., Ruff, S.W., Morris, R.V., Gellert, R., Klingelhoefer, G., Christensen, P.R., McCoy, T.J., Gosh, A., Moersch, J.M., Cohen, B.A., Rogers, A.D., Schroeder, C., Squyres, S.W., Crisp, J., Yen, A., 2008. Mineralogy of volcanic rocks in Gusev Crater, Mars: reconciling Moessbauer, Alpha Particle X-ray spectrometer, and miniature thermal emission spectrometer spectra. J. Geophys. Res. 113, E06S04. https://doi.org/10.1029/2007JE002970.

McSween, H.Y.J., Taylor, G.J., Wyat, M.B., 2009. Elemental composition of the martian crust. Science 324, 736–739.

Michalski, J.R., Reynolds, S.J., Sharp, T.G., Christensen, P.R., 2004. Thermal infrared analysis of weathered granitic rock compositions in the Sacaton Mountains, Arizona: Implications for petrologic classifications from thermal infrared remote-sensing data. J. Geophys. Res. Planets. 109. https://doi.org/10.1029/2003JE002197.

Michalski, J., Poulet, F.S., Bibring, J.-P., Mangold, N., 2010a. Analysis of phyllosilicate deposits in the Nili Fossae region of Mars: comparison of TES and OMEGA data. Icarus 206, 269–289.

Michalski, J.R., Cuadros, J., Bishop, J.L., Darby Dyar, M., Dekov, V., Fiore, S., 2015. Constraints on the crystal-chemistry of Fe/Mg-rich smectitic clays on Mars and links to global alteration trends. Earth Planet. Sci. Lett. 427, 215–225.

Michalski, J.R., Fergason, R.L., 2009. Composition and thermal inertia of the Mawrth Vallis region of Mars from TES and THEMIS data. Icarus 199, 25–48.

Michalski, J.R., Kraft, M.D., Sharp, T.G., Williams, L.B., Christensen, P.R., 2005. Mineralogical constraints on the high-silica martian surface component observed by TES. Icarus 174, 161–177.

Michalski, J.R., Kraft, M.D., Sharp, T.G., Williams, L.B., Christensen, P.R., 2006. Emission spectroscopy of clay minerals and evidence for poorly crystalline aluminosilicates on Mars from Thermal Emission Spectrometer data. J. Geophys. Res. 111, E03004. https://doi.org/10.1029/2005JE002438.

Michalski, J.R., Niles, P.B., Cuadros, J., Baldridge, A.M., 2013. Multiple working hypotheses for the formation of compositional stratigraphy on Mars: insights from the Mawrth Vallis region. Icarus 226, 816–840.

Michalski, J.R., Noe Dobrea, E.Z., 2007. Evidence for a sedimentary origin of clay minerals in the Mawrth Vallis region, Mars. Geology 35, 951–954.

Michalski, J.R., Poulet, F., Loizeau, D., Mangold, N., Noe Dobrea, E.Z., Bishop, J.L., Wray, J.J., McKeown, N.K., Parente, M., Hauber, E., Altieri, F., Carrozzo, F.G., Niles, P.B., 2010b. The Mawrth Vallis region of Mars: a potential landing site for the Mars Science Laboratory (MSL) mission. Astrobiology 10, 687–703.

Milliken, R.E., Grotzinger, J.P., Thomson, B.J., 2010. The paleoclimate of Mars as captured by the stratigraphic record in Gale Crater. Geophys. Res. Lett. 37, L04201. https://doi.org/10.1029/2009GL041870.

Milliken, R.E., Swayze, G.A., Arvidson, R.E., Bishop, J.L., Clark, R.N., Ehlmann, B.L., Green, R.O., Grotzinger, J., Morris, R.V., Murchie, S.L., Mustard, J.F., Weitz, C.M., 2008. Opaline silica in young deposits on Mars. Geology 36, 847–850. https://doi.org/10.1130/G24967A.1.

Ming, D.W., Archer, P.D., Glavin, D.P., Eigenbrode, J.L., Franz, H.B., Sutter, B., Brunner, A.E., Stern, J.C., Freissinet, C., Mcadam, A.C., Mahaffy, P.R., Cabane, M., Coll, P., Campbell, J.L., Atreya, S.K., Niles, P.B., Bell, J.F., Bish, D.L., Brinckerhoff, W.B., Buch, A., Conrad, P.G., Des Marais, D.J., Ehlmann, B.L., Fairén, A.G., Farley, K., Flesch, G.J., Francois, P., Gellert, R., Grant, J.A., Grotzinger, J.P., Gupta, S., Herkenhoff, K.E., Hurowitz, J.A., Leshin, L.A., Lewis, K.W., McLennan, S.M., Miller, K.E., Moersch, J., Morris, R.V., Navarro-González, R., Pavlov, A.A., Perrett, G.M., Pradler, I., Squyres, S.W., Summons, R.E., Steele, A., Stolper, E.M., Sumner, D.Y., Szopa, C., Teinturier, S., Trainer, M.G., Treiman, A.H., Vaniman, D.T., Vasavada, A.R., Webster, C.R., Wray, J.J., Yingst, R.A., Team, M.S., 2014. Volatile and organic compositions of sedimentary rocks in Yellowknife Bay, Gale Crater, Mars. Science. 343. https://doi.org/10.1126/science.1245267.

Ming, D.W., Mittlefehldt, D.W., Morris, R.V., Golden, D.C., Gellert, R., Yen, A.S., Clark, B.C., Squyres, S.W., Farrand, W.H., Ruff, S.W., Arvidson, R.E., Klingelhöfer, G., McSween Jr., H.Y., Rodionov, D.S., Schröder, C., De Souza, P.A., Jr. & Wang, A., 2006. Geochemical and mineralogical indicators for aqueous processes in the Columbia Hills of Gusev crater, Mars. J. Geophys. Res. 111, E02S12. https://doi.org/10.1029/2005JE002560.

Minitti, M.E., Mustard, J.F., Rutherford, M.J., 2002. The Effects of Glass Content and Oxidation on the Spectra of SNC-Like Basalts: Applications to Mars Remote Sensing. J. Geophys. Res. 107 (6), 1–16.

Moll, D.M., Vestal, J.R., 1992. Survival of microorganisms in smectite clays: implications for Martian exobiology. Icarus 98, 233–239.

Moore, D.M., Reynolds, R.C., 1997. X-ray Diffraction and the Identification and Analysis of Clay Minerals. Oxford Press, New York.

Morris, R.V., Klingelhöfer, G., Bernhardt, B., Schröder, C., Rodionov, D.S., De Souza Jr., P.A., Yen, A.S., Gellert, R., Evlanov, E.N., Foh, J., Kankeleit, E., Güttlich, P., Ming, D.W., Renz, F., Wdowiak, T.J., Squyres, S.W., Arvidson, R.E., 2004. Mineralogy at Gusev Crater from the Mössbauer spectrometer on the Spirit Rover. Science 305, 833–836.

Morris, R.V., Klingelhofer, G., Schroder, C., Fleischer, I., Ming, D.W., Yen, A., Gellert, R., Arvidson, R.E., Rodionov, D.S., Crumpler, L., Clark, B.C., Cohen, B.A., McCoy, T.J., Mittlefehldt, D.W., Schmidt, M.E., Desouza, P.A., Squyres, S.W., 2008. Iron mineralogy and aqueous alteration from Husband Hill through Home Plate at Gusev Crater, Mars: results from the Mössbauer instrument on the Spirit Mars Exploration Rover. J. Geophys. Res. 112. https://doi.org/10.1029/2008JE003201.

Murchie, S.L., Bibring, J.P., Arvidson, R.E., Bishop, J.L., Carter, J., Ehlmann, B.L., Langevin, Y., Mustard, J.F., Poulet, F., Riu, L., Seelos, K.D., Viviano-Beck, C.E., 2019. VSWIR spectral analyses of Mars from orbit using CRISM and OMEGA. In: Bishop, J.L., Bell III, J.F., Moersch, J.E. (Eds.), Remote Compositional Analysis: Techniques for Understanding Spectroscopy, Mineralogy, and Geochemistry of Planetary Surfaces. Cambridge University Press, Cambridge (Chapter 23).

Murchie, S.L., Mustard, J.F., Ehlmann, B.L., Milliken, R.E., Bishop, J.L., McKeown, N.K., Noe Dobrea, E.Z., Seelos, F.P., Buczkowski, D.L., Wiseman, S.M., Arvidson, R.E., Wray, J.J., Swayze, G.A., Clark, R.N., Des Marais, D.J., McEwen, A.S., Bibring, J.P., 2009a. A synthesis of Martian aqueous mineralogy after 1 Mars year of observations from the Mars Reconnaissance Orbiter. J. Geophys. Res. 114. https://doi.org/10.1029/2009JE003342.

Murchie, S.L., Roach, L.H., Seelos, F.P., Milliken, R.E., Mustard, J.F., Arvidson, R.E., Wiseman, S., Lichtenberg, K., Andrews-Hanna, J., Bibring, J.-P., Bishop, J.L., Parente, M., Morris, R.V., 2009b. Evidence for the origin of layered deposits in Candor Chasma, Mars, from mineral composition and hydrologic modeling. J. Geophys. Res. 114. https://doi.org/10.1029/2009JE003343.

Mustard, J., Poulet, F., Head, J.W., Mangold, N., Bibring, J.-P., Pelkey, S.M., Fassett, C., Langevin, Y., Neukum, G., 2007. Mineralogy of the Nili Fossae region with OMEGA/Mars Express data: 1. Ancient impact melt in the Isidis Basin and implications for the transition from the Noachian to Hesperian. J. Geophys. Res. 112. https://doi.org/10.1029/2006JE002834.

Mustard, J.F., Ehlmann, B.L., Murchie, S.L., Poulet, F., Mangold, N., Head, J.W., Bibring, J.-P., Roach, L.H., 2009. Composition, morphology, and stratigraphy of Noachian crust around the Isidis basin. J. Geophys. Res. 114. https://doi.org/10.1029/2009JE003349.

Mustard, J.F., Murchie, S.L., Pelkey, S.M., Ehlmann, B.L., Milliken, R.E., Grant, J.A., Bibring, J.-P., Poulet, F., Bishop, J.L., Noe Dobrea, E.Z., Roach, L.A., Seelos, F., Arvidson, R.E., Wiseman, S., Green, R., Hash, C., Humm, D., Malaret, E., McGovern, J.A., Seelos, K., Clancy, R.T., Clark, R.N., Des Marais, D., Izenberg, N., Knudson, A.T., Langevin, Y., Martin, T., McGuire, P., Morris, R.V., Robinson, M., Roush, T., Smith, M., Swayze, G.A., Taylor, H., Titus, T.N., Wolff, M., 2008. Hydrated silicate minerals on Mars observed by the Mars Reconnaissance Orbiter CRISM instrument. Nature 454, 305–309. https://doi.org/10.1038/nature07097.

Mustard, J.F., Tarnas, J.D., 2008. In: Hydrogen production in the upper 15 km of martian crust via serpentinization: implications for ancient and modern habitability. Astrobiology Science Conference, Mesa, AZ. Abstract #3426.

Nealson, K., 1997. The limits of life on Earth and searching for life on Mars. J. Geophys. Res. 102, 23675–23686.

Neukum, G., Jaumann, R., Hoffmann, H., Hauber, E., Head, J.W., Basilevsky, A.T., Ivanov, B.A., Werner, S.C., Van Gasselt, S., Murray, J.B., McCord, T., Teamthe, H.C.-I., 2004. Recent and episodic volcanic and glacial activity on Mars revealed by the High Resolution Stereo Camera. Nature 432, 971–979.

Noe Dobrea, E.Z., Bishop, J.L., McKeown, N.K., Fu, R., Rossi, C.M., Michalski, J.R., Heinlein, C., Hanus, V., Poulet, F., Arvidson, R., Mustard, J.F., Ehlmann, B.L., Murchie, S., McEwen, A.S., Swayze, G., Bibring, J.-P., Malaret, J.F.E., Hash, C., 2010. Mineralogy and Stratigraphy of Phyllosilicate-bearing and dark mantling units in the greater Mawrth Vallis/west Arabia Terra area: constraints on geological origin. J. Geophys. Res. 115. https://doi.org/10.1029/2009JE003351.

Noe Dobrea, E.Z., Wray, J.J., Calef, F.J., Parker, T.J., Murchie, S.L., 2012. Hydrated minerals on Endeavour Crater's rim and interior, and surrounding plains: new insights from CRISM data. Geophys. Res. Lett. 39. https://doi.org/10.1029/2012GL053180.

Papapanagiotou, P., Beaufort, D., Patrier, P., Traineau, H., 2003. Clay mineralogy of the >0.2 m rock formation of hte M1 drill hole of the geothermal field of Milos (Greece). Bull. Geol. Soc. Greece 28, 575–586.

Parente, M., Bishop, J.L., Bell, J.F., 2009. Spectral unmixing for mineral identification in Pancam images of soils in Gusev Crater, Mars. Icarus 203, 421–436.

Parfitt, R.L., 2009. Allophane and imogolite: role in soil biogeochemical processes. Clay Miner. 44, 135–155.

Parfitt, R.L., Furkert, R.J., Henmi, T., 1980. Identification and structure of two types of allophane from volcanic ash soils and tephra. Clay Clay Miner. 28, 328–334.

Petit, S., Madejova, J., Decarreau, A., Martin, F., 1999. Characterization of octahedral substitutions in kaolinites using near infrared spectroscopy. Clay Clay Miner. 47, 103–108.

Petit, S., Martin, F., Wiewiora, A., De Parseval, P., Decarreau, A., 2004. Crystal-chemistry of talc: a near infrared (NIR) spectroscopy study. Am. Mineral. 89, 319–326.

Pierazzo, E., Chyba, C.F., 1999. Amino acid survival in large cometary impacts. Meteorit. Planet. Sci. 34, 909–918.

Pinnavaia, T.J., 1983. Intercalated clay catalysts. Science 220, 365.

Post, J.L., 1984. Saponite from near Ballarat, California. Clay Clay Miner. 32, 147–152.

Post, J.L., Noble, P.N., 1993. The near-infrared combination band frequencies of dioctahedral smectites, micas, and illites. Clay Clay Miner. 41, 639–644.

Poulet, F., Bibring, J.-P., Mustard, J.F., Gendrin, A., Mangold, N., Langevin, Y., Arvidson, R.E., Gondet, B., Gomez, C., Omega, 2005. Phyllosilicates on Mars and implications for early Martian climate. Nature 438, 623–627.

Poulet, F., Carter, J., Bishop, J.L., Loizeau, D., Murchie, S.M., 2014. Mineral abundances at the final four curiosity study sites and implications for their formation. Icarus 231, 65–76.

Quantin, C., Carter, J., Thollot, P., Broyer, J., Lozach, L., Davis, J., Grindrod, P.M., Pajola, M., Baratti, E., Rossato, S., Allemand, P., Bultel, B., Leyrat, C., Fernando, J., Ody, A., 2016. In: Oxia Planum is a wide noachian clay bearing plain hosting younger fluvio-deltaic deposits which has been chosen by ESA as the landing site for Exomars 2018.Lunar Planet. Sci. Conf. XLVII, The Woodlands, TX. Abstract #2863.

Quantin, C., Flahaut, J., Clenet, H., Allemand, P., Thomas, P., 2012. Composition and structures of the subsurface in the vicinity of Valles Marineris as revealed by central uplifts of impact craters. Icarus 221, 436–452.

Rampe, E.B., Kraft, M.D., Sharp, T.G., Golden, D.C., Ming, D.W., Christensen, P.R., 2012. Allophane detection on Mars with Thermal Emission Spectrometer data and implications for regional-scale chemical weathering processes. Geology, 40. https://doi.org/10.1130/G33215.1.

Rampe, E.B., Kraft, M.D., Sharp, T.G., 2013. Deriving chemical trends from thermal infrared spectra of weathered basalt: Implications for remotely determining chemical trends on Mars. Icarus 225, 749–762.

Ramsey, M.S., Christensen, P.R., 1998. Mineral abundance determination: quantitative deconvolution of thermal emission spectra. J. Geophys. Res. 103, 577–596.

Rasmussen, C., Dahlgren, R.A., Southard, R.J., 2010. Basalt weathering and pedogenesis across an environmental gradient in the southern Cascade Range, California, USA. Geoderma 154, 473–485.

Rice, M.S., Bell III, J.F., Cloutis, E.A., Wang, A., Ruff, S.W., Craig, M.A., Bailey, D.T., Johnson, J.R., De Souza Jr, P.A., Farrand, W.H., 2010. Silica-rich deposits and hydrated minerals at Gusev Crater, Mars: Vis-NIR spectral characterization and regional mapping. Icarus 205, 375–395.

Roach, L.H., Mustard, J.F., Swayze, G.A., Milliken, R., Bishop, J.L., Murchie, S.L., Lichtenberg, K.A., 2010. Hydrated mineral stratigraphy of Ius Chasma, Valles Marineris. Icarus 206, 253–268.

Rogers, A.D., Christensen, P.R., 2007. Surface mineralogy of Martian low-albedo regions from MGS-TES data: implications for upper crustal evolution and surface alteration. J. Geophys. Res. 112, E01003. https://doi.org/10.1029/2006JE002727.

Ruff, S.W., 2004. Spectral evidence for zeolite in the dust on Mars. Icarus 168, 131–143.

Ruff, S.W., Farmer, J.D., 2016. Silica deposits on Mars with features resembling hot spring biosignatures at El Tatio in Chile. Nat. Commun. 713554. https://doi.org/10.1038/ncomms13554.

Saito, M.A., Sigman, D.M., Morel, F.M.M., 2003. The bioinorganic chemistry of the ancient ocean: the co-evolution of cyanobacterial metal requirements and biogeochemical cycles at the Archean-Proterozoic boundary? Inorg. Chim. Acta 356, 308–318.

Salisbury, J.W., 1993. Mid-infrared spectroscopy: laboratory data. In: Pieters, C.M., Englert, P.A.J. (Eds.), Remote Geochemical Analysis: Elemental and Mineralogical Composition. Cambridge University Press, Cambridge, pp. 79–98.

Salisbury, J.W., Walter, L.S., Vergo, N., D'aria, D.M., 1991. Infrared (2.1–25 μm) Spectra of Minerals. Johns Hopkins University Press, Baltimore.

Schon, S.C., Head, J.W., Fassett, C.I., 2012. An overfilled lacustrine system and progradational delta in Jezero crater, Mars: implications for Noachian climate. Planet. Space Sci. 67, 28–45.

Scudder, N.A., Horgan, B., Rutledge, A.M., Rampe, E.B., 2017. In: Using composition to trace glacial, fluvial, and aeolian sediment transport in a Mars-analog glaciated volcanic system.Lunar Planet. Sci. Conf. XLVIII, The Woodlands, TX. Abstract #2625.

Seelos, K.D., Seelos, F.P., Viviano-Beck, C.E., Murchie, S.L., Arvidson, R.E., Ehlmann, B.L., Fraeman, A.A., 2014. Mineralogy of the MSL Curiosity landing site in Gale crater as observed by MRO/CRISM. Geophys. Res. Lett. 41, 2014GL060310. https://doi.org/10.1002/2014GL060310.

Singer, R.B., 1985. Spectroscopic observation of Mars. Adv. Space Res. 5, 59–68.

Skok, J.R., Mustard, J.F., Ehlmann, B.L., Milliken, R.E., Murchie, S.L., 2010a. Silica deposits in the Nili Patera caldera on the Syrtis Major volcanic complex on Mars. Nat. Geosci. 3. https://doi.org/10.1038/NGEO990.

Skok, J.R., Mustard, J.F., Murchie, S.L., Wyatt, M.B., Ehlmann, B.L., 2010b. Spectrally distinct ejecta in Syrtis Major, Mars: evidence for environmental change at the Hesperian-Amazonian boundary. J. Geophys. Res. 115. https://doi.org/10.1029/2009JE003338.

Smith, D.E., Neumann, G.A., Arvidson, R.E., Guinness, E., Slavney, S., 2003. Mars Global Surveyor Laser Altimeter Mission Experiment Gridded Data Record. NASA Planetary Data System.

Summons, R.E., Amend, J.P., Bish, D., Buick, R., Cody, G.D., Des Marais, D.J., Dromart, G., Eigenbrode, J.L., Knoll, A.H., Sumner, D.Y., 2011. Preservation of Martian organic and environmental records: Final report of the Mars Biosignature Working Group. Astrobiology 11, 157–184.

Squyres, S.W., Arvidson, R.E., Bollen, D., Bell III, J.F., Brückner, J., Cabrol, N.A., Calvin, W.M., Carr, M.H., Christensen, P.R., Clark, B.C., Crumpler, L., Des Marais, D.J., D'uston, C., Economou, T., Farmer, J., Farrand, W.H., Folkner, W., Gellert, R., Glotch, T.D., Golombek, M., Gorevan, S., Grant, J.A., Greeley, R., Grotzinger, J., Herkenhoff, K.E., Hviid, S., Johnson, J.R., Klingelhöfer, G., Knoll, A.H., Landis, G., Lemmon, M., Li, R., Madsen, M.B., Malin, M.C., McLennan, S.M., McSween, H.Y., Ming, D.W., Moersch, J., Morris, R.V., Parker, T., Rice Jr., J.W., Richter, L., Rieder, R., Schroder, C., Sims, M., Smith, M., Smith, P., Soderblom, L.A., Sullivan, R., Tosca, N.J., Wänke, H., Wdowiak, T., Wolff, M., Yen, A., 2006. Overview of the opportunity Mars exploration rover mission to Meridiani Planum: eagle crater to purgatory ripple. J. Geophys. Res. 111. https://doi.org/10.1029/2006JE002771.

Tanaka, K.L., Skinner Jr., J.A., Hare, T.M., 2005. Geologic Map of the Northern Plains of Mars, Map 2888. U.S. Geol. Surv., Flagstaff, AZ

Thollot, P., Mangold, N., Ansan, V., Le Mouélic, S., Milliken, R.E., Bishop, J.L., Weitz, C.M., Roach, L.H., Mustard, J.F., Murchie, S.L., 2012. Most Mars minerals in a nutshell: various alteration phases formed in a single environment in Noctis Labyrinthus. J. Geophys. Res. 117E00J06. https://doi.org/10.1029/2011je004028.

Thorpe, M.T., Hurowitz, J.A., Dehouck, E., 2017. In: A frigid terrestrial analog for the paleoclimate of Mars.Lunar Planet. Sci. Conf. XLVIII, The Woodlands, TX Abstract #2599.

Tornabene, L.L., Moersch, J.E., McSween, H.Y., Hamilton, V.E., Piatek, J.L., Christensen, P.R., 2008. Surface and crater-exposed lithologic units of the Isidis Basin as mapped by coanalysis of THEMIS and TES derived data products. J. Geophys. Res. 113. https://doi.org/10.1029/2007JE002988.

Tosca, N.J., Milliken, R.E., Michel, F.M., 2008. Smectite formation on early mars: experimental constraints.Workshop on Martian Phyllosilicates: Recorders of Aqueous Processes, Paris. Abstract #7030.

Toulmin III, P., Baird, A.K., Clark, B.C., Keil, K., Rose, J.H.J., Christian, R.P., Evans, P.H., Kelliher, W.C., 1977. Geochemical and mineralogial interpretation of the Viking inorganic chemical results. J. Geophys. Res. 82, 4625–4634.

Treiman, A.H., 1998. The history of Allan Hills 84001 revised: multiple shock events. Meteorit. Planet. Sci. 33, 753–764.

Treiman, A.H., Barrett, R.A., Gooding, J.L., 1993. Preterrestrial aqueous alteration of the Lafayette (SNC) meteorite. Meteoritics 28, 86–97.

Treiman, A.H., Bish, D.L., Vaniman, D.T., Chipera, S.J., Blake, D.F., Ming, D.W., Morris, R.V., Bristow, T.F., Morrison, S.M., Baker, M.B., Rampe, E.B., Downs, R.T., Filiberto, J., Glazner, A.F., Gellert, R., Thompson, L.M., Schmidt, M.E., Le Deit, L., Wiens, R.C., Mcadam, A.C., Achilles, C.N., Edgett, K.S., Farmer, J.D., Fendrich, K.V., Grotzinger, J.P., Gupta, S., Morookian, J.M., Newcombe, M.E., Rice, M.S., Spray, J.G., Stolper, E.M., Sumner, D.Y., Vasavada, A.R., Yen, A.S., 2016. Mineralogy, provenance, and diagenesis of a potassic basaltic sandstone on Mars: CheMin X-ray diffraction of the Windjana sample (Kimberley area, Gale Crater). J. Geophys. Res. Planets 121, 75–106.

Treiman, A.H., Morris, R.V., Agresti, D.G., Graff, T.G., Achilles, C.N., Rampe, E.B., Bristow, T.F., Ming, D.W., Blake, D.F., Vaniman, D.T., Bish, D.L., Chipera, S.J., Morrison, S.M., Downs, R.T., 2014. Ferrian saponite from the Santa Monica Mountains (California, U.S.A., Earth): characterization as an analog for clay minerals on Mars with application to Yellowknife Bay in Gale Crater. Am. Mineral. 99, 2234–2250.

Vaniman, D.T., Bish, D.L., Ming, D.W., Bristow, T.F., Morris, R.V., Blake, D.F., Chipera, S.J., Morrison, S.M., Treiman, A.H., Rampe, E.B., Rice, M., Achilles, C.N., Grotzinger, J.P., McLennan, S.M., Williams, J., Bell, J.F., Newsom, H.E., Downs, R.T., Maurice, S., Sarrazin, P., Yen, A.S., Morookian, J.M., Farmer, J.D., Stack, K., Milliken, R.E., Ehlmann, B.L., Sumner, D.Y., Berger, G., Crisp, J.A., Hurowitz, J.A., Anderson, R., Des Marais, D.J., Stolper, E.M., Edgett, K.S., Gupta, S., Spanovich, N., Team, M.S., 2014. Mineralogy of a Mudstone at Yellowknife Bay, Gale Crater, Mars. Science. 343. https://doi.org/10.1126/science.1243480.

Velde, B., 1995. Origin and Mineralogy of Clays. Springer-Verlag, Berlin.

Viviano, C.E., Moersch, J.E., 2012. A technique for mapping Fe/Mg-rich phyllosilicates on Mars using THEMIS multispectral thermal infrared images. J. Geophys. Res. Planets. 117. https://doi.org/10.1029/2011JE003985.

Viviano, C.E., Moersch, J.E., McSween, H.Y., 2013. Implications for early hydrothermal environments on Mars through the spectral evidence for carbonation and chloritization reactions in the Nili Fossae region. J. Geophys. Res. Planets 118, 1858–1872.

Viviano-Beck, C.E., Murchie, S.L., Beck, A.W., Dohm, J.M., 2017. Compositional and structural constraints on the geologic history of eastern Tharsis Rise, Mars. Icarus 284, 43–58.

Wada, K., Henmi, T., Yoshinaga, N., Patterson, S.H., 1972. Imogolite and allophane formed in saprolite of basalt on Maui, Hawaii. Clay Clay Miner. 20, 375–380.

Weitz, C.M., Bishop, J.L., 2016. Stratigraphy and formation of clays, sulfates, and hydrated silica within a depression in Coprates Catena, Mars. J. Geophys. Res. 121, 805–835.

Weitz, C.M., Bishop, J.L., Baker, L.L., Berman, D.C., 2014a. Fresh exposures of hydrous Fe-bearing amorphous silicates on Mars. Geophys. Res. Lett. 41, 8744–8751.

Weitz, C.M., Bishop, J.L., Thollot, P., Mangold, N., Roach, L.H., 2011. Diverse mineralogies in two troughs of Noctis Labyrinthus, Mars. Geology 39, 899–902. https://doi.org/10.1130/G32045.1.

Weitz, C.M., Milliken, R.E., Grant, J.A., McEwen, A.S., Williams, R.M.E., Bishop, J.L., Thomson, B.J., 2010. Mars Reconnaissance Orbiter observations of light-toned layered deposits and associated fluvial landforms on the plateaus adjacent to Valles Marineris. Icarus 205, 73–102. https://doi.org/10.1016/j.icarus.2009.04.017.

Weitz, C.M., Noe Dobrea, E., Wray, J.J., 2015. Mixtures of clays and sulfates within deposits in western Melas Chasma, Mars. Icarus 251, 291–314.

Weitz, C.M., Noe Dobrea, E.Z., Lane, M.D., Knudson, A.T., 2012. Geologic relationships between gray hematite, sulfates, and clays in Capri Chasma. J. Geophys. Res. Planets. 117. E00J09. https://doi.org/10.1029/2012JE004092.

Weitz, C.M., Noe Dobrea, E.Z., Wray, J.J., 2014b. Mixtures of clays and sulfates within deposits in western Melas Chasma. Icarus. https://doi.org/10.1016/j.icarus.2014.04.009.

Wiseman, S.J., Arvidson, R.E., Morris, R.V., Poulet, F., Andrews-Hanna, J.C., Bishop, J.L., Murchie, S.L., Seelos, F.P., Des Marais, D., Griffes, J.L., 2010. Spectral and stratigraphic context of hydrated sulfate and phyllosilicate deposits in Northern Sinus Meridiani, Mars. J. Geophys. Res. 115, E00D18. https://doi.org/10.1029/2009JE003354.

Wray, J.J., 2013. Gale Crater: the Mars Science Laboratory/Curiosity Rover Landing Site. Int. J. Astrobiol. 12, 25–38.

Wray, J.J., Ehlmann, B.L., Squyres, S.W., Mustard, J.F., Kirk, R.L., 2008. Compositional stratigraphy of clay-bearing layered deposits at Mawrth Vallis, Mars. Geophys. Res. Lett. 35. https://doi.org/10.1029/2008GL034385.

Wray, J.J., Noe Dobrea, E.Z., Arvidson, R.E., Wiseman, S.M., Squyres, S.W., McEwen, A.S., Mustard, J.F., Murchie, S.L., 2009. Phyllosilicates and sulfates at Endeavour Crater, Meridiani Planum, Mars. Geophys. Res. Lett. 36. https://doi.org/10.1029/2009GL040734.

Wray, J.J., Squyres, S.W., Roach, L.H., Bishop, J.L., Mustard, J.F., Noe Dobrea, E.Z., 2010. Identification of the Ca-sulfate bassanite in Mawrth Vallis, Mars. Icarus 209, 416–421.

Wyatt, M.B., Hamilton, V.E., McSween, H.Y., Jr., Christensen, P.R., Taylor, L.A., 2001. Analysis of terrestrial and martian volcanic compositions using thermal emission spectroscopy, 1. Determination of mineralogy, chemistry, and classification strategies. J. Geophys. Res. 106, 14,711–14,732.

Yang, K., Browne, P.L., Huntington, J.F., Walshe, J.L., 2001. Characterising the hydrothermal alteration of the Broadlands-Ohaaki geothermal system, New Zealand, using short-wave infrared spectroscopy. J. Volcanol. Geotherm. Res. 106, 53–65.

CHAPTER 4

Martian Habitability as Inferred From Landed Mission Observations

Raymond E. Arvidson, Jeffrey G. Catalano

Contents

4.1 INTRODUCTION

The United States has successfully operated four landers (Viking Landers 1 and 2, Pathfinder, and Phoenix) and four rovers (Sojourner, Spirit, Opportunity, and Curiosity) on the surface of Mars (Table 4.1 and Fig. 4.1). When combined with the array of observations from orbiters, the data collected from these landed missions provide unique perspectives about the geologic evolution of Mars, including the role of water on and beneath the surface, and the implications for habitability and life. Lateral mobility has proven to be key to understanding Mars. This is evident in the Spirit, Opportunity, and Curiosity rovers' abilities to traverse thousands of meters to reach and characterize outcrops, in part directed to specific locations based on the analysis of orbiter-based imaging and spectral data. For the Phoenix Lander, which touched down above the northern

From Habitability to Life on Mars
https://doi.org/10.1016/B978-0-12-809935-3.00004-9

Table 4.1 Mars landed mission summary

Mission	Lifetime	Location (lat and lon)	Location source	Instrumentation
Mutch Memorial Station (Viking Lander 1)	1976–82	22.2715°, 312.0486° Chryse Planitia	Kuchynka et al. (2014)	Imaging, seismology, meteorology, arm-based delivery of soil to GCMS, XRFS, three biology experiments (Soffen, 1977)
Soffen Memorial Station (Viking Lander 2)	1976–79	47.6698°, 134.2803° Utopia Planitia	Kuchynka et al. (2014)	Same as Viking Lander 1
Sagan Memorial Station (Mars Pathfinder)	1997	19.0998°, 326.7458° Chryse Planitia	Kuchynka et al. (2014)	Lander-based imaging, rover-based APXS (Matijevic and Shirley, 1997)
Spirit Rover	2004–10	−14.5719°, 175.4785° Gusev Crater	Arvidson et al. (2004a)	Imaging, emission spectrometry, arm-based microscopic imager, APXS, Mössbauer spectrometry, Rock Abrasion Tool (Squyres et al., 2003)
Opportunity Rover	2004–current	−1.9483°, 354.4742° Meridiani Planum	Arvidson et al. (2004b)	Same as Spirit
Phoenix Lander	2008	68.2184°, 234.2487° Vastitas Borealis	Smith et al. (2009)	Imaging, meteorology, atmospheric laser sounder, arm-based scoop and rasp with icy soil delivery to evolved gas analyzer, wet chemistry, arm-based imaging, soil conductance and conductivity probes (Smith et al., 2009)
Curiosity Rover	2012–current	−4.5895°, 137.4417° Gale Crater	Vasavada et al. (2014)	Imaging, laser-induced breakdown and neutron spectrometers, meteorology package, radiation detectors, arm-based imager, APXS, brush, drill, and drill powder delivery to transmission XRD and GCMS with scanning laser spectrometer (Grotzinger et al., 2012)

Locations are given in areocentric coordinates with longitudes increasing toward the east. See text for references concerning instrumentation.

Latitude and longitude values are in areocentric coordinates. Values for VL1, VL2, and MPF are recent estimates of landing locations based on the analysis of radiometric tracking of the Opportunity Rover, while it was stationary during Martian winter. This analysis provided a more accurate Mars rotation model not yet adopted by IAU. Previous estimates for these lander locations were published in Folkner et al. (1997).

(From Arvidson, R.E., 2016. Aqueous history of Mars as inferred from landed mission measurements of rocks, soils, and water ice. J. Geophys. Res. doi:10.1002/2016JE005079.)

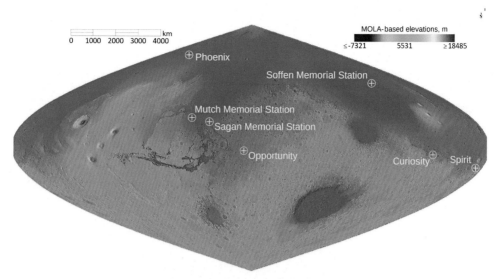

Fig. 4.1 Mars Global Surveyor's Mars Orbital Laser Altimeter (MOLA, Zuber et al., 1992) gridded data were used to generate this sinusoidal equal area shaded relief map overlain with color-coded elevations relative to the MOLA-based areoid equipotential surface. The projection is centered at 0 degrees in both longitude and latitude with the north pole shown at the top. The locations for the seven successful landed missions are shown. The Mutch Memorial Station is the name given to Viking Lander 1, the Soffen Memorial Station to Viking Lander 2, and the Sagan Memorial Station to the Mars Pathfinder Lander and its Sojourner rover.

Arctic Circle, vertical mobility using its robotic arm scoop and rasp to expose and collect subsurface icy soil proved to be just as important as lateral mobility for meeting this mission's specific scientific objectives.

In this chapter, key observations and inferences are reviewed that pertain to the presence of components needed for habitability on the surface and in the subsurface of Mars for the Spirit, Opportunity, and Curiosity rovers and the Phoenix Lander (Fig. 4.2). The focus of this paper, in part drawn from a previous survey of landed missions to Mars (Arvidson, 2016), is on more recent missions because the scientific objectives have been tightly focused on the extent to which water has interacted with the surface and interior of Mars and the implications for habitability and life. Also, these missions were directed to locations and included payloads that were optimized, within cost constraints, to capitalize on the inferences derived from the analysis of relatively recent and sophisticated orbital observations. These include, for example, Mars Global Surveyor Thermal Emission Spectrometer-based detections of crystalline hematite at Meridiani Planum (Christensen et al., 2001), which Opportunity characterized in great detail; Mars Reconnaissance Orbiter Compact Reconnaissance Imaging Spectrometer for Mars (CRISM, Murchie et al., 2007)-based detection of smectites and carbonate minerals within the Columbia Hills region explored by Spirit

Spirit and Opportunity

Curiosity

Phoenix

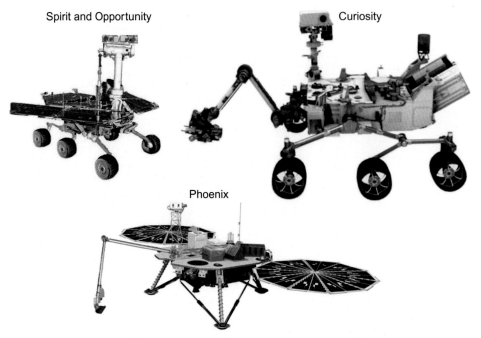

Fig. 4.2 Images for the three classes of landed missions are shown at correct relative scale. The Spirit and Opportunity rovers both landed in 2004, the Phoenix Lander in 2008, and the Curiosity rover in 2012. *(From Arvidson, R.E., 2016. Aqueous history of Mars as inferred from landed mission measurements of rocks, soils, and water ice. J. Geophys. Res. https://doi.org/10.1002/2016JE005079.)*

(Carter and Poulet, 2012); smectite localities examined by Opportunity on the rim of Endeavour Crater (Wray et al., 2009; Noe Dobrea et al., 2012; Arvidson et al., 2014; Fox et al., 2016); and crystalline hematite (Fraeman et al., 2013), smectites, and hydrated sulfate minerals within Gale Crater (Milliken et al., 2010; Fraeman et al., 2016).

This chapter proceeds by first providing a broad overview of all successful landed missions on Mars, organized as a function of when the spacecraft operated on the surface. Needs and challenges for habitability and life are then briefly described as a precursor to consideration of what has been found from landed missions. Highlights from the four key missions with direct relevance to understanding habitability are then presented in the order of increasing age of the soils and rock outcrops examined, with the first discussion focused on the icy soil examined by the Phoenix Lander (as explained in Fig. 4.3). Organization by increasing age is taken to understand how the nature of water interactions with the Martian surface and interior has changed over time. Prospects are then drawn with regard to habitability and life.

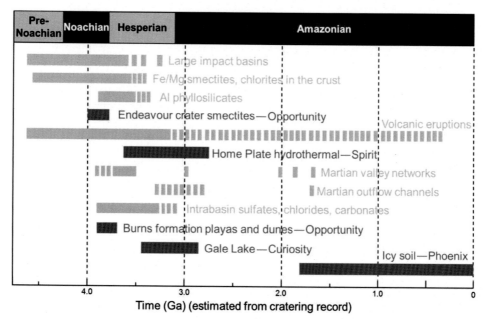

Fig. 4.3 Schematic showing major activity on Mars as gray bars as a function of time, largely inferred from analysis of orbital data. Also shown as *dark bars* are approximate times and key results from the four landed missions discussed in detail in this paper. The timing of the smectites on Endeavour rim is meant to be coincident with the formation of this Noachian-age crater. The timing of hydrothermal activity in the vicinity of Home Plate, Inner Basin, Columbia Hills, is assumed to be coincident with the timing of Gusev Crater volcanic plains emplacement, ~3.7 Ga (Greeley et al., 2005), with a long time shown toward younger ages given the uncertainty of the ages of features in the Inner Basin. The timing of deposition of the Burns formation is given by crater age dates from Arvidson et al. (2006a) and for the fluvial-deltaic-lacustrine deposits at the base of the stratigraphic column in Gale by Grant et al. (2014). The age of the polygonal plains and icy soil at the Phoenix landing site is Amazonian, and a long time span is shown to indicate likely orbitally controlled icy soil deposits and associated landforms at this latitude. *(Adapted from Ehlmann, B.L., Edwards, C.S., 2014. Mineralogy of the Martian surface. Annu. Rev. Earth Planet. Sci. 42(1), 291–315. doi:10.1146/annurev-earth-060313-055024. From Arvidson, R.E., 2016. Aqueous history of Mars as inferred from landed mission measurements of rocks, soils, and water ice. J. Geophys. Res. https://doi.org/10.1002/2016JE005079.)*

4.2 SUMMARY OF LANDED MISSIONS

Two Viking Landers were placed on the surface of Mars in 1976, both to begin the search for life on Mars and to celebrate the bicentennial of the United States. Viking Lander 1, subsequently renamed the Thomas A. Mutch Memorial Station, touched down on the Hesperian-age cratered volcanic plains of Chryse Planitia (Soffen, 1977). The original landing site was close to the mouth of a large, complex channel system emanating from the southern cratered highlands. The final landing site was moved further "downstream" to avoid rough elements detected from Earth-based radar observations. Viking Lander 2, subsequently renamed the Gerald Soffen Memorial Station, touched down in the high northern latitudes on polygonal ground located in Utopia Planitia (Soffen, 1977). Utopia Planitia was selected based on a very high atmospheric humidity and thus the best chance of finding evidence for extant life. Both landers were equipped with identical instrument payloads and robotic arms with scoops that sampled and delivered soil to three biology-related experiments, a gas chromatograph mass spectrometer to test for the presence of organic molecules and an X-ray fluorescence instrument to determine soil composition (Table 4.1). The focus on detecting organic materials and evidence for life within the soil deposits was a first step in the search for evidence concerning Martian habitability and life. The soils were found to have a basaltic composition, with evidence for minor cohesion induced by sulfate salts, perhaps by kieserite ($MgSO_4 \cdot H_2O$) (Clark et al., 1976). No in situ organic molecules were detected (Biemann et al., 1977), and results from the three biology experiments were equivocal at best (Klein, 1977), with a main result that found that the soils include a strong oxidant, perhaps one or more perchlorate-bearing salts (Quinn et al., 2013).

The next mission to successfully operate on the surface of Mars was Pathfinder (Matijevic and Shirley, 1997), touching down in 1997, and subsequently renamed the Carl Sagan Memorial Station. Pathfinder landed at the distal end of Ares Vallis, with the intent of using its lander-based imaging system and rover, Sojourner, to test the hypothesis that the landing site was scoured by floodwaters. Results were positive in that numerous large boulders were found that were inferred to have been transported during one or more flood events (Golombek et al., 1999). In addition, Sojourner included an Alpha Proton X-ray Spectrometer (APXS), and soil measurements again demonstrated a basaltic composition (Rieder et al., 1997). Rock measurements by APXS suggested a somewhat evolved composition toward basaltic andesites (Rieder et al., 1997). The greatest legacy of the Pathfinder mission was the demonstration that rovers can be remotely operated on Mars and can provide the critically needed lateral mobility to traverse to key soils and rocks for measurements needed to meet mission objectives. Sojourner traversed ~100 m across the landing site, keeping in sight of the lander for two-way communication with Earth.

The realization that lateral mobility and capable instrumentation were needed to access and characterize the array of key features at a landing site and beyond led to the development and implementation of the Mars Exploration Rover (MER) Mission (Squyres et al., 2003). The two MER rovers, Spirit and Opportunity, touched down in 2004. Both rovers have identical scientific payloads that consist of a mast-based multispectral stereo imagers (Pancam) and a miniature thermal emission spectrometer (Mini-TES), together with an arm-based microscopic imager (MI), Alpha Particle X-ray and Mössbauer Spectrometers (APXS and MB), and a Rock Abrasion Tool (RAT) to clean dust and coatings from rock surfaces (Table 4.1).

Spirit's landing site was on the Hesperian-age plains of Gusev Crater (Fig. 4.4) with the expectation that the deposits beneath the plains formed within a fluvial-deltaic-lacustrine environment that once filled the ~165 km-wide Noachian-age crater (Squyres et al., 2004a). The plains turned out to be covered with olivine-bearing basalt flows churned up by repeated impact events (Squyres et al., 2004a; Arvidson et al., 2006a). The relatively pristine nature of the basaltic rocks examined on the Gusev plains is consistent with long-term and relatively weak water-rock interactions that extend to the present, for example, only weak dissolution of olivine under low water-to-rock ratios (Hurowitz et al., 2006). Soon after landing on the plains, the rover was directed to the ~3 km distant Columbia Hills, an island of older rock on the floor of the crater. Spirit made landfall on the West Spur portion of the Columbia Hills in 2004 and found evidence for goethite ($FeO(OH)$), a mineral formed in aqueous environments

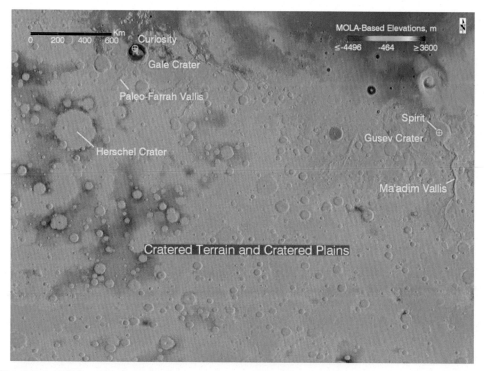

Fig. 4.4 A MOLA gridded product was used to generate the shaded relief map, color-coded with elevations relative to the MOLA-based areoid, as defined in the caption to Fig. 4.1. The location of the Spirit Rover in Gusev and the Curiosity rover in Gale Craters are shown. Both craters are at the margins between the heavily cratered terrains to the south and the plains to the north. Gusev Crater has been cut by the Ma'adim Vallis, and Gale Crater is the site of a ~5 km high interior mountain called Aeolis Mons or informally Mount Sharp. A large drainage system labeled as Paleo-Farrah Vallis and sourced in Herschel Crater was truncated by the formation of Gale Crater. The map extends from ~122°E, 2.75°N to 179.75°E, 42.75°S, planetocentric coordinates.

(Morris et al., 2006a). Traversing onto the Husband Hill portion of the Columbia Hills led to the discovery of ferrous iron carbonates, perhaps siderite ($FeCO_3$) (Morris et al., 2010), followed by descent into the Inner Basin, and the exploration of the Home Plate volcanoclastic feature, which led to the discovery of sulfate minerals and opal-A ($SiO_2 \cdot nH_2O$) deposits produced in hydrothermal environments (Squyres et al., 2008; Yen et al., 2008; Arvidson et al., 2010; Ruff et al., 2011). Contact with Spirit was lost in 2010 while the rover was embedded in soft sulfate-rich sands that were deposited next to Home Plate (Arvidson et al., 2010). During its mission, Spirit traversed ~7.7 km while exploring the floor of Gusev Crater.

Opportunity was directed to and successfully landed on the late Noachian- to Hesperian-age plains in Meridiani Planum (Fig. 4.5), with the site chosen based on the detection of crystalline hematite (Fe_2O_3) from the analysis of Mars Global Surveyor Thermal Emission Spectrometer data (Christensen et al., 2001). Hematite can form at

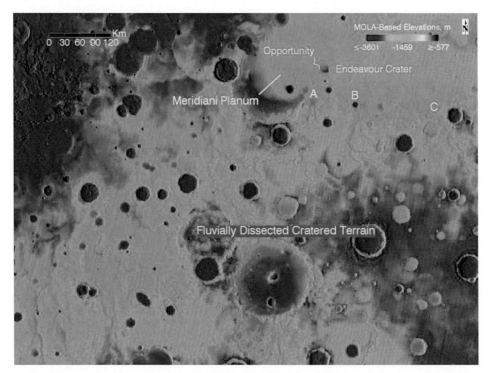

Fig. 4.5 A MOLA gridded product was used to generate the shaded relief map color-coded with elevations relative to the MOLA-based areoid, as defined in the caption to Fig. 4.1. The location of the Opportunity rover on the Meridiani plains and Endeavour Crater are shown. Note that fluvial channels that are buried by what are now known as Burns formation deposits that underlie the plains, with the embayment locations denoted by letters A, B, and C. White line shows 45 km of Opportunity's traverses from the plains' landing site to the rim of Endeavour Crater. The map extends from ~344.65°E, 0.34°S to 359.75°E, 11.63°S, planetocentric coordinates.

low temperatures and in aqueous environments, depending on the acidity and the presence of oxidants, and thus, Meridiani was an obvious choice for a landing site. The rover, encapsulated within its landing system, by chance bounced into one of the few craters (subsequently named Eagle) on the Meridiani Plains (Squyres et al., 2004b). The crater is ~22 m wide, and the rover was commanded to traverse to a nearby outcrop on the crater wall. The outcrop was found to consist of sulfate-rich sandstones, with ripple patterns indicative of sand transport in a fluvial environment (Squyres et al., 2004c). Hematitic concretions were found within the outcrop and interpreted to have been generated by diffusion-based reactions with rising groundwater environments after deposition of the sandstones. The mission became one of the crater hoppings to investigate stratigraphic sections exposed on crater walls of what became known as the Burns formation. The ensemble of data led to a self-consistent model of deposition of sulfate-rich muds in shallow evaporitic playas, with subsequent reworking to produce sulfate-rich sandstones (e.g., Grotzinger et al., 2005; McLennan et al., 2005).

Opportunity was directed later in its mission to the rim of the ~22 km-wide Noachian-age Endeavour Crater, with a change in objectives to search for evidence for aqueous alteration associated with these ancient rocks. Impact breccias, subsequently named the Shoemaker formation, were found to dominate the rim outcrops (Squyres et al., 2012; Crumpler et al., 2015; Mittlefehldt et al., 2018). Enhanced zinc concentrations were found on the Tisdale rock target located on the southern side of the Cape York rim segment and were interpreted to have formed while hydrothermal systems were active just after Endeavour Crater formed (Squyres et al., 2012). On the eastern side of Cape York, Opportunity data showed that finely layered outcrops, subsequently named the Matijevic formation, unconformably underlie the Shoemaker formation breccias (Arvidson et al., 2014). Synergistic analysis of CRISM hyperspectral images and Opportunity observations show that these older deposits have been isochemically altered to contain ferric iron smectites, with Opportunity observations showing extensive leaching toward an aluminum-rich smectite composition in fracture zones, where enhanced fluid flow would be expected (Arvidson et al., 2014; Clark et al., 2016).

On the Murray Ridge rim segment, Opportunity's wheels serendipitously excavated and overturned two small rocks from a fracture zone. Extensive measurements using Opportunity's instruments demonstrate that the rocks have coatings that are highly enriched in sulfate minerals and manganese oxides (Arvidson et al., 2016). The coatings are inferred to have formed in a subsurface evaporative environment, with the presence of manganese oxides indicative of the introduction of a strong oxidant toward the end of the coating formation period. Opportunity was then directed to Marathon Valley on the Cape Tribulation rim segment due to the CRISM-based discovery of iron-magnesium smectites (Fox et al., 2016). Opportunity's exploration of the highly fractured Marathon Valley floor showed extensive evidence for aqueous alteration, including strings of hematite-rich pebbles within sulfate-rich fractures, germanium enrichments for selected rocks, and likely isochemical alteration to produce the observed iron-magnesium smectite signature evident in CRISM data (Fox et al., 2016; Farrand et al., 2017; Mittlefehldt et al., 2018; Stein et al., 2018). As of 10 August 2017 (sol 4815), Opportunity had traversed 45 km during its exploration of Meridiani Planum and Endeavour Crater.

The Phoenix Lander touched down on the high northern plains in 2008 (Smith et al., 2009). This relatively low-cost Scout-class mission focused on the use of its robotic arm for rasping and scooping icy soil for analytic measurements using onboard wet chemistry, microscopy, electrochemistry, and conductive analyzer (MECA), and thermal and evolved gas analysis (TEGA) instrumentation (Table 4.1). Soil containing water ice in pores was expected to be just beneath the surface based on Mars Odyssey orbiter neutron and gamma-ray spectroscopy measurements (Boynton et al., 2002; Feldman et al., 2004). Other instrumentations on the Phoenix Lander included soil probes (also part of MECA), a descent imager (MARDI), mast- and arm-based multispectral imagers (SSI and RAC), and meteorology experiments (ASI/MET and Telltale) (Table 4.1). Icy soil was found in each of the dozen trenches excavated by the robotic arm (Arvidson et al., 2009), demonstrating that vertical mobility to depths of 10–20 cm was an appropriate match to mission objectives (Smith et al., 2009). Carbonates were detected in the crusty basaltic soil samples based on evolved gas analyzer results (Boynton et al., 2009), and perchlorates were detected using the wet chemistry capabilities (Hecht et al., 2009).

The Mars Science Laboratory Curiosity rover touched down in 2012 on the northern plains in Gale Crater (Fig. 4.4) and began its exploration of the ~5 km high set of sedimentary strata that underlie a mountain (Aeolis Mons, informally named Mount Sharp) located in the crater center (Grotzinger et al., 2012; Vasavada et al., 2014). Curiosity has on board the most extensive scientific instrumentation sent to date to the surface of Mars (Table 4.1). This includes a descent camera (MARDI); mast-based multispectral cameras (Mastcam); a laser-induced breakdown spectrometer for remotely determining compositions of soils and rocks (ChemCam); body-mounted meteorology, neutron spectrometer, and radiation experiments (REMS, DAN, and RAD); an arm with a color imaging system (MAHLI), brush, and drill; an Alpha Particle X-ray Spectrometer (APXS); and in situ instruments that accept drill powders for X-ray diffractometer (CheMin) and/or gas chromatograph, mass spectrometer, and laser spectrometer observations (SAM). The rover's exploration while crossing the plains and ascending the lower slopes of Mount Sharp uncovered unequivocal evidence for ancient fluvial-deltaic-lacustrine systems prograding from Gale's rim and walls toward the crater floor (Grotzinger et al., 2014, 2015). In addition, powders recovered from initial drilling into the lacustrine outcrops include smectites (Vaniman et al., 2013; Rampe et al., 2017) and low concentrations of chlorinated hydrocarbons (Freissinet et al., 2015; Mahaffy et al., 2015). As of 10 August 2017 (sol 1781), Curiosity had driven 16.4 km during its exploration of Gale Crater and its sedimentary deposits.

4.3 NEEDS AND CHALLENGES FOR HABITABILITY AND LIFE

Landed missions on Mars provide unique datasets for evaluating whether or not conditions were conducive to life at specific locations and times. Extensive past reviews have outlined in general the key components for habitability (e.g., Hoehler, 2007; McKay, 2014; Cockell et al., 2016) and on early Mars specifically (e.g., Knoll and Grotzinger, 2006; Des Marais, 2010; Cockell, 2014). The essential requirements to sustain life are generally grouped into four major categories: (1) a solvent, that is, water; (2) a source of energy; (3) an availability of essential chemical building blocks; and (4) a favorable environmental conditions. Critical aspects of these habitability requirements are discussed below, with a focus on features that are potentially observable by landed missions on Mars.

The availability of liquid water has been a primary focus of the search for past habitable conditions on Mars. Ionic and polar neutral molecules readily dissolve in water (Finney, 2004), providing a solvent medium through which bioessential chemical components can be made available to organisms. Compartmentalizing cells and obtaining energy via chemical gradients across cell membranes are also aided by water as the primary solvent. Mars has long been a focus in the search for habitable environments because of clear geomorphic evidence for the past surface water (Lasue et al., 2013). Exploration of the Martian surface thus seeks evidence for both the occurrence and duration of liquid water. One further challenge is that liquid water needs to be of low to moderate salt content to be favorable for life (Tosca et al., 2008a), as a low activity of water greatly limits biological activity (Grant, 2004).

Life also needs a source of energy to support critical biological functions. Sources of chemical energy provided by redox disequilibrium have been identified as key indicators of a habitable environment (Hoehler et al., 2007). The coexistence of reductants and oxidants, especially those with slow reaction kinetics, is potentially observable by landed missions in the Martian rock record. Whereas light provides an alternative energy source that is available in any nonshadowed surface environments, the evolution of photosynthesis substantially postdates the origin of life on Earth (Des Marais, 2000; Blankenship, 2010), and it is unlikely that evidence would be found of phototrophic life on Mars in the absence of chemotrophic predecessors. In addition, ionizing radiation in the form of sunlight and cosmic rays degrades organic compounds (Pavlov et al., 2012), and sunlight can also produce reactive oxidants that decompose carbon-bearing molecules (Court et al., 2006; Davila et al., 2008). These challenges for phototrophic life suggest that identifying environments that provided chemical energy is essential to investigating the habitability of Mars.

Habitable environments also must provide an adequate supply of elements essential to the formation of biomolecules: C, H, N, O, P, and S. The presence of organic carbon may come from meteoritic infall (Benner et al., 2000), although this would require protection from abiotic degradation. Fixed nitrogen (e.g., nitrate and ammonium) may be produced by impact processes on Mars (Manning et al., 2009). Detecting the presence of many of these elements in forms usable to life requires mass spectrometer measurements by landed missions, although P and S can be detected by X-ray fluorescence-based instruments (e.g., APXS), and ChemCam on MSL can also detect many of the elements essential to life (Wiens et al., 2013).

Finally, clement physical and chemical conditions are also key components of habitability. Temperatures of surface environments can be constrained by evidence of liquid water, and subsurface paleotemperatures can be inferred from mineral assemblages with known temperature-dependent stabilities. Whereas life on Earth appears to have a temperature range of roughly −20 to 120°C, the extremes represent challenging environments except for the most highly adapted organisms (Clarke et al., 2013; Kashefi and Lovley, 2003). Given the current climate on Mars, the lower temperature limit is likely of greatest relevance to many locations explored by landed missions. Whereas brines can preserve liquid water down to very low temperatures (e.g., Chevrier et al., 2009), the low activity of water in such fluids poses a substantial challenge to life (see earlier discussion). Extremes of pH also pose substantial challenges to life (Krulwich, 1995; Baker-Austin and Dopson, 2007) that favor circumneutral conditions. Radiation, radicals, and other oxidants, together with toxic compounds, are examples of additional factors that negatively impact habitability (Hoehler, 2007; Cockell et al., 2016). Finally, habitable conditions must persist beyond brief instances in time in order for life to gain a foothold and sustain itself (Cockell et al., 2016). Prolonged habitability is thus a key environmental factor to assess when investigating the Martian surface and subsurface.

4.4 INDICATORS OF HABITABILITY FROM LANDED MISSIONS

4.4.1 Phoenix: Northern Latitude Soils and Water Ice

4.4.1.1 Overview

The Phoenix Lander touched down above the Arctic Circle on Amazonian-age plains dominated by polygonal ground, with a wide variety of polygon sizes formed by repeated contraction and expansion of icy soils as seasons and orbital conditions changed during the Amazonian Period (Figs. 4.1 and 4.6) (Mellon et al., 2008, 2009a). The Phoenix mission was designed to explore key features of the geologically modern Martian water cycle, one dominated largely by vapor-ice dynamics. As noted in the landed mission summary section of this chapter, the primary objective for the Phoenix Lander mission was to sample icy soils predicted from orbital data to be present just a few centimeters beneath the surface at high-latitude sites. In fact, orbital observations show the presence of permanent water ice outcrops on the north-facing slopes within the nearby Heimdal Crater (Figs. 4.6 and 4.7). Phoenix Lander was equipped with a 2.4 m long, 4 degree of freedom arm,

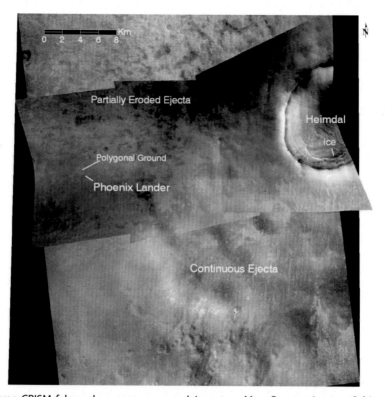

Fig. 4.6 Three CRISM false-color scenes are overlain onto a Mars Reconnaissance Orbiter's Context Camera (CTX) image (Malin et al., 2007) to show the location of the Phoenix landing site on the partially eroded ejecta from the Heimdal impact crater. This site was chosen because the ubiquitous high rock population characteristic of the high northern plains was reduced during ejecta emplacement, thus increasing the probability of landing without encountering mission-ending large rocks. The *blue colors* in Heimdal are associated with H_2O ice protected from sunlight and associated sublimation on the interior north-facing slopes. An ice spectrum from CRISM data is shown in Fig. 4.7, along with a typical spectrum of the polygonal ground on which Phoenix landed. For the CRISM data, RGB colors are assigned to single scattering albedos centered at 1.08, 1.51, and 2.53 μm. CRISM scenes from left to right are FRT0000BAAE, FRT0000B6D4, and FRT0000B5D2, all acquired during the northern summer season. CTX image is P21_009145_2484_XI_69N125W.img, acquired simultaneously with FRT0000B6D4.

with a scraper, scoop, and rasp (Fig. 4.2). This arm excavated a dozen trenches (Figs. 4.8 and 4.9) and delivered both ice-free and icy soil to onboard instruments that performed microscopic imaging and wet chemistry and evolved gas analyses. A few centimeters of crusty basaltic soil were encountered over very hard icy soil, with both pore ice and thin slabs of ice evident in the two polygonal mounds that were excavated (Smith et al., 2009; Arvidson et al., 2009; Mellon et al., 2009b).

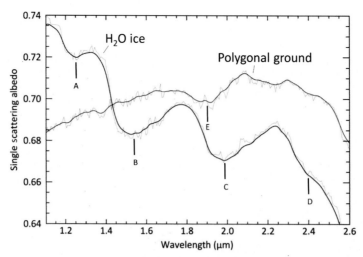

Fig. 4.7 Spectra are shown using a mean of 100 pixels (12 m/pixel) for the H_2O ice on the north-facing slopes within Heimdal Crater and polygonal ground near the Phoenix landing site. Values are shown as single scattering albedos, which are free from lighting and viewing conditions. Thin lines represent spectra before noise suppression. Absorptions A–D are diagnostic of H_2O ice, whereas E for the polygonal ground is indicative of adsorbed, absorbed, and/or H_2O bound in minerals (e.g., see Clark, 1999, for absorption vibrational assignments).

Fig. 4.8 Surface Stereo Imager (SSI, Lemmon et al., 2008) enhanced color image of the trenches dug by the Phoenix robotic arm, extending to the horizon to show the polygonal ground that dominates the plains' landing site. Slab H_2O ice was encountered on the left side trench (Dodo-Goldilocks), whereas pore H_2O ice was found on the right-side trenches (Snow White). These two trench complexes were excavated in polygonal mounds. RGB corresponds to bands centered at 0.600, 0.530, and 0.480 μm. Product available through the NASA Planetary Photojournal as PIA12105.

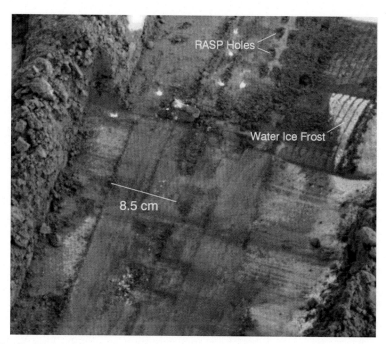

Fig. 4.9 SSI color image of the Snow White trenches excavated on the right side of the digging volume. This trench complex was scraped using the robotic arm scoop, and holes were excavated using the rasp on the scoop to be able to collect H_2O icy soil. H_2O frost can be seen accumulating on the shadowed portions of the trench during the northern fall season when the image was acquired. Product available through the NASA Planetary Photojournal as PIA11724.

4.4.1.2 Perchlorates and Other Salts

A key finding from the wet chemistry experiment was the presence of dissolved salts leached from the soil samples (Hecht et al., 2009; Kounaves et al., 2010). There were no differences in soluble chemical species observed from surface soils down to soils excavated from just above the icy soil interface. A moderately alkaline pH was observed when the soils were introduced into water within the wet chemistry cells, consistent with a carbonate-buffered solution. The data indicate the presence of hydrated magnesium sulfate salts, which are widely found on Mars, and the detection of 0.4–0.6 wt% perchlorate (Hecht et al., 2009; Toner et al., 2014). Other ions identified include calcium, potassium, sodium, and low concentrations of ammonia (NH_4^+). The presence of a hydrated magnesium perchlorate salt in excavated soil was indicated using multispectral Surface Stereo Imager data that detected an H_2O absorption feature unique to this mineral phase (Cull et al., 2010). Perchlorate salts may also have been excavated by the descent rockets, deliquescing to form droplets of brine on the lander struts (Rennó et al., 2009; Mehta et al., 2011).

The discovery of perchlorate salts is important because perchlorate is a strong oxidant in addition to being a freezing-point depressant. As noted in the landed mission summary of this chapter, it may also have been the strong oxidant in the soils sampled by the Viking Landers (e.g., Quinn et al., 2013). The presence of hydrated perchlorate salts may also explain how brines can be present under the modern and very cold conditions. For example, hydrated perchlorate salts have been identified using CRISM spectra in recurring slope lineae that emanate from cliffs in the midlatitudes when the summer season arrives and insolation is directly onto the slopes (Ojha et al., 2015).

4.4.1.3 Low Water Activity Environment

One of the last images received from Phoenix covers one part of the Snow White trench complex after arm work was finished (Fig. 4.9). The data were acquired during the northern fall season at 11:32 local solar time, when water ice frost was on the surface. This image is a fitting reminder that Mars does have a modern water cycle, one governed primarily by ice-vapor conditions, although the slab ice encountered on the Goldilocks trench complex indicates at least some minor liquid water migration (Mellon et al., 2009b; Cull et al., 2010). This is also consistent with the discovery of a minor amount of carbonate-bearing mineral or minerals in the Phoenix crusty soil, thought to have formed by reactions with thin films of water (Boynton et al., 2009). Thin films of water on grains are also consistent with the ubiquitous $3\,\mu m$ absorption fundamental due to OH and H_2O vibrations seen in orbital OMEGA data, with deeper absorptions at higher latitudes (Jouglet et al., 2007; Milliken et al., 2007). In fact, thermodynamically, liquid water should be present as thin films, migrating to grain-to-grain contacts, carrying dissolved ions that would precipitate on saturation to form salts (Boxe et al., 2012). Of issue for habitability and life would be the magnitude of water activity in these films, particularly for highly concentrated brines.

4.4.2 Curiosity: Fluvial-Deltaic-Lacustrine Deposits in Gale Crater

4.4.2.1 Overview

Curiosity's landing site on the plains within the $\sim 154\,km$-wide Gale Crater was selected to be as close as possible to the $\sim 5\,km$ stack of Aeolis Mons sedimentary strata that cover the center of the crater (Fig. 4.10) (Golombek et al., 2012). Gale Crater is inferred to have formed during the Hesperian Era, $\sim 3.61\,Ga$ ago (Le Deit et al., 2013), with deposition of sediments within the crater extending to $\sim 2.9\,Ga$, based on impact crater densities on the plains covering the deposits (Grant et al., 2014) (Fig. 4.3). Curiosity's examination of deposits within Gale Crater fits as the next oldest landforms and deposits examined by the three rovers and the Phoenix Lander.

Analysis of CRISM data covering Gale Crater indicates that the lower Mount Sharp strata include smectite exposures that transition upward to hydrated sulfate deposits (Milliken et al., 2010; Fraeman et al., 2016). A major erosional unconformity separates

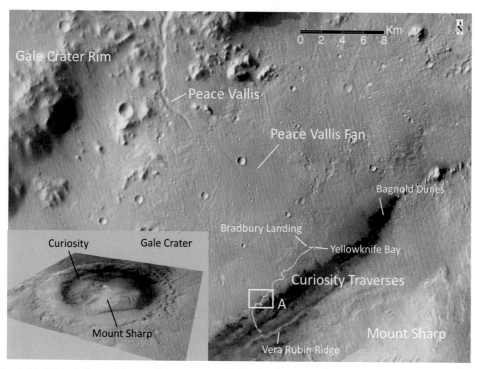

Fig. 4.10 Orbital view of the northwestern portion of Gale Crater, showing the rim, the plains, and a portion of Mount Sharp using a visible wavelength mosaic from the Odyssey orbiter's Thermal Emission Imaging System (THEMIS, Christensen et al., 2004) images and overlain with color-coded predawn THEMIS-based IR radiance values. *Red colors* indicate warmer areas likely because of an increase number of rocks and/or more exposed bedrock. The landing site and the traverses for the Curiosity rover are shown to just north of the Vera Rubin Ridge. Note the Peace Vallis channel that has cut the crater rim and deposited alluvial fan material onto the plains. Box A delineates area shown in detail in Fig. 4.12. Inserted on lower left is a simulated perspective view of Gale Crater looking to the northwest that shows the ~5 km high Mount Sharp located within the crater.

the top of the hydrated sulfate section from overlying clinoform deposits that are interpreted to be anhydrous dust stones (Anderson and Bell, 2010; Milliken et al., 2010). In addition, there is evidence for a hematite-capped ridge (Vera Rubin Ridge) on the lower stratigraphic section of Mount Sharp (Fraeman et al., 2013). The deposits that underlie Mount Sharp provide a Hesperian-age record for the period of time when Mars was undergoing significant changes in surface conditions, from formation of smectites and hydrated sulfates to anhydrous deposits, following a global model for environmental changes posed by Bibring et al. (2006a,b) on the basis of Mars Express OMEGA hyperspectral imaging observations. For this chapter, we focus on Curiosity's exploration of outcrops exposed on the plains and lower slopes of Mount Sharp that are stratigraphically below the rocks with smectite and hydrated sulfate signatures (Fig. 4.10).

4.4.2.2 Prolonged Surface Water

Analysis of MRO HiRISE images (McEwen et al., 2007) and associated digital elevation maps, combined with detailed assessment of the stratigraphy, rock textures, primary sedimentary features, compositions, and mineralogy from Curiosity's measurement campaigns, demonstrates that outcrops exposed on the plains and lower portions of Mount Sharp provide compelling evidence for a prograding fluvial-deltaic-lacustrine environment (Grotzinger et al., 2014, 2015). This was during a time of a prolonged and rising lake level within Gale Crater (Fig. 4.11). The evidence includes the initial observations of what have been mapped as Bradbury group fluvial gravels on the hummocky plains seen just after landing (Williams et al., 2013), together with the Sheepbed mudstone and Gillespie sandstone members of the Yellowknife Bay formation, located to

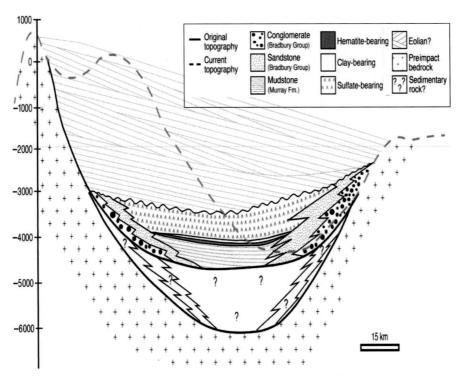

Fig. 4.11 Interpretative Gale Crater stratigraphic section sketch provides an overview of the evolution of the sedimentary deposits that underlie the plains and Mount Sharp. Bradbury group fluvial gravels and sandstones, together with deltaic sandstones, prograded over Murray formation lacustrine sandstones, all sourced from the crater rim. These were succeeded by sulfate-bearing strata and, after a period of erosion, large-scale clinoform dust stones. Not shown are the Stimson formation windblown sandstones deposited unconformably on top of eroded Murray formation deposits. (Based on Grotzinger, J.P., et al., 2015. Deposition, exhumation, and paleoclimate of an ancient lake deposit, Gale crater, Mars. Science 350(6257). https://doi.org/10.1126/science.aac7575.)

the northeast of the Bradbury Landing (Grotzinger et al., 2014, 2015). The Bradbury group deposits are for the most part older than the Murray formation mudstones (part of the Mount Sharp group) examined by Curiosity after leaving Yellowknife Bay and heading to the southwest toward Mount Sharp. Some interfingering of the Bradbury group and Murray formation deposits is inferred. Both units are unconformably overlain by Stimson formation windblown sandstones that are interpreted to have formed after deposition and exhumation of the Mount Sharp strata, implying a significant shift toward aridity relative to the time of deposition of the fluvial-deltaic-lacustrine deposits (Grotzinger et al., 2015).

A particularly illustrative example of the sedimentary facies associated with the lower Mount Sharp stratigraphic section is found in the Pahrump Hills and areas just to the southwest of these regions (Figs. 4.11–4.16). Murray formation mudstones, interpreted to be lacustrine deposits, cover the southern portion of a wide valley floor within the

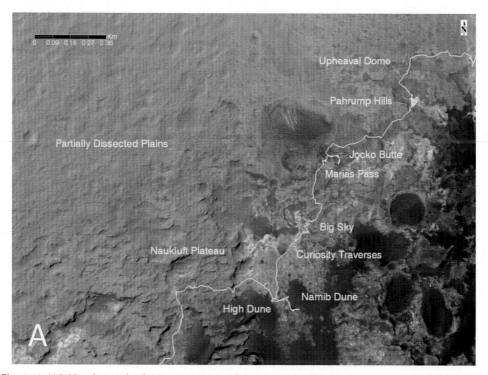

Fig. 4.12 HiRISE enhanced color image mosaic of traverses taken by Curiosity from Upheaval Dome to the Naukluft Plateau. Pahrump Hills and Marias Pass are locations where numerous observations were conducted and four drilled samples were delivered to the CheMin and SAM instruments. Also shown are the measurement stops at the High and Namib Dunes, part of the extensive Bagnold dune field shown in Fig. 4.10. RGB assigned to data covering wavelengths at 0.8–1.0, 0.55–0.850, and 0.40–0.60 μm.

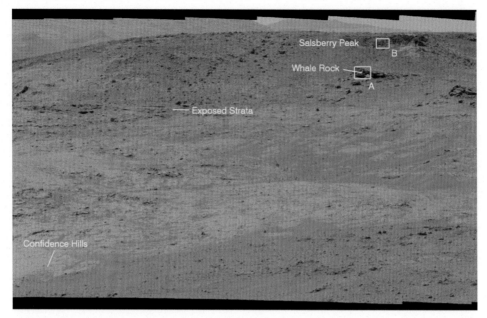

Fig. 4.13 Mastcam color mosaic (Malin et al., 2010) acquired looking to the south at the Pahrump Hills and the Murray formation mudstones, together with Whale Rock and Salsberry Peak sandstone outcrops. Letters A and B denote locations of close-up views of the sandstone strata. Whale Rock is ~2 m high. RGB corresponds to Mastcam wavelengths centered at 0.753, 0.535, and 0.432 μm. Mosaic CX 00753ML 0420852 F464328792VA is available through the PDS Geosciences Node Curiosity Analyst's Notebook.

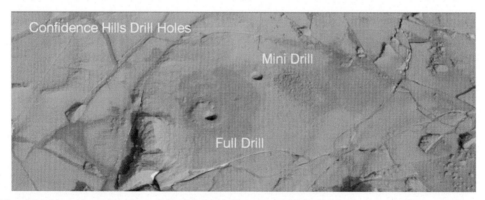

Fig. 4.14 Mastcam color view of Confidence Hills mini and full drill holes into flat-lying mudstone lacustrine facies of the Murray formation. The mini drill hole was done to test the ability to successfully drill into these particular rocks, whereas samples for CheMin and SAM were obtained from the full drill hole. A coarser grained, likely sandstone facies of the Murray formation rocks can be seen as a topographic high to the lower left of the full drill hole. Drill holes are ~1.6 cm in diameter. Image is frame 0762 MR 003273001 0403806 EOI DRCL, available through the PDS Geosciences Node Curiosity Analyst's Notebook.

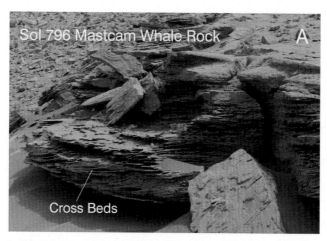

Fig. 4.15 Portion A of the Mastcam mosaic shown in Fig. 4.13, enlarged to show the side of Whale Rock with fluvial cross bedding in sandstone and dips indicative of flow to the south. The geometric configuration of Whale Rock and the fact that it is surrounded by mudstones are consistent with a fluvial channel in a fluvial-deltaic-lacustrine complex prograding toward the south. This cross-bedded section is ~25 cm high.

Fig. 4.16 Portion B of Mastcam mosaic covering the Salsberry Hills outcrop and showing coarse-grained rocks overlain by well-stratified rocks that are interpreted to be sandstones. Both the Salsberry Hills and Whale rock outcrops are interpreted to be gravel and sandstone fluvial deposits that interfinger with the lacustrine mudstones. Larger coarse-grained rock is a block that is ~20 cm wide.

Pahrump Hills area (Figs. 4.12 and 4.13). The Confidence Hills drill sample was obtained from these mudstones (Fig. 4.14). A lens of cross-bedded fluvial sandstone named Whale Rock is situated within the mudstone strata (Fig. 4.15). This sandstone outcrop is interpreted to have been deposited as a stream system or a bottom-hugging flow within a deltaic complex, with the cross beds dipping in the southerly direction, that is, the flow direction (Grotzinger et al., 2015). Continued rise of the lake level and associated submergence led to an upward transition in which the sandstones were covered by mudstones. The top of the section, called Salsberry Peak (Fig. 4.16), represents a distal fluvial unit, with a gravel base, implying continued southward progradation of the fluvial-deltaic-lacustrine deposits.

4.4.2.3 Varying Redox Conditions

As of sol 1298, Curiosity had successfully drilled nine holes, with delivery of rock powders to CheMin and SAM. The fourth drill target (Confidence Hills) was in the Murray formation mudstones at Pahrump Hills (Fig. 4.14). CheMin transmission XRD data for powder from this target indicate the presence of an amorphous or glassy phase, a phyllosilicate, likely a collapsed smectite or illite, hematite, minor amount of jarosite, feldspars, pyroxene, and a minor amount of magnetite, with an overall basaltic composition (Cavanagh et al., 2015; Rampe et al., 2017). Analyses of Confidence Hills samples, two additional drill powders (Mojave and Telegraph Peak) collected in Pahrump Hills mudstones located stratigraphically above the Confidence Hills target, and a sample collected in Marias Pass (Buckskin, youngest target, ~12 m above Confidence Hills) (Fig. 4.12) show an upsection decrease in hematite-jarosite and an increase in magnetite-crystalline silica contents (Rampe et al., 2017). This implies either temporal variations in oxic and anoxic lake conditions (Hurowitz et al., 2017) or variations in the extent of acid-sulfate aqueous weathering (Rampe et al., 2017).

4.4.2.4 Organic Carbon in Mudstones

SAM organic analyses of the drill samples discussed in the previous paragraph are ongoing. On the other hand, the drill sample from the Cumberland drill hole in the Sheepbed mudstone (in Yellowknife Bay, part of the Bradbury group) has been thoroughly analyzed and indicates the presence of low concentrations of chlorinated hydrocarbons (Freissinet et al., 2015; Mahaffy et al., 2015). The chlorinated hydrocarbons are interpreted to be the reaction products of Martian chlorine and organic carbon and are consistent with igneous, hydrothermal, atmospheric, exogenous, or biological sources (Freissinet et al., 2015). The lack of Martian organic compounds found in scooped samples from the modern windblown basaltic sands trapped within a cluster of rocks (Rocknest) and examined in detail by Curiosity (Glavin et al., 2013) strongly supports the hypothesis that preservation of organic materials will be largely in ancient bedrock deposits that have not been highly oxidized.

This is certainly consistent with the lack of Martian organic compounds in soils analyzed by the Viking Landers (Biemann et al., 1977).

4.4.2.5 Detection of Nitrate

An oxidized nitrogen species was detected by SAM in both modern windblown basaltic sands found at the Rocknest area and in the Cumberland and John Klein mudstone drill samples collected in Yellowknife Bay (Stern et al., 2015). Detailed analysis revealed 0.01–0.1 wt% nitrate (NO_3^-) in these samples. The distribution of nitrate in both windblown sediments and mudstones suggests that some form of fixed nitrogen was perhaps produced on early Mars during thermal impact shocks or lighting associated with volcanic activity. The nitrate-to-perchlorate ratio on Mars is lower than on Earth, suggesting both a lack of biological nitrogen fixation and distinctly different perchlorate formation mechanisms on the two planets (Stern et al., 2017).

4.4.2.6 Alteration by Ground Water

A surface-charged hydrologic system was required to generate the sedimentary deposits observed by Curiosity, and the data point to environments conducive to preservation of organic molecules and ones that were habitable, at least in terms of the sustained presence of water. In addition, the strata encountered by Curiosity also commonly have veins and fracture-controlled alteration zones, implying postdeposition aqueous alteration and thus sustained subsurface water availability. This includes the discovery of manganese oxide-rich veins along bedding planes (Lanza et al., 2016), raised ridges with fracture-filling cements (Léveillé et al., 2014), calcium sulfate-rich veins that cut across strata (Vaniman et al., 2013; Nachon et al., 2014), and silica-rich halos centered about fractures (Frydenvang et al., 2016; Yen et al., 2016).

For this chapter, we focus on implications for the silica-rich halos associated with fractures in the Stimson formation sandstones on the Naukluft Plateau. As noted, the Stimson formation sandstones were deposited by wind after the Murray formation was eroded and lie unconformably on this older unit. Curiosity's ChemCam and APXS observations show that fractures that cut across the Stimson outcrops have high-silica (~70% SiO_2) values close to the fracture centers, transitioning outward to typical basaltic compositions (Frydenvang, et al. 2016; Yen et al., 2016). This is interpreted to be a consequence of enhanced flow of postdepositional aqueous fluids along fracture zones and associated high hydraulic conductivities. Given that the Stimson formation sandstones were deposited after Mount Sharp strata were deposited and were subsequently exhumed, these halos indicates a sustained presence of groundwater in Mount Sharp.

4.4.2.7 Evolution of Water Availability

Analysis of the Cumberland sample by SAM also shows that the smectite-bearing hydrogen has a D/H approximately three times Earth's standard mean ocean water and 50%

lower than the Martian atmosphere, consistent with extended desiccation of the planet by solar wind stripping (Mahaffy et al., 2015). Again, this suggests that water-laid sediments such as those evident for the Yellowknife Bay and Pahrump Hills areas were largely confined to early geologic times, that is, from Noachian to Hesperian ages.

4.4.3 Opportunity: Burns Formation Sandstones

4.4.3.1 Overview

Opportunity's multiyear observations of Burns formation included outcrops on the plains and the use of impact craters to probe the local stratigraphy (Fig. 4.17). The Burns formation consists of sulfate-rich sandstones that were unconformably deposited onto the dissected Noachian-age cratered landscape produced by extensive fluvial erosion and transport of sediment to the northwest (Hynek and Phillips, 2001; Arvidson et al., 2006b). Impact crater densities imply a depositional age for the Burns formation of late Noachian to early Hesperian (Arvidson et al., 2006b). A number of hypotheses have been presented for the environment that produced the Burns formation sandstones, including aqueous alteration of impact ejecta deposits (Knauth et al., 2005), polar dust deposits that

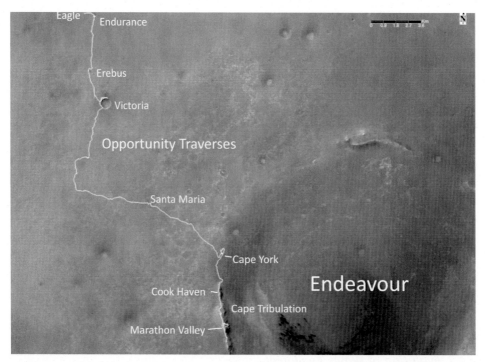

Fig. 4.17 Opportunity traverses are shown in white overlain onto a HiRISE-based image mosaic. Key craters used by Opportunity to show vertical sections of the Burns formation include Eagle through Santa Maria. The rim of Endeavour Crater has been examined in detail, including Marathon Valley.

formed under different orbital conditions (Niles and Michalski, 2009), and volcanic deposits (McCollom and Hynek, 2005). The hypothesis that best explains the detailed observations made by Opportunity is the formation of sulfate-rich muds by evaporation in shallow playa environments, reworking by wind and water to generate sand-sized particles, followed by alteration and cementation during periods of rising groundwater, and finally wind erosion and cratering to produce the current landscape and outcrops (e.g., Squyres and Knoll, 2005).

4.4.3.2 From Playa Muds to Sandstones

Observations of the compositions (McLennan et al., 2005; McLennan, 2012), mineralogy (Glotch et al., 2006; Morris et al., 2006b), grain textures, primary sedimentary features (e.g., ripple patterns) (Grotzinger et al., 2006; Edgar et al., 2014), diagenetic features such as hematitic concretions and veins (Knoll et al., 2008), and detailed local stratigraphic cross sections (Grotzinger et al., 2005) provide the definitive evidence for an evolving environment of deposition for Burns formation rocks that began with playa muds that were transformed into sulfate-rich sandstones. The Burns formation depositional setting thus contrasts with the Hesperian-age fluvial-deltaic-lacustrine system in Gale Crater that required significant surface runoff and associated erosion and transport of sediment.

The driving mechanism for the formation of the Burns formation is interpreted to be the regional-scale groundwater hydrostatic head that was sourced from the southern highlands, with the Meridiani Planum region as a prime regional-scale lowland for groundwater upwelling (Andrews-Hanna et al., 2010). As noted, this would have existed after the region underwent massive fluvial erosion and transport of sediment to the northwest earlier during the Noachian Period (Hynek and Phillips, 2001). The hydrologic system thus shifted from surface-dominated erosion to a groundwater-dominated depocenter, with recharge from the southern highlands. A plausible scenario for the formation of the Burns formation rocks would start with groundwater reacting with basaltic rocks during its transit to Meridiani Planum lowlands. As a consequence, these waters would have transported dissolved cations and anions until reaching the surface in Meridiani Planum. Evaporation of these waters in the presence of sulfur from aerosols or dissolution from reactions with subsurface rocks led to production of sulfate-rich muds in shallow, ephemeral playa environments. Outcrops exposed in the ~100 m-diameter Santa Maria crater are interpreted to represent these original muds (Edgar et al., 2014) (Figs. 4.18 and 4.19). Elsewhere, the Burns formation outcrops encountered by Opportunity consist of sulfate-rich sandstones interpreted to have formed as playa muds and subsequently reworked in fluvial and windblown environments to produce sands (Grotzinger et al., 2005, Metz et al., 2009) (Figs. 4.20 and 4.21). Rising groundwater events cemented the deposits and generated hematitic concretions and veins (Knoll et al., 2008). The hematite signature in Mars Global Surveyor Thermal Emission spectra

Fig. 4.18 Portion of a HiRISE view of the relatively fresh Santa Maria Crater with Opportunity traverses shown in white. Note the bright crater rim and windblown ripples on the crater floor. Luis de Torres and Ruiz Garcia are two targets for which Opportunity acquired in situ measurements. HiRISE frame PSP_009141_1780_red.

Fig. 4.19 Pancam false-color mosaic is shown for the ~100 m-wide Santa Maria Crater. Also shown in the lower left is a Microscopic Imager-based mosaic (Herkenhoff et al., 2003) acquired on the boulder Ruiz Garcia. Arrows show direction of migration of windblown ripples on the crater floor, likely a direction that keeps the rocks on the southeast side of the crater relatively clean of dust. The MI image shows evidence for mud clasts (Edgar et al., 2014), and APXS data show this rock to be sulfate rich. Luis de Torres is a Burns formation contact science sulfate-rich target with a sandstone texture. Terreros has a bluish color similar to Ruiz Garcia and is likely another mud clast target. RGB shown as Pancam left-eye bands L257 in this and subsequent Pancam images, with wavelengths centered at 0.753, 0.535, and 0.432 μm (Bell et al., 2003). Pancam mosaic is SOL2512B_P2290_L257F.tif, and the MI inset is B2527RuizGracia_4_raw.tif.

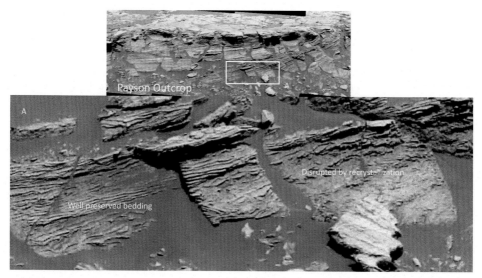

Fig. 4.20 View of the Payson outcrop on the western side of the ~300 m wide, highly eroded Erebus Crater. This outcrop displays evidence for what is interpreted to be interdune wet conditions, including the presence of shallow surface water flow based on ripple patterns. Also shown in A is evidence for bedding disruption by recrystallization soon after deposition of the sulfate-rich deposits (Metz et al., 2009). Mosaic available on Planetary Photojournal as PIA02696.

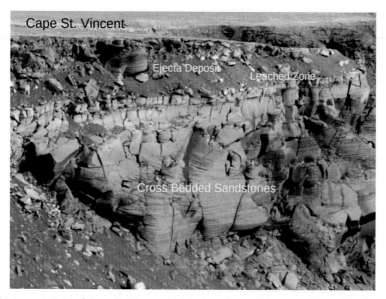

Fig. 4.21 Pancam mosaic of Cape St. Vincent located on the ~10 m high wall of Victoria Crater. Large-scale cross beds are indicative of a dune environment. Opportunity's measurements on other portions of Victoria's walls show that the bedrock is a sulfate-rich sandstone. A bright zone at the top of the intact section is interpreted to be due to leaching from surface waters migrating downward. The ejecta deposit can be seen at the top of the section. Mosaic available through the PDS Geosciences Node as SOL1167B_P2419_L257_F.tif.

that led to a landing on Meridiani Planum has thus been shown by Opportunity to be due to hematitic concretions generated by diagenetic alteration of the sandstones and concentrated as a relatively immobile lag deposit on the surface as the more friable sandstones were eroded by wind (Jerolmack et al., 2006).

Large-scale cross bedding in the Burns formation sandstones is particularly well exposed on the walls of the ~800 m-wide Victoria Crater and is interpreted to have formed in a windblown dune environment during a particularly dry period when the Burns formation rocks were being emplaced (Squyres et al., 2009) (Fig. 4.21). In addition, the in-place wall rocks exhibit a bright band near the top of the section that Opportunity measurements have shown to have reduced concentrations of chlorine relative to the immediately underlying strata (Arvidson et al., 2011). This is interpreted to be evidence for leaching of more soluble species after deposition of the Burns formation and before the impact event that produced Victoria Crater. This leaching was likely top down to depths of several meters, implying at least one period of minor surface water recharge during or after deposition of Burns formation strata.

4.4.3.3 Acidic Conditions

Mineralogical investigation of the Burns formation by Mössbauer spectroscopy revealed the presence of jarosite in the sulfate-rich sandstones (Klingelhöfer et al., 2004; Morris et al., 2006b). Jarosite requires an acidic environment (pH 2–4) to form, and its occurrence has been taken as evidence that the waters that generated the evaporite components of this unit were highly acidic (King and McSween, 2005; Squyres and Knoll, 2005; McLennan et al., 2005). This acidity could have come solely from oxidation of ferrous iron contained in upwelling groundwater, which would have been an anoxic fluid with a circumneutral pH (Hurowitz et al., 2010). This configuration established a redox interface, with a chemical gradient between shallow ferrous iron-bearing waters and an evaporative, oxidized surface. An alternative model where the anoxic groundwater initially evaporates to form a mixture of magnesium, calcium, and ferrous iron sulfates, followed by later oxidation and acidification during subsequent eolian reworking or diagenesis, would also generate a jarosite-bearing mineral assemblage (Tosca et al., 2008b).

4.4.4 Spirit: Aqueous Activity in the Columbia Hills
4.4.4.1 Overview

Next in the order of increasing age is the discovery from Spirit observations of volcanoclastic activity and associated sulfate-rich sands and silica deposits adjacent to Home Plate, Inner Basin, Columbia Hills (Figs. 4.22 and 4.23). The age of these deposits must be younger than the Columbia Hills rocks, which are uplifted portions of the floor of the Noachian-age Gusev Crater. We prefer an age coincident with the emplacement of the extensive volcanic flows on the floor of Gusev Crater during the Hesperian Period, ~3.7 Ga (Greeley et al., 2005), although the deposits could conceivably be Noachian in age.

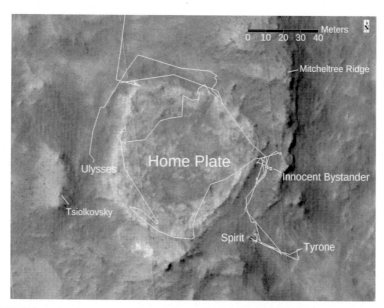

Fig. 4.22 HiRISE-based image is shown covering Home Plate with Spirit's traverses overlain. Home Plate is interpreted to be a partially eroded volcanoclastic construct. Innocent Bystander is a silica-rich target. Sulfate-rich soils excavated by the wheels were identified in a number of locations, including Ulysses, where Spirit was attempting to move to the southeast and was halted because of incipient embedding. Frame ID ESP_0013499_1650_red.

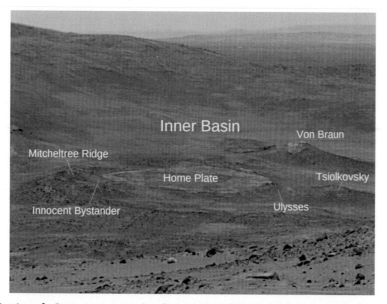

Fig. 4.23 Portion of a Pancam mosaic taken from Husband Hill, located to the north of the Inner Basin and the Home Plate. Home Plate and von Braun are interpreted to be partially eroded volcanic constructs. Innocent Bystander is high-silica target encountered by Spirit, and Ulysses is the sulfate-rich sand location that led to embedding of Spirit. Product available through the NASA Planetary Photojournal as PIA17760.

Home Plate, based on the presence of layered strata and a bomb sag (Lewis et al., 2008), is interpreted to be a partially eroded volcanoclastic construct. This inference is consistent with nearby constructs that are also thought to also be of volcanoclastic origin due to their conical shape (von Braun, Fig. 4.23) or caldera-like appearance (Goddard to the south of Home Plate).

4.4.4.2 Sulfate-Rich Sands

In the rolling hills immediately to the north of Home Plate, Spirit's wheels excavated bright soils during one of its traverses. The excavated soils were found to be dominated by sulfate-rich sands, with associated enrichment in silica relative to the more common basaltic soils found in Gusev Crater (Yen et al., 2008; Wang et al., 2008). Home Plate and the surrounding areas were then extensively explored, and the valley to the east of Home Plate was found to be underlain by sulfate-rich soils, with friable silica-rich outcrops exposed that were in part disaggregated during wheel passages. The sulfate-rich sands have been interpreted to have been deposited in fumarolic and/or hydrothermal systems when volcanism was active within the Inner Basin.

The most studied exposure of sulfate-rich deposits is also the location of Spirit's final measurement campaign. Traversing south along the valley bordering the western side of Home Plate, Spirit became embedded in soft ferric iron sulfate-rich sands when the left front wheel broke through a thin soil crust (Figs. 4.24 and 4.25). Subsequent measurements by Spirit showed that the crust is enriched in hematite and one or more calcium sulfate phases relative to the subsurface ferric iron sulfate-rich sands (Arvidson et al., 2010). The distribution of minerals with depth in the soil is consistent with aqueous activity. Hematite and calcium sulfate(s) found at the surface are relatively insoluble and would have been left as residual deposits when a downward percolating fluid preferentially dissolved the soluble ferric iron sulfates (Arvidson et al., 2010). Given that Spirit encountered sulfate sands within the Inner Basin only when excavated by the wheels, and given an active windblown environment that should uncover these materials, the soil profile development must be a consequence of a relatively modern hydrologic cycle. This could be associated with periodic snow cover, minor melting, and water percolation into the underlying soils during times of relatively high spin axis obliquity (Arvidson et al., 2010; Kite et al., 2013).

4.4.4.3 Silica-Rich Deposits

As noted above, Spirit discovered silica-rich outcrops while traversing the valley just to the east of Home Plate (Figs. 4.26 and 4.27). The silica-rich deposits have up to 91% SiO_2 concentrations (Squyres et al., 2008), and analysis of Mini-TES emission spectra shows that these deposits are dominated by the mineral opal-A ($SiO_2 \cdot nH_2O$) (Ruff et al., 2011). These deposits have been interpreted as rocks highly leached by acid-sulfate-rich fluids in which low-pH fluids dissolved basaltic rocks, leaving behind a silica residuum

Fig. 4.24 Front Hazcam image looking to the north within the valley to the west of Home Plate. Bright soils were excavated by the left front wheel during this backward drive. The ridges within the tracks indicate a stick-slip motion and associated high slip. The right front wheel drive actuator was inoperative, and the rover dragged it through the soil. A number of MI and APXS targets are labeled, and measurements provided detailed information on the surface and subsurface soils. Hazcam data are available through the PDS Geosciences Node as 2F292473825FFLB188P1212R0M1.img. *(From Arvidson, R.E., et al., 2010. Spirit Mars Rover Mission: overview and selected results from the northern Home Plate Winter Haven to the side of Scamander crater. J. Geophys. Res. 115. https://doi.org/10.1029/2010JE003633.)*

(Ming et al., 2006). The silica-rich deposits have also been interpreted to be hot spring sinters associated with Home Plate volcanic activity, produced as hot waters cooled upon reaching the surface (Ruff and Farmer, 2016). In either case, the deposits are evidence for significant fluid flow, likely resulting from hydrothermal activity during the formation of Home Plate.

4.4.5 Opportunity: Fracture-Related Aqueous Processes on Endeavour Crater's Rim

4.4.5.1 Overview

Opportunity's investigation of the rim of the Noachian-age Endeavour Crater characterized arguably the oldest extensive outcrops studied by the landed missions discussed in this chapter (Fig. 4.3). Exploration of Cape York, Murray Ridge, and Cape Tribulation rim segments was designed to search for and characterize the structure, stratigraphy,

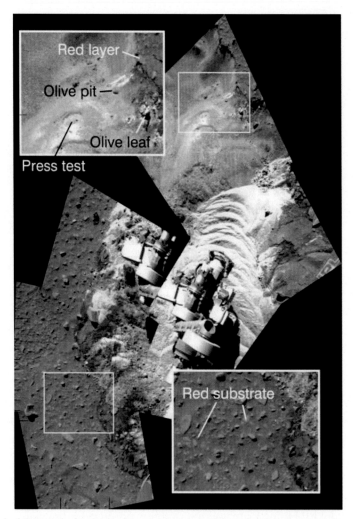

Fig. 4.25 The last Pancam enhanced color mosaic acquired from Spirit is shown and was acquired just before contact with Spirit was lost. The rover became embedded as the left wheels dug deeply into the bright soft soil known after numerous measurements to consist of ferric sulfates mixed with basaltic sands. The mosaic was acquired after attempting a backward drive maneuver in which the left front wheel was rotated azimuthally before the drive actuator moved the rover backward, thus creating the waves of sulfate-rich sands. Detailed measurements of the soils show horizons with relatively insoluble species at the top, indicating continued water percolating into the soil. *Small white boxes* show locations of enlarged image views of the scene. Olive pit and leaf are sites in which extensive measurements were conducted. Press test was the site where the Mössbauer spectrometer was pushed into the soil as part of a physical properties experiment. Pancam mosaic is available through the PDS Geosciences Node as SOL2163_P2397_L257_F_Full.tif.

Fig. 4.26 Navcam mosaic of the high-silica targets Norma Luker and Innocent Bystander located in the valley between Home Plate and Mitcheltree Ridge. *White box* shows the location of false-color view of the two high-silica targets shown in Fig. 4.27. Innocent Bystander was broken by dragging the right front wheel across the silica-rich deposits. Mosaic available through PDS Geosciences Node as Spirit_left_navcam_site_130_pos_37_cyl.jpg.

Fig. 4.27 Pancam false-color view of the Norma Luker and Innocent Bystander high-silica targets. Note the debris dragged to the lower right by the rover's front wheel while breaking the rocks apart. Data available through the Planetary Data System Geosciences Node as ID Sol1234A_P2378_L257_F.jpg.

and lithology associated with this ~22 km-wide degraded crater (Crumpler et al., 2015; Mittlefehldt et al., 2018). A particular focus was finding and documenting evidence for formation or alteration of rocks in the presence of aqueous environments. As noted in the summary of landed mission results, traverses taken by Opportunity were in part directed based on findings from CRISM observations, including detections of ferric iron smectites on the eastern side of Cape York (Arvidson et al., 2014) and iron-magnesium smectites in a gentle swale (Marathon Valley) on Cape Tribulation (Fox et al., 2016). In this section, several examples are presented that indicate enhanced aqueous activity along fractures in order to elucidate how subsurface fluids have interacted with Noachian-age impact crater rims and the extent to which habitable environments may have been present.

4.4.5.2 Matijevic Formation and the Espérance Fracture

CRISM data indicated the presence of ferric iron smectites (Fig. 4.17) in the eastern portion of Cape York, which is located approximately midway along the strike of this rim segment (Arvidson et al., 2014). Opportunity was directed to turn into Cape York outcrops at this location (subsequently named Matijevic Hill) and found that the smectite detections are associated with planar outcrops (Matijevic formation) that have numerous thin, interleaved dark and bright layers, each ~0.5 mm thick, and dominated by cemented silt to sand-sized grains (Arvidson et al., 2014) (Fig. 4.28). These outcrops sit

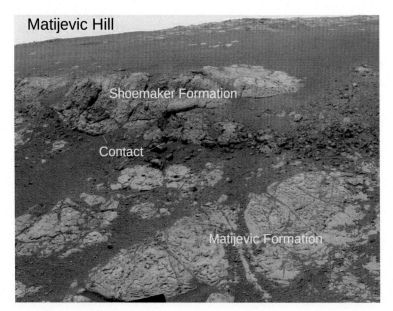

Fig. 4.28 Pancam false-color mosaic acquired looking to the west on Matijevic Hill showing the bright and dark layers exposed on the Matijevic formation and the overlying Shoemaker formation impact breccias. Also shown is the contact between the two units. Product available through the NASA Planetary Photojournal as PIA16704.

stratigraphically just above the rocks that are laden with ~2–3 mm-diameter spherules slightly enriched in hematite relative to the rock matrix. The finely layered outcrops are unconformably overlain by the Shoemaker formation impact breccias that dominate Endeavour's rim. Both the Matijevic and Shoemaker formation rocks are cut by calcium sulfate-rich veins, interpreted to be gypsum ($CaSO_4 \cdot 2H_2O$) based on a spectral reflectance downturn at ~1 μm in Pancam-based spectra and assigned to bound H_2O-related vibrations (Squyres et al., 2012). The spherule-dominated and finely layered Matijevic formation deposits have basaltic compositions and are interpreted to be pre-Endeavour rocks that were uplifted during the Endeavour impact event. The Matijevic formation rocks are slightly enriched in soluble elements, including zinc, sulfur, chlorine, and bromine as compared with the Shoemaker formation breccias, with no discernable bulk compositional differences between the bright and the dark strata. The banded nature of the strata is consistent with deposition by a number of mechanisms in which environmental conditions shifted back and forth, including lacustrine environments. This is certainly not a unique environmental interpretation, given the limited lateral and vertical outcrop exposures. For example, the dark layers have also been interpreted to be bedding plane veins (e.g., Clark et al., 2016; Mittlefehldt et al., 2018).

The Espérance fracture zone cutting the Matijevic formation outcrops was examined in detail to test the hypothesis that enhanced subsurface flow would have leached the rocks within the fracture (Fig. 4.29). Opportunity's RAT was commanded to grind repeatedly into this fracture zone, and MI and APXS measurements were acquired between individual grinds. A clear trend was found toward an aluminum-rich smectite composition (e.g., montmorillonite) with increasing grind depths (Arvidson et al., 2014; Clark et al., 2016) (Fig. 4.29). This result suggests that the interior of the fracture zone was subjected to significant leaching and alteration by enhanced fluid flow along fractures. This leaching contrasts sharply with the mild (i.e., low water-to-rock ratio) isochemical alteration inferred from the basaltic compositions of the finely layered Matijevic formation strata that exhibit the CRISM-based ferric iron smectite signature. The extensive alteration and evidence for fluid flow focused through fracture systems is consistent with a long-lived hydrothermal system of the type envisioned by Newsom et al. (2001).

4.4.5.3 Sulfates and Manganese Oxides on the Island Rocks

Opportunity spent one of its winter season campaigns in Cook Haven in the Murray Ridge rim segment of Endeavour Crater (Fig. 4.17). Along the route into this shallow valley, the rover excavated Pinnacle and Stuart Island rocks from a soil-filled fracture (Fig. 4.30). The surfaces of these two small rocks showed both bright and dark coatings, which were subsequently investigated in detail using Pancam multispectral observations, MI imaging, and overlapping APXS observations (Arvidson et al., 2016). Detailed analysis of the data showed that the bright coatings are dominated by magnesium and iron sulfates, whereas the dark coatings, inferred to be on top of the bright coatings, were shown to be dominated

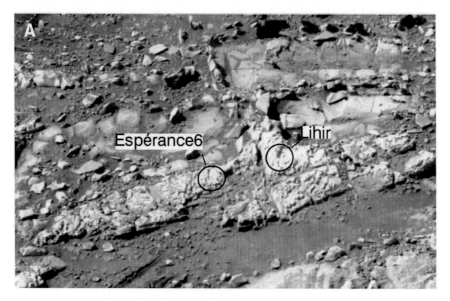

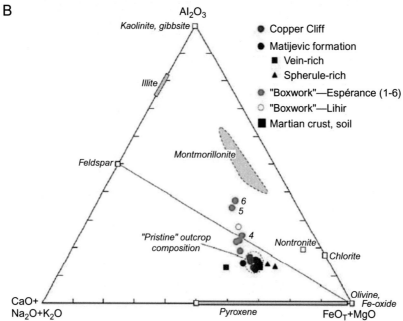

Fig. 4.29 (A) Pancam false-color view showing the box work fracture and examined in detail by Opportunity. *Circle* indicates the targets Espérance 6 (which was abraded with the RAT) and Lihir. Approximate scale across the scene is 70 cm. (B) Ternary plot of mole fraction Al_2O_3-(CaO + Na_2O + K_2O)-(FeO_T + MgO) for selected rocks from Matijevic Hill and other materials. Mineral compositions based on an idealized stoichiometry, the field for montmorillonite based on structural formulae of 25 natural montmorillonites, and average Martian crust and soils. Opportunity APXS-based compositions trend toward montmorillonite. *(From Arvidson, R.E., et al., 2014. Ancient aqueous environments at Endeavour Crater, Mars. Science 343(6169), 1248097–1248097. https://doi.org/10.1126/science.1248097.)*

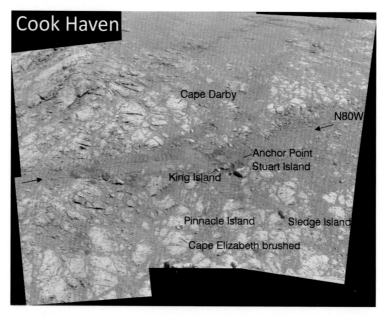

Fig. 4.30 Pancam false-color mosaic from Cook Haven looking south after excavating Pinnacle and Stuart Island rocks, which were the targets of extensive in situ observations. Also shown are in situ bedrock targets Cape Darby and Cape Elizabeth (brushed by the RAT). Anchor Point in situ targets are soils excavated from the soil-filled fracture, King Island is another target with a bright coating and perhaps the mate of Pinnacle Island, and Sledge Island may or may not have existed as an erratic rock before Opportunity arrived. For reference, Pinnacle Island is ~3.5 cm wide, and Stuart Island is ~12 cm in its longest dimension. *(From Arvidson, R.E., et al., 2014. Ancient aqueous environments at Endeavour Crater, Mars. Science 343(6169), 1248097–1248097. https://doi.org/10. 1126/science.1248097.)*

by one or more manganese oxides. The inference is that sulfate-rich fluids moved through the fracture and evaporated to coat the rocks, followed by a final episode in which manganese oxides formed. The presence of manganese oxides indicates that a strong oxidant reacted with dissolved Mn^{2+} and implies that a redox gradient once existed at this location. Note that what have been interpreted to be bedding plane manganese oxide deposits have also been found by Curiosity in the Murray formation, although these deposits are quite rare (Lanza et al., 2016). The fact that manganese oxides produced by subsurface aqueous processes have been found at two widely separated locations does indicate that strong oxidizers have been present on at least a regional-scale basis.

4.4.5.4 Smectites, Hematite, and Sulfates in Marathon Valley

Opportunity explored in detail Marathon Valley on the Cape Tribulation rim segment, guided by the discovery from CRISM data of iron-magnesium smectites on the valley floor (Fox et al., 2016). Opportunity's observations show that the flat, polygonally fractured

Fig. 4.31 Pancam enhanced color mosaic acquired from the floor of Marathon Valley and looking toward the northern wall named Hinners Point. Center of mosaic has a decorrelation stretch overlay where rock and soil colors have been exaggerated. Note the flat-lying nature of the outcrops, which have been cut by polygonal fractures. A string of red pebbles is evident, along with soil-filled fractures. Opportunity scuffed one of the soil-filled fractures located to the south of the location from which the mosaic was acquired. Mosaic available on the Planetary Photojournal as PIA19820.

Shoemaker formation breccias that dominate the floor outcrops are the carrier of the spectral signature. Opportunity APXS data show that the alteration occurred in an isochemical system (Fox et al., 2016) (Fig. 4.31). These breccias exhibit a fine-grained matrix with embedded clasts. On the other hand, the extensive fractures cutting across the floor were found to be accompanied by rows of red pebbles, some of which, based on Pancam 13F observations, show evidence for crystalline hematite (Farrand et al., 2017). In addition, Opportunity deliberately used one of its wheels to scuff one of the soil-filled fractures with red pebbles on its side and uncovered evidence for magnesium and other sulfate deposits just beneath the surface (Stein et al., 2018). These observations indicate that leaching by both a sulfate-rich fluid and an iron oxidation occurred within the fracture zone.

4.5 OUTLOOK FOR HABITABILITY AND LIFE ON MARS

4.5.1 Overview

The successful Mars landed missions have acquired remote sensing and in situ observations that both complement observations acquired from orbit and provide data that could not have been acquired without operating on the surface. This is especially true given the Mars exploration focus of understanding the role of water interacting with the surface and subsurface and the implications for habitability and life. The Mars Exploration Rovers Spirit and Opportunity and the Mars Science Laboratory Curiosity, with their ability to traverse thousands of meters to access and analyze key rock outcrops and soils, have been particularly important in providing the data needed to infer the nature and extent of water activity. The following discussion reviews the current state of knowledge about the sites of the post-2000 landed missions.

4.5.2 Sustained Water

The Phoenix landing site contains limited liquid water, occurring primarily as thin water films that likely have high salt contents. In contrast, the Gale Crater site explored by Curiosity provides evidence of sustained liquid water in a lacustrine setting, although further work is needed to provide a quantitative estimate of lake duration. There is also ample evidence for subsurface diagenetic fluids, which extend the timing of water availability beyond that of surface water inundation as these fluids were postdepositional. The Burns formation in Meridiani Planum explored by Opportunity clearly shows evidence of formation from liquid water. Whereas the evaporite mineral components indicate that the water present was highly saline, the estimated volume of water needed to produce the observed deposits (Hurowitz et al., 2010) suggests that the source fluid was less concentrated. The subsurface hydrologic system feeding the playa system thus may have had a water activity more amenable to life, although the highly saline and low-pH surface environment would have been problematic (Knoll et al., 2005). Waters in the Columbia Hills explored by Spirit likely existed in the form of springs that were elevated in temperature associated with volcanic activity and may or may not have occurred over a prolonged period. Exploration of the Endeavour Crater rim by Opportunity shows clear evidence of subsurface water flow through fractures, possibly from postimpact hydrothermal circulation. Flow may also have occurred through these fractures during formation of the overlying Burns formation, as this would introduce the abundant sulfate component, suggesting multiple periods of subsurface water activity.

4.5.3 Chemical Energy

Observations at the Phoenix site provide little evidence of past redox disequilibrium needed to provide chemical energy. Atmospheric O_2 and soil perchlorate salts are oxidants that can be coupled with reductants by terrestrial microorganisms (Coates and Achenbach, 2004) to provide energy; however, any evidence of reductants is lacking at this site. On the other hand, the Gale Crater site shows abundant evidence of reduced inorganic species that can serve as electron donors when coupled to an oxidant, including iron sulfide minerals, magnetite, and ferrous iron in phyllosilicates and primary mafic silicate minerals (Vaniman et al., 2013; Rampe et al., 2017; Hurowitz et al., 2017). The site also shows clear evidence of ferrous iron oxidation, with hematite, jarosite, and akaganéite (Vaniman et al., 2013) all observed. In addition, at least some of the iron in phyllosilicates is oxidized, and the magnetite may be in a partially oxidized nonstoichiometric form, which can be produced by microbial redox iron cycling (Byrne et al., 2015). The occurrence of manganese oxides (Lanza et al., 2016) also indicates that oxidation of Mn^{2+} occurred, as the localized concentration of manganese requires transport in its more soluble reduced state. This record of reduced species, along with oxidation processes, indicates that the site had chemical energy available during the deposition and diagenesis of the sedimentary deposits. The proposed redox stratification of the lake in Gale Crater (Hurowitz et al., 2017) would have provided an optimal setting for life to obtain energy. Oxidation of Fe^{2+} also occurred during the deposition or diagenesis of the Burns formation. Whether this could be utilized by life depends on the kinetics of abiotic ferrous iron oxidation and the source of oxidant. Photooxidation stimulated by UV light (Hurowitz et al., 2010) may not have provided energy accessible to organisms; terrestrial microorganisms utilize ferrous iron faster than abiotic chemical oxidation only at acidic pH or under microaerophilic conditions (Weber et al., 2006). The presence of ferric-bearing minerals at the surface around Home Plate in the Columbia Hills suggests that oxidation of ferrous iron occurred at the site, but observations to date do not allow for constraints on the coexistence of oxidized and reduced species. It is thus unclear whether ample chemical energy was once available at this location. In the rim of Endeavour Crater, the presence of ferric oxides and phyllosilicates in proximity to materials with a largely basaltic composition (and thus ferrous iron-bearing silicates) suggests that oxidant(s) penetrated into the subsurface. The occurrence of manganese oxides in a fracture requires that Mn^{2+} was leached from the surrounding rock and transported in fluid until it was oxidized. These observations indicate that the fractured rocks in the rim of Endeavour Crater were settings where chemical energy was available in the subsurface.

4.5.4 Essential Elements

For the purpose of the present review, organic carbon is the most critical essential element, followed by fixed nitrogen and then other required elements. The detection of organic carbon and nitrate by Curiosity in Gale Crater suggests that this site likely contained key building blocks for life. Note that carbon dioxide is abundant in the atmosphere and was certainly present at all sites explored by landed missions. In addition, Phoenix provided evidence of carbonate minerals in the soil. However, the biological availability of carbon from carbon dioxide or carbonates at all sites is uncertain because this requires organisms capable of autotrophy. Biological carbon fixation was acquired early in the evolution of life on Earth and may even have been a key step in the origin of life (Fuchs, 2011), and similar expectation of biological capabilities exists for possible past or extant life on Mars. Water-rock interactions enhance phosphorus availability through weathering processes, although phosphorus has low solubility in the presence of calcium carbonates (Arai and Sparks, 2007) and may have limited availability at the Phoenix site. Sulfur is widely abundant at all sites as sulfate salts and is expected to have high biological availability.

4.5.5 Favorable Environmental Conditions

The pH at the Phoenix site was found to be circumneutral to slightly alkaline in nature, but it is unclear whether the site experienced favorable temperatures in the past and water may have only existed as brines. Gale Crater provided the strongest evidence for prolonged conditions with temperatures supporting liquid water, both on the surface and in subsurface fractures. The lake's pH was inferred to be circumneutral for much of its duration, and subsurface fluids flowed through materials that would also promote mild pH conditions. Water flow was required to produce the Burns formation at Meridiani Planum, indicating that temperatures supported the stability of liquid water in the near subsurface, although it is unclear how long this water persisted on the surface, where substantial evaporation occurred. The pH was clearly acidic during the formation of jarosite, owing to the acidity generated by iron oxidation, but the subsurface may have maintained a pH more favorable for habitability. The Columbia Hills clearly once contained highly acidic waters, as indicated by ferric sulfate minerals, and likely had elevated temperatures. It is possible that the waters that formed Home Plate silica-rich deposits were more neutral in nature if they are in fact sinter deposits formed in a hot spring setting. The rim of Endeavour Crater saw subsurface water that may have originated from hydrothermal fluids associated with the impact event, and these were likely in temperature ranges amenable to life, given the crater size (Newsom et al., 2001). Some alteration assemblages are isochemical in nature, suggesting circumneutral to alkaline pH conditions, but leaching and the presence of sulfate minerals observed within fractures on Endeavour's rim show that other fluids were at least slightly acidic.

4.5.6 Hazards to Habitability

The landed missions have also identified a number of potential hazards to life. Perchlorate formation likely also generates other, more reactive oxychlorine species (Carrier and Kounaves, 2015), including hypochlorite (ClO^-) and chlorite (ClO_2^-), key chemical components in bleach and chlorinated drinking water. Such species would clearly be detrimental to life and thus may pose habitability challenges for all shallow soils on Mars today. Chlorinated hydrocarbons can be toxic to organisms (Hanschloer, 1994) and are known to form from the reaction of oxychlorine species with organic compounds (Richardson et al., 2007). Oxychlorine species on Mars thus poses both direct and indirect chemical hazards to life today, and resolving the timing of their widespread

Habitability potential

Mission	Age estimate	Sustained solvent	Chemical energy	Essential elements	Favorable conditions	Hazards
Phoenix	Amazonian					
Curiosity	Hesperian					
Opportunity—Burns Fm.	Noachian-Hesperian					
Spirit—Home Plate	Noachian-Hesperian					
Opportunity—Endeavour	Noachian					

Fig. 4.32 Fever chart summarizing the current state of knowledge of the habitability of the landed sites discussed in detail in this chapter. Each site is evaluated based on the four established habitability factors, along with consideration of potential hazards to life. In this diagram, *green* indicates clear evidence of favorable condition, *red* indicates unfavorable conditions, and *yellow* indicates either a lack of information needed to assess habitability or locations where only some aspects of habitability are met.

formation on Mars is needed to better assess past habitability. Direct ultraviolet and cosmic ray exposure and the radicals they generate also likely posed hazards for much of the planet's history, favoring environments protected from exposure, including the subsurface systems and the lower portions of the lake in Gale Crater.

4.5.7 Overall Assessment

This combination of factors provides for an overall assessment of habitability potential at the landing sites based on the current state of knowledge (Fig. 4.32). For many sites, substantial uncertainties remain regarding key aspects of habitability because key components, especially the presence of organic carbon, fixed nitrogen, and other essential elements, could not be detected. However, the landed missions provided constraints on the duration of liquid water, available chemical energy, and environmental conditions, and together, these indicate that Gale Crater has the greatest habitability potential of sites visited to date, with the rim of Endeavour Crater also indicating the presence of some key attributes required for a habitable subsurface environment.

REFERENCES

Anderson, R., Bell III, J.F., 2010. Geologic mapping and characterization of Gale Crater and implications for its potential as a Mars Science Laboratory landing site. Mars J. 5, 76–128. https://doi.org/10.1555/mars.2010.0004.

Andrews-Hanna, J.C., Zuber, M.T., Arvidson, R.E., Wiseman, S.M., 2010. Early Mars hydrology: meridiani playa deposits and the sedimentary record of Arabia Terra. J. Geophys. Res. 115(E6). https://doi.org/10.1029/2009JE003485.

Arai, Y., Sparks, D.L., 2007. Phosphate reaction dynamics in soils and soil components: a multiscale approach. Adv. Agron. 94, 135–179. https://doi.org/10.1016/S0065-2113(06)94003-6.

Arvidson, R.E., 2016. Aqueous history of Mars as inferred from landed mission measurements of rocks, soils, and water ice. J. Geophys. Res. 121. https://doi.org/10.1002/2016JE005079.

Arvidson, R.E., et al., 2004a. Localization and physical properties experiments conducted by Spirit at Gusev Crater. Science 305 (5685), 821–824. https://doi.org/10.1126/science.1099922.

Arvidson, R.E., et al., 2004b. Localization and physical property experiments conducted by Opportunity at Meridiani Planum. Science 306 (5702), 1730–1733. https://doi.org/10.1126/science.1104211.

Arvidson, R.E., et al., 2006a. Overview of the Spirit Mars Exploration Rover Mission to Gusev crater: landing site to Backstay Rock in the Columbia Hills. J. Geophys. Res. 111(E2). https://doi.org/10.1029/2005JE002499.

Arvidson, R.E., et al., 2006b. Nature and origin of the hematite-bearing plains of Terra Meridiani based on analyses of orbital and Mars Exploration rover data sets. J. Geophys. Res. 111(E12). https://doi.org/10.1029/2006JE002728.

Arvidson, R.E., et al., 2009. Results from the Mars Phoenix Lander Robotic Arm experiment. J. Geophys. Res. 114. https://doi.org/10.1029/2009JE003408.

Arvidson, R.E., et al., 2010. Spirit Mars Rover Mission: overview and selected results from the northern Home Plate Winter Haven to the side of Scamander crater. J. Geophys. Res. 115. https://doi.org/10.1029/2010JE003633.

Arvidson, R.E., et al., 2011. Opportunity Mars Rover Mission: overview and selected result from purgatory ripple to traverses to Endeavour Crater. J. Geophys. Res. 116E00F15. https://doi.org/10.1029/2010JE003746.

Arvidson, R.E., et al., 2014. Ancient aqueous environments at Endeavour Crater, Mars. Science 343 (6169), 1248097. https://doi.org/10.1126/science.1248097.

Arvidson, R.E., et al., 2016. High concentrations of manganese and sulfur in deposits on Murray Ridge, Endeavour Crater, Mars. Am. Mineral. 101. https://doi.org/10.2138/am-2016-5599.

Baker-Austin, C., Dopson, M., 2007. Life in acid: pH homeostasis in acidophiles. Trends Microbiol. 15 (4), 165–171. https://doi.org/10.1016/j.tim.2007.02.005.

Bell, J.F., et al., 2003. Mars Exploration Rover Athena Panoramic Camera (Pancam) investigation. J. Geophys. Res. 108(E12). https://doi.org/10.1029/2003JE002070.

Benner, S.A., Devine, K.G., Matveeva, L.N., Powell, D.H., 2000. The missing organic molecules on Mars. Proc. Natl. Acad. Sci. U. S. A. 97 (6), 2425–2430. https://doi.org/10.1073/pnas.040539497.

Bibring, J.-P., Squyres, S.W., Arvidson, R.E., 2006a. Merging views on Mars. Science 313 (5795), 1899–1901. https://doi.org/10.1126/science.1132311.

Bibring, J.-P., et al., 2006b. Global mineralogical and aqueous Mars history derived from OMEGA/Mars Express data. Science 312 (5772), 400–404. https://doi.org/10.1126/science.1122659.

Biemann, K., et al., 1977. The search for organic substances and inorganic volatile compounds at the surface of Mars. J. Geophys. Res. 82, 4641–4658. https://doi.org/10.1029/JB082i028p04641.

Blankenship, R.E., 2010. Early evolution of photosynthesis. Plant Physiol. 154 (2), 434–438. https://doi.org/10.1104/pp.110.161687.

Boxe, C.S., Hand, K.P., Nealson, K.H., Yung, Y.L., Yen, A.S., Saiz-Lopez, A., 2012. Adsorbed water and thin liquid films on Mars. Int. J. Astrobiol. 11 (03), 169–175. https://doi.org/10.1017/S1473550412000080.

Boynton, W.V., Feldman, W.C., Squyres, S.W., Prettyman, T.H., Bruckner, J., Evans, L.G., Reedy, R.C., Starr, R., Arnold, J.R., 2002. Distribution of hydrogen in the near surface of Mars: evidence for subsurface ice deposits. Science 297 (5578), 81–85. https://doi.org/10.1126/science.1073722.

Boynton, W.V., et al., 2009. Evidence for calcium carbonate at the Mars Phoenix landing site. Science 325 (5936), 61–64. https://doi.org/10.1126/science.1172768.

Byrne, J.M., Klueglein, N., Pearce, C., Rosso, K.M., Appel, E., Kappler, A., 2015. Redox cycling of Fe(II) and Fe(III) iron in magnetite by Fe-metabolizing bacteria. Science 347 (6229), 1473–1476. https://doi.org/10.1126/science.aaa4834.

Carrier, B., Kounaves, P., 2015. The origins of perchlorate in the Martian soil. Geophys. Res. Lett. 42 (10), 3739–3745. https://doi.org/10.1002/2015gl064290.

Carter, J., Poulet, F., 2012. Orbital identification of clays and carbonates in Gusev Crater. Icarus 219, 250–253.

Cavanagh, P.D., et al., 2015. Confidence Hills mineralogy and CheMin results from base of Mt. Sharp, Pahrump Hills, Gale Crater, Mars. 46th Lunar and Planetary Science Conference. Lunar and Planetary Institute, Houston (p. Abstract #2735).

Chevrier, V.F., Hanley, J., Altheide, T.S., 2009. Stability of perchlorate hydrates and their liquid solutions at the Phoenix landing site, Mars. Geophys. Res. Lett. 36(10), GL037497. https://doi.org/10.1029/2009GL037497.

Christensen, P.R., Morris, R.V., Lane, M.D., Bandfield, J.L., Malin, M.C., 2001. Global mapping of Martian hematite mineral deposits: remnants of water-driven processes on early Mars. J. Geophys. Res. 106 (E10), 23873–23885. https://doi.org/10.1029/2000JE001415.

Christensen, P., et al., 2004. The Thermal Emission Imaging System (THEMIS) for the Mars 2001 Odyssey mission. Space Sci. Rev. 110 (1–2), 85–130.

Clark, R.N., 1999. Spectroscopy of rocks and minerals, and principles of spectroscopy, in manual of remote sensing. In: Rencz, A.N. (Ed.), Remote Sensing for the Earth Sciences. In: vol. 3. John Wiley and Sons, New York, pp. 3–58 16-5561 (Chapter 1).

Clark, B.C., Baird, A.K., Rose, H.J., Toulmin, P., Keil, K., Castro, A.J., Kelliher, W.C., Rowe, C.D., Evans, P.H., 1976. Inorganic analyses of Martian surface samples at the Viking landing sites. Science 194 (4271), 1283–1288. https://doi.org/10.1126/science.194.4271.1283.

Clark, B.C., et al., 2016. Esperance: multiple episodes of aqueous alteration involving fracture fills and coatings at Matijevic Hill, Mars. Am. Mineral. 101. https://doi.org/10.2138/am-20.

Clarke, A., Morris, G.J., Fonseca, F., Murray, B.J., Acton, E., Price, H.C., 2013. A low temperature limit for life on earth. PLoS One. 8(6), e66207. https://doi.org/10.1371/journal.pone.0066207.

Coates, J.D., Achenbach, L.A., 2004. Microbial perchlorate reduction: rocket-fuelled metabolism. Nat. Rev. Microbiol. 2 (7), 569–580. https://doi.org/10.1038/nrmicro926.

Cockell, C.S., 2014. Trajectories of Martian habitability. Astrobiology 14 (2), 182–203. https://doi.org/10.1089/ast.2013.1106.

Cockell, C.S., et al., 2016. Habitability: a review. Astrobiology 16 (1), 89–117. https://doi.org/10.1089/ast.2015.1295.

Court, R.W., Sephton, M.A., Parnell, J., Gilmour, I., 2006. The alteration of organic matter in response to ionising irradiation: chemical trends and implications for extraterrestrial sample analysis. Geochim. Cosmochim. Acta 70 (4), 1020–1039. https://doi.org/10.1016/j.gca.2005.10.017.

Crumpler, L.S., Arvidson, R.E., Bell, J., Clark, B.C., Cohen, B.A., Farrand, W.H., Gellert, R., Golombek, M., Grant, J.A., Guinness, E., et al., 2015. Context of ancient aqueous environments on Mars from in situ geologic mapping at Endeavour Crater. J. Geophys. Res. 120, 538–569. https://doi.org/10.1002/2014JE004699.

Cull, S.C., Arvidson, R.E., Catalano, J.G., Ming, D.W., Morris, R.V., Mellon, M.T., Lemmon, M., 2010. Concentrated perchlorate at the Mars Phoenix landing site: evidence for thin film liquid water on Mars. Geophys. Res. Lett. 37(22). https://doi.org/10.1029/2010GL045269.

Davila, A.F., Fairén, A.G., Gago-Duport, L., Stoker, C., Amils, R., Bonaccorsi, R., Zavaleta, D., Lim, D.S.-M., McKay, C.P., 2008. Subsurface formation of oxidants on Mars and implications for the preservation of organic biosignatures. Earth Planet. Sci. Lett. 272 (1), 456–463. https://doi.org/10.1016/j.epsl.2008.05.015.

Des Marais, D.J., 2000. When did photosynthesis emerge on earth? Science 289 (5485), 1703–1705. https://doi.org/10.1126/science.289.5485.1703.

Des Marais, D.J., 2010. Exploring Mars for evidence of habitable environments and life. Proc. Am. Philos. Soc. 154 (4), 402–421.

Edgar, L.A., Grotzinger, J.P., Bell, J.F., Hurowitz, J.A., 2014. Hypotheses for the origin of fine-grained sedimentary rocks at Santa Maria crater. Meridiani Planum 234, 36–44. https://doi.org/10.1016/j.icarus.2014.02.019.

Farrand, W.H., et al., 2017. Pancam multispectral and APXS chemical examination of rocks and soils in Marathon Valley and points south along the rim of Endeavour Crater. Lunar and Planet. Sci. Conf. 48. (Abstract 2453).

Feldman, W.C., et al., 2004. Global distribution of near-surface hydrogen on Mars. J. Geophys. Res. 109 (E9). https://doi.org/10.1029/2003JE002160.

Finney, J.L., 2004. Water? What's so special about it? Philos. Trans. R. Soc. Lond. Ser. B Biol. Sci. 359, 1145–1165.

Folkner, W.M., Yoder, C.F., Yuan, D.N., Standish, E.M., Preston, R.A., 1997. Interior structure and seasonal mass redistribution of Mars from radio tracking of Mars Pathfinder. Science 278 (5344), 1749–1752. https://doi.org/10.1126/science.278.5344.1749.

Fox, V.K., Arvidson, R.E., Guinness, E.A., McLennan, S.M., Catalano, J.G., Murchie, S.L., Powell, K.E., 2016. Smectite deposits in Marathon Valley, Endeavour Crater, Mars, identified using CRISM hyperspectral reflectance data. Geophys. Res. Lett. https://doi.org/10.1002/2016GL069108.

Fraeman, A.A., et al., 2013. A hematite-bearing layer in Gale Crater, Mars: mapping and implications for past aqueous conditions. Geology 41 (10), 1103–1106. https://doi.org/10.1130/G34613.1.

Fraeman, A.A., et al., 2016. The stratigraphy and evolution of lower Mount Sharp from spectral, morphological, and thermophysical orbital data sets. J. Geophys. Res. https://doi.org/10.1002/2016JE005095.

Freissinet, C., et al., 2015. Organic molecules in the Sheepbed Mudstone, Gale Crater, Mars: detection of organics in Martian sample. J. Geophys. Res. 120 (3), 495–514. https://doi.org/10.1002/2014JE004737.

Frydenvang, J., et al., 2016. Discovery of silica-rich lacustrine and eolian sedimentary rocks in Gale Crater. Mars, 47th Lunar and Planetary Science Conference (Abstract 2349.pdf).

Fuchs, G., 2011. Alternative pathways of carbon dioxide fixation: insights into the early evolution of life? Annu. Rev. Microbiol. 65, 631–658. https://doi.org/10.1146/annurev-micro-090110-102801.

Glavin, D.P., et al., 2013. Evidence for perchlorates and the origin of chlorinated hydrocarbons detected by SAM at the Rocknest aeolian deposit in Gale Crater. J. Geophys. Res. 118. https://doi.org/10.1002/jgre.20144.

Glotch, T.D., Bandfield, J.L., Christensen, P.R., Calvin, W.M., McLennan, S.M., Clark, B.C., Rogers, A.D., Squyres, S.W., 2006. Mineralogy of the light-toned outcrop at Meridiani Planum as seen by the Miniature Thermal Emission Spectrometer and implications for its formation. J. Geophys. Res. 111(E12). https://doi.org/10.1029/2005JE002672.

Golombek, M.P., et al., 1999. Overview of the Mars Pathfinder Mission: launch through landing, surface operations, data sets, and science results. J. Geophys. Res. 104 (E4), 8523–8553. https://doi.org/10.1029/98JE02554.

Golombek, M., et al., 2012. Selection of the Mars Science Laboratory landing site. Space Sci. Rev. 170 (1–4), 641–737. https://doi.org/10.1007/s11214-012-9916-y.

Grant, W.D., 2004. Life at low water activity. Philos. Trans. R. Soc. Lond. Ser. B Biol. Sci. 359 (1448), 1249–1266. https://doi.org/10.1098/rstb.2004.1502.

Grant, J.A., Wilson, S.A., Mangold, N., Calef, F., Grotzinger, J.P., 2014. The timing of alluvial activity in Gale Crater, Mars. Geophys. Res. Lett. 41 (4), 1142–1149. https://doi.org/10.1002/2013GL058909.

Greeley, R., et al., 2005. Fluid lava flows in Gusev crater, Mars. J. Geophys. Res. 110(E5)https://doi.org/10.1029/2005JE002401.

Grotzinger, J.P., et al., 2005. Stratigraphy and sedimentology of a dry to wet eolian depositional system, Burns formation, Meridiani Planum, Mars. Earth Planet. Sci. Lett. 240 (1), 11–72. https://doi.org/10.1016/j.epsl.2005.09.039.

Grotzinger, J., et al., 2006. Sedimentary textures formed by aqueous processes, Erebus Crater, Meridiani Planum, Mars. Geology 34 (12), 1085. https://doi.org/10.1130/G22985A.1.

Grotzinger, J.P., et al., 2012. Mars Science Laboratory mission and science investigation. Space Sci. Rev. 170 (1–4), 5–56. https://doi.org/10.1007/s11214-012-9892-2.

Grotzinger, J.P., et al., 2014. A habitable fluvio-lacustrine environment at Yellowknife Bay, Gale Crater, Mars. Science 343 (6169), 1242777. https://doi.org/10.1126/science.1242777.

Grotzinger, J.P., et al., 2015. Deposition, exhumation, and paleoclimate of an ancient lake deposit, Gale Crater, Mars. Science. 350(6257). https://doi.org/10.1126/science.aac7575.

Hanschloer, D., 1994. Toxicity of chlorinated organic compounds: effects of the introduction of chlorine in organic molecules. Angew. Chem. Int. Ed. Eng. 33 (19), 1920–1935. https://doi.org/10.1002/anie.199419201.

Hecht, M.H., et al., 2009. Detection of perchlorate and the soluble chemistry of Martian soil at the Phoenix Lander site. Science 325 (5936), 64–67. https://doi.org/10.1126/science.1172466.

Herkenhoff, K.E., et al., 2003. Athena microscopic imager investigation. J. Geophys. Res. 108(E12). https://doi.org/10.1029/2003JE002076.

Hoehler, T.M., 2007. An energy balance concept for habitability. Astrobiology 7 (6), 824–838. https://doi.org/10.1089/ast.2006.0095.

Hoehler, T.M., Amend, J.P., Shock, E.L., 2007. A "follow the energy" approach for astrobiology. Astrobiology 7 (6), 819–823. https://doi.org/10.1089/ast.2007.0207.

Hurowitz, J.A., Fischer, W.W., Tosca, N.J., Milliken, R.E., 2010. Origin of acidic surface waters and the evolution of atmospheric chemistry on early Mars. Nat. Geosci. 3 (5), 323–326. https://doi.org/10.1038/ngeo831.

Hurowitz, J.A., McLennan, S.M., Tosca, N.J., Arvidson, R.E., Michalski, J.R., Schrőder, C., Squyres, S.W., 2006. In situ and experimental evidence for acidic weathering of rocks and soils on Mars. J. Geophys. Res. 111. https://doi.org/10.1029/2005JE002515.

Hurowitz, J.A., et al., 2017. Redox stratification of an ancient lake in Gale Crater, Mars. Science. 356(6341). https://doi.org/10.1126/science.aah6849.

Hynek, B.M., Phillips, R.J., 2001. Evidence for extensive denudation of the Martian highlands. Geology 29 (5), 407–410. https://doi.org/10.1130/0091-7613(2001)029<0407:EFEDOT>2.0.CO;2.

Jerolmack, D.J., Mohrig, D., Grotzinger, J.P., Fike, D.A., Watters, W.A., 2006. Spatial grain size sorting in eolian ripples and estimation of wind conditions on planetary surfaces: application to Meridiani Planum, Mars. J. Geophys. Res. 111(E12). https://doi.org/10.1029/2005JE002544.

Jouglet, D., Poulet, F., Milliken, R.E., Mustard, J.F., Bibring, J.-P., Langevin, Y., Gondet, B., Gomez, C., 2007. Hydration state of the Martian surface as seen by Mars Express OMEGA: 1. Analysis of the 3 μm hydration feature. J. Geophys. Res. 112(E8). https://doi.org/10.1029/2006JE002846.

Kashefi, K., Lovley, D.R., 2003. Extending the upper temperature limit for life. Science 301 (5635), 934. https://doi.org/10.1126/science.1086823.

King, P.L., McSween, H.Y., 2005. Effects of H_2O, pH, and oxidation state on the stability of Fe minerals on Mars. J. Geophys. Res. 110(E12)E12S10. https://doi.org/10.1029/2005JE002482.

Kite, E.S., Halevy, I., Kahre, M.A., Wolff, M.J., Manga, M., 2013. Seasonal melting and the formation of sedimentary rocks on Mars, with predictions for the Gale Crater mound. Icarus 223 (1), 181–210. https://doi.org/10.1016/j.icarus.2012.11.034.

Klein, H.P., 1977. The Viking biological investigation: general aspects. J. Geophys. Res. 82 (28), 4677–4680. https://doi.org/10.1029/JS082i028p04677.

Klingelhöfer, G., et al., 2004. Jarosite and hematite at Meridiani Planum from Opportunity's Mössbauer Spectrometer. Science 306 (5702), 1740–1745. https://doi.org/10.1126/science.1104653.

Knauth, L.P., Burt, D.M., Wohletz, K.H., 2005. Impact origin of sediments at the Opportunity landing site on Mars. Nature 438 (7071), 1123–1128. https://doi.org/10.1038/nature04383.

Knoll, A.H., Carr, M., Clark, B., Des Marais, D.J., Farmer, J.D., Fischer, W.W., Grotzinger, J.P., Hayes, A., McLennan, S.M., Malin, M., Schröder, C., Squyres, S., Tosca, N.J., Wdowiak, T., 2005. An astrobiological perspective on Meridiani Planum. Earth Planet. Sci. Lett. 240, 179–189.

Knoll, A.H., Grotzinger, J., 2006. Water on Mars and the prospect of Martian life. Elements 2 (3), 169–173. https://doi.org/10.2113/gselements.2.3.169.

Knoll, A.H., et al., 2008. Veneers, rinds, and fracture fills: relatively late alteration of sedimentary rocks at Meridiani Planum, Mars. J. Geophys. Res. 113(E6). https://doi.org/10.1029/2007JE002949.

Kounaves, S.P., Hecht, M.H., Kapit, J., Gospodinova, K., DeFlores, L., Quinn, R.C., Boynton, W.V., Clark, B.C., Catling, D.C., Hredzak, P., Ming, D.W., Moore, Q., Shusterman, J., Stroble, S., West, S.J., Young, S.M.M., 2010. Wet Chemistry experiments on the 2007 Phoenix Mars Scout Lander mission: data analysis and results. J. Geophys. Res. 115. E00E10.

Krulwich, T.A., 1995. Alkaliphiles: 'basic' molecular problems of pH tolerance and bioenergetics. Mol. Microbiol. 15 (3), 403–410. https://doi.org/10.1111/j.1365-2958.1995.tb02253.x.

Kuchynka, P., Folkner, W.M., Konopliv, A.S., Parker, T.J., Park, R.S., Maistre, S.L., Dehant, V., 2014. New constraints on Mars rotation determined from radiometric tracking of the Opportunity Mars Exploration Rover. Icarus 229, 340–347. https://doi.org/10.1016/j.icarus.2013.11.015.

Lanza, N.L., Wiens, R.C., Arvidson, R.E., et al., 2016. Oxidation of manganese in an ancient aquifer, Kimberley formation, Gale crater, Mars. Geophys. Res. Lett. https://doi.org/10.1002/2016GL069109.

Lasue, J., Mangold, N., Hauber, E., Clifford, S., Feldman, W., Gasnault, O., Grima, C., Maurice, S., Mousis, O., 2013. Quantitative assessments of the Martian hydrosphere. Space Sci. Rev. 174 (1), 155–212. https://doi.org/10.1007/s11214-012-9946-5.

Le Deit, L., Hauber, E., Fueton, F., Pondrelli, M., Pio Rossi, A., Jaumann, R., 2013. Sequence of infilling events in Gale Crater, Mars: results from morphology, stratigraphy, and mineralogy. J. Geophys. Res. 118. https://doi.org/10.1002/2012JE004322.

Lemmon, M.T., et al., 2008. In: The Phoenix Surface Stereo Imager (SSI) investigation. 39th Lunar and Planetary Science Conference. Lunar and Planetary Institute, Houston (Abstract #2156).

Léveillé, R.J., et al., 2014. Chemistry of fracture-filling raised ridges in Yellowknife Bay, Gale Crater: Window into past aqueous activity and habitability on Mars: chemistry of raised ridges, Gale Crater. J. Geophys. Res. 119 (11), 2398–2415. https://doi.org/10.1002/2014JE004620.

Lewis, K.W., Aharonson, O., Grotzinger, J.P., Squyres, S.W., Bell, J.F., Crumpler, L.S., Schmidt, M.E., 2008. Structure and stratigraphy of Home Plate from the Spirit Mars Exploration Rover. J. Geophys. Res. 113(E12). https://doi.org/10.1029/2007JE003025.

Mahaffy, P.R., et al., 2015. The imprint of atmospheric evolution in the D/H of Hesperian clay minerals on Mars. Science 347 (6220), 412–414. https://doi.org/10.1126/science.1260291.

Malin, M.C., et al., 2007. Context Camera Investigation on board the Mars. Reconnaissance Orbiter. J. Geophys. Res. 112E05S04. https://doi.org/10.1029/2006JE002808.

Malin, M.C., et al., 2010. In: The Mars Science Laboratory (MSL) Mast-mounted cameras (Mastcams) flight instruments. 41st Lunar and Planetary Science Conference. Lunar and Planetary Institute, Houston (p. Abstract #1123).

Manning, C.V., Zahnle, K.J., McKay, C.P., 2009. Impact processing of nitrogen on early Mars. Icarus 199 (2), 273–285. https://doi.org/10.1016/j.icarus.2008.10.015.

Matijevic, J., Shirley, D., 1997. The mission and operation of the Mars Pathfinder microrover. Control. Eng. Pract. 5 (6), 827–835. https://doi.org/10.1016/S0967-0661(97)00067-1.

McCollom, T.M., Hynek, B.M., 2005. A volcanic environment for bedrock diagenesis at Meridiani Planum on Mars. Nature 438 (7071), 1129–1131. https://doi.org/10.1038/nature04390.

McEwen, A.S., et al., 2007. Mars reconnaissance orbiter's high resolution imaging science experiment (HiRISE). J. Geophys. Res. 112(E5). https://doi.org/10.1029/2005JE002605.

McKay, C.P., 2014. Requirements and limits for life in the context of exoplanets. Proc. Natl. Acad. Sci. U. S. A. 111 (35), 12628–12633. https://doi.org/10.1073/pnas.1304212111.

McLennan, S.M., 2012. Geochemistry of sedimentary processes on Mars. In: Grotzinger, J.P., Milliken, R.E. (Eds.), Sedimentary Geology of Mars. SEPM Society for Sedimentary Geology, Tulsa, OK, pp. 119–138.

McLennan, S.M., et al., 2005. Provenance and diagenesis of the evaporite-bearing Burns formation, Meridiani Planum, Mars. Earth Planet. Sci. Lett. 240 (1), 95–121. https://doi.org/10.1016/j.epsl.2005.09.041.

Mehta, M., et al., 2011. Explosive erosion during the Phoenix landing exposes subsurface water on Mars. Icarus 211 (1), 172–194. https://doi.org/10.1016/j.icarus.2010.10.003.

Mellon, M.T., Arvidson, R.E., Marlow, J.J., Phillips, R.J., Asphaug, E., 2008. Periglacial landforms at the Phoenix landing site and the northern plains of Mars. J. Geophys. Res. 113. https://doi.org/10.1029/2007JE003039.

Mellon, M.T., Malin, M.C., Arvidson, R.E., Searls, M.L., Sizemore, H.G., Heet, T.L., Lemmon, M.T., Keller, H.U., Marshall, J., 2009b. The periglacial landscape at the Phoenix landing site. J. Geophys. Res. 114. https://doi.org/10.1029/2009JE003418.

Mellon, M.T., et al., 2009a. Ground ice at the Phoenix landing site: stability state and origin. J. Geophys. Res. 114. https://doi.org/10.1029/2009JE003417.

Metz, J.M., et al., 2009. Sulfate-rich eolian and wet interdune deposits, Erebus crater, Meridiani Planum Mars. J. Sediment. Res. 79, 247–264.

Milliken, R.E., Grotzinger, J.P., Thomson, B.J., 2010. Paleoclimate of Mars as captured by the stratigraphic record in Gale Crater. Geophys. Res. Lett. 37(4). https://doi.org/10.1029/2009GL041870.

Milliken, R.E., Mustard, J.F., Poulet, F., Jouglet, D., Bibring, J.-P., Gondet, B., Langevin, Y., 2007. Hydration state of the Martian surface as seen by Mars Express OMEGA: 2. H_2O content of the surface. J. Geophys. Res. 112(E8). https://doi.org/10.1029/2006JE002853.

Ming, D.W., et al., 2006. Geochemical and mineralogical indicators for aqueous processes in the Columbia Hills of Gusev crater, Mars. J. Geophys. Res. 111(E2). https://doi.org/10.1029/2005JE002560.

Mittlefehldt, D.W., et al., 2018. Compositions of diverse lithologies from the rim of Noachian-aged Endeavour Crater, Meridiani Planum Mars. J. Geophys. Res. https://doi.org/10.1002/2017JE005474.

Morris, R.V., et al., 2006a. Mössbauer mineralogy of rock, soil, and dust at Gusev crater, Mars: spirit's journey through weakly altered olivine basalt on the plains and pervasively altered basalt in the Columbia Hills. J. Geophys. Res. 111(E2). https://doi.org/10.1029/2005JE002584.

Morris, R.V., et al., 2006b. Mössbauer mineralogy of rock, soil, and dust at Meridiani Planum, Mars: Opportunity's journey across sulfate-rich outcrop, basaltic sand and dust, and hematite lag deposits. J. Geophys. Res. 111(E12). https://doi.org/10.1029/2006JE002791.

Morris, R.V., et al., 2010. Identification of carbonate-rich outcrops on Mars by the Spirit Rover. Science 329 (5990), 421–424. https://doi.org/10.1126/science.1189667.

Murchie, S., et al., 2007. Compact reconnaissance imaging spectrometer for Mars (CRISM) on Mars reconnaissance orbiter (MRO). J. Geophys. Res. 112(E5). https://doi.org/10.1029/2006JE002682.

Nachon, S.M., et al., 2014. Calcium sulfate veins characterized by ChemCam/Curiosity at Gale Crater, Mars. J. Geophys. Res. 119. https://doi.org/10.1002/2013JE004588.

Newsom, H.E., Hagerty, J.J., Thorsos, I.E., 2001. Location and sampling of aqueous and hydrothermal deposits in Martian impact craters. Astrobiology 1 (1), 71–88. https://doi.org/10.1089/153110701750137459.

Niles, P.B., Michalski, J., 2009. Meridiani Planum sediments on Mars formed through weathering in massive ice deposits. Nat. Geosci. 2 (3), 215–220. https://doi.org/10.1038/ngeo438.

Noe Dobrea, E.A., et al., 2012. Hydrated minerals on Endeavour Crater's rim and interior, and surrounding plains: new insights from CRISM data. Geophys. Res. Lett. 39L23201. https://doi.org/10.1029/2012GL053180.

Ojha, L., Wilhelm, M.B., Murchie, S.L., McEwen, A.S., Wray, J.J., Hanley, J., Massé, M., Chojnacki, M., 2015. Spectral evidence for hydrated salts in recurring slope lineae on Mars. Nat. Geosci. 8 (11), 829–832. https://doi.org/10.1038/ngeo2546.

Pavlov, A.A., Vasilyev, G., Ostryakov, V.M., Pavlov, A.K., Mahaffy, P., 2012. Degradation of the organic molecules in the shallow subsurface of Mars due to irradiation by cosmic rays. Geophys. Res. Lett. 39(13) L13202. https://doi.org/10.1029/2012GL052166.

Quinn, R.C., Martucci, H.F.H., Miller, S.R., Bryson, C.E., Grunthaner, F.J., Grunthaner, P.J., 2013. Perchlorate radiolysis on Mars and the origin of Martian soil reactivity. Astrobiology 13 (6), 515–520. https://doi.org/10.1089/ast.2013.0999.

Rampe, E.B., et al., 2017. Mineralogy of an ancient lacustrine mudstone succession from the Murray formation, Gale Crater, Mars. Earth Planet. Sci. Lett. 471. https://doi.org/10.1016/j.epsl.2017.04.021.

Rennó, N.O., et al., 2009. Possible physical and thermodynamical evidence for liquid water at the Phoenix landing site. J. Geophys. Res. 114. https://doi.org/10.1029/2009JE003362.

Richardson, S.D., Plewa, M.J., Wagner, D., Schoeny, R., DeMarini, D.M., 2007. Occurrence, genotoxicity, and carcinogenicity of regulated and emerging disinfection by-products in drinking water: a review and roadmap for research. Mutat. Res. Rev. Mutat. Res. 636 (1–3), 178–242. https://doi.org/10.1016/j.mrrev.2007.09.001.

Rieder, R., Economou, T., Wanke, H., Turkevich, A., Crisp, J., Breckner, J., Dreibus, G., McSween Jr, H.Y., 1997. The chemical composition of Martian soil and rocks returned by the mobile Alpha Proton X-ray Spectrometer: preliminary results from the X-ray mode. Science 278 (5344), 1771–1774. https://doi.org/10.1126/science.278.5344.1771.

Ruff, S.W., Farmer, J.D., 2016. Silica deposits on Mars with features resembling hot spring biosignatures at El Tatio in Chile. Nat. Commun. 7(2016)13554. https://doi.org/10.1038/ncomms13554.

Ruff, S.W., et al., 2011. Characteristics, distribution, origin, and significance of opaline silica observed by the Spirit rover in Gusev crater, Mars. J. Geophys. Res. 116. https://doi.org/10.1029/2010JE003767.

Smith, P.H., et al., 2009. H$_2$O at the Phoenix landing site. Science 325 (5936), 58–61. https://doi.org/10.1126/science.1172339.

Soffen, G.A., 1977. The Viking project. J. Geophys. Res. 82 (28), 3959–3970. https://doi.org/10.1029/JS082i028p03959.

Squyres, S.W., Knoll, A.H., 2005. Sedimentary rocks at Meridiani Planum: origin, diagenesis, and implications for life on Mars. Earth Planet. Sci. Lett. 240 (1), 1–10. https://doi.org/10.1016/j.epsl.2005.09.038.

Squyres, S.W., et al., 2003. Athena Mars rover science investigation. J. Geophys. Res. 108(E12). https://doi.org/10.1029/2003JE002121.

Squyres, S.W., et al., 2004a. The Spirit Rover's Athena science investigation at Gusev Crater, Mars. Science 305 (5685), 794–799. https://doi.org/10.1126/science.1100194.

Squyres, S.W., et al., 2004b. The Opportunity Rover's Athena science investigation at Meridiani Planum, Mars. Science 306 (5702), 1698–1703. https://doi.org/10.1126/science.1106171.

Squyres, S.W., et al., 2004c. In situ evidence for an ancient aqueous environment at Meridiani Planum, Mars. Science 306 (5702), 1709–1714. https://doi.org/10.1126/science.1104559.

Squyres, S.W., et al., 2008. Detection of silica-rich deposits on Mars. Science 320 (5879), 1063–1067. https://doi.org/10.1126/science.1155429.

Squyres, S.W., et al., 2009. Exploration of Victoria Crater by the Mars Rover Opportunity. Science 324 (5930), 1058–1061. https://doi.org/10.1126/science.1170355.

Squyres, S.W., et al., 2012. Ancient impact and aqueous processes at Endeavour Crater, Mars. Science 336 (6081), 570–576. https://doi.org/10.1126/science.1220476.

Stein, N.T., et al., 2018. Retrieval of compositional endmembers from Mars Exploration Rover Opportunity observations in a soil-filled fracture in Marathon Valley, Endeavour Crater rim. J. Geophys. Res. 123. https://doi.org/10.1002/2017JE005339.

Stern, J.C., Sutter, B., Jackson, W.A., Navarro-González, R., McKay, C.P., Ming, D.W., Archer, P.D., Mahaffy, P.R., 2017. The nitrate/(per)chlorate relationship on Mars. Geophys. Res. Lett. 44, 2643–2651.

Stern, J.C., et al., 2015. Evidence for indigenous nitrogen in sedimentary and aeolian deposits from the Curiosity rover investigations at Gale Crater, Mars. Proc. Natl. Acad. Sci. U. S. A. 112, 4245–4250.

Toner, J.D., Catling, D.C., Light, B., 2014. Soluble salts at the Phoenix Lander site, Mars: a reanalysis of the Wet Chemistry Laboratory data. Geochim. Cosmochim. Acta 136, 142–168. https://doi.org/10.1016/j.gca.2014.03.030.

Tosca, N.J., Knoll, A.H., McLennan, S.M., 2008a. Water activity and the challenge for life on early Mars. Science 320 (5880), 1204–1207. https://doi.org/10.1126/science.1155432.

Tosca, N.J., McLennan, S.M., Dyar, M.D., Sklute, E.C., Michel, F.M., 2008b. Fe oxidation processes at Meridiani Planum and implications for secondary Fe mineralogy on Mars. J. Geophys. Res. 113(E5), E05005. https://doi.org/10.1029/2007JE003019.

Vaniman, D.T., et al., 2013. Mineralogy of a mudstone at Yellowknife Bay, Gale Crater, Mars. Science. 1243480. https://doi.org/10.1126/science.1243480.

Vasavada, A.R., et al., 2014. Overview of the Mars Science Laboratory mission: Bradbury Landing to Yellowknife Bay and beyond. J. Geophys. Res. 119 (6), 1134–1161. https://doi.org/10.1002/2014JE004622.

Wang, A., et al., 2008. Light-toned salty soils and coexisting Si-rich species discovered by the Mars Exploration Rover Spirit in Columbia Hills. J. Geophys. Res. 113(E12). https://doi.org/10.1029/2008JE003126.

Weber, K.A., Achenbach, L.A., Coates, J.D., 2006. Microorganisms pumping iron: anaerobic microbial iron oxidation and reduction. Nat. Rev. Microbiol. 4 (10), 752–764.

Wiens, R.C., et al., 2013. Pre-flight calibration and initial data processing for the ChemCam laser-induced breakdown spectroscopy instrument on the Mars Science Laboratory rover. Spectrochim. Acta B At. Spectrosc. 82, 1–27. https://doi.org/10.1016/j.sab.2013.02.003.

Williams, R.M.E., et al., 2013. Martian fluvial conglomerates at Gale Crater. Science 340 (6136), 1068–1072. https://doi.org/10.1126/science.1237317.

Wray, J.J., Noe Dobrea, E.Z., Arvidson, R.E., Wiseman, S.M., Squyres, S.W., McEwen, A.S., Mustard, J.F., Murchie, S.L., 2009. Phyllosilicates and sulfates at Endeavour Crater, Meridiani Planum, Mars. Geophys. Res. Lett. 36(21)L21201. https://doi.org/10.1029/2009gl040734.

Yen, A.S., et al., 2008. Hydrothermal processes at Gusev Crater: an evaluation of Paso Robles class soils. J. Geophys. Res. 113(E6). https://doi.org/10.1029/2007JE002978.

Yen, A.S., et al., 2016. Cementation and aqueous alteration of a sandstone unit under acidic conditions in Gale Crater, Mars.47th Lunar and Planetary Science Conference. (Abstract 1649.pdf).

Zuber, M.T., et al., 1992. The Mars Observer laser altimeter investigation. J. Geophys. Res. https://doi.org/10.1029/92JE00341.

FURTHER READING

Ehlmann, B.L., Edwards, C.S., 2014. Mineralogy of the Martian surface. Annu. Rev. Earth Planet. Sci. 42 (1), 291–315. https://doi.org/10.1146/annurev-earth-060313-055024.

CHAPTER 5

Archean Lakes as Analogues for Habitable Martian Paleoenvironments

David T. Flannery, Roger E. Summons, Malcolm R. Walter

Contents

5.1 INTRODUCTION

In the early 1970s, the Mariner 9 and Viking missions sent back the first high-resolution imagery of the surface of Mars. Fluvial features were clearly visible in these images, and they immediately reignited a long-running debate centering on the possible existence of a Martian hydrologic cycle. Suddenly, hypotheses featuring rivers, deltas, crater lakes, and a global ocean started to gain momentum. Such ideas were debated during a multidecadal data drought caused by a series of misfortunes involving spacecraft sent to Mars by several nations and the shifting priorities of NASA. This drought was finally broken in 1996 by the arrival of Mars Global Surveyor—a NASA orbiter mission that was closely followed by several highly successful orbiter and rover missions that encountered abundant evidence for flowing liquid water in the past. Exactly when and for how long standing bodies of liquid water persisted on the surface of Mars remains a subject of vigorous debate, but a general consensus has been reached in regard to the presence of habitable lacustrine environments early in Mars' history (e.g., Grotzinger et al., 2014; Hurowitz et al., 2017).

From Habitability to Life on Mars
https://doi.org/10.1016/B978-0-12-809935-3.00005-0

Several lines of evidence, including crater counts and a radiometric date (Farley et al., 2014) generated by the SAM instrument aboard the Mars Science Laboratory (MSL), suggest water flowed into closed basins during the Late Noachian/Early Hesperian, around the same time that Earth's oldest surviving sedimentary rocks were deposited. Now that NASA has achieved its goal of identifying past habitable environments on Mars, attention is likely to shift toward detecting evidence for organisms that may have lived in these settings (Mustard et al., 2013). Future Mars surface missions will search for and attempt to characterize putative microbial biosignatures in ancient sedimentary rocks. The scientists tasked with designing, operating, and interpreting the science investigations of these missions will use the terrestrial geologic record, our only available analogue, as their guide (McMahon et al., 2018).

Where then to study the residue of life that flourished in Mars analogue lake systems? One approach is to visit modern lakes in order to observe relevant biological and sedimentologic processes operating today. However, such an approach must mitigate a problem that is well known to Precambrian geologists; the principle of uniformitarianism often fails when we attempt to compare geobiological processes across deep time. A case in point is that of stromatolites—the enduring sedimentologic expressions of microbial communities—which form stromatolites through the accumulation of layer upon layer of microbial mats. Stromatolites are undoubtedly the best known macroscopic biosignature of the Archean, and they are found in practically all Precambrian shallow-water environments featuring chemical sediments. In contrast, living stromatolites are rarely seen today. Their precipitous decline in the Late Neoproterozoic is generally attributed to the evolution of grazing metazoans, which disrupt laminar mat fabrics. The sedimentologic expression of microbial ecosystems has thus changed through time, and if we are correct in assuming that, as on Earth, ancient Martian lake life was prokaryotic, we are compelled to focus our studies on "extreme" modern environments that exclude or inhibit the growth of eukaryotes. Alternatively, this problem can be overcome by focusing on lakes that existed prior to the evolution of complex life on Earth, although this approach comes with its own problems, as discussed below.

Ideally, the mineralogy and depositional facies of a Mars analogue lake will be similar to known settings on Mars. Siliciclastic rocks deposited in aeolian, fluvial, and lacustrine settings have been well characterized by Mars rover missions, and many components of these systems are visible from orbit. However, whereas terrestrial clastic sedimentary rocks are typically quartz-bearing, rocks characterized from Mars orbit and by rovers on the ground are dominated by basaltic minerals. Since basalt weathers relatively rapidly in a terrestrial atmosphere, basaltic sedimentary rocks, and, accordingly, river and lake systems dominated by basaltic sediment, are uncommon on Earth. Conversely, carbonate rocks, particularly sedimentary carbonate precipitated in shallow water environments (which typically preserves macroscopic microbial biosignatures on earth), have yet to be discovered on Mars. Well-characterized Martian sedimentary rocks are instead dominated by oxidized iron and sulfur phases (e.g., Squyres et al., 2004; Vaniman et al., 2013; Hurowitz et al., 2017). The latter is known to play a similar role to carbonate in preserving microbialites, as is halite and silica, although the terrestrial geologic record preserves comparatively few examples of this. Diagenesis poses an additional conundrum. Modern analogues have not experienced the taphonomic changes that occur during billions of years of burial, and, on the other hand, plate tectonic activity has ensured that most ancient terrestrial analogues have experienced metamorphism and deformation far in excess of that expected for Mars. The Martian radiation environment and its effects on chemical biosignatures exposed at the near surface is an additional taphonomic factor that is difficult to replicate on Earth.

For the reasons described above, locating the optimum combination of depositional facies, microbial biosignatures, age, mineralogy, diagenesis, and radiation environment in a Mars analogue lake system is extremely challenging. The Archean lake systems that are the focus of this chapter tick several, but not all, of these boxes. Due to their position at the center of a stable, deep-rooted craton, they have endured billions of years of diagenesis without ever having experienced deep burial or substantial deformation. Continental rifting occurring during their deposition led to extensive subaerial volcanism and to the formation of basaltic plateaus that were outwardly similar to ancient flood basalt provinces visible from Mars orbit. Over a period of several tens of millions of years, these plateaus developed ephemeral rivers and lakes that hosted microbially dominated ecosystems. River-borne basaltic and quartz-rich sediment, and chemical sedimentary rocks including carbonate, accumulated in these closed basins. Stromatolites accreted under an anoxic or suboxic, greenhouse gas–rich atmosphere and were frequently subjected to the ecological and sedimentologic effects of catastrophic volcanic eruptions. Multiple episodes of flooding and the complete evaporation of lakes that formed the distal expressions of massive braided river systems are recorded by cyclic lithofacies changes in ancient lake sediments. Similar stratigraphic sections may exist for Martian lakes that were subjected to extreme climatic shifts related to major changes in planetary obliquity. Critically, some of these Archean lake deposits came to host an interrelated suite of microbial biosignatures, including microbialites, isotope fractionations, organic molecules and microfossils, that has survived to the present day.

5.2 ARCHEAN LAKES

One of the consequences of the constant recycling of rocks in the process of plate tectonics is that very little outcrop has survived from the first half of Earth's history. The majority of surviving sedimentary rocks postdate the Precambrian (Blatt and Jones, 1975), and those that have endured are generally highly metamorphosed. Not only do sedimentary rocks make up <5% of all Precambrian rocks, which are themselves uncommon, but the vast majority of this small fraction were deposited in marine environments.

Extensive and well-preserved Archean rocks of lacustrine origin are known from just two sedimentary sequences: the Fortescue Group in the Pilbara Craton, Western Australia, and the Ventersdorp Supergroup, Kaapvaal Craton, South Africa (Tables 5.1). The former includes laterally extensive outcrop that was deposited between ~2775 and ~2630 Ma and metamorphosed to prehnite-pumpellyite facies (see Thorne and Trendall, 2001, for a review). The chronology of the Ventersdorp Supergroup is not as well constrained, but it was likely deposited after ~2765 Ma, the maximum U-Pb radiometric date obtained for volcanic rocks of the underlying Witwatersrand Group (England et al., 2001), and prior to ~2664 Ma, the youngest of several dates obtained for the upper Ventersdorp Supergroup (Barton et al., 1995). Lacustrine rocks of the Ventersdorp Group outcrop less extensively than those of the Fortescue Group and have been metamorphosed to greenschist facies. Both successions are extremely well preserved by Archean standards, and both consist of several stratigraphic kilometers of subaerial lava flows intercalated with sedimentary rocks overlying Early Archean basement. The age, lithostratigraphic, paleomagnetic, and chemical similarities have led some to suggest that both groups may have been deposited in the same basin on the supercontinent "Vaalbara," a portmanteau of the Kaapvaal and Pilbara cratons (e.g., Grobler and Meakins, 1988; de Kock et al., 2009; Altermann and Lenhardt, 2012). Less extensive lacustrine

Table 5.1 Significant Archean lake deposits

Age (Ma)	Unit	Craton	Metamorphic grade	Lithology
~2690	Woodiana Member, Jeerinah Formation, Fortescue Group	Pilbara	Prehnite-pumpellyite facies	Silicilastic rocks with stromatolites
~2740	Ventersdorp Group	Kaapvaal	Greenschist facies	Siliciclastic and carbonate rocks with stromatolites
~2720	Kuruna Member, Kylena Formation, Fortescue Group	Pilbara	Prehnite-pumpellyite facies	Siliciclastic and carbonate rocks with stromatolites
~2720	Meentheena Member, Tumbiana Formation, Fortescue Group	Pilbara	Prehnite-pumpellyite facies	Extensive siliciclastic and carbonate rocks with stromatolites
~2775	Mt. Roe Basalt, Fortescue Group	Pilbara	Prehnite-pumpellyite facies	Siliciclastic rocks

facies are reported from the slightly older (~2985–2780 Ma) Witwatersrand Group, which consists largely of siliciclastic rocks deposited in braided stream and marine environments metamorphosed to greenschist facies or higher.

Compendia of Archean lake deposits are slimmed further by the difficulty of unambiguously identifying lacustrine units. Geologists have historically relied primarily on fossils when distinguishing marine from lacustrine deposits. This is an effective approach for Phanerozoic rocks where each setting possesses a distinctive biota, but cannot be applied to Precambrian units deposited prior to the rise of complex life. As a consequence, interpreting the depositional setting of Archean rocks is notoriously difficult. Isotopic and trace element data may rule out certain hypotheses, but because alteration is typically more severe than in younger rocks, and so little is known about the composition of the Archean atmosphere and oceans, these data may also be ambiguous. The few sedimentologic "smoking guns" that do exist are associated with marine settings (e.g., tidally generated sedimentary features; even these features can be difficult to interpret due to the generation of seiches in large lakes). Thus, when a lacustrine setting is suspected, the interpretation is generally based on multiple lines of circumstantial evidence, and an element of doubt is unavoidable.

Of the few Archean lakes known, the Tumbiana Formation of the Fortescue Group has received the most attention. Prior to the advent of radiometric dating, the Tumbiana Formation and other rocks of the Fortescue Group were thought to be of Phanerozoic age, owing largely to their excellent preservation. The curious absence of fossils was noted by several researchers, but early descriptions lacked detailed paleoenvironmental interpretations (e.g., Gregory and Gregory, 1884; Woodward, 1891; Maitland, 1904, 1915; Talbot, 1919). Later studies focusing solely on the Tumbiana Formation resulted in lacustrine interpretations that were based on stratigraphic and sedimentologic observations (Walter, 1972, 1983; Lipple, 1975). Recently, studies incorporating trace element data: stable C, O, and Sr isotope values: and detailed analyses of stratigraphy, mineralogy and sedimentology, reached similar conclusions (Buick, 1992; Bolhar and van Kranendonk, 2007; Awramik and Buchheim, 2009; Coffey et al., 2013; Flannery et al., 2016). Notable exceptions to this consensus include the interpretations of Packer (1990), Thorne and Trendall (2001), and Sakurai et al. (2005), who proposed a marine environment for the Tumbiana Formation. Diagnostic marine sedimentary features have not been reported. The observation of herringbone cross stratification made by Sakurai et al. (2005) has not been confirmed by several subsequent studies. Fluvial and lacustrine units occurring elsewhere in the northern Fortescue Group (reviewed in Flannery et al., 2016) include numerous examples of shallow lakes developed within braided river systems. They are generally thinly bedded and stratigraphically bounded by fluvial sandstone and subaerial lava flows. Lacustrine facies in the Ventersdorp Group are probably the least studied of all, but despite this, there has been little to no debate as to their origin (e.g., Winter, 1963; Buck, 1980; Karpeta, 1989).

5.3 FORTESCUE GROUP REGIONAL GEOLOGIC SETTING

The Hamersley Basin hosts the Mount Bruce Supergroup, an exceptionally well-preserved succession of supracrustal volcanic and sedimentary rocks deposited during the Late Archean and Early Proterozoic upon Early Archean basement in remote north-western Australia. The oldest rocks of the Mount Bruce Supergroup are the terrestrial lavas and sedimentary rocks of the northern Fortescue Group, which were deposited between ~2.77 and ~2.63 Ga. The Fortescue Group is conformably overlain by off-shore, presumably marine, siliciclastic rocks of the ~2.63–2.45 Ga Hamersley Group. These are in turn overlain by shallow-water siliciclastic and carbonate rocks of the ~2.45–2.22 Ga Turee Creek Group.

Thorne and Trendall (2001) divided the Hamersley Basin into four subbasins. We focus here on the northern subbasins, where the Fortescue Group consists predominantly of subaerial lava flows, fluvial sandstone, and minor stromatolitic carbonate and shale (Fig. 5.1). The northern Fortescue Group outcrop extends approximately 500 km along the southern boundary of the Pilbara Craton, where the group reaches a maximum local thickness of 6 km. The southern subbasin lies to the south of the Fortescue River, and has been more severely folded and metamorphosed. It contains the supposed offshore lateral equivalents of Fortescue Group rocks preserved in the northern subbasins. These rocks include the Boongal Formation (proposed equivalent to the Kylena Formation), Pyradie Formation (Tumbiana Formation), and Bunjinah Formation (Maddina Formation). Preservation is best in the north, where the metamorphic grade was limited to prehnite-pumpellyite facies and dips average <10° (Smith et al., 1982; Hoshino et al.,

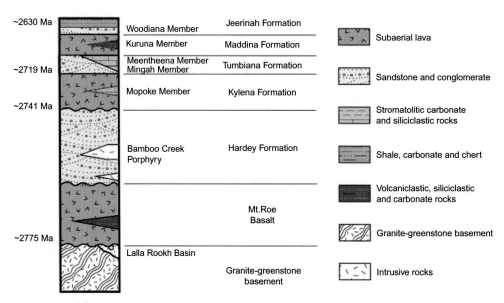

Fig. 5.1 Simplified Fortescue Group stratigraphy.

2014). The metamorphic grade of southern subbasin rocks is greenschist facies or greater (Smith et al., 1982). Limited geochronological data (Arndt et al., 1991) indicate that deposition of at least the lower portion of the southern subbasin succession was contemporaneous with deposition of the Mt. Roe Basalt and the Hardey Formation in the northern subbasins. Rocks of the northern Fortescue Group are thought to have been deposited during one or more periods of continental rifting associated with extensive volcanism (Blake, 1984; Thorne and Trendall, 2001). Available paleomagnetic and geochronological data suggest that the Pilbara Craton was located at ∼50° south during this interval and may have been close to or contiguous with the Kaapvaal Craton, South Africa, and lacustrine rocks of the similarly aged Ventersdorp Supergroup (Nelson et al., 1992; Strik et al., 2003; de Kock et al., 2009).

5.4 SEDIMENTARY ENVIRONMENTS

Fluviolacustrine facies of the Fortescue Group occur within the Mt. Roe Basalt, Hardey Formation, Kylena Formation, Maddina Formation and lower Jeerinah Formation. We focus here on the Meentheena Member (Tumbiana Formation), which preserves laterally extensive stromatolitic carbonate and siliciclastic rocks stratigraphically bounded by kilometer-thick subaerial lava flows. The Tumbiana Formation is divided into the Mingah Member, which consists of fluvial sandstone, conglomerate, volcanic tuff, and lenticular beds of stromatolitic carbonate, and the conformably overlying Meentheena Member, which consists of lacustrine stromatolitic carbonate, siltstone, shale and tuff. A simplified sedimentary facies division for the Meentheena Member is shown in Table 5.2. The Mingah Member is typically 50–150m thick. Debris flows at the base

Table 5.2 Facies divisions of the Meentheena Member

Facies	Lithology	Sedimentary features	Biosignatures
Fluvial	Sandstone	Trough and tabular cross stratification, asymmetrical ripples, desiccation cracks	Kerogen; C, N, S isotope fractionations
Nearshore, very shallow-water lake	Grainstone/calcisiltite	Symmetrical ripples, desiccation cracks, stone rosettes, tepee structures, ooids, storm event beds, fenestrae	Stromatolites; ooids; oncolites; MISS; microfossils; kerogen; C, N, S isotope fractionations
Shallow-water lake	Calcilutite/calcisiltite	Soft-sediment deformation; storm event beds	Stromatolites; roll-up structures; kerogen; C, N, S isotope fractionations
Deep-water lake	Siltstone/shale	Planar lamination	Kerogen; C, N, S isotope fractionations

of the unit are reported from Mingah Member outcrop in the northeast subbasin (Walter, 1983; Sakurai et al., 2005; Coffey et al., 2013). Trough and tabular cross stratified vol-caniclastic sandstone forms the vast majority of outcrop. Intercalated subaerial lava flows and desiccation cracks suggest periods of subaerial exposure during deposition, likely within a large (>700 km) braided river system. Limited paleocurrent data reported by Coffey et al. (2013) suggest this system was likely transporting sediment in a northeasterly direction—toward carbonate lakes recorded by the overlying Meentheena Member—which probably represents the distal expression of this endorheic river system. A cyclic lithofacies succession in the Meentheena Member was first identified by Awramik and Buchheim (2009) (Fig. 5.2). This cycle consists of edgewise conglomerate, cross stratified calcarenite, wavy laminated calcilutite/siltite, and laminated shale featuring desiccation cracks at upper contacts. The base of transgressive cycles is typically represented by detri-tal and authigenic carbonate rocks bearing abundant evidence for very shallow-water deposition, ground water upwelling, and regular subaerial exposure. A transition to deep-water conditions in overlying units is indicated by the increasing scarcity of coarse sediment particles, cross stratification, desiccation cracks, tepee structures and edgewise conglomerate, and the increasing prevalence of flat to wavy laminated calcisiltite/lutite and large stromatolites with considerable synoptic relief. Storm-related event beds are occasionally observed in this deep-water facies, which is typically overlain by organic rich, planar-laminated siltstone/shale. The absence of current-generated sedimentary features suggests the siltstone/shale facies was deposited in the deepest environments of the Meentheena Member lakes, below storm wave base and below the zone of car-bonate precipitation. Desiccation cracks occurring at the upper contacts of shale units mark the complete drying of the lake. Overlying carbonate rocks represent the initiation of a new transgressive cycle.

5.4.1 Fluvial Facies

Although the Meentheena Member consists predominantly of stromatolitic carbonate and deep-water shale, it also contains fluvial siliciclastic rocks, including migrating bar bedforms and trough cross stratified gravel. These units are lithologically and sedimen-tologically similar to siliciclastic rocks of the underlying Mingah Member (Fig. 5.3A). They occur intermittently in the upper Meentheena Member, where they are interca-lated with shallow water carbonate units interpreted here as lake shore deposits. Fluvial facies are also common in the Hardey Formation, Kuruna Member (Maddina Forma-tion), and lower Jeerinah Formation, although they are not well studied from a paleon-tological perspective. Ancient soil horizons reported from the Mt. Roe Basalt are a notable exception. The close association of organic matter in paleosols and fluviolacus-trine facies in the Mt. Roe Basalt led Rye and Holland (2000) to suggest organic matter accumulated in ephemeral ponds. Fluvial processes are thought to have then reworked

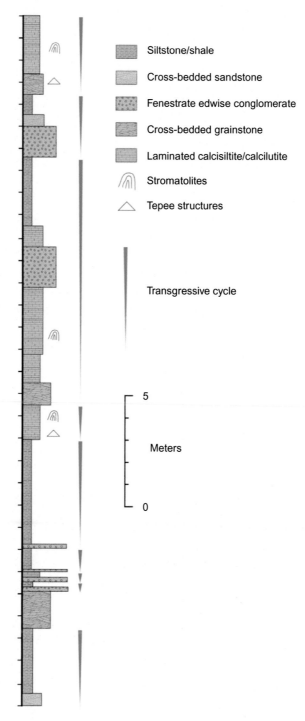

Fig. 5.2 Simplified representative stratigraphic section measured in the Meentheena Member showing a repeating succession of common lithofacies.

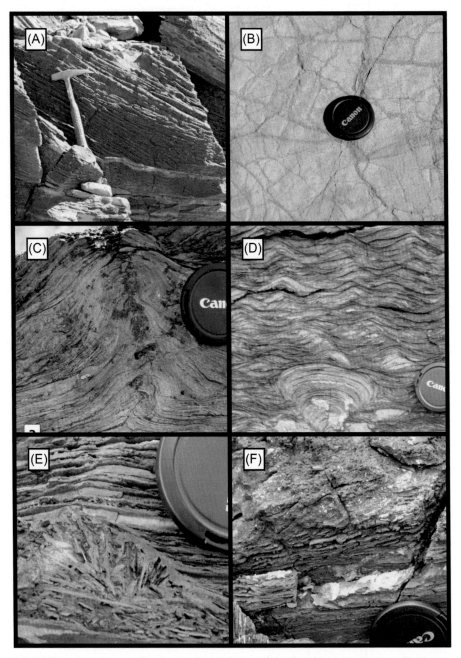

Fig. 5.3 (A) Trough cross stratification in fluvial facies of the Tumbiana Formation. (B) Desiccation cracks in near shore lake facies of the Meentheena Member. (C) Vertical cross section through a stromatolite encrusted, fluid escape tepee structure in near shore lake facies of the Meentheena Member. (D) Symmetrical (wind wave-driven) ripple cross stratification surrounding a turbinate columnar stromatolite in the near shore lake facies of the Meentheena Member. (E) Stone rosette formed by the reworking of edgewise clasts by wind waves in shallow water in near shore lake facies of the Meentheena Member. (F) Unit preserving fenestrae ("birds eye" calcite) associated with an event bed in near shore lake facies of the Meentheena Member.

organic matter that is now preserved within sericitic sandstone deposited on basaltic lava plains. Elsewhere, carbonate stromatolites are reported from decicentimeter-thick lacustrine deposits intercalated with fluvial sandstone of the Mingah Member (Flannery et al., 2016). Macroscopic microbial features (e.g., roll-up structures) are notably not reported from fluvial or deltaic facies of the Fortescue Group.

5.4.2 Near Shore Lake Facies

A very shallow water depositional environment is well represented in the Meentheena Member by an association of desiccation cracks, symmetrical (wind-generated) ripple cross stratification, stone rosettes, and tepee structures. Stromatolites of diverse scales and morphologies are common in this facies. Kerogen is visible in thin section and occurs predominantly within micritic stromatolitic laminae and in stylolites.

5.4.2.1 Desiccation Cracks

Desiccation cracks occur at the tops of regressive lithofacies cycles in the Meentheena Member. They are typically present at the contact between underlying, deep water shale and overlying, shallow water calcitic rocks hosting symmetrical ripple cross stratification (Fig. 5.3B).

5.4.2.2 Tepee Structures

Tepee structures are ridges formed by the buckling of lake or seafloor pavement, which creates sedimentary features that appear tepee-shaped when viewed in vertical cross section. They are known to form due to a variety of processes involving either ground water upwelling, desiccation, or syndepositional tectonism. Meentheena Member tepee structures are raised areas of calcitic pavement characterized by laminar crusts and oncolites (Flannery et al., 2016), and are thus closest to the peritidal and lacustrine tepee types defined by Kendall and Warren (1987). They are typically associated with features suggestive of very shallow water deposition (e.g., symmetrical ripple marks, stone rosettes, and desiccation cracks). Topographic relief during deposition is indicated by onlapping sediment. They range from 20 to 50 cm in height and are tepee-shaped in vertical cross section only—tepees seen in bedding plane exposures appear as elongate ridges. In vertical cross section, Meenthena Member tepees typically display a central fracture and a brecciated central zone, suggesting the synsedimentary upwelling of fluid. Stromatolites often drape the areas of raised topography provided by Meentheena Member tepee structures (Fig. 5.3C). Secondary replacement of calcite by silica in the brecciated zone and in draping stromatolites may be related to the movement of silica-rich groundwater. Modern examples of tepee structures that are texturally similar to those described from the Meentheena Member are reported from Lake Macleod in Western Australia (Handford et al., 1984) and from coastal salinas of South Australia (Warren, 1982;

Ferguson et al., 1988), where they form through the disruptive action of periodically discharging groundwater in nearshore environments.

5.4.2.3 Symmetrical Ripple Cross-Stratification

Symmetrical ripple cross stratification (Fig. 5.3D) is ubiquitous in the nearshore lake facies. This type of cross stratification forms through the reworking of sandy sediments by wind-driven standing waves in shallow water. The predominant dip direction recorded by some laminae likely reflects deposition within shoaling waves, with the predominant dip direction indicating the direction of wave propagation (Allen, 1979; Martel and Gibling, 1991). Climbing sets reported by Flannery et al. (2016) may reflect episodes of waning sediment suspension generated by shoaling wind-generated waves (Reineck and Singh, 1972).

5.4.2.4 Edgewise Conglomerate and Stone Rosettes

Tabular clasts that have originated in underlying calcitic crusts typically occur in lenses of edgewise conglomerate. These clasts are occasionally arranged in "rosettes" (Fig. 5.3E). Such stone rosettes are common components of younger marine and lacustrine environments (Kazmierczak and Goldring, 1978; Ricketts and Donaldson, 1979; Myrow et al., 2003, 2004), where they are formed in shallow water settings through the reworking of clasts by wind-generated waves.

5.4.2.5 Fenestrae

Decacentimeter-scale beds of fenestrate (birds eye) limestone (Fig. 5.3F) occur in cross stratified calcarenite, typically in chaotic accumulations of laminated allochthonous carbonate clasts. Fenestrae also occur under large tabular clasts of edgewise conglomerate and in the apical zones of micritic stromatolites. Fenestrae are 1–5 mm wide and are commonly filled with sparry calcite and surrounded by a matrix of sand-sized carbonate, tuff, and siliciclastic grains. Multiple hypotheses exist for the origin of fenestrae in Precambrian carbonates, including the repeated wetting and drying of microbial mats, the generation of gas bubbles through the decomposition of organic matter, oxygen evolution during oxygenic photosynthesis, and the expansion of microbial mats during growth (Shinn, 1968; Logan et al., 1974; Monty, 1976; Bosak et al., 2009). Fenestrae in the Tumbiana Formation are found in a wide variety of contexts, and more than one process may be responsible for their generation. However, the concentration of fenestrae in storm event beds favors an explanation involving the decomposition of disrupted microbial mats.

5.4.2.6 Ooids

Ooids are accretionary carbonate grains that are generally thought to require water saturated in $CaCO_3$, regular agitation, and, potentially, biological processes in order to form

(Davies et al., 1978; Summons et al., 2013; Diaz et al., 2015). Micritic ooids occurring in the Meentheena Member are the oldest well-preserved ooids known. Nuclei include detrital pyrite and quartz, as well as volcaniclastic and carbonate clasts (Buick, 1992; Flannery et al., 2016). Meentheena Member ooids retain fine ($<5\,\mu m$) laminar fabrics (Fig. 5.4A), and are mineralogically and texturally similar to some modern ooids currently accreting in temperate lakes (Swirydczuk et al., 1979; Wilkinson et al., 1985; Davaud and Girardclos, 2001; Plee et al., 2008).

5.4.2.7 Near-shore Stromatolites

Domical and columnar stromatolites in the nearshore facies are characterized by very low synoptic relief ($<3\,cm$) and the erosion of column bases (Fig. 5.4B). Onlapping sediment typically consists of ripple cross laminated calcarenite. Progradation of this material across stromatolites commonly appears to have initiated a morphological shift toward greater lamina convexity and the branching of columns, before terminating stromatolite growth altogether (Fig. 5.4C). Stromatolitic laminae are composed of neomorphic micrite and microspar with a high proportion of sand-sized detrital grains. Stylolites are common in stromatolites and generally follow the same plane of lamination that is created by alternating layers of spar and micrite. Oncolites formed around edgewise clasts derived from calcitic pavement (Fig. 5.4D and E).

5.4.3 Shallow Water Lake Facies

The shallow water lake facies is dominated by flat to wavy-laminated calcilutite/calcisiltite hosting large domical stromatolites. The absence of symmetrical ripple cross lamination and the general scarcity of current-generated sedimentary features suggest that deposition occurred below wave base. Periodic high energy events, most likely storms, are recorded by current generated cross stratification and displaced stromatolitic material. The depositional setting of this facies is thus likely to have been between wave base and storm wave base. Stromatolitic micrite preserves organic matter in concentrations averaging $\sim0.1\%$ TOC in both outcrop and drill core samples (Eigenbrode and Freeman, 2006; Thomazo et al., 2009; Flannery et al., 2016).

5.4.3.1 Soft Sediment Deformation

Soft sediment deformation structures are common at the contacts between calcilutite and calcisiltite units. Examples include flame structures, ball and pillow structures, and dish and pillar structures (Fig. 5.4F).

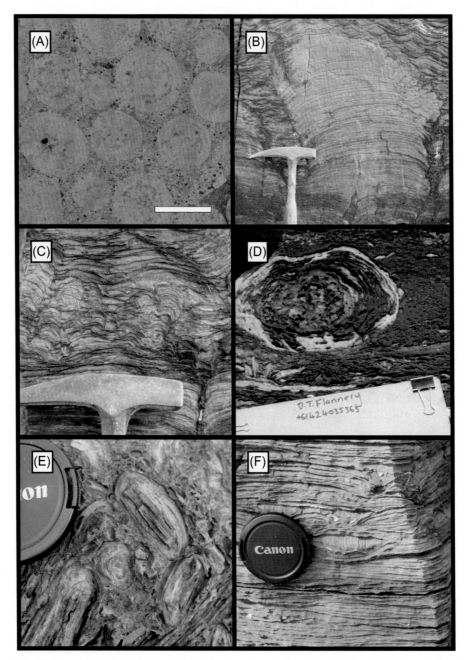

Fig. 5.4 (A) Transmitted light thin section photomicrograph showing micritic ooids in near shore lake facies of the Meentheena Member. Scale bar = 1 mm. (B) Columnar stromatolite incorporating a high proportion of sand-sized detrital grains in near shore lake facies of the Meentheena Member. (C) Low-relief columnar stromatolites shifting to branching and increasingly convexly laminated forms within a matrix of symmetrically ripple cross stratified calcarenite in the near shore lake facies of the Meentheena Member. (D) Oncolite in an event bed related to storm activity in the shallow water lake facies of the Meentheena Member. The clip board is 30 cm in width. (E) Oncolites nucleated on edgewise clasts in near-shore lake facies of the Meentheena Member. (F) Calcisiltite/calcilutite featuring soft sediment deformation in shallow-water lake facies of the Meentheena Member.

5.4.3.2 Event Beds

Laterally extensive event beds (Fig. 5.5A), most likely related to storm activity, consist of tabular clast edgewise conglomerate; displaced and contorted stromatolitic material; and pebble-sized clasts of carbonate, lithic, and volcaniclastic material. A matrix of trough cross stratified sand sized carbonate grains, often with abundant fenestrae, is typically present.

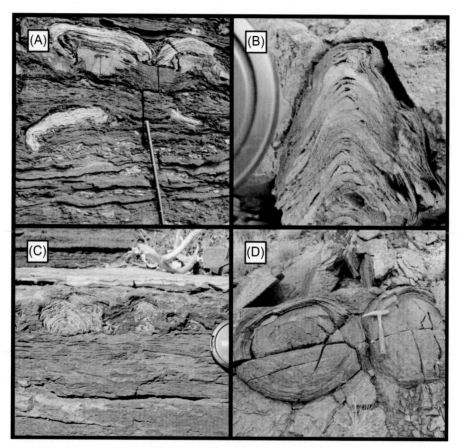

Fig. 5.5 (A) Storm-induced event beds featuring disrupted stromatolites in the shallow water lake facies of the Meentheena Member. Gradations on the Jacob's Staff are 10 cm. (B) Coniform stromatolite with a shifting axial zone and detrital material entrained within steep stromatolite flanks in shallow-water lake facies of the Meentheena Member. Growth appears to have been terminated by an overlying unit of volcanic tuff. (C) Light toned unit of volcanic tuff that appears to have terminated the growth of domical Meentheena Member stromatolites in the shallow-water lake facies. (D) Large domical stromatolites in shallow water lake facies of the Meentheena Member.

5.4.3.3 Shallow Water Stromatolites

Well preserved conical and domical stromatolites occur in laminated calcilutite/ calcisiltite. Among all Fortescue Group microbialites, these examples display the highest diversity of macroscopic lines of evidence for biogenicity, such as roll-up structures, conical morphologies, axial zones, detrital material entrained on steep stromatolite flanks, and tufted and reticulate morphologies. These microbialites accreted in a relatively quiet, deep water environment that was less influenced by physical environmental factors such as wave action and desiccation. Laminae are typically composed of neomorphic micrite and spar but are locally rich in siliciclastic and volcaniclastic grains. Macrolamination generally derives from alternating micritic and sparitic laminae bordered by stylolites and/or silicified tuffaceous laminae.

Conformly laminated columnar stromatolites (Fig. 5.5B) are comparable to *Conophyton* described from younger rocks. The examples reported here are among the oldest known conical stromatolites exhibiting axial zones: a thickening and contortion of laminae along the central axis of the stromatolite that is generally interpreted as representing the vertical migration of filamentous microbes (Walter et al., 1976). Millimeter-scale "tufts" are morphologically similar to textures described from modern microbial mats (Flannery and Walter, 2011) and grade laterally into reticulate structures that are also associated with microbial motility (Shepard and Sumner, 2010). In some cases, "tufts" grade vertically into decacentimeter-scale, conically laminated stromatolites composed of macrolaminated micrite and diagenetic silica. Coniform stromatolites are surrounded by a matrix of onlapping, massive to faintly laminated micrite, volcanic glass, and intraformational clasts. The paucity of current generated sedimentary features is consistent with the accretion of these stromatolites in a low energy, deep water setting, as has been reported for the vast majority of Precambrian *Conophyton* (e.g., Donaldson, 1976). The macrolamination that is visible in outcrop is related to destructive diagenetic silica that has produced laminae that are more resistant to weathering. The mean inclination of laminae from horizontal is 55°. This is beyond the angle of repose for sand sized grains, some of which have been incorporated into the steeply inclined flanks of conical stromatolites (cf. Allwood et al., 2015), suggesting the former presence of a gelatinous microbial mat (cf. Frantz et al., 2015). An influx of volcaniclastic material appears to have terminated stromatolitic growth in several instances.

In addition to coniform stromatolites, a large variety of domical stromatolite morphologies occur in the shallow water lake facies (Fig. 5.5C and D). Synoptic relief of stromatolitic laminae ranges from less than a centimeter to up to 40 cm, which is the highest relief reported for stromatolites of the Fortescue Group, and is consistent with a comparatively deep water setting.

5.4.4 Deep Water Lake Facies

The deep water lake facies consists of planar laminated siltstone/shale. The absence of current generated sedimentary structures in this facies suggests deposition occurred below storm wave base, possibly below the zone of carbonate precipitation in the deepest parts of the Meentheena Member lake system. Shale units are up to several tens of meters thick but outcrop very poorly compared with prominently weathering carbonate and chert. Hand samples acquired from outcrop are highly fissile and pervasively oxidized. Shale obtained from drill core preserves macromolecular organic matter in relatively high concentrations ($\sim$0.1%–0.4% TOC; Eigenbrode and Freeman, 2006; Thomazo et al., 2006, 2009).

5.5 BIOSIGNATURE PRESERVATION

5.5.1 Microbialites

Lacustrine facies of the Fortescue Group preserve extraordinarily diverse stromatolite morphologies. In many cases, multiple lines of evidence for biogenicity are available. Such evidence includes organic matter and microfossils trapped within stromatolitic laminae, wrinkly laminar fabrics, detrital grains adhered to steeply inclined stromatolite flanks, roll-up structures, and complex morphologies associated with microbial motility (Walter, 1983; Awramik and Buchheim, 2009; Flannery and Walter, 2011; Flannery et al., 2016). The extremely fine-grained micritic mineralogy of Meentheena and Kuruna Member stromatolites and the abundance of detrital grains in these stromatolites preclude an origin in abiotic mineral crusts or fans. Micritic microfabrics lack mineral pseudomorphs or ghost fabrics indicative of an isopachous precursor, but are well-enough preserved to retain micron-scale laminae and detrital grains of carbonate and volcanic glass. Laminae pinch and swell, and are often wrinkly, which is inconsistent with an origin in abiotic crusts, but comparable to microbial textures reported from modern environments (e.g., Hagadorn and Bottjer, 1997). Roll-up structures in the Meentheena Member demonstrate that some stromatolitic laminae remained flexible after accretion. Generally speaking, the Tumbiana Formation is anomalously devoid of evidence for the precipitation of mineral crusts and aragonite fans. In contrast, these are common components of similarly aged marine units (e.g., Sumner, 1997). In addition to stromatolites, microbially induced sedimentary structures (cf. Gerdes, 2007) are reported from shallow-water lacustrine marl in the Meentheena Member (Flannery, 2013). All microbialites reported from lacustrine facies of the Fortescue Group are preserved in shallow-water/nearshore environments.

5.5.2 Microfossils

Schopf and Walter (1983) reported filamentous microfossils from limestone stromatolites of the Meentheena Member. Similar features have not been relocated by subsequent studies, although palimpsest fabrics first reported by Walter (1972) are widespread and may represent highly degraded microbial filaments. Unlike the majority of Precambrian microfossils, which are preserved in microcrystalline quartz matrixes that allow for extremely fine-scale preservation of morphological features, the microfossils reported by Schopf and Walter (1983) are described from a micritic matrix. Microcrystalline quartz is relatively common in both the Meentheena and Kuruna members, where it commonly replaces microbial carbonate in a style that is reminiscent of younger, microfossil-bearing units. However, microfossils are not reported from siliceous rocks of either member. Typically, the long-term preservation of microfossils within stromatolites is linked to the very early replacement of a chemical sediment with silica, which must occur prior to *postmortem* degradation of microbial cells. SiO_2 concentrations during this window must be high enough to induce silica precipitation, at least within the microenvironments created by sediment pore space and extracellular polymeric substances associated with biofilms. The rarity of such conditions is often invoked to account for the general paucity of Archean microfossil deposits, despite the prevalence of chert in the early geologic record. Ephemeral lakes in volcanic settings are perhaps more likely to host this taphonomic window due to the abundance of syndepositional silica supplied by volcanic ash, fluctuating water levels, and silica-rich groundwaters upwelling at lake peripheries (cf. Kremer et al., 2012). However, despite the clear presence of these advantageous environmental factors in many of the Fortescue Group lakes, taphonomic conditions do not appear to have favored microfossil preservation.

5.5.3 Chemical and Isotopic Biosignatures

Interest in the organic geochemistry of ancient lacustrine sediments was sparked by the discovery of anomalous C isotope signatures for organic carbon, in the form of kerogen, preserved in sediments of the Tumbiana Formation. It was observed (Hayes, 1983) that $\delta^{13}C_{org}$ values in the range −41.2‰ to −51.9‰ PDB, together with coeval carbonate C isotope values ($\delta^{13}C_{carb}$) close to 0‰, could not be explained as being solely the result of fractionations associated with photoautotrophic carbon fixation. Rather, it was proposed that such signatures were more likely recording a significant component of carbon assimilation by microaerobic methanotrophic bacteria (Hayes, 1983, 1994; Hayes et al., 1983). Such extreme depletions recorded during a $\sim$ 150-million-year interval in the Fortescue Group and in black shales of other Neoarchean sedimentary settings were proposed to be indicative of an Archean carbon cycle that was fundamentally distinct from that which operates in contemporary aquatic environments (Des Marais, 1997, 2001). Rather, a world with low but nonzero levels of O_2 in the ocean atmosphere system could sustain "a globally prominent methanogen-methylotroph cycle" (Des Marais, 1997). Later, these ideas were modified when it was suggested that recycling of carbon by anaerobic microbial communities could also have led to significant ^{13}C depletion of kerogens and that aerobic CH_4 cycling need not be invoked as the sole cause (Des Marais, 2001). A subsequent study (Eigenbrode and Freeman, 2006) identified acetate recycling as one process that could drive strong ^{13}C depletion in sedimentary organic C, although these authors also asserted that either aerobic or anaerobic methanotrophy would be a simpler explanation. This is because methane oxidation processes ultimately require electron acceptors that are coupled to oxygenic photosynthesis, with the possible exceptions of sulfate reduction using the sulfate formed from photolysis of volcanogenic SO_2, or Fe^{3+} produced through photochemical reactions or as a consequence of ferrophototrophy. N isotope data acquired from Tumbiana Formation sediments show extremely positive values (as high as +8.6‰) and a robust anticorrelation with the organic C isotope data for the same samples. These findings have been interpreted as marking the onset of nitrification coupled to its continuous consumption by denitrification, an inescapable indicator of O_2 availability (Thomazo et al., 2011).

Whereas the lacustrine kerogens of the Tumbiana Fm. have afforded informative isotopic data and their syngenicity is uncontested, they have proved resistant to molecular characterization that might be biogeochemically revealing. Bitumens, on the other hand, have been the focus of molecular work that originally appeared to support the isotopic findings. Methylhopane biomarker hydrocarbons, thought to infer the availability of molecular O_2, were detected in samples collected from the Tumbiana Formation and other sedimentary rocks of the Mount Bruce Supergroup (Brocks et al., 1999; Eigenbrode et al., 2008). However, in a test of prior reports of Archean biomarkers, clean drilling, sampling, and handling protocols were applied when redrilling Tumbiana Formation sediments at the eastern edge of the Pilbara Craton, a region thought to be most prospective in terms of metamorphic grade and potential for preservation of molecular biosignatures. A careful reanalysis of these sediments showed that the previously reported biomarkers were contaminants overprinting mixtures of noninformative polyaromatic and diamondoid hydrocarbons (French et al., 2015).

5.5.4 Biosignatures That Are Notably Absent

Biosignatures that are absent from, or yet to be discovered in, Archean lake deposits includes microbial borings in volcanic glass, impact glass, or carbonate grains. Controversial examples of endolithic microbial trace fossils are reported from Early Archean pillow lavas (McLoughlin et al., 2008). Similar microtubules are reported from Tumbiana Formation volcanic glass by Lepot et al. (2011), who interpreted their origin as abiotic, or uncertain in some cases. Ooids are another common substrate for endolithic microbes in modern environments. Morphologically complex microborings are reported from several examples of well-preserved Precambrian ooids (e.g., Campbell, 1982; Knoll et al., 1986; Zhang and Golubic, 1987). The absence of obvious microborings in exceptionally well preserved Meentheena Member ooids, glassy lava flow tops and abundant volcanic glass shards, suggests that endolithic microbes were not present during deposition of the unit. Macroscopic microbial features in siliciclastic sediment (e.g., roll-up structures) are notably not reported from fluvial or deltaic facies of the Fortescue Group.

5.6 LESSONS FOR MARTIAN PALEOLAKE EXPLORATION

Although there are obvious differences and a great many unknowns (e.g., the likelihood of photosynthetic organisms inhabiting ancient Martian lakes), several features of the lacustrine units within the Neoarchean Fortescue Group make them excellent analogues for some Martian paleolakes. In particular, the regional depositional settings, which feature large braided river systems that fed extensive, ephemeral lakes developed upon basaltic lava plateaus, are broadly similar to the settings of many Martian examples. The Curiosity rover has traversed along the erosional remnants of a thick sequence of fluvial and nearshore sandstone, conglomerate, and lacustrine mudstone (e.g., Hurowitz et al., 2017). A number of depositional settings and sedimentary structures discovered by Curiosity, such evidence for regular subaerial exposure and burial diagenesis (e.g., shrinkage cracks and nodules), are familiar features of the Fortescue Group. The great age of the Fortescue Group, the prevalence of microbially dominated ecosystems, and the minimal deformation and metamorphism that have occurred since deposition, also contribute to the utility of these rocks as a Mars analogue. Significantly, evidence for ancient microbial life is widespread in the northern Fortescue Group. All of the lacustrine facies defined here preserve at least one type of biosignature, and many preserve multiple lines of evidence for biological processes that are visible in outcrop using instrumentation that is likely to be carried by future Mars rovers (e.g., textural biosignatures featuring spatially resolved distributions of elements, minerals, and organics that can be mapped using rastering microfocus X-ray fluorescence and UV Raman/fluorescence spectroscopy. See for example Flannery et al., 2018a,b). Below, based on our observations of Fortescue Group lake systems, we tentatively propose several lessons in relation to which elements of these paleolakes would offer the best prospects for in situ biosignature characterization using the suite of contact science instruments selected for the Mars 2020 rover mission:

(1) *Biosignatures, and particularly macroscopic textural biosignatures, are comparatively rare in fluvial sections.* Macroscopic biosignatures such as stromatolites and roll-up structures are not reported from fluvial systems feeding the Fortescue Group lakes. Fluvial facies consist primarily of laterally extensive, thickly bedded, cross stratified sandstone that represents a suboptimal prospecting ground for biosignatures that are easily identified in outcrop. Hand samples of coarse-grained siliciclastic sediment investigated in the laboratory have thus far yielded values for average total organic carbon content that are low compared to lacustrine carbonate rocks and unweathered lacustrine shale.

(2) *Deep water siliciclastic rocks preserve organic matter, but only at depth from current outcrop, and they do not host macroscopic biosignatures.* Drill core obtained from below the modern weathering horizon (>100 m) includes shale preserving relatively abundant macromolecular organic matter (mean TOC = ~0.3%), as well as sulfides and other minerals recording biogenic isotope fractionations. Macroscopic biosignatures are not reported from deep water siliciclastic units. Shale is generally pervasively weathered near the surface and outcrops very poorly. Surface weathering has oxidized shale components, remobilized organic matter, and typically led to extreme friability. Investigation of surface expressions of this facies in the field with the goal of characterizing biosignatures using the Mars 2020 contact science instrument suite would thus be challenging.

(3) *The greatest abundance and diversity of biosignatures is found at the upper contacts of regressive lithostratigraphic cycles and at lake peripheries.* In shallow water environments at lake peripheries, and during times when the siliciclastic sediment flux was reduced or water levels fell and ions were concentrated, chemical sediments were precipitated. Carbonate has preserved macroscopic biogenic textures (i.e., stromatolites), microbial body fossils, and laminar distributions of organic matter. Zones of groundwater upwelling at lake peripheries, recorded in the Meentheena Member by tepee structures, provided chemical disequilibria that could have been harnessed by metabolic processes, and also created a taphonomic window in the form of early diagenetic silica, a mineral known for its excellent preservation of microfossils. Minor mineral phases involved in key biogeochemical cycles can also be "locked in" by chemical sediments precipitated at lake peripheries. For example, pyrite survives in some carbonate rock outcrop, whereas as pyrite in shale outcrop has generally been pervasively oxidized. TOC values obtained for carbonate rocks are comparable to those obtained for deep water siliciclastic units sampled in drill core (Thomazo et al., 2009; Flannery et al., 2016). In the case of the Fortescue Group lakes, chemical sediment is composed predominantly of micritic carbonate. However, halite, gypsum, silica, and other chemical sediments precipitated in lakes through similar processes are also known to host stromatolites, microfossils, and other microbial biosignatures (cf. Allwood et al., 2013). As in the Meentheena Member, these sediments are typically precipitated at lake peripheries and at contacts recording transitions between regressive and transgressive cycles. An exploration model focused on chemical sediments precipitated in these facies can be applied to many such systems, including Martian examples.

(4) *Volcaniclastic horizons can be traced laterally to well-preserved microbialites in shallow water facies.* Ongoing volcanism during deposition of the Fortescue Group resulted in airfall tuff regularly settling in shallow water lake environments and terminating the growth of stromatolites. The bases of extremely fine-grained tuffaceous units reliably preserve molds and casts of micrometer/millimeter-scale microbial textures and

other sedimentary features in exquisite detail. Tuffaceous units located in deeper water siliciclastic settings can typically be traced laterally to shallow water equivalents featuring chemical sediments.

REFERENCES

Allen, J., 1979. A model for the interpretation of wave ripple-marks using their wavelength, textural composition, and shape. J. Geol. Soc. Lond. 136, 673–682.

Allwood, A., Burch, I., Rouchy, J., Coleman, M., 2013. Morphological biosignatures in gypsum: diverse formation processes of Messinian (∼ 6.0 Ma) gypsum stromatolites. Astrobiology 13 (9), 870–886.

Allwood, A., Clark, B., Flannery, D., Hurowitz, J., Wade, L., Elam, T., Foote, M., Knowles, E., 2015. Texture-specific elemental analysis of rocks and soils with PIXL: the planetary instrument for X-ray lithochemistry on Mars 2020. Aerospace Conference, 2015. IEEE, pp. 1–13.

Altermann, W., Lenhardt, N., 2012. The volcano-sedimentary succession of the Archean Sodium Group, Ventersdorp Supergroup, South Africa: volcanology, sedimentology and geochemistry. Precambrian Res. 214–215, 60–81.

Arndt, N.T., Nelson, D.R., Compston, W., Trendall, A.F., Thorne, A.M., 1991. The age of the Fortescue Group, Hamersley Basin, Western Australia, from ion microprobe zircon U-Pb results. Aust. J. Earth Sci. Int. Geosci. J. Geol. Soc. Aust. 38 (3), 261–281.

Awramik, S.M., Buchheim, H.P., 2009. A giant, Late Archean lake system: The Meentheena Member (Tumbiana Formation; Fortescue Group), Western Australia. Precambrian Res. 174 (3–4), 215–240.

Barton, J., Blignaut, E., Salnikova, E., Kotov, A., 1995. The stratigraphical position of the Buffelsfontein Group based on field relationships and chemical and geochronological data. S. Afr. J. Geol. 98 (4), 386–392.

Blake, T.S., 1984. The lower Fortescue Group of the northern Pilbara Craton: stratigraphy and paleogeography. In: Muhling, J.R., Groves, D.I., Blake, T.S. (Eds.), Archaean and Proterozoic Basins of the Pilbara, Western Australia—Evolution and Mineralization Potential. University of Western Australia, Geology Department and University Extension, Western Australia, pp. 123–143. Publication no. 9.

Blatt, H., Jones, R.L., 1975. Proportions of exposed igneous, metamorphic, and sedimentary rocks. Geol. Soc. Am. Bull. 86 (8), 1085–1088.

Bolhar, R., van Kranendonk, M.J., 2007. A non-marine depositional setting for the northern Fortescue Group, Pilbara Craton, inferred from trace element geochemistry of stromatolitic carbonates. Precambrian Res. 155 (3–4), 229–250.

Bosak, T., Liang, B., Sim, M.S., Petroff, A.P., 2009. Morphological record of oxygenic photosynthesis in conical stromatolites. Proc. Natl. Acad. Sci. 106 (27), 10939–10943.

Brocks, J.J., Logan, G.A., Buick, R., Summons, R.E., 1999. Archean molecular fossils and the early rise of eukaryotes. Science 285, 1033–1036.

Buck, S.G., 1980. Stromatolite and ooid deposits within the fluvial and lacustrine sediments of the Precambrian Ventersdorp Supergroup of South Africa. Precambrian Res. 12 (1–4), 311–330.

Buick, R., 1992. The antiquity of oxygenic photosynthesis: evidence from stromatolites in sulphate-deficient Archaean lakes. Science 255 (5040), 74–77.

Campbell, S.E., 1982. Precambrian endoliths discovered. Nature 299 (5882), 429–431.

Coffey, J., Flannery, D., Walter, M., George, S., 2013. Sedimentology, stratigraphy and geochemistry of a stromatolite biofacies in the 2.72 Ga Tumbiana Formation, Fortescue Group, Western Australia. Precambrian Res. 236, 282–296.

Davaud, E., Girardclos, S., 2001. Recent freshwater ooids and oncoids from western lake Geneva (Switzerland): indications of a common organically mediated origin. J. Sediment. Res. 71 (3), 423–429.

Davies, P.J., Bubela, B., Ferguson, J., 1978. The formation of ooids. Sedimentology 25 (5), 703–730.

de Kock, M.O., Evans, D.A.D., Beukes, N.J., 2009. Validating the existence of Vaalbara in the Neoarchean. Precambrian Res. 174 (1–2), 145–154.

Des Marais, D.J., 1997. Isotopic evolution of the biogeochemical carbon cycle during the Proterozoic Eon. Org. Geochem. 27, 185–193.

Des Marais, D.J., 2001. Isotopic evolution of the biogeochemical carbon cycle during the Precambrian. Rev. Mineral. Geochem. 43, 555–578.

Diaz, M.R., Swart, P.K., Eberli, G.P., Oehlert, A.M., Devlin, Q., Saeid, A., Altabet, M.A., 2015. Geochemical evidence of microbial activity within ooids. Sedimentology 62 (7), 2090–2112.

Donaldson, J.M., 1976. Paleoecology of *Conophyton* and associated stromatolites in the Precambrian dismal lake and Rae groups. In: Walter, M.R. (Ed.), Stromatolites. Developments in Sedimentology. Elsevier, Amsterdam, Oxford, New York, pp. 523–534.

Eigenbrode, J.L., Freeman, K.H., 2006. Late Archean rise of aerobic microbial ecosystems. Proc. Natl. Acad. Sci. 103 (43), 15759–15764.

Eigenbrode, J.L., Freeman, K.H., Summons, R.E., 2008. Methylhopane biomarker hydrocarbons in Hamersley Province sediments provide evidence for Neoarchean aerobiosis. Earth Planet. Sci. Lett. 273, 323–331.

England, G.L., Rasmussen, B., McNaughton, N.J., Fletcher, I.R., Groves, D.I., Krapez, B., 2001. SHRIMP U-Pb ages of diagenetic and hydrothermal xenotime from the Archaean Witwatersrand Supergroup of South Africa. Terra Nova 13 (5), 360–367.

Farley, K.A., Malespin, C., Mahaffy, P., Grotzinger, J.P., Vasconcelos, P.M., Milliken, R.E., Malin, M., Edgett, K.S., Pavlov, A.A., Hurowitz, J.A., Grant, J.A., Miller, H.B., Arvidson, R., Beegle, L., Calef, F., Conrad, P.G., Dietrich, W.E., Eigenbrode, J., Gellert, R., Gupta, S., Hamilton, V., Hassler, D.M., Lewis, K.W., McLennan, S.M., Ming, D., Navarro-González, R., Schwenzer, S.P., Steele, A., Stolper, E.M., Sumner, D.Y., Vaniman, D., Vasavada, A., Williford, K., Wimmer-Schweingruber, R.F., 2014. In situ radiometric and exposure age dating of the martian surface. Science 343, 1247166. https://doi.org/10.1126/science.1247166.

Ferguson, J., Chambers, L.A., Donnelly, T.H., Burne, R.V., 1988. Carbon and oxygen isotopic composition of a recent megapolygon-spelean limestone, fisherman bay, South Australia. Chem. Geol. Isot. Geosci. 72 (1), 63–76.

Flannery, D.T., 2013. Palaeobiology of the Neoarchean Fortescue Group (doctoral thesis). University of New South Wales.

Flannery, D.T., Walter, M.R., 2011. Archean tufted microbial mats and the Great Oxidation Event: new insights into an ancient problem. Aust. J. Earth Sci. 59 (1), 1–11.

Flannery, D., Allwood, A., Van Kranendonk, M.J., 2016. Lacustrine facies dependence of highly 13C-depleted organic matter during the global age of methanotrophy. Precambrian Res. 285, 216–241.

Flannery, D., et al., 2018. Spatially-resolved isotopic study of carbon trapped in 3.43 Ga Strelley Pool Formation stromatolites. Geochimica et Cosmochimica Acta 223, 21–35.

Flannery, D., Allwood, A., Hodyss, R., Summons, R., Tuite, M., Walter, M., Williford, K., 2018. Microbially-influenced formation of Neoarchean ooids, Geobiology (in press).

Frantz, C.M., Petryshyn, V.A., Corsetti, F.A., 2015. Grain trapping by filamentous cyanobacterial and algal mats: implications for stromatolite microfabrics through time. Geobiology 13 (5), 409–423.

French, K.L., Hallmann, C., Hope, J.M., Schoon, P.L., Zumberge, J.A., Hoshino, Y., Peters, C.A., George, S.C., Love, G.D., Brocks, J.J., Buick, R., Summons, R.E., 2015. Reappraisal of hydrocarbon biomarkers in Archean rocks. Proc. Natl. Acad. Sci. 112, 5915–5920.

Gerdes, G., 2007. Structures left by modern microbial mats in their host sediments. In: Schieber, J., Bose, P.K., Eriksson, P.G., Banerjee, S., Sarkar, S., Altermann, W., Catuneau, O. (Eds.), Atlas of Microbial Mat Features Preserved Within the Siliciclastic Rock Record. Elsevier, Amsterdam, pp. 5–38.

Gregory, A.C., Gregory, F.T., 1884. North-West Coast. J. Aust. Explor.

Grobler, N., Meakins, A., 1988. Comparison between the Fortescue Group and the Ventersdorp Supergroup. Extended Abstracts of Geocongress, vol. 88, pp. 211–214.

Grotzinger, J.P., Sumner, D.Y., Kah, L.C., Stack, K., Gupta, S., Edgar, L., Rubin, D., Lewis, K., Schieber, J., Mangold, N., Milliken, R., 2014. A habitable fluvio-lacustrine environment at Yellowknife Bay, Gale Crater, Mars. Science 343 (6169), 1242777. https://doi.org/10.1126/science.1242777pmid:24324272.

Hagadorn, J.W., Bottjer, D.J., 1997. Wrinkle structures: microbially mediated sedimentary structures common in subtidal siliciclastic settings at the Proterozoic-Phanerozoic transition. Geology 25 (11), 1047–1050.

Handford, C.R., Kendall, A.C., Prezbindowski, D.R., Dunham, J.B., Logan, B.W., 1984. Salina-margin tepees, pisoliths, and aragonite cements, Lake MacLeod, Western Australia: their significance in interpreting ancient analogs. Geology 12 (9), 523–527.

Hayes, J.M., 1983. Geochemical evidence bearing on the origin of aerobiosis, a speculative hypothesis. In: Schopf, J.W. (Ed.), Earth's Earliest Biosphere: Its Origin and Evolution. Princeton University Press, Princeton, NJ, pp. 291–301.

Hayes, J.M., 1994. Global methanotrophy at the archean-proterozoic transition. In: Bengston, S. (Ed.), Early Life on Earth (Nobel Symposium). Columbia University Press, New York, pp. 200–236.

Hayes, J.M., Kaplan, I.R., Wedeking, K.W., 1983. Precambrian organic geochemistry: preservation of the record. In: Schopf, J.W. (Ed.), Earth's Earliest Biosphere: Its Origin and Evolution. Princeton Univ. Press, Princeton, pp. 93–135.

Hoshino, Y., Flannery, D., Walter, M., George, S., 2014. Hydrocarbons preserved in a$\sim$ 2.7 Ga outcrop sample from the Fortescue Group, Pilbara Craton, Western Australia. Geobiology 13 (2), 99–111.

Hurowitz, J.A., Grotzinger, J.P., Fischer, W.W., McLennan, S.M., Milliken, R.E., Stein, N., Vasavada, A.R., Blake, D.F., Dehouck, E., Eigenbrode, J.L., Fairén, A.G., Frydenvang, J., Gellert, R., Grant, J.A., Gupta, S., Herkenhoff, K.E., Ming, D.W., Rampe, E.B., Schmidt, M.E., Siebach, K.L., Stack-Morgan, K., Sumner, D.Y., Wiens, R.C., 2017. Redox stratification of an ancient lake in Gale crater, Mars. Science 356, eaah6849. https://doi.org/10.1126/science.aah6849.

Karpeta, W., 1989. Bedded cherts in the Rietgat Formation, Hartbeesfontein, South Africa; a late Archaean to early Proterozoic magadiitic alkaline playa lake deposit? S. Afr. J. Geol. 92 (1), 29–36.

Kazmierczak, J., Goldring, R., 1978. Subtidal flat-pebble conglomerate from the Upper Devonian of Poland; a multiprovenant high-energy product. Geol. Mag. 115 (5), 359–366.

Kendall, C., Warren, J., 1987. A review of the origin and setting of tepees and their associated fabrics. Sedimentology 34 (6), 1007–1027.

Knoll, A.H., Golubic, S., Green, J., Swett, K., 1986. Organically preserved microbial endoliths from the late Proterozoic of East Greenland. Nature 321 (6073), 856–857.

Kremer, B., Kazmierczak, J., Łukomska-Kowalczyk, M., Kempe, S., 2012. Calcification and silicification: fossilization potential of cyanobacteria from stromatolites of Niuafo'ou's Caldera Lakes (Tonga) and implications for the early fossil record. Astrobiology 12 (6), 535–548.

Lepot, K., Benzerara, K., Philippot, P., 2011. Biogenic versus metamorphic origins of diverse microtubes in 2.7 Gyr old volcanic ashes: multi-scale investigations. Earth Planet. Sci. Lett. 312 (1–2), 37–47.

Lipple, L.S., 1975. Definitions of New and Revised Stratigraphic Units of the Eastern Pilbara Region. Western Australia Geological Survey.

Logan, B.W., Hoffman, P., Gebelein, C.D., 1974. Algal Mats, Cryptalgal Fabrics, and Structures. Hamelin Pool, Western Australia.

Maitland, A.G., 1904. Preliminary report on the geological features and mineral resources of the Pilbara Goldfield. Bull. Geol. Surv. W. Aust. 15, p. 1906.

Maitland, A.G., 1915. The geology of Western Australia. Bull. geol. Surv. West. Aust. 64, 79–91.

Martel, A., Gibling, M., 1991. Wave-dominated lacustrine facies and tectonically controlled cyclicity in the Lower Carboniferous Horton Bluff Formation, Nova Scotia, Canada. Lacustrine Facies Analysis: International Association of Sedimentologists Special Publication 13, 223–243.

McLoughlin, N., Furnes, H., Banerjee, N.R., Staudigel, H., Muehlenbachs, K., de Wit, M., van Kranendonk, M.J., 2008. Micro-bioerosion in volcanic glass: extending the ichnofossil record to Archaean basaltic crust. In: Current Developments in Bioerosion. Springer, Heidelberg, pp. 371–396.

McMahon, S., Bosak, T., Grotzinger, J.P., Milliken, R.E., Summons, R.E., Daye, M., Newman, S.A., Fraeman, A., Williford, K.H., Briggs, D.E.G., 2018. A Field Guide to Finding Fossils on Mars. JGR Planet. (in Review).

Monty, C., 1976. The origin and development of cryptalgal fabrics. In: Walter, M.R. (Ed.), Stromatolites. Developments in Sedimentology, Amsterdam, Oxford, New York, pp. 193–249.

Mustard, J., Adler, M., Allwood, A., Bass, D., Beaty, D., Bell, J., Brinckerhoff, W., Carr, M., Des Marais, D., Drake, B., Edgett, K., Grant, J., Milkovich, S., Ming, D., Murchie, S., Onstott, T., Ruff, S., Sephton, M., Steele, A., Treiman, A., 2013. Report of the Mars 2020 Science Definition Team. Mars Explor. Progr. Anal. Gr., 155–205.

Myrow, P.M., Taylor, J.F., Miller, J.F., Ethington, R.L., Ripperdan, R.L., Allen, J., 2003. Fallen arches: dispelling myths concerning Cambrian and Ordovician paleogeography of the Rocky Mountain region. Geol. Soc. Am. Bull. 115 (6), 695–713.

Myrow, P.M., Tice, L., Archuleta, B., Clark, B., Taylor, J.F., Ripperdan, R.L., 2004. Flat-pebble conglomerate: its multiple origins and relationship to metre-scale depositional cycles. Sedimentology 51 (5), 973–996.

Nelson, D.R., Trendall, A.F., de Laeter, J.R., Grobler, N.J., Fletcher, I.R., 1992. A comparative study of the geochemical and isotopic systematics of late archaean flood basalts from the pilbara and kaapvaal cratons. Precambrian Res. 54 (2–4), 231–256.

Packer, B.M., 1990. Sedimentology, Paleontology and Isotope-Geochemistry of Selected Formations in the 2.7 Billion-Year-Old Fortescue Group, Western Australia. University of California, Los Angeles.

Plee, K., Ariztegui, D., Martini, R., Davaud, E., 2008. Unravelling the microbial role in ooid formation—results of an in situ experiment in modern freshwater Lake Geneva in Switzerland. Geobiology 6 (4), 341–350.

Reineck, H.-E., Singh, I.B., 1972. Genesis of laminated sand and graded rhythmites in storm-sand layers of shelf mud. Sedimentology 18 (1–2), 123–128.

Ricketts, B.D., Donaldson, J.A., 1979. Stone rosettes as indicators of ancient shorelines; examples from the Precambrian Belcher Group, Northwest Territories. Can. J. Earth Sci. 16, 1887–1891.

Rye, R., Holland, H.D., 2000. Life associated with a 2.76 Ga ephemeral pond? Evidence from Mount Roe #2 paleosol. Geology 28 (6), 483–486.

Sakurai, R., Ito, M., Ueno, Y., Kitajima, K., Maruyama, S., 2005. Facies architecture and sequence-stratigraphic features of the Tumbiana Formation in the Pilbara Craton, northwestern Australia: implications for depositional environments of oxygenic stromatolites during the Late Archean. Precambrian Res. 138 (3–4), 255–273.

Schopf, J.W., Walter, M., 1983. Archean microfossils: new evidence of ancient microbes. In: Schopf, J.W. (Ed.), Earth's Earliest Biosphere: It's Origins and Evolution. Princeton University Press, Princeton, pp. 214–239.

Shepard, R.N., Sumner, D.Y., 2010. Undirected motility of filamentous cyanobacteria produces reticulate mats. Geobiology 8 (3), 179–190.

Shinn, E.A., 1968. Practical significance of birdseye structures in carbonate rocks. J. Sediment. Res. 38 (1), 215–223.

Smith, R.E., Perdrix, J.L., Parks, T.C., 1982. Burial metamorphism in the Hamersley Basin, Western Australia. J. Petrology 23 (1), 75–102.

Squyres, S.W., Arvidson, R.E., Bell, J.F., Brückner, J., Cabrol, N.A., Calvin, W., Carr, M.H., Christensen, P.R., Clark, B.C., Crumpler, L., Des Marais, D.J., 2004. The opportunity Rover's Athena science investigation at Meridiani Planum, Mars. Science 306 (5702), 1698–1703.

Strik, G., Blake, T.S., Zegers, T.E., White, S.H., Langereis, C.G., 2003. Palaeomagnetism of flood basalts in the Pilbara Craton, Western Australia: Late Archaean continental drift and the oldest known reversal of the geomagnetic field. J. Geophys. Res. 108 (B12), 2551.

Summons, R., Bird, L., Gillespie, A., Pruss, S., Roberts, M., Sessions, A., 2013. Lipid biomarkers in ooids from different locations and ages: evidence for a common bacterial flora. Geobiology 11 (5), 420–436.

Sumner, D.Y., 1997. Late Archean calcite-microbe interactions: two morphologically distinct microbial communities that affected calcite nucleation differently. PALAIOS 12 (4), 302–318.

Swirydczuk, K., Wilkinson, B.H., Smith, G.R., 1979. The Pliocene Glenns Ferry oolite; lake-margin carbonate deposition in the southwestern Snake River plain. J. Sediment. Res. 49 (3), 995–1004.

Talbot, H.W.B., 1919. Notes on the geology and mineral resources of parts of the North-West, Central and Eastern Divisions. Western Australia Department of Mines, Annual Report for 1918, 83–93.

Thomazo, C., Ader, M., Philippot, P., 2006. Organic δ13C negative excursion at 2.7 Ga viewed along a 100m depth drill-core profile (Tumbiana Formation Western Australia). Geophys. Res. Abstr. 8, 09013.

Thomazo, C., Ader, M., Farquhar, J., Philippot, P., 2009. Methanotrophs regulated atmospheric sulfur isotope anomalies during the Mesoarchean (Tumbiana Formation, Western Australia). Earth Planet. Sci. Lett. 279 (1–2), 65–75.

Thomazo, C., Ader, M., Philippot, P., 2011. Extreme 15N-enrichments in 2.72-Gyr-old sediments: evidence for a turning point in the nitrogen cycle. Geobiology 9, 107–120.

Thorne, A., Trendall, A.F., 2001. Geology of the Fortescue Group, Pilbara Craton, Western Australia. 144 Geological Survey of Western Australia, Department of Minerals and Energy.

Vaniman, D.T., Bish, D.L., Ming, D.W., Bristow, T.F., Morris, R.V., Blake, D.F., Chipera, S.J., Morrison, S.M., Treiman, A.H., Rampe, E.B., Rice, M., 2013. Mineralogy of a mudstone at Yellowknife Bay, Gale crater, Mars. Science, 1243480.

Walter, M.R., 1972. Stromatolites and the biostratigraphy of the Australian Precambrian and Cambrian. Special Papers in Palaeontology, The Palaeontological Association, London.

Walter, M.R., 1983. Archean stromatolies: evidence of the Earth's earliest benthos. In: Schopf, J.W. (Ed.), Earth's Earliest Biosphere. Princeton University Press, Princeton.

Walter, M.R., Bauld, J., Brock, T.D., 1976. Microbiology and morphogenesis of columnar stromatolites (Conophyton, Vacerrila) from hot springs in Yellowstone National Park. In: Walter, M.R. (Ed.), Stromatolites. Elsevier, New York, pp. 273–310.

Warren, J.K., 1982. The hydrological significance of Holocene tepees, stromatolites, and boxwork limestones in coastal salinas in South Australia. J. Sediment. Res. 52 (4), 1171–1201.

Wilkinson, B.H., Owen, R.M., Carroll, A.R., 1985. Submarine hydrothermal weathering, global eustasy, and carbonate polymorphism in phanerozoic marine oolites. J. Sediment. Res. 55 (2), 171–183.

Winter, H., 1963. Algal structures in the sediments of the Ventersdorp system. Trans. Geol. Soc. S. Afr. 66, 115.

Woodward, H.P., 1891. Annual Report of the Government Geologist, 1890. Geological Survey of Western Australia.

Zhang, Y., Golubic, S., 1987. Endolithic microfossils (cyanophyta) from early Proterozoic stromatolites, Hebei, China. Acta Micropaleontol. Sin. 4, 1–12.

FURTHER READING

Ferralis, N., Matys, E.D., Knoll, A.H., Hallmann, C., Summons, R.E., 2016. Rapid, direct and non-destructive assessment of fossil organic matter via microRaman spectroscopy. Carbon 108, 440–449.

CHAPTER 6

Evolution of Altiplanic Lakes at the Pleistocene/Holocene Transition: A Window Into Early Mars Declining Habitability, Changing Habitats, and Biosignatures

Nathalie A. Cabrol, Edmond A. Grin, Pierre Zippi, Nora Noffke, Diane Winter

Contents

6.1 OVERVIEW

Lakes on early Mars could have provided favorable environments for the development of microbial life and for its preservation in the fossil record. While they ultimately disappeared as atmospheric changes were taking place, the evolution of their habitability would have controlled the ability of any life present to transition to the newly developing environments (i.e., by adapting to more protected habitats that offered radiation refuge and moisture retention). A better understanding of the magnitude and impact of environmental stress on lake microbial habitats during climate transition is thus an essential component in the development of exploration strategies for future missions that will search for traces of biosignatures on Mars, like NASA's Mars 2020 and ESA's ExoMars.

Here, the investigation of Laguna Verde in south Bolivia documents environmental change over a period of ~7000 years at the Pleistocene/Holocene transition and provides

From Habitability to Life on Mars
https://doi.org/10.1016/B978-0-12-809935-3.00006-2

a window in time into analog processes that could have taken place at the Noachian/ Hesperian transition in martian lakes. Radiocarbon dating of microbialites and other geologic samples collected on the paleoterraces and shores constrains the timing of peak water volume during the Late Pleistocene, followed by increased water loss from 8230 to 6300 BP. Episodic interactions between volcanic/hydrothermal and lacustrine activity and changes in the chemical and physical characteristics of the water column are recorded in the morphology, geochemistry, and composition of the samples and in the fossil record.

6.2 INTRODUCTION

Lakes are sentinels of climate change (Williamson et al., 2009). Over time, sequences of sediments may record both environmental tipping points at planetary scale and subtle watershed fluctuations. Rapid burial of dead organisms in sediments provides anoxic conditions favorable to the preservation of a fossil record. Within that context and using an analogy to Earth, lakes on early Mars would have been favorable sites for the inception and the development of life and for the potential preservation of its record (see *Lakes on Mars*, Cabrol and Grin, eds., and references therein). For these reasons, ancient lakes have been considered priority candidate landing sites for rover missions characterizing the past habitability of Mars and for upcoming missions such as Mars 2020 and ExoMars. In support of the preparation for these missions to seek traces of ancient life, various aspects of extreme lakes, aqueous environments, and biosignatures have been studied in terrestrial analogs over the years (e.g., Farmer and Des Marais, 1999; Cady et al., 2003; Knoll and Grotzinger, 2006; Cabrol et al., 2003, 2005, 2007a,b, 2009; Cabrol and Grin, 2010 and references therein; Noffke 2009, 2014; Fairén et al., 2010; Davila and Schulze-Makuch, 2016; Hays et al., 2017; Pontefract et al., 2017; Koo et al., 2017a,b). Recently, a study on macroscopic morphologies preserved in the >3.7 Ga-old Gillespie Lake Member on Mars demonstrated the possible steps of exploration for features presenting analogies with microbially induced sedimentary structures (MISS, Noffke, 2014). On Earth, such microbial structures are caused by interaction of benthic microbes with physical sediment dynamics (erosion and deposition of sediment). This interaction generates meter- to decimeter-scale sedimentary structures. In contrast to microbialites, where chemical processes play a role in their formation, the microbial-physical origin of the MISS makes them prime candidates for life exploration.

Here, while Chapter 5 showed how Neoarchean lakes can inform us on ancient environmental conditions and habitability of early Earth and Mars, in this chapter, we illustrate the evolution of Laguna Verde, an altiplanic lake that provides a modern analog to early martian lake habitats. Its investigation allows us to examine the impact of environmental change between the Late Pleistocene and the Early Holocene as a wet climate transitioned into arid conditions in a relatively short period of time. Data show how major and minor changes were encapsulated into a stratigraphic section of ancient terraces covering 7000 years and how some of the physical, geochemical, environmental, and biological signatures may provide relevant insights into the Noachian/Hesperian transition on Mars.

6.3 ENVIRONMENTAL SETTING

The altiplano is a high plateau reaching 3500–4200 m altitude. Its arid core (18–28S) is characterized by a volcanic, sulfur-rich environment; hydrothermal springs and deposits; evaporating lakes; playas; dune fields; and other aeolian deposits. Dry lakes (salars) and evaporating lakes (lagunas) give an opportunity to measure in situ the impact of rapid climate change on microbial life and on the morphology, mineralogy, and chemistry of microbial habitats. Because of elevation, aridity, and latitude, the environmental shift that started at the end of the last glacial maximum (LGM) took place in a thin, unstable atmosphere and extreme ultraviolet (UV) radiation—the highest UV recorded at the surface of our planet to date (Cabrol et al., 2014; Feister et al., 2015). This environment is also characterized by strong and sudden daily temperature (T) fluctuations that generate high UV/T ratios (Table 6.1). Both high UV and sharp temperature swings would have been common with the thinning atmosphere on Mars at the Noachian/Hesperian transition.

Table 6.1 Physical and environmental characteristics
Environmental characteristics

Elevation (m)	4340
Atmospheric pressure (mb)	550
Air temperature (°C)	−30/+13
Relative humidity (%)	9–25
Snow precipitation (mm/year)	100
Wind (km/h)	0–100
UV index max[a]	29 (4300 m), 43.3 (5914 m)
Volcanic environment	Basaltic andesite
Sedimentary environment	Carbonate, sulfate
Hydrothermal springs (°C)	12–36

Lake physical characteristics	Laguna Blanca	Laguna Verde
Water temperature (°C)	+14	+15
pH	7.2–8.42	8.19–9.0
Freezing point of water (°C)	−5	−25
Maximum depth (m)	0.5	5
Ice cover	Variable year round	Rare
Total dissolved solids (TDS, mg/L)[b]	22,400	~ 50,000

[a]Based on yearly measurements from Eldonet UV dosimeters and environmental stations placed on the shore (Cabrol et al., 2009, 2014).
[b]Geochemistry from water samples (Hock et al., 2005).

Most altiplanic lakes were formed at the end of the LGM, as demonstrated by the sedimentary record and climate studies performed in the Cordillera Real in northern Bolivia (e.g., Gouze et al., 1986; Seltzer 1990, 1992; Francou et al., 1995; Seltzer et al., 1995; Valero-Garces et al., 1996; Abbott et al., 1997; Clapperton et al., 1997; Baucom and Rigsby 1999; Grosjean et al., 2000), near Lake Titicaca and Lake Poopo (e.g., Servant and Fontes, 1978; Wirrmann and Fernando De Oliveira Almeida, 1987; Rondeau, 1990; Wirrmann et al., 1990; Risacher and Fritz, 1991; Roche et al., 1992; Wirrmann and Mourguiart, 1995; Seltzer et al., 2002), and in the Central and South Central Andes (Risacher and Fritz, 1991; Rouchy et al., 1996; Blodgett et al., 1997; Clayton et al., 1997; Grosjean et al., 1998; Grosjean 2001; Fornari et al., 2001).

Comparatively, the southern part of Bolivia is less documented (e.g., Rondeau, 1990; Risacher and Fritz, 1991; Sylvestre et al., 1999; Cabrol et al., 2009, 2010a,b, 2014; Cabrol and Grin, 2010; Hock et al., 2005; Demergasso et al., 2010; Fleming and Prufert-Bebout, 2010). In addition to glacial melting, most lakes formed under a precipitation regime of 400–500 mm/year (e.g., Messerli et al., 1993; Wirrmann and Mourguiart, 1995; Abbott et al., 1997; Baucom and Rigsby, 1999; Sylvestre et al., 1999; Vuille et al., 2000; Grosjean, 2001). They are currently receding under negative water balance (e.g., Laguna Colorada, P/Ev = 160/500 mm/year—unpublished data, Bolivian National Park Services).

Located in the Potosi region (south Bolivia), Laguna Verde is typical of the evaporating lakes of this modern arid altiplanic regime. Because of evaporation and topography, the original ($\sim$27 × 11 km) Late Pleistocene lake is now $\sim$10 × 4 km and in the process of dividing into two bodies of water with distinct physicochemical characteristics: Laguna Blanca to the east (22°47.00′S/67°47.00′W) and Laguna Verde to the west (22°47.32′S/67°49.16′W); see Fig. 6.1. Both lakes are still connected by a channel a few meters long and $\sim$1 m wide. Laguna Verde is 5 m deep and Laguna Blanca 0.50 m. The elevation of the earliest lacustrine deposits and a survey of the modern lake through sonar sounding show that the paleolake was at least 45 m deeper than the residual lakes and thus close to 50 m (Cabrol et al., 2003, 2009).

Despite the connecting channel, both lakes present notable differences, which affect the diversity and abundance of present-day life in each basin. The physicochemical and biological characteristics of active lakes in this region, including Laguna Blanca and Laguna Verde, were documented as part of the NASA Astrobiology Institute (NAI)-funded High Lakes Project (e.g., Acs et al., 2003; Cabrol et al., 2007a,b, 2009, 2010a,b, 2014; Dorador et al., 2009; Demergasso et al., 2010; Fleming and Prufert-Bebout, 2010; Feister et al., 2015). Here, we discuss unpublished data collected during the investigation of a stratigraphic section of Laguna Verde that records the end of the Pleistocene and the beginning of the Holocene.

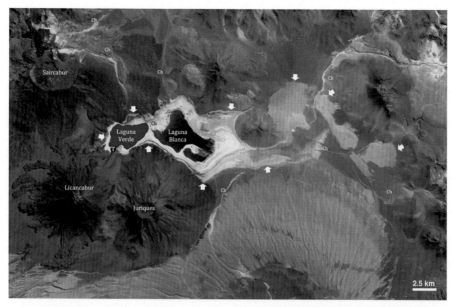

Fig. 6.1 Regional context—the interaction between lacustrine and volcanic activity is illustrated in this advanced spaceborne thermal emission and reflection radiometer (ASTER) image. The margins of the Late Pleistocene Laguna Verde are marked by *white arrow* and covered ~ 27 km from west (left) to east. Many of the early lake sediments are covered by more recent volcanic and alluvial fan deposits on the east shore, while the west shore is characterized by an intense erosion of the bases of volcanic constructs (e.g., Licancabur, Sairecabur, and Juriques). *Ch*, main channels draining into the current lakes; *red dot*, location of the transect. *(Credit image: NASA.)*

6.4 VOLCANIC/HYDROTHERMAL ACTIVITY

Hydrothermal springs and a limited amount of snow precipitation control the water level of Laguna Blanca and Laguna Verde (Table 6.1 and Cabrol et al., 2009). A strong negative water balance has led to a noticeable retreat of the shorelines for both lakes every year during the time of the study. At current rate and without engineering or modification in the input of hydrothermal springs, the two lakes could become totally separated in just a few years.

Hydrothermal activity is generated through interactions between aquifers and ongoing local and regional magmatic and volcanic activity. Small, closed, and now dry pools on the ancient terraces indicate that hydrothermal springs contributed to the lake water balance during the early Holocene. The combined basin is surrounded by volcanoes towering over 6000 m elevation, many of them dormant (e.g., Licancabur) and some still active today. Located ~ 70 km to the south, Lascar is one of the most active volcanoes of the Andes. Eruptions take place nearly every 7 years (the latest in 2016), and the volcano has daily displays of water vapor and sulfur emissions. About 60 km to the north, Putana has continuous geyser activity. After 10,000 years of inactivity, a new hot spot was identified in the ASTER satellite imagery over Chiliquès in 2002 (unpublished data, JPL news release). Chiliquès belongs to the same volcanic chain as Licancabur, Lascar, and Putana. Regionally, geothermal centers (e.g., El Tatio in Chile that extends into Bolivia at Sol de Mañana), hydrothermal springs, geysers, and mud pots is evidence of abundant residual heat (Fig. 6.2). Additional signs of continuous volcanic/hydrothermal activity at Laguna Verde include new fumaroles and hydrothermal springs observed nearshore at the foot of the Sairecabur volcano and bubbling in the summit lake of the Licancabur volcano in 2006, with water in the south shore reaching +14°C, while the water temperature of the lake is 4°C on average (Cabrol et al., 2009).

Fig. 6.2 Hydrothermal and volcanic activity—the region surrounding Laguna Verde is characterized by intense hydrothermal and volcanic activity. (A) Spring mound ($\sim$ 1 m high) in the geothermal field of El Tatio, in Chile, which is the third largest geothermal center in the world after Yellowstone (the United States), and Dolina Giezerov (Russia). (B) The energy center that supplies El Tatio extends to the east in Bolivia, where most of the activity occurs as mud pots. (C) Hydrothermal springs on the north shore of Laguna Verde. (D) Emissions of water vapor and sulfur from the Lascar volcano in November 2016. *(Credit photos: NASA Astrobiology Institute, the SETI Institute, and Victor Robles, SETI Institute/Campoalto (D).)*

6.5 STRATIGRAPHIC RECORD

Ancient microbialites cover the Pleistocene shores and terraces over an area of $\sim 100\,km^2$, while modern cyanobacteria-dominated colonies are still forming small ($\sim \leq 10\,cm$) structures and mats in shallow waters (Cabrol et al., 2009; Fleming and Prufert-Bebout, 2010; Demergasso et al., 2010). Various sizes, shapes, and ages of these Late Pleistocene to Early Holocene microbialites mark changes in the dynamics and physico-chemical evolution of the paleolake environment over time (Fig. 6.3). They present remarkable similarities with the modern stromatolite systems described in Shark Bay, Australia (Suosaari et al., 2016), in morphogenesis, size, composition, and microbial systems (see next sections). They are distributed in distinct provinces around the paleolake, suggesting that location (shoreward to lakeward) and currents may have affected the microbial mat types involved and their growth pattern.

The boundaries of the Late Pleistocene lake were identified with the 15–90 m/pixel resolution ASTER Vis/NIR/TIR 15-channel satellite imagery (Fig. 6.1) and global

Fig. 6.3 Microbialites—the paleoshores of what are now Laguna Verde and Laguna Blanca are covered with microbialites of distinct shapes and sizes distributed in provinces around the paleolake. Biostructures similar to (A–C) in size and shape have been described in Shark Bay, Australia, by Suosaari et al. (2016), as characteristics of a shoreward to lakeward distribution. The microbialites shown in (A) and (B) are about 40 cm wide, and the elongated structures in (C) can reach 3 m long and 50–70 cm high. (D) Lava flows capping the remains of microbialites at the foot of the Juriques volcano. (E) Remnants of dome structures (50–70 cm high on average) on the east paleoshore. *(Credit photos: High Lakes Project, NASA Astrobiology Institute, the SETI Institute.)*

positioning system (GPS). A transect was performed on the paleoterraces of Laguna Verde south shore at a location where stratigraphy was well preserved. The earliest record of lacustrine activity was found at 4378 ± 13 m elevation, where deposits come in contact with lava flows. The stratigraphic sequence down to the most recent material (4340 ± 5 m) consists mostly of fossiliferous carbonates and calcareous sediments intercalated with basaltic andesite. Other samples were collected at Laguna Blanca at discrete sites along the paleoshores in order to compare the evolution of rock major element oxides and correlate species and environment across the paleolake at radiocarbon ages comparable with those of Laguna Verde (Tables 6.2 and 6.3).

Terrace elevations and ^{14}C show a maximum water level of ~50 m at the end of Pleistocene (13,270 BP) and a decline of the lake level in mid-Holocene (8290–6300 BP), with an average loss of 24 m in 2000 years (1.2 mm/year) compared with 0.28 mm/year 13,240–8290 BP. The focus of the stratigraphic analysis on the main terraces did not allow the identification of small-scale environmental fluctuations within that trend, but data support previous studies showing a transition from a Late Pleistocene humid phase to an increasingly arid climate from the early Holocene to the present time (e.g., Messerli et al., 1993; Abbott et al., 1997; Sylvestre et al., 1999).

6.6 GEOSIGNATURES

Petrographic analysis was used to determine the textural and mineralogical diversity, to identify silicates, carbonates, oxides, and hydroxides and to make a first-order characterization of the fossil record. Volcanic samples consisted of basaltic andesite, which is typical of the local volcanism and regional deposits (Marinovic and Lahsen, 1984). Fossiliferous carbonates were the most common rock types (68%), with calcareous sediments making 16% of the samples and then andesite (12%) and agglomerates mixing carbonate and volcanic fragments. Samples from both Laguna Blanca and Laguna included relatively coarse (up to 600 μm) clastic sediments intercalated with fossiliferous carbonates. Mineralogy was predominantly that of angular quartz, lesser plagioclase, amphibole, pyroxene, biotite, oxide, and hydroxide (magnetite and hematite).

Overall, data from the stratigraphic column and the petrographic micrographs confirm the interaction between volcanic and lacustrine activity in a nearshore environment (Fig. 6.4) especially at the time when the paleolake had reached its peak volume 13,270 BP. The fossil record (diversity and abundance) shows comparable changes in habitat over time between what are now Laguna Blanca and Laguna Verde (Fig. 6.5). Fluctuations occurred in both sample series at comparable ages, consistent with an overall response of the large lake to environmental changes, indicating that the paleolake had not undergone significant physicochemical separation yet during that time frame.

Table 6.2 Age and characteristics of the stratigraphic column at Laguna Verde

Radiocarbon Age[a] (years, BP)	Sample weight (mg) and type	Sample ID	Description
Laguna Verde Series			
		LV1	Volcanic material
		LV2	Volcanic material
		LV4	Mixed volcanic/lacustrine unit. Higher concentration of small (1–10 cm) volcanic rocks, gravel, and sand. Two shorelines with no compositional difference
13,270 ± 100	527 microbialite	LV6	Volcanic material exposure: three to four exposed flows, possible shoreline over one vertical meter
12,250 ± 90	667 microbialite	LV9	Five-meter terrace descending toward current shoreline. High density of fossil microbialites, very eroded, dark brown to light tone. Lake–bed material
12,070 ± 160	899 fossiliferous carbonate	LV8	Lake bed with submillimeter- to millimeter-scale layering. Pitted material, light-toned to yellow/brown. Mottled maize-colored material, similar to that of LV5. Major terrace (2 m). Small microbialite structure
11,470 ± 90	591 microbialite	LV5	Mixed volcanic/lacustrine unit. More sand. Mottled, patterned, maize-colored lake–bed material. Material similar in morphology, tone, and color to that observed in Laguna Blanca
8290 ± 70	723 microbialite	LV3	Dark volcanic material. Blocks embayed by lacustrine material (light-toned). Shorelines
10,320 ± 90	1000 microbialite	LV10	One-meter–high terrace. Microbialites present but less abundant. Exposure of dark lava and mud-like material. Terrace extends to 25 m in the direction of the current shoreline
10,270 ± 80	577 microbialite	LV7	Platy, lacustrine material; light-toned, granular; gypsum. Some curved or domed structures. Evidence of secondary mineralization
6300 ± 70	6900 calcareous sediment	LV11	Level connecting with the current level of Laguna Verde. Volcanics and salt deposits. Microbialites are less abundant. Fossil algae

[a]Note that the sample ages are quoted as uncalibrated conventional radiocarbon dates in before present (BP) using the Libby ^{14}C mean life of 8033 years (Reimer et al., 2004). All solutions, with a probability of 50% or greater for the calibrated age of resulting radiocarbon ages, were calculated from the dendrocalibration data. The 68% and 95% confidence intervals, which are the 1σ and 2σ limits for a normal distribution, were also calculated. A probability of 100% was determined when the radiocarbon data intersected the dendrocalibration curve at this age. All results are rounded to the nearest multiple of 5 years.

Table 6.3 Age and characteristics of Laguna Blanca samples

Radiocarbon Age[a] (years, BP)	Sample weight (mg) and type	Sample ID	Description
Laguna Blanca series			
11,540 ± 110	293 microbialite	LB2	Dome structures 0.45 m high, round, a few elongated (0.40–0.70 m). High density, light-toned
10,650 ± 110	257 microbialite	LB4	Margin of volcanic and lake shore material. Horizontal laminations over 10 cm. Light-toned to lightly reddish. Associated with boulders
10,590 ± 100	464 microbialite	LB9	Microbialite
9960 ± 100	335 microbialite	LB5	Dome structures; highly eroded and disintegrating into sand- to granule-size particles. Laminations well preserved ranging from 1 to 7 mm in thickness. Light-toned to reddish
9210 ± 100	338 microbialite	LB7	Elongated structures, poorly preserved. High density; size <1 m long and 0.50 m wide; 0.50 m high; mixed with volcanic rocks and anchored on volcanic material; brownish–gray texture, pitted
8440 ± 80	4900 microbialite	LB1	Tall (1.2 m) and large structures (15 × 5 m) on average, elongated, eroded, high density, brown
8250 ± 90	279 microbialite	LB6	Large, elongated, coalescent structures, laminated. Average height is 1.2 m, well preserved, high density, light-toned with some reddish layers
8140 ± 90	319 microbialite	LB8	Tabular, elongated structures up to 5 m long and 0.50 m high, massive terrace of microbialites, brownish to light-toned, a few eroded round structures (1 m) exposed partly, 50–60 cm wide, dark brown color
7520 ± 90	396 microbialite	LB3	Low density. A few very large specimen (5 m diameter and 1. 2 m high), light-toned

[a]Note that the sample ages are quoted as uncalibrated conventional radiocarbon dates in before present (BP) using the Libby ^{14}C mean life of 8033 years (Reimer et al., 2004). All solutions, with a probability of 50% or greater for the calibrated age of resulting radiocarbon ages, were calculated from the dendrocalibration data. The 68% and 95% confidence intervals, which are the 1σ and 2σ limits for a normal distribution, were also calculated. A probability of 100% was determined when the radiocarbon data intersected the dendrocalibration curve at this age. All results are rounded to the nearest multiple of 5 years.

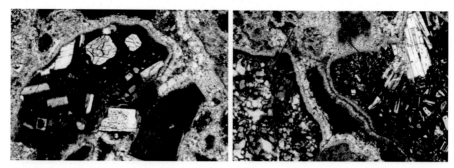

Fig. 6.4 Volcanic/carbonate interactions—two regions of sample LV6 (13,270 ± 100 BP). Left: Agglomerate consisting of volcanic fragments and mineral clasts cemented and intercalated with subrounded, framboidal carbonate fragments containing minute relict algae. The fine-grained carbonate is rimmed by overgrowth of more clear carbonate with growth bands. Right: Basaltic andesite and fine-grained carbonates. The field of view for both micrographs is 2.3 mm. *(Credits: High Lakes Project/The SETI Institute and NASA Astrobiology Institute/Biostratigraphy.com.)*

6.7 CHEMICAL AND ISOTOPIC SIGNATURES

Chemical composition and isotopic characteristics are markers of geologic, climatic, and/ or biological changes. Major and trace elements were obtained from fusion-inductively coupled plasma (FUS-ICP), total digestion ICP multielement analytic procedure, and rock analysis through X-ray fluorescence (XRF) and XRF pressed pellets.

A high CaO concentration reflects a majority of carbonate rocks on the paleoshores with the exception of the Late Pleistocene/Holocene transition, where a significant peak in silica and other oxides is observed. Previous climate models and lava-flow studies (Kull et al., 2003; Figueroa et al., 2008) have suggested that the Licancabur volcano was mostly built up after the LGM (12–10 ka). Our data independently confirm that volcanic activity related to Licancabur was taking place at the end of LGM as the oldest lacustrine sediments from Laguna Verde (13,270 ± 100) are capped by its lava flows. Another, stronger, peak in SiO_2 is observed at 12,070 BP but, unlike the first one, is limited to silica with no substantial variations in other oxides. No older deposits were observed, which either suggests that there was no lacustrine activity before that time (which would provide a boundary age for the formation of Laguna Verde) or more ancient lake sediments were eroded away or buried by subsequent deposits.

The influence of the Licancabur volcano buildup on the west paleoshore environment is further revealed by the evolution over time of the rock major element oxides. Although none of the Laguna Blanca samples are as old as Laguna Verde's oldest, both series can be compared after 11,540 BP. The youngest sample in Laguna Verde shows the highest percentage of SiO_2 (8.75 wt.%) of the Holocene samples. While not as high as the Late Pleistocene values that we attribute to Licancabur's activity (16.6–20.20 wt.%), it represents double the average seen in the Holocene samples (average 5.26 wt.%). The

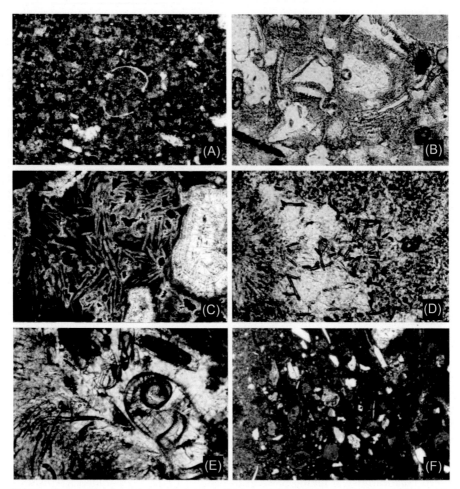

Fig. 6.5 Fossil record in changing environment: (A) curved fossil fragment in *dark*, murky carbonate matrix in deposits that show evidence of a high-energy environment and no seasonality (12,250 BP). (B) Carbonate replacement of fossil fragments (12,250 BP). (C) Carbonized fossil fragments including ostracods during a desiccation event (10,270 BP). (D) Algae fragments in clear recrystallized carbonate in an environment of stagnant waters (8440 BP). (E) Samples dominated by cyanobacteria filaments in growth bands (8290 BP). (F) Silicate clasts in murky carbonate matrix in a high-energy lake (6300 BP). See also Table 6.4 for summary of changes in lake dynamics and environment. Field of view: 2.3 mm (A and C) and 0.45 mm (B, D, and E). *(Credits: High Lakes Project/The SETI Institute and NASA Astrobiology Institute/Biostratigraphy.com.)*

same increase is noted at Laguna Blanca, where it started earlier in the Holocene (9960 BP), with temporal fluctuations (Fig. 6.6).

Major element oxide ratios significantly differ prior to 10,000 BP, while they have remained comparable for both Laguna Verde and Laguna Blanca since that period. After the end of the Pleistocene, levels of SiO_2 are on average four times lower in the samples (5 wt.%). At Laguna Blanca, the average concentration is slightly higher (6–9 wt.%)

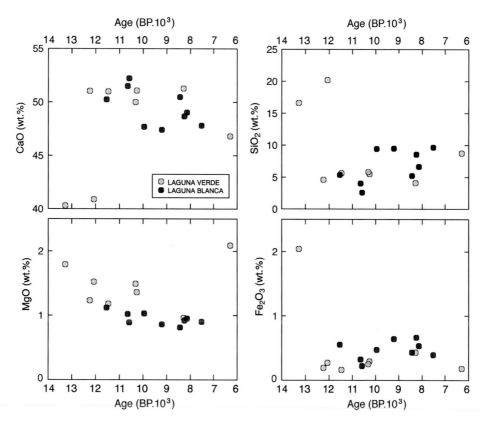

Fig. 6.6 Major element oxides—carbonates dominate the sample geochemistry. High values in SiO_2 and Fe_2O_3 during the end of the Pleistocene are attributed to volcanic activity related to the Licancabur volcano. Excepted for these peaks, similar trends over time on samples collected on what are the shores of Laguna Verde and Laguna Blanca in present times show that the early Holocene lake was still reacting to change in the basin as one unit. Higher silica content in more recent times for the LB sample series is attributed to a higher abundance of hydrothermal springs on the east shore compared with the west shore, as evidenced in the geologic record of the ancient terraces.

throughout the stratigraphic column than the average for Laguna Verde. Except for LB4 (10,650 BP), no direct contact with volcanic material was observed for the samples in that series, and in this case, the variability of SiO_2 might be related to variations in hydrothermal and biological activity over time. Hydrothermal springs are present around both Laguna Blanca and Laguna Verde, and diatoms are observed in both fossil records. Currently, they are significantly more abundant and diverse in Laguna Blanca (Acs et al., 2003; Cabrol et al., 2007a). Diatoms use SiO_2 as ballast by depositing it in their cell walls to sink more rapidly (e.g., Honjo et al., 1995). They also use silica for growth (e.g., Tilman and Kilham, 1976; Volcani, 1981), and fluctuations in their population over time may have affected the geochemical record (Fig. 6.7).

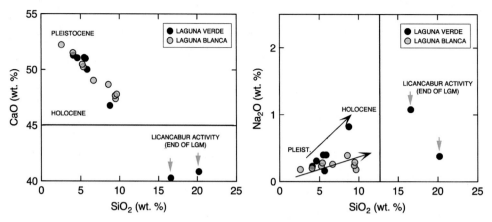

Fig. 6.7 Sample major element oxide ratios—the Pleistocene samples associated with the buildup of Licancabur appear as outliers in otherwise comparable trends for both sample series. With decreasing regional volcanic activity during the Holocene, the increase in SiO_2 versus CaO might be related to hydrothermal and/or biological activity (e.g., changes in diatom populations). The trend observed for SiO_2 versus Na_2O suggests variability in the weathering of volcanic rocks that could be associated with fluctuations in lake dynamics over time.

6.8 CHANGES IN LAKE HABITAT AND BIOSIGNATURES

The impact of environmental changes on microbial habitat and biosignatures was characterized by comparing thin sections from both lakes at close radiocarbon ages. Palynology (pollen, spores, and algae), diatoms, and other organic residues provide a window into environmental changes through the stratigraphy. Organic material from cyanobacteria in the microbialites was preserved only as relicts, mostly as brownish, organic-rich carbonate, which made the identification of main groups nearly impossible.

In the fossil record (Fig. 6.8), cuspate-shaped fragments resembling ostracods (12,070 BP) and abundant broken diatom valves (12,250 BP, Fig. 6.8A and B) show that the Late Pleistocene was dominated by a high-energy environment. Volcanic activity and the weathering of volcanic material transported into the lake are recorded at 13,270 BP and once again at the end of the LGM (11,470 BP). The abundance of clay, the intense oxidation, and the almost complete replacement of mafic minerals by amorphous iron-hydroxide suggest wet episodes and/or extensive underwater residence of the volcanic material. Layered tufted mats (Gerdes and Krumbein, 1994; Gerdes et al., 2000; Noffke, 2010) covered by coccoid cyanobacteria populations at Laguna Blanca (Fig. 6.8C) record the increasing aridity at the beginning of the Holocene (10,590 BP). Within a few-hundred-year interval (10,270 BP) on the opposite shore, fossil types had also changed at Laguna Verde when compared with the previous periods. Carbonates made of dark, microcrystalline murky domains and light clear domains (or recrystallized and second-

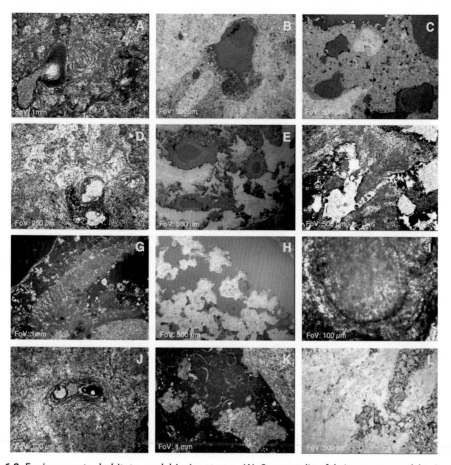

Fig. 6.8 Environments, habitats, and biosignatures: (A) Stromatolite fabric constructed by in situ networks of cyanobacterial filaments. These networks include many broken diatom skeletons, showing a high-energy environment. No seasonal variation recorded. (B) Details of diatom fragments baffled, trapped, and bound by cyanobacteria. (C) Network of in situ lithified cyanobacteria visible in the *brownish*, organic-rich carbonate and in the older, fully lithified, and lighter portion of the stromatolite. (D) The stromatolite shows baffling, trapping, and binding of fine sand material (e.g., quartz grains). In situ lithification of the cyanobacteria can be seen within the *brownish-colored*, organic-rich carbonate portion. No seasonal variation recorded. (E) Seasonality recorded in the growth pattern of the stromatolite. Two mat generations are distinguished, one composed of filamentous cyanobacteria and the other of vertically oriented filaments. This suggests that growth stopped at times during which the cyanobacterial filaments oriented themselves parallel to the surface. (F and G) Growth pattern (from E) with a tufted mat (F) and thick, vertically oriented filaments (G). (H) The stromatolites are formed by coccoid and filamentous cyanobacteria. Several generations of cements are visible. The pattern of lithification records seasonal variations in the chemical composition of the lake. Here, the cement layers are of even thickness, thus recording the formation of the stromatolite in almost stagnant water. (I) No main direction of water current is visible. Episodic desiccation events created the *reddish-brownish* seams visible in the carbonate matrix. The stromatolite is very porous, which documents a high gas production by the cyanobacteria. Gas bubbles are visible within the brownish, organic-rich carbonates and in the fully lithified portions of the stromatolite. Most of these gas bubbles contained CO_2. (J) Filamentous cyanobacteria. (K) The development of the stromatolite came to an end. (L) This stromatolite was formed in an environment of stronger water currents. The carbonate fabric is constructed of fragments of coarse material in hollows (e.g., ostracod shell fragments). The sample shows baffling, trapping, and binding of the coarse material.

generation carbonates) emphasize these changing conditions and their impact of microbial ecosystems: In this environment, sinuous algae are missing from the fossil record, while worm burrows are now abundant. Evidence of baffling, trapping, and binding of fine sand material, such as oriented quartz grains (Noffke et al., 1997), is found 9960 BP, with in situ lithification of cyanobacteria (genus *Lyngbya*) within the brownish-colored, organic-rich carbonate portion of the sample (Fig. 6.8D).

Seasonality is detected for the first time 9210 BP (Fig. 6.8E) in the growth pattern of the microbialites through the presence of two generations of mats both in the organic-rich carbonate margins and in the fully lithified fabrics: one is composed of a tufted mat made of filamentous cyanobacteria forming bundled pinnacles and the other of thick, vertically oriented filaments. Fig. 6.8F and G shows both mat generations. Cyanobacterial filaments oriented parallel to the surface indicate that the microbialite growth stopped at times (e.g., Browne et al., 2000; Noffke et al., 2003).

Over a thousand years later (8440 BP), the environment has changed to almost stagnant waters evidenced by cement layers of even thickness in the microbialites. No main direction of water current could be inferred from the sample. Thin section through the structure shows several generations of cement with a pattern of lithification recording seasonal variation in the chemical composition of the water (Fig. 6.8H). Reddish to brownish seams in the carbonate matrix reveal episodic desiccation, and the high porosity of the microbialite is an indication of abundant gas production by the metabolic activity of cyanobacteria (e.g., Sumner, 2000).

This time period is also represented by three other samples in the stratigraphic column (Tables 6.2 and 6.3). They all have similar characteristics, strengthening the case for seasonality and fluctuating environmental conditions. Two separate carbonate generations are observed at 8290 BP (Fig. 6.4), showing more than one deposition event: One is characterized by dark, murky microcrystalline carbonate and represents the original depositional environment; the other is a light, fine-grained carbonate with radiating starburst-like and spherulitic carbonates, which suggests a later and cleaner depositional environment.

In more recent times (Fig. 6.8I–L), filamentous algae are randomly oriented in the microbialites. Rare grains of magnetite and hematite suggest little oxidation. The youngest sample in the stratigraphic column 6300 BP is made of carbonate fragments (95%) cemented by amorphous carbonate matrix. The remainder of the sample consists of clastic material (quartz, plagioclase, hornblende, clinopyroxene, biotite, and magnetite). Silicates are derived in part from local volcanic rocks. The abundance and broken morphology of fossils and the presence of clastic volcanic material may indicate the resurgence of a higher energy environment at that time (wind and waves) and possibly more erosion of the volcanic material on the shore.

The summary of observations and inferred environments is shown in Table 6.4.

Table 6.4 Summary of microbialite main characteristics and inferred paleolake environments

Epoch	Year (BP)	Sample main characteristics	Lake dynamics/environment
Late Pleistocene	13,270	Fragment of basaltic andesite rimmed by carbonate in agglomerate	Volcanic activity and/or weathering of volcanic material. Significant oxidation and abundance of clay. Increased precipitation
	12,250	In situ lithified cyanobacteria	High hydraulic energy. No evidence of seasonality
	12,070	Dark, murky, poorly crystalline, amorphous carbonate. Clear carbonates outline walls of vesicles and form rims of carbonate–altered fossil fragment. Many fossil fragments	High hydraulic energy. No evidence of seasonality
Transition to Holocene	11,470	Matrix replaced by red-brown clay	Extensive oxidation wet episode and/or extensive residence of rock underwater
	10,590	Two layers of tufted mats. Tops are covered by dark coccoids	Potential desiccation events
Holocene	9960	Cyanobacteria probably belonging to the genus *Lyngbya*. In situ lithification of the cyanobacteria	No seasonality observed
	9210	Organic-rich carbonate margins and lithified fabric. Two mat generations: filamentous cyanobacteria forming tufted bundles and thick filaments	Clear evidence of seasonality recorded for the first time in growth pattern
	8840	Coccoid and filamentous cyanobacteria. Cements of even thickness	Stagnant waters. Fabric suggests seasonal variation in the water composition. Episodes of desiccation
	8290	Two separate carbonate generations	Seasonal variations
	8250	Dark and light carbonate. Poorly preserved algae and ostracods shells. Growth bands	Oxidation. Growth band related to seasonality
	8140	Growth bands	Comparable with 8250 BP
	7520	Microcrystalline to cryptocrystalline carbonate in small domains. Abundance of fossil fragments. Fossil algae are randomly oriented	No significant oxidation
Transition to mid-Holocene	6300	Conglomerate cemented by amorphous matrix. Many broken fossils. The presence of volcanic clasts. Quartz does not belong to local volcanic material	Broken fossils and volcanic may indicate higher hydraulic regime (locally?) and erosion. Silica could signal the input of a hydrothermal source

Diatoms confirm depositional environments of littoral to lake margin suggested by the distribution and shapes of the microbialites, with a dry, sandy to rocky lakeshore; elevated lake salinity; and somewhat alkaline water chemistry. Environmental stress, such as climate change, acidification, and/or eutrophication of the lake, was further tracked with a transfer function to chart changes in diatom assemblages between samples. An initial principal component analysis (PCA) reveals species and assemblage associations in the better-preserved samples (Fig. 6.9). The salinity index (after van Dam et al., 1994) suggests that all samples were formed in a lake with an average elevated salinity. The Late Pleistocene lake was marked by brackish to fresh waters (salinity index, SI, 1.99–1.90; see Table 6.5). A decrease in the salinity index around 9960 BP may be indicative of a wetter climate phase with increased runoff, groundwater input, hydrothermal spring, and/or change in precipitation regime. In the same geologic timescale, Abbott et al. (1997) note an increase in organic matter in high-altitude lakes in the Cordillera Real (northern Bolivia) that they attributed to an increase in precipitation. Species that are commonly associated with finer sediments or quiet water (e.g., *Nitzschia*, *Amphora*, and *Gyrosigma*) suggest a slow distal stream discharging into

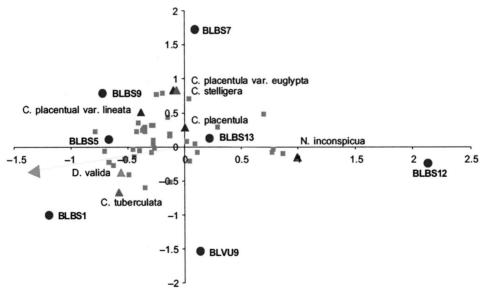

Fig. 6.9 Principal component analysis of diatom populations and salinity index—the PCA illustrates sites and species data with the salinity index passively shown as the *green arrow*. The *circles* represent the seven samples; *the squares and triangles* represent the species and the position for all species in relation to each other. The differently *colored triangles* are the dominant species in the samples, with their names near the symbols. The principal axis appears to be most closely tied to the salinity preferences of the species and the secondary axis to the environment (i.e., pelagic vs benthic). Dominant species varied with samples.

Table 6.5 Summary of paleoenvironmental analyses[a]

Year (BP)	Rock texture	Acid-insoluble residue	Thin section	Palynology	Diatoms[b]
12,250	Radiating bladed/botryoidal	Abundant filaments[c]	Stronger water currents, no seasonality	Shallow alkaline water	SI = 1.99—brackish salinity, benthic mats, quiet water, fine sediments
10,590	Layered, vuggy, compact	Dark filament	Desiccation, seasonality	Shallow alkaline, saline water	SI = 1.90—rare diatoms, saline-brackish
9960	Layered	Lenticular insoluble minerals	*Lyngbya*. In situ lithification, no seasonality	Shallow alkaline, saline water, algal crusts	SI = 1.57—fresh-brackish warm open water, fine sediments
9210	Layered, vuggy	Fine insoluble minerals	Seasonality	Alkaline, saline water, fluctuating level	SI = 1.95—saline deep water; fine sediments
8840	Granular	Insoluble minerals and fine organics	Coccoid/filamentous cyanobacteria, seasonality	Fluctuating water level	SI = 2.13—brackish pelagic, deeper, or more open warm water
LB12[d]	Botryoidal, vuggy	Clear filament	No seasonality, continuously subaquatic	More expansive, alkaline, saline water, algal crusts	SI = 1.23—fresh-brackish water
LB13[d]	Radiating botryoidal	Abundant filaments	Stronger water currents	Transitional shoreline	SI = 1.87—higher salinity-brackish

[a]From thin-section analysis of the stromatolite microfabric, analysis of the siliceous diatom content, and palynological analysis of the organic content.
[b]SI = salinity index (range 1–3; 1 = fresh, 2 = fresh to brackish, and 3 = purely brackish).
[c]Organic filamentous sheaths of probable cyanobacterial origin are present in most samples.
[d]Supplement samples with no radiocarbon age. Data from palynology and petrography suggest an Early Holocene age.

the lake. The sometimes large numbers of *N. inconspicua* might coincide with periodic high water discharge from this river.

More climate fluctuations are recorded throughout the stratigraphic column (Table 6.5). High salinity is a trend that continues today. Laguna Verde is three times saltier than seawater and Laguna Blanca 10 times saltier than average continental waters (Acs et al., 2003). However, while Laguna Verde remains highly alkaline (Cabrol et al., 2009), the pH of Laguna Blanca is mostly neutral with local variations (Table 6.1), reflecting the chemical characteristics and relative abundance of the many active hydrothermal thermal springs currently flowing into the lake compared with Laguna Verde. It also confirms the physicochemical separation between the two bodies of water.

The high abundance of the *Botryococcus* algae in some of the samples underscores elevated water alkalinity and the occurrence of a changing lake dynamics, possibly related to new wind regimes, with wind and waves concentrating the buoyant alga to form crusts along the lakeshore. Modern observations of this wind concentration mechanism in major saline lakes and ocean embayments around the world are summarized in Zippi (1998). High winds (40–100 km/h) are a daily constant in current climate conditions in the study area, and the accumulation of *Botryococcus* may be evidence for the timing of their initiation.

6.9 CONCLUSION

Sample analysis within the stratigraphic survey at Laguna Verde provides insights into the evolution of microbial habitats and biogeosignatures in a changing planetary environment, particularly into their rapid response to change. Some of the datasets used for this reconstruction (e.g., diatoms, zooplankton, and pollen) have no relevance to Mars. However, the transition from a wet to arid climate and its impact on hydrogeologic dynamics and physicochemical and geochemical signatures offer suitable geosignature analogies to the Noachian/Hesperian transition on Mars and possibly biogeosignatures as well (e.g., microbialites, especially MISS). This analogy also bears important lessons for a potential evolution of life on Mars (Cabrol, 2018) and for the upcoming exploration of biosignatures.

The environmental transformation observed at Laguna Verde can be considered rapid at geologic scale, with fluctuations over several thousands of years, and a dramatic acceleration today. Yet, whether during the Late Pleistocene/Holocene transition (this chapter) or in the present-day Laguna Verde and Laguna Blanca (Cabrol et al., 2007a,b, 2009, 2010a,b), the response of the lacustrine habitats could be immediately observed in the seasonality captured in the samples (or the lack thereof), in the composition of the sediments, and in the changes in the fossil record. This reemphasizes the value of ancient lakes as landing sites when contextual environmental data at major and minor scales are critically needed to reconstruct the biological and geologic evolution of a site, as that will be the case for the upcoming missions searching for biosignatures. One example is the array of potential MISS in the lake deposits of the Gillespie Lake Member on Mars, where surface structures could be consistent with the reaction of microbial mats to the fluctuations of the local climate, including humidity and changes in water level of the ancient lake (Noffke, 2014).

Laguna Verde also carries an important message about how we should think about the exploration of Mars (scale and resolution) as it transitions from the characterization of habitability to the search for biosignatures. The loss of the magnetosphere 4.1 Ga and that of the atmosphere by solar-wind sputtering possibly as early as 3.9/3.8 Ga (Jakosky et al., 2017) implies that by terrestrial standards, the surface of Mars became a polyextreme environment very early on (e.g., extreme UV, cold, aridity, and thin atmosphere). In many ways, the altiplanic lakes are evolving in a similar polyextreme environment. At high elevation and in a thin atmosphere, the spatiotemporal impact of environmental stress on altiplanic lakes may reflect what was taking place on early Mars. In this analogous environment, while the dynamics of the lakes results from regional aridification trends, the habitats currently forming out of the separation of Laguna Verde into two distinct bodies of water demonstrate that in polyextreme environments, local—not global—processes dominate (Cabrol, 2018). Although Laguna Verde and Laguna Blanca are still connected today by a small channel, individual basin topography, geology, and abundance and physicochemical characteristics of hydrothermal springs feeding into each lake are generating different water chemistries, weathering and deposition processes, sediment texture, composition, and ultimately ecosystems with different internal feedback loops and biosignature preservation potential.

Understanding ancient martian lake habitats that may have formed during the Noachian/Hesperian transition requires, therefore, shifting the scope of exploration from planetary habitability (global trends) to that of (local) microbial habitat potential. Currently, the resolution and scale of the morphological, geologic, mineralogical, and compositional data that would matter for the evaluation of the microbial habitats and biosignature potential for candidate landing-site selection are not available, with the exception of Gusev and Gale Craters and Meridiani. Unless a return to one of these landing sites is planned, both Mars 2020 and ExoMars will have to spend a substantial amount of mission time deciphering habitat and biosignature potential from the ground before they can first understand where to acquire the most valuable samples.

The issue is compounded on Mars by the fact that it is unknown whether biological processes, if any, had time to accumulate enough organic matter to be detectable by mission instruments or can be recognized as such. Further, terrestrial studies have demonstrated that as environmental pressure increases and opportunities for adaptation decrease, life tends to reduce its footprint to micrometer-meter niches (Cabrol, 2018 and references therein). In this respect, the continuous study of terrestrial analogs integrating orbital, field, and laboratory data is imperative in support to the upcoming missions and to future sample return, in order to provide guidance on instrument payloads and detection thresholds.

ACKNOWLEDGMENTS

The data presented in this chapter were collected as part of the High Lakes Project (HLP), a project supported by the National Aeronautics and Space Administration Astrobiology Institute (NAI) between 2002 and 2009 under grant no. NNA04CC05A. This investigation continues today in the Chilean Altiplano and Andes with the NAI-funded project entitled Changing Planetary Environments and the Fingerprints of Life (2015–19) that characterizes biosignatures relevant to the Mars 2020 and ExoMars missions under grant no. NNX15BB01A.

REFERENCES

Abbott, M., Brenner, M.W., Kelts, K.R., 1997. A 3500 ^{14}C yr high-resolution record of water-level changes in Lake Titicaca, Bolivia/Peru. Quat. Res. 47, 169–180.

Acs, E., Cabrol, N.A., Grigorszky, I., Friedmann, I., Kiss, A., Szabó, K., Kiss, K.T., 2003. In: Dombos, M., Lakner, G. (Eds.), Similarities and dissimilarities in biodiversity of three high-altitude mountain lakes (Andes, Bolivia). Sixth Hungarian Ecological Congress. St. Stephan University Publishers, Godollo (Abstract No. 305).

Baucom, P.C., Rigsby, C.A., 1999. Climate and lake-level history of the northern Altiplano, Bolivia, as recorded in Holocene sediments of the Rio Desaguadero. J. Sed. Res. 69 (3), 597–611.

Blodgett, T.A., Lenters, J.D., Isacks, B.L., 1997. Constraints on the origin of paleolake expansions in the central Andes. Earth Interact. 1 (1), 1–28.

Browne, K.M., Golubuc, S., Seong-Joo, L., 2000. Shallow-marine microbial carbonate deposits. In: Microbial Sediments. Springer, Berlin, Heidelberg, pp. 233–249.

Cabrol, N.A., 2018. The coevolution of life & environment on mars: an ecosystem perspective on the robotic search for biosignatures. Astrobiology 18 (1), 1–26.

Cabrol, N.A., Grin, E.A. (Eds.), 2010. Lakes on Mars. Elsevier Science, Amsterdam. 390 pp.

Cabrol, N.A., et al., 2009. The high lakes project. J. Geophys. Res. Biogeosci. High Lakes Project, special issue, https://doi.org/10.1029/2008JG000818.

Cabrol, N.A., Grin, E.A., Friedmann, R., De Vore, E., McKay, C.P., Murbach, M., Friedmann, I., Chong, G., Demergasso, C., Tambley, C., Escudero, L., Kisse, K., Grigorszky, I., Fike, D., Hock, A., Grigsby, B., 2003. Licancabur: exploring the limits of life in the highest lake on Earth. NASA/TM-2003-211862pp. 64–67.

Cabrol, N.A., Grin, E.A., Bebout, L., Rothschild, L., Hock, A.N., The Mars Underwater Science Team, 2005. Field and diving exploration of the highest lakes on Earth: analogy of environment and habitats with early Mars and life adaptation strategies to UV.NAI 2005 Biennal Meeting, University of Colorado, Boulder. Center for Astrobiology, No. 627.

Cabrol, N.A., Grin, E.A., Kiss, K.T., Ács, E., Grigorszky, I., Szabò, K., Tóth, B., Fike, D.A., Hock, A.N., Demergasso, C., Escudero, L., Chong, G., Galleguillos, P., Grigsby, B.H., Zambrana Román, J., McKay, C.P., Tambley, C., 2007a. Signatures of habitats and life in earth's high-altitude lakes: clues to noachian aqueous environments on Mars. In: Chapman, M.G., Skilling, I. (Eds.), The Geology of Mars: Evidence from Earth-Based Analogues. Cambridge University Press, New York, NY.

Cabrol, N.A., Grin, E.A., Hock, A.N., 2007b. Mitigation of environmental extremes as a possible indicator of extended habitat sustainability for lakes on early Mars. Proc. SPIE. 6694. https://doi.org/10.1117/12.731506.

Cabrol, N.A., Grin, E.A., Chong, G., Hock, A.N., Häder, D.P., et al., 2010a. Declining lake habitat in rapid climate change. In: Cabrol, N.A., Grin, E.A. (Eds.), Lakes on Mars. Elsevier, Amsterdam, pp. 347–369 (Chapter 13).

Cabrol, N.A., Andersen, D.T., Stoker, C.R., Lee, P., McKay, C.P., Wettergreen, D.S., 2010b. Chapter 10: other analogs to mars: high altitude, subsurface, desert, and polar environments. In: Doran, P.T., Bery Lyons, W., McKinght, D.M. (Eds.), Life in Antarctic Deserts and Other Cold Dry Environments: Astrobiological Analogues. Cambridge University Press, Cambridge Astrobiology, New York, pp. 258–305.

Cabrol, N.A., Feister, U., Häder, D.-P., Piazena, H., Grin, E.A., Klein, A., 2014. Record solar UV irradiance in the tropical Andes. Front. Environ. Sci. 2. https://doi.org/10.3389/fenvs.2014.00019.

Cady, S.L., Farmer, J.D., Grotzinger, J.P., Schopf, J.W., Steele, A., 2003. Morphological biosignatures and the search for life on Mars. Astrobiology 3, 351–368.

Clapperton, C.M., Clayton, J.D., Benn, D.I., Marden, C.J., Argollo, J., 1997. Late quaternary glacier advances and palaeolake highstands in the Bolivian Altiplano. Quat. Internat. 38/39, 49–59.

Clayton, J., Clapperton, D., Chalmers, M., 1997. Broad synchrony of a late-glacial glacier advance and the highstand of palaeolake Tauca in the Bolivian Altiplano. J. Quat. Sci. 12 (3), 169–182.

Davila, A.F., Schulze-Makuch, D., 2016. The last possible outposts for life on Mars. Astrobiology 16, 159–168.

Demergasso, C., Dorador, C., Meneses, D., Blamey, J., Cabrol, N.A., Escudero, L., Chong, G., 2010. Prokaryotic diversity pattern in high-altitude ecosystems of the Chilean Altiplano. J. Geophys. Res. 115, G00D09. https://doi.org/10.1029/2008JG000836.

Dorador, C., Meneses, D., Urtuvia, V., Demergasso, C., Vila, I., Witzel, K.-P., Imhoff, J.F., 2009. Diversity of *Bacteroidetes* in high-altitude saline evaporitic basins in northern Chile. J. Geophys. Res. Biogeosci. G00D05, 1–11. https://doi.org/10.1029/2008JG000837.

Fairén, A.G., Davila, A.F., Lim, D., Bramall, N., Bonaccorsi, R., Zavaleta, J., Uceda, E.R., Stoker, C., Wierzchos, J., Dohm, J.M., Amils, R., Andersen, D., McKay, C.P., 2010. Astrobiology through the ages of Mars: the study of terrestrial analogues to understand the habitability of Mars. Astrobiology 10, 821–843.

Farmer, J.D., Des Marais, D.J., 1999. Exploring for a record of ancient Martian life. J. Geophys. Res. Planets 104, 26977–26995. https://doi.org/10.1029/1998je000540.

Feister, U., Cabrol, N., Häder, D., 2015. UV irradiance enhancements by scattering of solar radiation from clouds. Atmosphere 6, 1211–1228. https://doi.org/10.3390/atmos6081211.

Figueroa, O., Déruelle, B., Demaiffe, D., 2008. Genesis of adakite-like lavas of Licancabur volcano (Chile-Bolivia, Central Andes). Compt. Rendus Geosci. 341, 310–318.

Fleming, E.D., Prufert-Bebout, L., 2010. Characterization of cyanobacterial communities from high-elevation lakes in the Bolivian Andes. J. Geophys. Res. Biogeosci. G00D09, 1–14. https://doi.org/10.1029/2008JG000817.

Fornari, M., Risacher, F., Feraud, G., 2001. Dating of paleolakes in the central Altiplano of Bolivia. Palaeogeo. Palaeoclim. Palaeoecol. 172, 269–282.

Francou, B., Ribstein, R., Semiond, H., Rodriquez, A., 1995. Balances de glaciares y clima en Bolivia y Peru: Impactos de los eventos ENSO. Bull. Inst. Fr. Et. And. 24, 661–670.

Gerdes, G., Krumbein, W.E., 1994. Peritidal potential stromatolites—a synopsis. In: Bertrand-Sarfati, J., Monty, C. (Eds.), Phanerozoic Stromatolites II. Kluwer, Dordrecht, pp. 103–130.

Gerdes, G., Noffke, N., Klenke, T., Krumbein, W.E., 2000. Microbial signatures in peritidal sediments—a catalogue. Sedimentology 47, 279–308.

Gouze, P., Argollo, J., Saliege, J.-F., Servant, M., 1986. Interpretation paleoclimatique des oscillations des glaciers au cours des 20 derniers millénaires dans les regions tropicales: Exemple des Andes Boliviennes. C. R. Acad. Sci. II 303, 219–233.

Grosjean, M., 2001. Mid-Holocene climate in the south-central Andes: humid or dry? Science 292, 2391–2392.

Grosjean, M., Geyh, M.A., Messerli, B., Schreier, H., Veit, H., 1998. A late-Holocene (<2600 BP) glacial advance in the south-central Andes (29 degrees S), northern Chile. The Holocene 8, 473–479.

Grosjean, M., van Leeuwen, J.F.N., van der Knapp, W.O., Geyh, M.A., Ammann, B., Tanner, W., Messerli, B., Nunez, L.A., Valero-Garce's, B.L., Veit, H., 2000. A 22,000 14C year BP sediment and pollen record climate change from Laguna Miscanti. Global Planet. Chang. 28, 35–51.

Hays, L.E., Graham, H.V., Des Marais, D.J., Hausrath, E.M., Horgan, B., McCollom, T.M., Parenteau, M.N., Potter-McIntyre, S.L., Williams, A.J., Lynch, K.L., 2017. Biosignature preservation and detection in Mars analog environments. Astrobiology 17 (1), 17363–17400.

Hock, A.N., Cabrol, N.A., Grin, E.A., Kovacs, G.T., Rothschild, R.L., Parazynski, S.E., Prufert-Bebout, L., The Mars Underwater Science Team, 2005. In: Mars-relevant conditions at the lakes of Licancabur volcano, Bolivia.2005 AGU Fall Meeting, San Francisco. (Abstract P41D-06).

Honjo, S., Dymond, J., Collier, R., Manganini, S.J., 1995. Export production of particles to the interior of the equatorial Pacific Ocean during the 1992 EqPac experiment. Deep Sea Res. II 44, 831–870.

Jakosky, B.M., Slipski, M., Benna, M., Mahaffy, P., Elrod, M., Yelle, R., Stone, S., Alsaeed, N., 2017. Mars' atmospheric history derived from upper-atmosphere measurements of ^{38}Ar/^{36}Ar. Science 355, 1408–1410.

Knoll, A.H., Grotzinger, J., 2006. Water on Mars and the prospect of martian life. Elements 2, 169–173.

Koo, H., Mojib, N., Hakim, J.A., Hawes, I., Tanabe, Y., Andersen, D.T., Bej, A.K., 2017a. Microbial communities and their predicted metabolic functions in growth laminae of a unique large conical mat from Lake Untersee, East Antarctica. Front. Microbiol. 8 (1347), 1–15. https://doi.org/10.3389/fmicb.2017.01347.

Koo, H., Hakim, J.A., Morrow, C.D., Eipers, P.G., Davila, A., Andersen, D.T., Bej, A.K., 2017b. Comparison of two bioinformatics tools used to characterize the microbial diversity and predictive functional attributes of microbial mats from Lake Obersee, Antarctica. J. Microbiol. Methods 140, 15–22.

Kull, C., Hänni, F., Grosjean, M., Veit, H., 2003. Evidence of an LGM cooling in NW-Argentina derived from a glacier climate model. Quat. Int. 108, 3–11.

Marinovic, N., Lahsen, A., 1984. Hoja Calama. Carta Geologica de Chile No. 58, S.N.G.M. Santiago, Chile.

Messerli, B., Grosjean, M., Bonani, G., Bürgi, A., Geyh, M.A., Graf, K., Ramseyer, K., Romero, H., Schotterer, U., Schreier, H., Vuille, M., 1993. Climate change and dynamics of natural resources in the Altiplano of northern Chile during Late Glacial and Holocene time. First synthesis. Mount. Res. Dev. 13 (2), 117–127.

Noffke, N., 2009. The criteria for the biogeneicity of microbially induced sedimentary structures (MISS) in Archean and younger, sandy deposits. Earth Sci. Rev. 96, 173–180.

Noffke, N., 2010. Microbial Mats in Sandy Deposits From the Archean to the Present Time. Springer, Heidelberg. 196 p.

Noffke, N., 2014. Ancient sedimentary structures in the <3.7 Ga Gillespie lake member, Mars, that resemble macroscopic morphology, spatial associations, and temporal succession in terrestrial microbialites. Astrobiology 15 (2), 169–192.

Noffke, N., Gerdes, G., Klenke, T., Krumbein, W.E., 1997. A microscopic sedimentary succession indicating the presence of microbial mats in siliciclastic tidal flats. Sed. Geol. 110, 1–6.

Noffke, N., Gerdes, G., Klenke, T., 2003. Benthic cyanobacteria and their influence on the sedimentary dynamics of peritidal depositional systems (siliciclastic, evaporitic salty and evaporitic carbonatic). Earth Sci. Rev. 62/1–2, 163–176.

Pontefract, A., Zhu, T.F., Walker, V.K., Hepburn, H., Lui, C., Zuber, M.T., Ruvkun, G., Carr, C.E., 2017. Microbial diversity in a hypersaline sulfate lake: a terrestrial analog of ancient Mars. Front. Microbiol. https://doi.org/10.3389/fmicb.2017.01819.

Reimer, P.J., Baillie, M.G.L., Bard, E., Bayliss, A., Beck, J.W., Bertrand, C., Blackwell, P.G., Buck, C.E., Burr, G., Cutler, K.B., Damon, P.E., Edwards, R.L., Fairbanks, R.G., Friedrich, M., Guilderson, T.P., Hughen, K.A., Kromer, B., McCormac, F.G., Manning, S., Bronk Ramsey, C., Reimer, R.W., Remmele, S., Southon, J.R., Stuiver, M., Talamo, S., Taylor, F.W., van der Plicht, J., Weyhenmeyer, C.E., 2004. IntCal04 terrestrial radiocarbon age calibration, 26-0 ka BP. Radiocarbon 46, 1029–1058.

Risacher, F., Fritz, B., 1991. Geochemistry of Bolivian salars, Lipez, southern Altiplano: origin of solutes and brine evolution. Geochim. Cosmochim. Acta 55, 687–705.

Roche, M.A., Bourges, J.C., Mattos, R., 1992. Climatology and hydrology of the Lake Titicaca basin. In: Dejoux, C., Iltis, A. (Eds.), Lake Titicaca: A Synthesis of Limnological Knowledge. Kluwer Academic Publishers, Dordrecht, pp. 63–83.

Rondeau, B., 1990. Géochimie isotopique et géochronologie des stromatolites lacustres quaternaires de l'Altiplano bolivien. (Ph.D. thesis). Université du Québec, Montreal.100 p.

Rouchy, J.-M., Servant, M., Fournier, M., Causse, C., 1996. Extensive carbonate algal bioherms in upper Pleistocene saline lakes of the central Altiplano of Bolivia. Sedimentology 43, 973–993.

Seltzer, G.O., 1990. Recent glacial history and paleoclimate of the Peru-Bolivian Andes. Quat. Sci. Rev. 9, 137–152.

Seltzer, G.O., 1992. Late quaternary glaciation of the Cordillera Real, Bolivia. J. Quat. Sci. 7, 87–98.

Seltzer, G.O., Rodbell, D.T., Abbott, M.B., 1995. Andean glacial lakes and climate variability since the last glacial maximum. Bull. Inst. Fr. Et. And. 24, 539–550.

Seltzer, G.O., Rodbell, D.T., Baker, P.A., Fritz, S.C., Tapia, P.M., Rowe, H.D., Dunbar, R.B., 2002. Early warming of tropical South America at the last glacial-interglacial transition. Science 296, 1685–1686.

Servant, M., Fontes, J., 1978. Les lacs quaternaires des hauts plateaux des Andes Boliviennes: Premieres interpretations paleoclimatiques. Cahiers de l'ORSTOM, Ser. Geol. 10, 9–23.

Sumner, D., 2000. Microbial vs. environmental influences on the morphology of late archean fenestrate microbialites. In: Riding, R., Awramik, S.M. (Eds.), Microbial Sediments. Springer, Berlin, Heidelberg, pp. 307–314.

Suosaari, E.P., Reid, R.P., Playford, P.E., Foster, J.S., Stolz, J.F., Casaburi, G., Hagan, P.D., Chirayath, V., Macintyre, I.G., Planavsky, N.J., Eberli, G.P., 2016. New multi-scale perspectives on the stromatolites of Shark Bay, Western Australia. Sci. Rep. 6, 1–13. https://doi.org/10.1038/srep20557.

Sylvestre, F., Servant, M., Servant-Vildary, S., Causse, C., Fournier, C., 1999. Lake-level chronology on the southern Bolivian Altiplano (18 degrees-23 degrees S) during late-glacial time and the early Holocene. Quat. Res. 51, 54–66.

Tilman, D., Kilham, P., 1976. Sinking in freshwater phytoplankton: some ecological implications of cell nutrient status and physical mixing processes. Limnol. Oceanogr. 21, 409–417.

Valero-Garces, B.L., Grosjean, M., Schwalb, A., Geyh, M., Messerli, B., Kelts, K., 1996. Limnogeology of Laguna Miscanti: evidence for mid to late Holocene moisture changes in the Atacama Altiplano. J. Paleolimnol. 16, 1–21.

van Dam, H., Mertens, A., Sinkeldam, J., 1994. A coded checklist and ecological indicator values of freshwater diatoms from the Netherlands. Neth. J. Aquat. Ecol. 28 (1), 117–133.

Volcani, B.E., 1981. Cell wall formation in diatoms: morphogenesis and biochemistry. In: Simpson, T.L., Volcani, B.E. (Eds.), Silicon and Siliceous Structures in Biological Systems. Springer-Verlag, New York, pp. 157–200.

Vuille, M., Bradley, S., Keimig, F., 2000. Interannual climate variability in the central Andes and its relation to tropical Pacific and Atlantic forcing. J. Geophys. Res. 105 (12), 447.

Williamson, C.E., Saros, J.E., Schindler, D.W., 2009. Sentinels of change. Science 323 (5916), 887–888.

Wirrmann, D., Fernando De Oliveira Almeida, L., 1987. Low Holocene level (7700 to 3650 years ago) of Lake Titicaca (Bolivia). Palaeogeo. Palaeoclim. Palaeoecol. 59, 315–323.

Wirrmann, D., Mourguiart, P., 1995. Late quaternary spatio-temporal limnological variations in the Altiplano of Bolivia. Quat. Res. 43, 344–354.

Wirrmann, D., Mourguiart, P., Fernando De Oliveira Almeida, L., 1990. Holocene sedimentology and ostracods distribution of Lake Titicaca-paleohydrological interpretations. In: Rabassa, J. (Ed.), Quaternary of South America and Antarctic Peninsula. In: vol. 6. A.A. Balkema, Rotterdam, pp. 89–129.

Zippi, P., 1998. Freshwater algae from the Mattagami formation (Albian) Ontario: paleoecology, botanical affinities, and systematic taxonomy. Micropaleontology 44 (1), 1–78.

FURTHER READING

Mourguiart, P., 1995. Une approche nouvelle du problème posé par les reconstructions des paléoniveaux lacustres: utilisation d'une fonction de transfert basée sur les faunes d'ostracodes. Géodynamique 5, 151–166.

CHAPTER 7

Siliceous Hot Spring Deposits: Why They Remain Key Astrobiological Targets

Sherry L. Cady, John R. Skok, Virginia G. Gulick, Jeff A. Berger, Nancy W. Hinman

Contents

For nearly 40 years, hydrothermal deposits have been recognized as potential paleobiological repositories for astrobiological exploration of Mars. Here, we summarize the motivation for this astrobiological search strategy as it pertains to our current understanding of silica-depositing hot spring ecosystems and terrestrial siliceous hot spring deposits. We also discuss the rover and orbital observations of recently discovered hydrothermal opaline silica deposits on Mars—interpreted as evidence of hot spring activity. The opaline silica digitate sinters near Columbia Hills represent the strongest evidence to date for potential fossilized biosignatures on Mars. The high habitability and preservation potentials of hot spring deposits on Earth, along with their ability to reveal insight into the metabolic evolution of life, strengthen the rationale for targeting siliceous hot spring deposits as high-priority astrobiology sites for future Mars missions.

7.1 INTRODUCTION

If microbial life ever emerged on Mars, it is likely that it would have thrived in hot springs, given that hydrothermal systems would have been widespread throughout the planet's history (Walter, 1996; Farmer, 1996; Schultze-Makuch et al., 2007). In general, hydrothermal systems develop when subsurface fluid (meteoric, magmatic, connate),

From Habitability to Life on Mars
https://doi.org/10.1016/B978-0-12-809935-3.00007-4

heated by rising or impact-generated magma, circulates and interacts with subterranean rock during its ascent to the surface. Hydrothermal fluid reaches the surface via subterranean fractures and passes through hydrothermal spring and geyser effluents where it redeposits dissolved minerals and aqueous precipitates as it cools (Fournier, 1989; Henley and Ellis, 1983; Sillitoe, 2015). Hydrothermal systems that developed along mid-ocean ridges and terrestrial hot springs, whether they were initiated by volcanic, impact, or tectonic activity, represent some of the most ancestral niches for microorganisms on the early Earth (Walter, 1996; Henley, 1996; Nisbet and Sleep, 2001).

Relevant to an astrobiology search strategy for Mars is the high habitability and preservation potentials of hydrothermal ecosystems. Silica-depositing terrestrial hot springs, in particular, have the ability to serve as paleontological repositories on Earth, as well as on Mars or any other rocky planet that could have experienced sustained periods of hydrothermal activity and hosted life as we know it. Silica sinter[1] deposits can preserve a variety of microbial biosignatures that include body fossils (morphologically and chemically recognizable cellular remains), biofilm and microbial mat fabrics, and biosedimentary structures (stromatolites, microbialites, and microbially induced sedimentary structures), along with chemical fossils (biosynthetic molecules, biologically fractionated stable-isotope signatures, biominerals, and anomalous concentrations and combinations of elements and minerals) (e.g., Walter, 1976b; Cady and Farmer, 1996; Jones and Renaut, 1996; Jones et al., 2001; McKenzie et al., 2001; Hinman and Walter, 2005; Georgiou and Deamer, 2014; Campbell et al., 2015a,b; Siljeström et al., 2017). Recent discoveries of massive primary opaline silica deposits in two locations on Mars, hypothesized for different reasons to be hydrothermal in origin (Squyres et al., 2008; Ruff et al., 2011; Skok et al., 2010; Ruff and Farmer, 2016), strengthen the relevance of the astrobiology search strategy for hydrothermal spring deposits on Mars that was originally proposed nearly four decades ago (Walter and Des Marais, 1993), and has been pursued and subsequently reported in hundreds of publications.

The key attributes that make siliceous sinters a compelling astrobiological target on Mars are reviewed here. In the sections that follow, we discuss a number of topics that

---

[1] Although it was suggested recently to restrict the term sinter to describe sedimentary rock primarily composed of silica that precipitates from hot spring waters at the vents of high-temperature (high-enthalpy) hot springs and geysers and from cooled waters on their surrounding discharge (Renaut and Jones, 2011b), we maintain the traditional, more inclusive definition of the term sinter sensu lato to describe a sedimentary rock type composed primarily of hot spring precipitate. Note that the broader use of the term does not imply (a) the temperature of the fluid from which the sinter formed (from boiling to ambient); (b) the primary mineralogy of the sinter (e.g., silica, iron, manganese, and carbonate, though travertine is the term typically used to describe carbonate sinter); or (c) the diagenetic state of the sinter precipitate, which for silica could be primary opal-A, any of the other varieties of microcrystalline opals or fibrous quartz phases, or microcrystalline quartz (the microstructures of opal and fibrous quartz varieties are highly disordered but opals rather than quartz varieties are kinetically favored to precipitate from aqueous fluids, cf., Cady et al., 1996, 1998). In this chapter, the focus is on silica sinter, the relative geological age of which is specified by the primary mineralogy.

highlight the potential astrobiological importance of the recently discovered hydrothermal opaline silica deposits on the red planet.

7.2 HOT SPRING DEPOSITS AS ASTROBIOLOGY TARGETS

Silica-depositing hot spring ecosystems host a wide range of metabolic strategies that are concentrated in sequential, concentrically arranged zones around hot spring effluents and pools. When hot spring fluids flow out and away from near-boiling pools and geysers, the metabolism of the primary producers in the microbial communities changes from chemotrophic to anoxygenic phototrophic to oxygenic phototrophic. As shown in Fig. 7.1, the zonal distribution of distinct phototrophic microbial communities—visible because of differences in the colors of their dominant photosynthetic pigments—produces a distinctive bullseye pattern when viewed from above. The major controls on the occurrence of biofilm- and mat-forming communities in silica-depositing hot springs include temperature, pH, and H_2S and O_2 concentrations (e.g., Peary and Castenholz, 1964; Brock,

Fig. 7.1 Aerial photograph of Grand Prismatic hot spring, Yellowstone National Park, USA. Chemotrophic microbial communities appear in the field as light-colored, transparent biofilms and streamers that colonize sinters in and around hot spring pools like this one, so long as they are episodically bathed with near-boiling fluids. Anoxygenic phototrophic communities occur as red- and green-pigmented layered mats and green streamers that occur adjacent to the hyperthermophile chemotrophic communities, living downstream in slightly lower temperature fluids. Oxygenic phototrophic microbial communities, which can be characterized by brown, orange, green pigmented communities, live at lower-to-ambient temperatures downstream of the higher temperature anoxygenic phototrophs. *Photo taken on March 26, 2015, © Sean Beckett | Dreamstime.com, Image ID 64049867.*

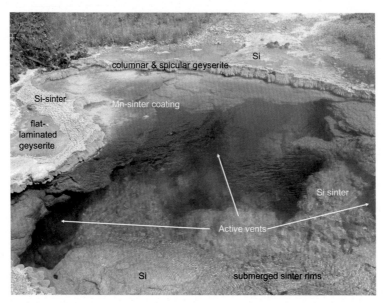

Fig. 7.2 Mixed Si- and Mn-depositing hot spring pool, Yellowstone National Park, USA. Deeper cavities in the sinter on the bottom of the pool reveal the position (and upflows, not visible in photo) of hydrothermal effluents of active (and likely former) vents. Submerged terraces just below the pool rim extend out over the vents, building up sinter from beneath as hydrothermal fluids bubble toward the surface and circulate convectively throughout the pool. The shallow submerged surfaces represent the ledges of sinter that build out over the pool surface during a prior period of quiescent flow. The columnar geyserite along the rims of the pool reveals periodic boiling-like activity due to vigorous degassing, at this site. The mixed chemical composition of the high-temperature sinter reveals evidence of multiple pathways for fluids of different composition.

1978; Castenholz and Pierson, 1995; Ward et al., 1998; Bryant et al., 2007; Boomer et al., 2009; Inskeep et al., 2010). In general, the steeper the thermal and geochemical gradients along the pool rims and edges of the outflow channels, the more abrupt the transition from one microbial community to the next and the more concentrated the biodiversity at such transition zones.

Hyperthermophilic chemotrophs live, by definition, in fluids that range in temperature from near-boiling to 80°C (Brock, 1978; Stetter, 1996). In terrestrial silica-depositing hot springs, like the one shown in Fig. 7.2, hyperthermophilic communities are dominated by filamentous bacteria (Blank et al., 2002; Inskeep et al., 2013). When filamentous hyperthermophilic biofilms colonize geyserite[2] surfaces, they form

---

[2] High-temperature silica sinters, known as geyserites (sensu White et al., 1964), precipitate within or immediately adjacent to thermal springs and geyser effluents from hydrothermal fluids ejected at or above surface boiling temperatures, waters that were once thought to be sterile (Allen, 1934) and capable only of producing abiotic sinters (Walter, 1976a).

very thin biofilms only a few cell-layers thick, which are typically not visible with the naked eye in the field (Cady and Farmer, 1996). In rapidly flowing outflow channels of near-boiling hot spring pools, filamentous hyperthermophiles can form long (a few to a few tens of centimeters long) filamentous "streamers" that flow freely from continuous (e.g., as on a stick or rock) or isolated attachment points (e.g., from a topographical high point a millimeter or two in diameter on a flat mat surface or on a loose piece of sinter) (Reysenbach et al., 1994). Cady and Farmer (1996) were the first to demonstrate that communities of hyperthermophilic biofilms colonize nearly any surface within and along the rims of near-boiling silica-depositing hot springs and that their presence influences the fabrics and the macrostructural characteristics of geyserites. The paleobiological relevance of geyserites was recently reviewed by Campbell et al. (2015a). Hyperthermophiles are considered by many to be the closest living relatives of microbes that occupied ancient ancestral hydrothermal niches at the ocean floor and on land (e.g., Shock, 1996; Farmer, 1998; Doolittle, 1999; Nisbet, 2000; Nisbet and Sleep, 2001; Rothschild and Mancinelli, 2001; Reysenbach and Cady, 2001; Schwartzman and Lineweaver, 2004; Stetter, 2006; Glansdorff et al., 2008; Deamer and Szostak, 2010; Deamer, 2012; Weiss et al., 2016; Soto et al., 2016; Cavalazzi et al., 2018; Strazzulli et al., 2017; Dai, 2017; Price et al., 2017). Recent chemical, geological, and biochemical computational evidence has reinforced the hypothesis that life could have originated in terrestrial hot springs (cf., Van Kranendonk et al., 2018; Westall et al., 2018, and references therein).

The transition from hyperthermophilic chemotrophic biofilms to filamentous thermophilic anoxygenic phototrophic communities occurs downstream, or in vertical microniches (Jones et al., 1997), where fluid temperatures drop below 80°C. Their visible light-harvesting bacterial chlorophyll pigments reveal that these microbial populations occupy the upper layers of stratified thermophilic mats that can reach several millimeters in thickness (Castenholz and Pierson, 1995; Boomer et al., 2009). Filamentous "streamers" of anoxygenic phototrophs can also develop in rapidly flowing outflow channels in this region of a hot spring system when they become intertwined with one another and with the highest-temperature oxygenic phototrophs (cf., Meyer-Dombard et al., 2011; Siljeström et al., 2017).

Different populations of cyanobacteria, which are oxygenic thermophilic phototrophs (Castenholz, 1969; Brock, 1978), produce an even wider variety of microbial mat types located downstream from the anoxygenic phototrophic communities at lower-to-ambient fluid temperatures. Conical, pinnacle, tufted, bubble, and pustular mats develop in hot spring outflow channels and across the discharge aprons of silica-depositing hot springs. Massive low-relief, yet laterally continuous, "carpets" of sheathed cyanobacteria often form where fluid depths thin to only a few millimeters thick (i.e., "sheet" flow) and temperatures drop to ambient. The combination of evaporation and cooling of hydrothermal fluids at the distal ends of silica-depositing hot spring ecosystems typically encapsulates sheathed cyanobacteria, which enhance their preservation potential (Farmer, 1999; Guido and Campbell, 2017).

The sequential distribution of distinct thermophilic communities and their corresponding lithified remains in silica-depositing thermal springs in Yellowstone National Park led Walter (1976b) to propose a correlative set of biological and lithological facies[3] for such systems. When mineralized, the different mat biofacies can be correlated with morphologically similar silica sinter deposits that Walter (1976b) described as sinter lithofacies (cf., Fig. 7.1, Cady and Farmer, 1996). This biofacies-lithofacies model provides a robust framework for reconstructing the paleoecology of these types of hot spring deposits, even after the primary opaline silica of the sinters transforms diagenetically to more thermodynamically stable silica phases (Rodgers et al., 2004; Lynne, 2012).

The biofacies-lithofacies model is highly relevant to astrobiology search strategies on Mars for three main reasons: (1) Lithofacies of silica hot spring/geyser deposits preserve a variety of biosignatures indicative of the primary microbial inhabitants of the ecosystem across multiple spatial scales. Multiple lines of evidence increase the probability of the biogenic origin of possible biosignatures (Mustard et al., 2013; Westall et al., 2015a; Hays, 2015; Horneck et al., 2016). For silica-depositing hot springs, the biosignatures with the highest morphological fidelity include microbial mat fabrics and fossilized microbial remains (Walter, 1976b; Walter et al., 1996; Hinman and Walter, 2005; Campbell et al., 2015b); (2) Recently discovered primary opaline silica deposits on Mars, interpreted as silica sinters (Squyres et al., 2008; Skok et al., 2010; Ruff et al., 2011), do not appear to have undergone even the earliest stage(s) of diagenesis, which implies that the deleterious effects of silica phase transformations, which cause subsequent, incremental loss of paleobiological information, would not have impacted microbial biosignatures if they were preserved in such deposits (Walter et al., 1996; Hinman and Walter, 2005; Guido and Campbell, 2017); (3) Lithofacies of hot springs and geysers preserve a paleobiological record of the biodiversity of its inhabitants. If the paleoecology of a hot spring ecosystem was decipherable in siliceous sinter deposits on Mars, the discovery of different lithofacies could reveal whether anoxygenic and oxygenic phototrophy ever evolved as key microbial metabolic strategies on another world. In addition to having implications for astrobiology, such a discovery would impact the selection and/or optimization of life-detection instruments on future Mars missions and the design of exploration strategies for other possible habitats (Parnell et al., 2007; Worms et al., 2009; Hays, 2015; Horneck et al., 2016; Vago et al., 2017).

---

[3] Facies are established to differentiate units of rock from adjacent units within a contiguous body of rock by physical, chemical, or biological means. For hot springs, biofacies were established to distinguish mappable microbial communities recognized by the color of their phototrophic pigments and the morphology of their mats; lithofacies were established as the mineralized equivalents of hot spring biofacies (cf., Walter, 1976b).

The ubiquitous occurrence on Earth of laminated microbial mat fabrics in silica-depositing hot spring systems, enhanced by pervasive in situ mineralization of biofilms and mats, strengthens the proposition for astrobiological exploration of sites where hydrothermal silica deposits have been located on Mars. The morphological and cellular fidelity of biosignature preservation in hot springs on Earth depends upon the intrinsic cellular characteristics of heat-loving organisms (e.g., Jones et al., 2001; Konhauser et al., 2001; Yee et al., 2003; Benning et al., 2004a,b; Amores and Warren, 2007, 2009; Hugo et al., 2010; Campbell et al., 2015b) and the extrinsic geochemical, hydrodynamic, and seasonal factors that influence silica sinter precipitation and accumulation (e.g., Hinman and Lindstrom, 1996; Braunstein and Lowe, 2001; Amores and Warren, 2007; Yee et al., 2003; Orange et al., 2013; Alleon et al., 2016). An example of how the intrinsic characteristics of particular cells can enhance their preservation is illustrated by the sheathed cyanobacterium *Calothrix*, the dominant cyanobacterial population in distal low-temperature regions of silica–depositing hot springs. These organisms tend to be preferentially preserved in the geological record due to their dynamic response to silicification (i.e., their sheath thickens (Benning et al., 2004b) and mineral precipitation can be localized on specific ultracellular structures in their sheaths (Hugo et al., 2010)) and the preferential preservation of their sheaths compared to other cellular components in siliceous sinter deposits (Farmer, 1999; Hinman and Walter, 2005; Guido and Campbell, 2011; Campbell et al., 2015a,b). An example of the impact of extrinsic factors on preservation is illustrated via a comparison of the effects of different modes of fossilization (cf., Fig. 9 in Cady and Farmer, 1996): cells rapidly and completely replaced by opaline silica can retain their morphological fidelity; cells completely entombed but only partially replaced by opaline silica can be permineralized (cf., Cady, 2002); and cells incompletely entombed and cemented while still viable in opaline silica typically lyse after death and their cellular contents are rapidly destroyed through oxidation. When the carbonaceous remains of cells and extracellular polymeric substances (EPS) are preserved in opaline silica, they can be characterized by multiple chemical biosignatures (e.g., Siljeström et al., 2017). The degree to which the effects of high UV and ionizing radiation and post-preservational oxidation affect the structural and chemical fidelity of microbial remains has only recently been explored. Theoretical considerations and experimental studies indicate that cosmic rays can destroy amino acids mixed with hydrated silica phases like opaline silica in <100 million years (Summons et al., 2011; Pavlov et al., 2012).

Comparative studies of modern and ancient thermal spring deposits have shown that the highest morphological fidelity of preservation is skewed toward sinter biofabrics and stromatolite structures (Walter, 1976b; Hinman and Walter, 2005; Handley et al., 2008; Campbell et al., 2015b; Westall et al., 2015b; Guido and Campbell, 2017). The ability to recognize macroscale stromatolite structures and especially the microscale biofabrics of the majority of hot spring and geyser lithofacies in older sinter deposits is remarkable given that, on Earth, the primary opal-A of sinter ultimately transforms to quartz (Rice et al., 1995, 2002; Walter et al., 1996, 1998; Herdianita et al., 2000; Campbell et al., 2001, 2015b; Trewin et al., 2003; Lynne and Campbell, 2003, 2004; Lynne et al., 2005, 2007; Guidry and Chafetz, 2003; Hinman and Walter, 2005; Guido and Campbell, 2011; Barbieri et al., 2014; Westall et al., 2015b; Djokic et al., 2017). The fine-scale (millimeter) laminations of macrostromatolites (columns, digitate structures) are preserved in the oldest known siliceous sinters, which were recently discovered in the Archean Dresser Formation of the Pilbara Craton in Western Australia (Djokic et al., 2017). On Earth, the long residence time of quartz at the Earth's surface is imperative for preservation of such morphological details in the geological record (Farmer and Des Marais, 1999). On Mars, the persistence of massive monomineralic opaline silica deposits that could have originated in hot springs suggests that the more delicate morphological features of microbial fossils and stromatolites can be expected to still exist in such deposits.

Relevant to interpretations of the source of the opaline silica deposits on Mars is the fact that economic geologists and geochemists have known for decades that both alkali-chloride and acid-sulfate-chloride fluids can source surficial hot springs that deposit silica sinters in terrestrial hydrothermal environments (cf., Ellis and Mahon, 1977; Henley and Ellis, 1983; Nicholson, 1993; Renaut and Jones, 2011a). In addition to sourcing hot springs, subsurface hydrothermal fluids contribute to the production of a variety of different types of silica deposits that have only recently been recognized as being distinct from sinters (Sillitoe, 2015). These may include silica residue and pseudosinter (e.g., silicified water table deposits, silicified travertines, silicified volcanics, and silicified volcaniclastics (Rodgers et al., 2004; Guido and Campbell, 2011, 2017; Sillitoe, 2015). The association of such deposits with hydrothermal processes can lead to their misidentification as sinter because they may display laminated fabrics and have a primary opaline silica mineralogy. The discovery of fossil evidence of filamentous fabrics in epithermal samples collected from hydrothermal deposits worldwide, archived in independent museum and researcher collections, suggests that mineralization of filamentous chemotrophic microbiota is a common process in low-temperature (<boiling) subterranean environments where silica-saturated fluids precipitated primary opaline silica on Earth (Hofmann and Farmer, 2000).

Walter and Des Marais (1993) emphasized that remote-sensing techniques could be used to distinguish surface sinters produced by hydrothermal activity due to their distinctive geomorphic features (typically mounds flanked by terraces, channels, and broad discharge aprons) and the geochemical and mineralogical differences between the nearly monomineralic sinters and their surrounding country rock. Hydrothermally altered ground in large hydrothermal fields, caused by fumarolic activity and hydrothermal fluid-rock interactions, can produce a range of geochemically predictable mineral assemblages (Buchanan, 1981; Tosdal et al., 2009). Another consideration relevant to remote detection of hydrothermal deposits is the possibility that surficial hot spring deposits may have been partly/completely removed by erosion, weathering, or impacts, leaving the exposed epithermal paleosurface of the more voluminous subsurface component of a hydrothermal system. This evidence for hydrothermal activity could also be detected remotely as local to regional-scale mineral alteration haloes, the geochemical nature of which could reveal insight with regard to the composition of the hydrothermal fluid if the composition of unaltered country rock can be determined (Henley, 1996; Sillitoe, 1993, 2015).

A final consideration relevant to an astrobiology search strategy for hydrothermal systems on Mars is the geological context of hydrothermal systems, which require a localized subsurface heat source, typically volcanic or impact-related, and subsurface water or ice (cf., Newsom, 1980; Gulick, 1998). Such features have been used to prospect for large-scale epithermal mineral deposits for decades (e.g., Huntington, 1996). Even on small rocky planets like Mars that never developed appreciable plate tectonic activity, hot spring deposits likely formed whenever impact formation, volcanism, magmatic intrusions, or tectonic events resulted in prolonged hydrothermal activity (Gulick, 1998; Osinski et al., 2013; Westall et al., 2015b).

Some of the earliest convincing evidence for hydrothermal activity on Mars were remote images of simple channel systems along the margins of impact craters that were found in data sets of images generated from mapping Mars globally at relatively low spatial resolution (~200m per pixel) during the Viking Orbital Missions. These investigations revealed that several major martian volcanic provinces contained possible hydrothermal features evidenced by the superposition of fluvial geomorphological features associated with extended volcanic activity (e.g., locations that include Apollinaris Mons, Alba Patera, Hadriaca Patera, Tyrrhena Patera, Hecates Tholus) (Gulick and Baker, 1989, 1990; Gulick, 1998, 2001; Schultze-Makuch et al., 2007; Osinski et al., 2013). In spite of the promising findings from remote imaging data sets, the remote and standoff acquisition of elemental and mineral spectra from the surface of Mars proved instrumental in the discovery of geochemical evidence of hydrothermal activity on Mars.

7.3 DETECTION OF SILICEOUS HYDROTHERMAL HOT SPRING DEPOSITS ON MARS

Once the technologically more advanced imaging spectrometers were flown on the recent Mars Global Surveyor (MGS) and Mars Reconnaissance Orbiter (MRO) missions, the discovery of possible hydrothermal features on Mars that were associated with fluvial morphology and a volcanic geological context steadily increased (Crumpler, 2003; Farrand et al., 2005; Crumpler et al., 2007; Rossi et al., 2007, 2008a,b; Raitala et al., 2008; Allen and Oehler, 2008). MGS returned high-resolution (~6m/pixel) narrow-angle camera coverage over targeted regions for nearly 10years (nearly 5 Mars years) after its arrival in 1997, and the MRO HiRISE camera imaged the surface with ~0.25 m/pixel resolution starting in late 2006. CRISM visible and near-infrared (VNIR) spectrometers, also onboard the MRO mission, acquired mineralogical data (~20 m/pixel highest resolution) that resolved surface and geochemical features on spatial dimensions of ~ 80 m or larger.[4]

Even with the combination of higher resolution imaging and spectroscopic data sets, there were still complications in making reliable identifications of hydrothermal fields from orbit, let alone identifying compositionally distinct individual hot spring deposits. Many areas that display the geomorphological evidence consistent with the presence of hydrothermal activity on the surface of Mars failed to produce the needed geological evidence in terms of mineralogical and surface landforms at higher resolution. This is largely because much of the hydrothermal activity associated with the formation of large volcanoes or impacts on Mars is ancient and took place between several hundred million and 3.5 billion years ago. Numerous subsequent resurfacing events and erosional and depositional processes would have erased much of the meters-to-tens-of-meters-scale landforms associated with hydrothermal activity. Large volcanoes, like the Tharsis Montes and the Elysium and Apollinaris Mons, are situated at high elevations where global dust storms can deposit material but where the atmosphere is too thin for subsequent winds to remove it. This creates recent, thick dust layers at high elevations that obscure spectral and geomorphic observations of small-scale hydrothermal features. The presence of a persistent dust cover is a major hindrance for reliable remote mineralogical identification. For example, in CRISM VNIR spectra, the mineralogy of only the top few tens of micrometers of the surface can be detected.

Fortunately, CRISM VNIR data from the volcanic complex of Syrtis Major—one of the lowest elevation volcanoes and most dust-free regions on Mars—revealed robust evidence of regional hydrothermal activity in a volcanic geological setting. Multiple distinct opaline silica mound structures identified at this site are similar in mineralogy and morphology to those associated with terrestrial hot springs on Earth (Skok et al., 2010).

7.4 MARS HOT SPRING DEPOSITS AT NILI PATERA

At the center of Syrtis Major is a series of nested caldera depressions, the best defined of which is a 50-km-wide depression in Nili Patera (Fig. 7.3). The caldera is unique among martian volcanic terrains in that it hosts evidence of both effusive and explosive volcanism, nearly monomineralic silica deposits, and compositional diversity that ranges from olivine-rich basalts to silica-enriched units (Fawdon et al., 2015).

---

[4] To distinguish mineralogically distinctive terranes on a Mars requires a spatial resolution of at least ~9 pixels at the ~ 20 m per pixel resolution of the CRISM instrument and the absence of a dust cover.

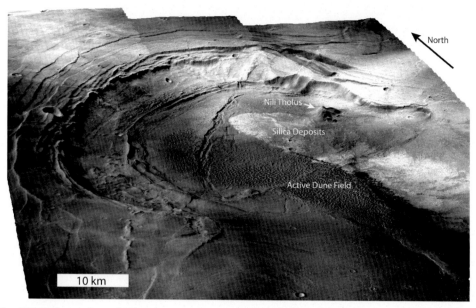

Fig. 7.3 Oblique view of Nili Patera caldera on Mars illustrates the variety of regional geological features associated with this setting. Silica sinter deposits and sinter-type mounds have been identified on and around the Nili Tholus volcanic cone (Skok et al., 2010).

Northeast of the caldera is the 300-m-high volcanic cone Nili Tholus. White-toned deposits were reported by Skok et al. (2010) as opaline silica, which was detected in CRISM spectral data on and around the cone (Fig. 7.4).

As shown in Fig. 7.5, CRISM data also revealed that—when normalized to the surrounding materials—the proximal white-tone deposits have a strong, asymmetric 2.21-μm VNIR absorption feature, which is caused by a combination vibrational absorption (OH stretch and SiOH bend) (Skok et al., 2010). These deposits also have weak (even for Mars) hydration absorptions at 1.4 and 1.9 μm (cf., Skok et al., 2010; Sun et al., 2016a,b). In combination, these spectral absorptions are consistent with an opal-A sinter composition, an interpretation strengthened by the geological context and distribution of the deposits on, and adjacent to, the Nili Tholus volcanic cone within the Nili Patera caldera. Since the initial report of the discovery of these silica sinter deposits around Nili Tholus, additional Si-OH spectral signatures have been identified in several areas within the volcanic flows. Deposits with similar absorptions were also detected in regions to the west and southwest of the cone, all lying on an evolved silica-enriched unit identified by thermal infrared observations with the use of THEMIS (Christensen et al., 2005).

In contrast to the discrete and point source nature of the near-cone mound deposits, opaline silica deposits to the west tend to be laterally more continuous (e.g., inset, Fig. 7.5). Their distribution may represent the remains of a transient hydrothermal system

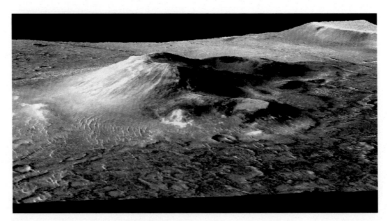

Fig. 7.4 Oblique CRISM colored HiRISE DEM project image of Nili Tholus volcanic cone. White-toned silica sinter deposits occur as mounds similar to those observed in large hydrothermal fields on Earth. Sinter deposits beneath the mound shown in the center of the image, located on the flank of Nili Tholus cone, radiates in a fan-shaped deposit oriented downslope, consistent with channelized outflow deposits from terrestrial hydrothermal effluents (Skok et al., 2010). *Image Credit: NASA/JPL/ MSSS/JHU-APL, CTX: P04_002427_1888_XI_08N292W, CRISM: FRT00010628.*

driven by volcanic heat flow and the release of subsurface volatiles. Alternatively, the laterally continuous nature of the westernmost silica deposits, in contrast to the distinct mound structures of the proximal deposits, could be the result of enhanced hydrothermal venting and subsequent opaline silica deposition along shallow subterranean fractures.

Hydrothermal silica deposits to the southwest of Nili Tholus lie within the circumference of a region that consists of ejecta of two ~100-m-wide craters. Such features are consistent with opaline silica deposition during active volcanism. Syrtis Major is estimated to have been active for about 100 million years and experienced multiple caldera–forming events (Hiesinger and Head III, 2004). While hydrothermal systems may have been active throughout this entire period in this region of Mars, only the events younger than any local volcanic resurfacing or dust burial would be visible from orbit. Such a long-lived, stable subterranean hydrothermal system could have generated numerous surface springs, each of which had the potential to support life and preserve it in paleobiological sinter repositories if Mars was ever inhabited by extremophilic bacteria.

7.5 OPALINE SILICA DEPOSITS AT COLUMBIA HILLS

Robust evidence for hot spring sinter deposits was discovered near Gusev crater during ground exploration by the Spirit rover during the twin Mars Exploration Rover (MER) missions. The Spirit rover landed in 2004 on relatively young, Hesperian, lava plains. In addition to other sedimentary rock types, hydrothermal deposits were predicted to occur

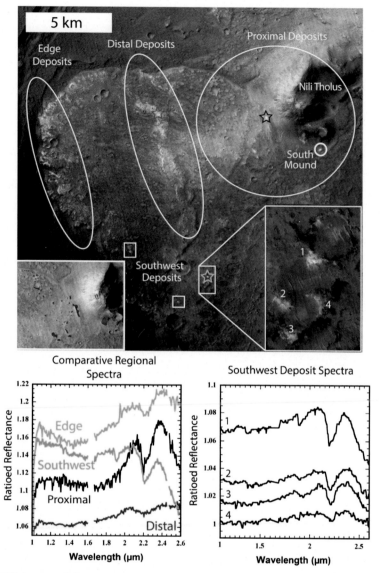

Fig. 7.5 HiRISE image of the regions analyzed by CRISM as part of NASA's Mars Reconnaissance Orbiter mission. Colored stars in the HiRISE image correspond to colored spectrum in the CRISM VNIR spectral plots, the latter of which are characteristic of the data acquired from the silica sinter mounds. All spectra display a broad asymmetrical spectral absorption feature at 2.21 μm, indicative of water in opaline silica, though the feature is notably weaker for the distal deposits. Left: HiRISE of Southwest deposits excavated by impact. Right: Proximal deposits of multiple mounds (Background HiRISE CTX: B05_011459_1891_XI_09N292W, Southwest inset: HiRISE: PSP_005684_1890).

within Gusev crater, likely sourced from exposed deposits that rimmed and surrounded smaller impacts (Cabrol et al., 2003). After exploring Gusev crater via multiple ground traverses, the Spirit rover traversed eastward for 6 months toward the Columbia Hills.

The terrain at the Columbia Hills was found to be remarkably different from the volcanic terrain located at the landing site, consisting of older Noachian terrains that lie topographically higher than the surrounding, younger volcanics. From the top of Husband Hill, a panoramic survey photographed by Spirit revealed a pentagon-shaped light-toned feature in the valley below. As shown in Fig. 7.6, this ~80-m-wide feature, named Home Plate, became a prime mission objective. The rover reached Home Plate on sol 744 and began a multiyear study of the surrounding terrain.

In the Eastern Valley, located between Home Plate and Mitcheltree Ridge, the results of multiple in situ instruments showed that the light-toned nodules found in association with hydrated ferric sulfates were enriched in Si. Alpha Particle X-Ray Spectrometer (APXS) analysis revealed their highly Si-enriched nature relative to the hydrated ferric sulfates located in nearby rocks and soils (Ming et al., 2008). Miniature Thermal Emission Spectrometer (Mini-TES) analysis of outcrop nodules known as the "Tyrone-nodules" (sols 1100 and 1101), named for its proximity to the Tyrone sulfate-rich soil deposit

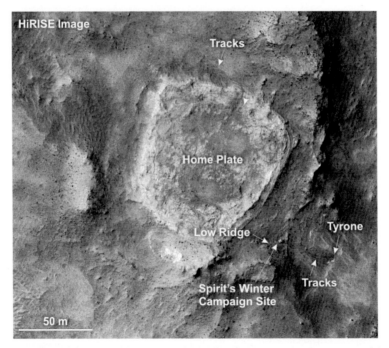

Fig. 7.6 HiRISE image of the Home Plate region visited by the Spirit rover during one of the twin MER rover missions. Most of the investigations were conducted on the eastern edge of the deposits with the excavated silica soils and digitate features identified near the Tyrone unit (NASA image).

(Yen et al., 2008; Wang et al., 2008), revealed that the light-toned materials produced an opaline silica spectral signature comparable with recent hydrothermal sinter deposits on Earth (e.g., Fig. 3 of Squyres et al., 2008; Ruff et al., 2011). On sol 1148, a nearby underlying patch of light-toned sediment (named Gertrude Wiese (e.g., Fig. 2 in Squyres et al., 2008 Navcam frame 2N233253342) was exposed after the Spirit rover drove over an encrusted, poorly consolidated unit that crumbled due to the resistance created when the broken wheel of the rover dragged across its surface. Subsequent APXS elemental analysis of this white-toned sediment indicated that it ranged in composition from ~65% to 92% wt% SiO_2 (Squyres et al., 2008), the latter concentration representing the highest silica content found to date on Mars. An example of the nodular opaline silica in outcrop is shown in Fig. 7.7.

Additional analytical measurements made via the Spirit rover on soils and outcrops around Home Plate over the next several months revealed opaline silica deposits in a variety of locations in the region. By sol 1220, 17 Mini-TES outcrop measurements that were consistent with the presence of opaline silica, though contaminated to varying degrees by silica-poor soil and dust, were acquired before analysis had to be stopped due to interferences created by the accumulation of airborne dust on the Mini-TES optics. Collectively, the silica-rich materials were interpreted by Squyres et al. (2008)

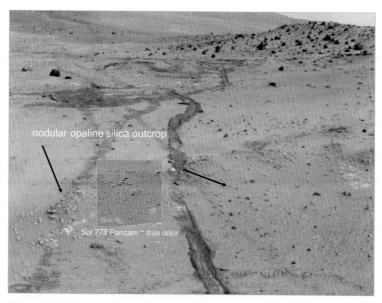

Fig. 7.7 Nodular opaline silica outcrops adjacent to Home Plate (Navcam mosaic, sol 1116, rover wheel tracks are ~1 m apart). Pancam approximate true color image of opaline silic nodules before rover traverse (ATC; sol 778, P2388). The approximately linear distribution of the opaline silica nodules (from lower left to middle right of Navcam image) is similar to runoff channel-like deposits at active silica-depositing hot springs found worldwide (NASA images).

as having formed under low-pH hydrothermal conditions, either as fumarole-related acid-sulfate leaching of basalts or precipitated as low-pH silica sinter. Low-pH hydrothermal conditions were favored because of the trends in major element enrichment (Si and Ti) and depletion (Fe, Na, Al) relative to other Gusev crater volcanic materials and the proximity of the opaline materials and the ferric sulfates, the latter of which were also interpreted as having a probable low-pH hydrothermal origin (Yen et al., 2008).

A critical examination of the characteristics and distribution of opaline silica led Ruff et al. (2011) to conclude that the silica-rich deposits around Home Plate originated as the erosion of a laterally persistent, stratigraphically restricted interval of silica sinter. They found no spectroscopic evidence to indicate that the opaline silica had been diagenetically matured beyond opal-A and no geochemical evidence in the deposits that ruled out the precipitation of the opaline silica by near-neutral pH thermal spring fluids.

Morphological comparisons of several features imaged in silica deposits at Home Plate with those of silica sinters associated with hot springs at El Tatio, Chile, strengthened the interpretation that the martian silica-rich deposits originated in hot spring discharge outflow channels (Ruff, 2015; Ruff and Farmer, 2016). The presence of centimeter-size pieces of nodular masses of opaline silica with different types of digitate protrusions of various length, shape, and orientation, along with their distribution in stratiform outcrops on the floors of local topographic lows, are remarkably consistent with the morphology and distribution of nodular sinters produced in outflow channels and debris aprons covered with water of various depth around some silica-depositing hot springs at El Tatio (Fig. 7.8). Spectroscopic analysis of halite-encrusted nodular and digitate silica structures

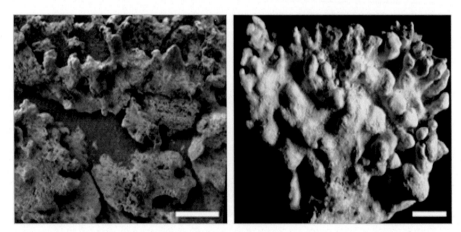

Fig. 7.8 The morphology and distribution of digitate sinters discovered near Home Plate on Mars (grayscale Microscopic Imager mosaic (sol 1157)) are remarkably similar to nodular sinters that formed in outflow channels and debris aprons covered with water of various depth around hot springs at El Tatio. White bar scale represents 1 cm in both images. Image credit: Ruff and Farmer (2016).

at El Tatio also produces infrared spectra that are most similar to the spectra obtained from morphologically similar structures in the siliceous sinter deposits at Home Plate (cf., Fig. 5a, Ruff and Farmer, 2016).

7.6 THE LIKELIHOOD OF FINDING MORE HOT SPRING DEPOSITS ON MARS

The question arises as to whether additional hydrothermal deposits will be found with further exploration, given that these two detections are the basis of our understanding of spring deposits on Mars. After the caldera spring deposits were found in Nili Patera, a significant effort was made to search the other martian calderas for orbital evidence of hot spring-like deposits characterized by similar morphology and composition. However, the concerted effort resulted in no new findings, and hence the question is whether Nili Patera is fundamentally unique, either in its geological setting or because the region is relatively dust free.

A comparative analysis of fundamental differences in the geological setting of the Nili Patera caldera and Syrtis Major volcanic region indicates that their geological setting is unique. The Nili Patera caldera surface lies in an $\sim$2-km depression that likely formed as a result of the crustal relaxation of the Isidis basin (Fig. 7.9). This setting contrasts that of most of the main Hesperian and Amazonian volcanics, which lie topographically higher than the volcanic terrain that surrounds Syrtis Major (estimated to be $\sim$500-m thick, Hiesinger and Head III, 2004). Regional crustal relaxation would also have created fractures that could have enhanced volcanism and caldera collapse. The latter would be expected to produce a longer-lived source of high-silica hydrothermal fluid and more voluminous spring sinter deposits at surface effluents (e.g., generated from a deeper part of the volcano closer to differentiating high-Si magma bodies). Syrtis Major is surrounded by hydrated Noachian crust, unlike all other calderas on Mars that are surrounded by their own volcanics. Such high concentrations of dissolved silica in crustal hydrothermal fluid could have sustained large-scale hydrothermal systems and made spring deposits common in this region.

The discovery of siliceous hot spring deposits at Home Plate was fortuitous, even though hydrothermal deposits were predicted as potential sedimentary rock in Gusev crater (Cabrol et al., 2003; Schwenzer et al., 2012). If widespread volcanism during this period on Mars drove hydrothermal systems that fed abundant spring systems across the surface of the planet (including possible deep-sea hydrothermal deposits, cf., Michalski et al., 2017), hot spring deposits like those found at Home Plate could be more common in Noachian terrains. They may not, however, be obvious; extensive erosion would have erased much of the volcanic context and global dust layers would have covered local deposits. Such factors may inhibit the remote detection of locally restricted outcrops

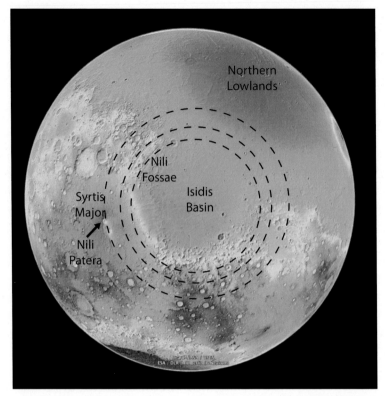

Fig. 7.9 Global map of Mars shows the location of Syrtis Major along the edge of the Isidis Basin. The eruption conduits and calderas may lie on a region of weakness formed by basin relaxation, which enhances the collapse.

of sinter from orbit, as they did at Home Plate. To test this hypothesis, further exploration of Noachian-age terrains is required.

7.7 GEOCHEMICAL CONSIDERATIONS

Given the necessary geochemical context (i.e., some type of heat source and subsurface water), it might be considered strange *not* to find silica-rich hydrothermal deposits distributed across Mars. The average martian crust is basaltic (Taylor and McLennan, 2010) with a relatively narrow range of silica content (45–52 wt%), inferred primarily on evidence from martian meteorites, orbital spectroscopic observations, and in situ rover analyses of rocks interpreted as igneous (including volcaniclastics). Orbital spectral signatures indicate that basalt and basaltic minerals (i.e., pyroxene, plagioclase, and olivine) dominate much of the martian surface. Analytical comparison of the geochemical composition of unconsolidated regolith located thousands of kilometers apart by the *Spirit* and *Opportunity* rovers and the Mars Science Laboratory (MSL) mission rover *Curiosity* indicates regional basaltic sources (Yen et al., 2008; O'Connell-Cooper et al., 2017). Dust measured by the three rovers is also basaltic, with enrichments in volatile and moderately volatile elements (S, Cl, and Zn) (Yen et al., 2008; Berger et al., 2016). Because global dust storms have been observed regularly (about every three martian years), and the dust has a uniform composition (like at these three rover sites), it is considered a global geological unit that represents an average sampling of the Mars surface. Indications of magmatic diversity have been discovered (e.g., Papike et al., 2009; Thompson et al., 2016; Treiman et al., 2016); however, evidence of high-silica lithologies that formed by igneous fractionation processes (e.g., andesite, trachyte, rhyolite, granite) is limited and often ambiguous, suggesting that this pathway to silica enrichment is not widespread on Mars.

On Earth, aquifers with abundant mafic minerals (pyroxene and olivine) usually have the highest silica concentrations. Pyroxene, olivine, and glass in basalt weathers readily, releasing silica into solution. Silica concentrations can be even higher in volcaniclastic deposits, where small irregular grain shape, small grain size, and increased porosity increase the surface area of the sediments to promote chemical and physical weathering processes. Another consequence of weathering is that groundwater in pyroxene- and olivine-rich volcaniclastic units can have ~5 times more dissolved silica than in average groundwater (Langmuir, 1997). Increasing temperature can increase silica solubility by up to two orders of magnitude, leading to concentrations ~5–30 times higher in hydrothermal systems on Earth. Thus, a hydrothermal system in a basaltic regime has a high potential for mobilizing and concentrating silica.

The buffering capacity of basalt must be considered when predicting and modeling silica solubility in hydrothermal fluids in a basaltic setting such as the martian crust. Silica solubility generally does not change with pH in circumneutral to acidic waters; however, silica solubility increases significantly in alkaline fluids (pH > 10). Thus, buffering of weathering reactions influences dissolved silica concentrations. Weathering of basalt by mildly acidic fluids will lead to the release of cations into solution, which buffers the pH to circumneutral or mildly alkaline levels. In rare ultramafic settings on Earth, this process can lead to a pH of 12 in Ca^{2+}-OH^- waters (Barnes et al., 1978). In open weathering systems on Earth, atmospheric CO_2 provides acidity via carbonic acid formation at pH <6, which promotes the formation of carbonates in basaltic settings. Atmospheric CO_2 is a plausible source of acidity for ancient Mars, but evidence of sulfate deposits greatly exceeds that of carbonates, an indication that sulfuric acid may have inhibited massive carbonate formation and dominated weathering. Alternative hypotheses suggest that carbonate deposits—in addition to those already detected (Ehlmann et al., 2008)—are deeply buried, or that they formed early and were dissolved by acidic fluids later in Mars' history. Nevertheless, at low water-to-rock ratios and in highly concentrated hydrothermal brines, the buffering capacity of basalt can be exceeded and acidic or alkaline solutions can evolve.

7.8 COMPETING HYPOTHESES FOR THE ORIGIN OF SILICA-RICH DEPOSITS ON MARS

The traverses of both MER *Spirit* and MSL *Curiosity* rovers led to the discovery of silica-rich materials that illustrate the diversity of silica-rich systems on Mars. In Gusev crater, the rover *Spirit* encountered highly localized silica enrichments in soils and rocks; whereas, at the base of Mt. Sharp in Gale crater, the rover *Curiosity* encountered localized high-silica rocks as well as a major unit with moderately elevated silica. A comparison of the occurrence, distribution, and potential source mechanisms of high-silica concentrations in rocks and soils analyzed at the two sites provides insights about silica-depositing systems on Mars.

Home Plate. The association of the silica-rich unconsolidated sediment (soil) and nodular outcrop (up to 92 wt% SiO_2) with volcanic tephra comprising Home Plate is consistent with formation under hydrothermal conditions. Two interpretations for the silica enrichment have been proposed: (1) the high silica is residue after the leaching of a basaltic precursor by acid-sulfate solutions (Squyres et al., 2008; Milliken et al., 2008) and (2) silica-saturated hydrothermal fluids precipitated opaline silica as sinter (Ruff et al., 2011).

Low pH solution leaching of the basaltic precursors in the Home Plate area is suggested by the compositional diversity of the silica-rich rocks (Squyres et al., 2008). Additionally, the apparent enrichment of Ti with the silica, both of which have low solubility in low pH fluids, indicates the Ti may be a residue derived from primary titanomagnetite in the basalt. Acid-sulfate solution weathering processes are also evident in the sulfate-rich soils of Gusev crater. These soils contain 4–33 wt% silica, and thus the sulfate and silica enrichments could have a common origin.

Precipitation of silica from high pH alkaline hydrothermal fluids can account for the nodular morphology and unique textures in the silica-rich rocks near Home Plate, which would be less likely to occur by way of alteration of basaltic rocks in the area (Ruff et al., 2011). Low pH acid-sulfate alteration commonly results in the addition of sulfur, but the silica-rich rocks and soils in Gusev crater have low sulfur content. In addition, Ti can be mobilized in alkali-chloride brines and precipitate as anatase (e.g., Kinsinger et al., 2010; Campbell et al., 2015b); thus, the Ti enrichment in opaline silica in Gusev crater is not necessarily an indicator of low pH conditions.

Gale crater. Silica-rich materials discovered within Gale crater have four different occurrences, two of which are grouped as altered sedimentary bedrock units (Rampe et al., 2017), and the third is in fracture-associated haloes (Yen et al., 2017). A fourth occurrence is a group of high-alkali mugearitic rocks (a class named Jake M). Silica enrichment in the Jake M rocks has been interpreted to be due to metasomatic and igneous processes (Stolper et al., 2013) and is less relevant to our discussion of the origin of primary hydrothermal silica deposits.

The altered sedimentary bedrock units have elevated silica concentrations (i.e., most in the 45–55 wt% range), though one tridymite-bearing unit named Buckskin has up to 75 wt% silica (Morris et al., 2016). The mineralogy of these units varies in detail, but crystalline SiO_2 and high-Si amorphous material comprise about 25–60 wt% of the samples analyzed by the Chemistry and Mineralogy (CheMin) X-ray diffractometer (XRD) instrument (Rampe et al., 2017). The crystalline silica phases in these units include, in variable fractions, cristobalite and tridymite, and microcrystalline opal-CT. Quartz is present at ∼1 wt% or less. The fracture-associated haloes have silica concentrations up to ∼70 wt%. In contrast with the microcrystalline-to-crystalline silica bedrock units, silica in the haloes is largely X-ray amorphous (Yen et al., 2017).

Conflicting interpretations have been proposed to account for silica enrichment in Gale crater. Several models suggest that acidic diagenetic fluids leached basaltic sedimentary bedrock of cations (Al, Ca, Mg, Fe, Mn, Ni, and Zn), while Si and Ti were retained as residue (Rampe et al., 2017; Yen et al., 2017; Berger et al., 2017). Jarosite, a sulfate that forms at low pH, was detected in CheMin XRD data, which support the acid-sulfate fluid alteration scenario (Rampe et al., 2017). Both low (diagenetic < 150°C) and high (hydrothermal > 150°C) temperature fluids have been invoked in geochemical leaching models. One interpretation is that silica enrichment, via precipitation from either circumneutral or alkaline diagenetic fluids, occurred via chemical weathering during erosion and transport of sediment into the crater (Frydenvang et al., 2017; Hurowitz et al., 2017). The addition of high-silica phases to the sediments analyzed in the crater, which includes detrital material such as tridymite in the Buckskin unit (a likely product of silicic volcanism), would enhance silica enrichment in Gale crater (Morris et al., 2016). Reconciling these different hypotheses is challenging, in part, because the provenance of the Gale crater sedimentary rocks is poorly constrained.

At present, the same conundrum faced in interpreting the origin and potential importance of the silica deposits at Home Plate near the Gusev crater is plaguing interpretation of the opaline silica detected at Gale crater. Are the deposits leached basalts, primary aqueous hydrothermal precipitates, or diagenetic precipitates, and what was the temperature and pH of the fluid from which the opaline silica precipitated?

7.9 SITE SELECTION CONSIDERATIONS RELEVANT TO THE RETURN TO MARS

On March 22, 2010, the Spirit rover stopped operations while still investigating the Home Plate region, leaving many unanswered questions about the formation of the hydrothermal opaline sinters and their potential to host biosignatures. These questions and the detailed exploration by the Spirit rover led to selection of the Home Plate target and Columbia Hills, considered representative of hydrothermal spring environments, as one of the final three options for the Mars 2020 rover-landing site. The other two landing site options include the river delta in Jezero crater (Goudge et al., 2017, 2018), considered representative of fluvial and lacustrine environments, and Northeast Syrtis (Bramble et al., 2017), considered representative of deep crustal environments and cold springs.

The Mars 2020 mission is designed to identify ancient environments capable of supporting microbial life, and seek signs of possible past microbial life in those habitable environments, particularly in specific rocks known to preserve signs of life over time. The rover will explore sites that could have supported life as we know it on Earth and then collect and cache samples that may be returned to Earth as part of a subsequent Mars sample return mission. Returned samples can be analyzed in greater detail with higher-resolution instruments in the laboratory to determine whether life established itself on Mars and if any unique biosignatures indicative of life are preserved in such materials.

As a landing-site candidate, Columbia Hills has had mixed support. Its strength as a sample return target is often seen as a weakness to many in the scientific community, i.e., that is, the site has already been explored. As one of the only three locations on Mars to have been explored by a long duration rover, extensive details are known about the target rocks and intended samples, including the presence of multiple sedimentary facies in the hydrothermal opaline silica deposits and the morphological similarities of macrostructures to hot spring sinters. Thus, the strength of this target is that more detailed exploration and analysis of a known site with potential biosignatures is likely to be critical for astrobiology missions. Earth-based studies have shown that it can take years to decades to understand the interplay of geochemical, hydrological, and biological processes that characterize specific habitats. Having ground-truthed data for a known site on Mars would result in significantly more strategic exploration of a potential ecosystem (Cabrol, 2018). For a mission designed to quickly sample and cache the rocks of interest, knowing the rocks and their location and context would be a significant advantage. For scientists eager to explore the diversity of ancient martian geology, there would be a missed opportunity if such a major rover and return mission was dedicated to incrementally advancing a previously explored location when so much about the planet is still unknown.

The case for returning to Columbia Hills is centered on the realization that opaline silica sinter at the site exhibits potential morphological biosignatures—millimeter-scale digitate features—that are texturally similar to those found in terrestrial hot spring systems, such as the El Tatio, Chile, geothermal field (Ruff, 2015; Ruff and Farmer, 2016). The high elevation, correspondingly lower pressure, high UV, and high evaporation rates at El Tatio (Nicolau et al., 2014) make it arguably the most Mars-like of the known terrestrial spring deposits. Geomorphic mapping has demonstrated macroscale and microscale similarities in the digitate deposits at El Tatio and those on Mars. This mapped, crossplanetary feature would be the primary target of a Mars 2020, or future, mission to Columbia Hills.

On Earth, digitate sinter features preserve abundant microscale evidence of the microbes that colonized their surfaces when the opaline silica structures formed (Cady and Farmer, 1996; Braunstein and Lowe, 2001; Jones and Renaut, 2003; Lowe and Braunstein, 2003; Handley et al., 2005, 2008). Digitate sinters on Earth are associated with hot spring features in which microbial life thrived throughout our planet's history. Whether the macroscale and microscale biogenic characteristics of digitate sinters are a requirement for their formation or just the result of having formed on the biologically ubiquitous Earth is still an active area of research.

The opaline silica digitate sinters on Mars represent some of the strongest evidence to date for potential biosignatures on the planet (Ruff and Farmer, 2016). Until a future mission returns to the Columbia Hills, or a rover traverses to Nili Patera, or encounters siliceous hot spring deposits at some other site on Mars, a number of key outstanding questions with regard to the potential for life to have gained a foothold and evolve on Mars remain unanswered. As discussed in this chapter, until sinter deposits on Mars are proven to be void of biosignatures, silica hot springs deposits on Mars remain one of the more compelling astrobiology targets known in our Solar System.

ACKNOWLEDGMENTS

SLC, VG, and NWH thank and acknowledge financial support from the NASA Astrobiology Institute under the SETI Institute NAI Team's grant NNX15BB01A and NASA Interagency NNA16BB06I. Additional financial support for SLC was provided by EMSL, a DOE Office of Science User Facility sponsored by the Office of Biological and Environmental Research. JRS thanks and acknowledges financial support from the NASA PSTAR Program under grant NNX15AJ38G. JB thanks the University of Guelph. Research by SCL within Yellowstone National Park has been supported by the National Park Service permit number 1994. The authors thank the referees for comments and suggestions that improved the chapter.

REFERENCES

Allen, E.T., 1934. The agency of algae in the deposition of travertine and silica from thermal waters. Am. J. Sci. 28, 373–389.

Allen, C.C., Oehler, D.Z., 2008. A case for ancient springs in Arabia Terra, Mars. Astrobiology 6, 1093–1122.

Alleon, J., Bernard, S., Le Guillou, C., Daval, D., Skouri-Panet, F., Pont, S., Delbe, L., Robert, F., 2016. Early entombment within silica minimizes the molecular degradation of microorganisms during advanced diagenesis. Chem. Geol. 437, 98–108.

Amores, D.R., Warren, L.A., 2007. Identifying when microbes biosilicify: the interconnected requirements of acidic pH, colloidal SiO_2 and exposed microbial surface. Chem. Geol. 240, 298–312.

Amores, D.R., Warren, L.A., 2009. Metabolic patterning of biosilicification. Chem. Geol. 268, 81–88.

Barbieri, R., Cavalazzi, B., Stivaletta, N., López-García, P., 2014. Silicified biota in high-altitude, geothermally-influenced ignimbrites at El Tatio Geyser Field, Andean Cordillera (Chile). Geomicrobiol J. 31, 493–508.

Barnes, I., O'Neil, J.R., Trescases, J.J., 1978. Present day serpentinization in New Caledonia, Oman and Yugoslavia. Geochim. Cosmochim. Acta 42, 144–145.

Benning, L.G., Phoenix, V.R., Yee, N., Tobin, M.J., 2004a. Molecular characterization of cyanobacterial silicification using synchrotron infrared micro-spectroscopy. Geochim. Cosmochim. Acta 68 (4), 729–741.

Benning, L.G., Phoenix, V.R., Yee, N., Konhauser, K.O., 2004b. The dynamics of cyanobacterial silicification: an infrared micro-spectroscopic investigation. Geochim. Cosmochim. Acta 68, 743–757.

Berger, J.A., Schmidt, M.E., Gellert, R., Campbell, J.L., King, P.L., Flemming, R.L., Ming, D.W.,
Clark, B.C., Pradler, I., Van Bommel, S.J.V., Minitte, M.E., Fairén, A.G., Boyd, N.I.,
Thompson, L.M., Perrett, G.M., Elliott, B.E., Desouza, E., 2016. A global Mars dust composition
refined by the alpha-particle X-ray spectrometer in Gale Crater. Geophys. Res. Lett. 43, 67–75.

Berger, J.A., Schmidt, M.E., Gellert, R., Boyd, N.I., Desouza, E.D., Flemming, R.L., et al., 2017. Zinc and
germanium in the sedimentary rocks of Gale Crater on Mars indicate hydrothermal enrichment followed
by diagenetic fractionation. J. Geophys. Res. Planets. 2017JE005290. https://doi.org/10.1002/
2017JE005290.

Blank, C.E., Cady, S.L., Pace, N.R., 2002. Microbial composition of near-boiling silica-depositing thermal
springs throughout Yellowstone National Park. Appl. Environ. Microbiol. 68, 5123–5135.

Boomer, S.M., Noll, K.L., Geesey, G.G., Dutton, B.E., 2009. Formation of multilayered photosynthetic
biofilms in an alkaline thermal spring in Yellowstone National Park, Wyoming. Appl. Environ. Microbiol. 75 (8), 2464–2475.

Bramble, M.S., Mustard, J.F., Salvatore, J.R., 2017. The geological history of Northeast Syrtis Major, Mars.
Icarus 293, 66–93.

Braunstein, D., Lowe, D.R., 2001. Relationship between spring and geyser activity and the deposition and
morphology of high temperature (>73°C) siliceous sinter, Yellowstone National Park, Wyoming,
U.S.A. J. Sediment. Res. 71, 747–763.

Brock, T.D., 1978. Thermophilic Microorganisms and Life at High Temperatures. Springer, New York
478 pp.

Bryant, D.A., Garcia Costas, A.M., Maresca, J.A., Chew, A.G., Klatt, C.G., Bateson, M.M., et al., 2007.
Candidatus Chloracidobacterium thermophilum: an aerobic phototrophic acidobacterium. Science
317, 523–526. https://doi.org/10.1126/science.1143236.

Buchanan, L.J., 1981. Precious metal deposits associated with volcanic environments in the southwest.
In: Dickson, W.R., Payne, W.D. (Eds.), Relations of Tectonics to Ore Deposits in the Southern
Cordillera, vol. 14. Arizona Geological Society Digest, pp. 237–262.

Cabrol, N., 2018. The coevolution of life and environment on Mars: an ecosystem perspective on the
robotic exploration of biosignatures. Astrobiology 18, 1–27.

Cabrol, N.A., Grin, E.A., Carr, M.H., Sutter, B., Moore, J.M., Farmer, J.D., Greeley, R., Kuzmin, R.O.,
Des Marais, D.J., Kramer, M.G., Newson, H., Berber, C., Thorsos, I., Tanaka, K.L., Barlow, N.G.,
Fike, D.A., Urquhart, M.L., Grigsby, B., Grant, F.D., de Goursac, O., 2003. Exploring Gusev crater
with Spirit: review of science objectives and testable hypotheses, J. Geophys. Res. 108 (E12), 8076.

Cady, S.L., 2002. Formation and preservation of bone-fide microfossils. In: Signs of Life: A Report Based on
the April, 2000 Workshop on Life-Detection Techniques. The National Academies Space Studies Board
and Board on Life Sciences, Carnegie Institution of Washington, National Academies Press, Washington,
DC, pp. 149–155.

Cady, S.L., Farmer, J.D., 1996. Fossilization processes in siliceous thermal springs: trends in preservation
along thermal gradients. In: Bock, G.R., Goode, J.A. (Eds.), Hydrothermal Ecosystems on Earth (and
Mars?), Ciba Foundation Symposium 202. John Wiley and Sons, Chichester, pp. 214–235.

Cady, S.L., Wenk, H.-R., Downing, K.H., 1996. HRTEM of microcrystalline opal in chert and porcelanite
from the Monterey Formation, California. Am. Mineral. 81, 1380–1395.

Cady, S.L., Wenk, H.-R., Sintubin, M., 1998. Microfibrous quartz varieties: characterization by quantitative
X-ray texture analysis and transmission electron microscopy. Contrib. Mineral. Petrol. 130, 320–335.

Campbell, K.A., Sannazzaro, K., Rodgers, K.A., Herdianita, N.R., Browne, P.R.L., 2001. Sedimentary
facies and mineralogy of the Late Pleistocene Umukuri silica sinter, Taupo Volcanic Zone, New Zealand.
J. Sediment. Res. 71, 727–746.

Campbell, K.A., Guido, D.M., Gautret, P., Foucher, F., Ramboz, C., Westall, F., 2015a. Geyserite in hot-
spring siliceous sinter: window on Earth's hottest terrestrial (paleo) environment and its extreme life.
Earth Sci. Rev. 148, 44–64.

Campbell, K.A., Bridget, Y., Lynne, K.M., Handley, S.J., Farmer, J.D., Guido, D.M., Foucher, F.,
Turner, S., Perry, R.S., 2015b. Tracing biosignature preservation of geothermally silicified microbial textures into the geological record. Astrobiology 15 (10), 858–882.

Castenholz, R.W., 1969. Thermophilic blue-green algae and the thermal environment. Bacteriol. Rev. 33 (4), 476–504.

Castenholz, R.W., Pierson, B.K., 1995. Ecology of thermophilic anoxygenic phototrophs. In: Blankenship, R.E., Madigan, M.T., Bauer, C.E. (Eds.), Anoxygenic Photosynethic Bacteria. Kluwer Academic Publishers, The Netherlands, pp. 87–103.

Cavalazzi, B., Glamocljia, M., Barbieri, R., Cady, S.L., 2018. Exobiology, life formation and planetary exploration. In: Rossi, A.P., van Gasselt, S. (Eds.), Planetary Geology. Springer International Publishing, Cham, Switzerland, pp. 347–367.

Christensen, P.R., McSween Jr., H.Y., Bandfield, J.L., Ruff, S.W., Rogers, A.D., Hamilton, V.E., Gorelick, N., Wyatt, M.B., Jakosky, B.M., Kieffer, H.H., Malin, M.C., Moersch, J.E., 2005. Evidence for magmatic evolution and diversity on Mars from infrared observations. Nature 436, 504–509.

Crumpler, L.S., 2003. Physical characteristics, geologic setting, and possible formation processes of spring deposits on Mars based on terrestrial analogs. Sixth International Conference on Mars, Abstract #3228.

Crumpler, L.S., Aubele, J.C., Zimbelman, J.R., 2007. Volcanic features of New Mexico analogous to volcanic features on Mars: Mars analogs. In: Chapman, M.G. (Ed.), The Geology of Mars, Evidence for Earth-Based Analogs. (Chapter 10). Cambridge University Press, Cambridge, pp. 95–125.

Dai, J., 2017. New insights into a hot environment for early life. Environ. Microbiol. Rep. 9, 203–210.

Deamer, D., 2012. First Life. Discovering the Connections Between Stars, Cells, and How Life Began. University of California Press, Berkeley and Los Angeles, CA, USA. 288 pp.

Deamer, D.W., Szostak, J.W. (Eds.), 2010. The Origins of Life. Cold Spring Harbor Laboratory Press, Cold Spring Harbor, NY.

Djokic, T., Van Kranendonk, M.J., Campbell, K.A., Walter, M.R., Ward, C.R., 2017. Earliest signs of life on land preserved in ca. 3.5 Ga hot spring deposits. Nat. Commun 8, 15263.

Doolittle, W.F., 1999. Phylogenetic classification and the universal tree. Science 284, 2124–2129.

Ehlmann, B.L., Mustard, J.F., Murchie, S.L., Poulet, F., Bishop, J.L., Brown, A.J., Calvin, W.M., Clark, R.N., Des Marais, D.J., Milliken, R.E., Roach, L.H., Roush, T.L., Swayze, G.A., Wray, J.J., 2008. Orbital identification of carbonate-bearing rocks on Mars. Science 322 (5909), 1828–1832.

Ellis, A.J., Mahon, W.A.J., 1977. Chemistry and Geothermal Systems. Academic Press. 392 pp.

Farmer, J.D., 1996. Hydrothermal systems on Mars: an assessment of present evidence. In: Bock, G.R., Goode, J.A. (Eds.), Hydrothermal Ecosystems on Earth (and Mars?), Ciba Foundation Symposium 202. John Wiley and Sons, Chichester, pp. 273–299.

Farmer, J.D., 1998. Thermophiles, early biosphere evolution and the origin of life on Earth: implications for the expbiological exploration of Mars. J. Geophys. Res. 103, 28457–28461.

Farmer, J.D., 1999. Taphonomic modes in microbial fossilization. Size Limits of Very Small Microorganisms: Proceedings of a Workshop. Washington National Academy Press, Washington, DC.

Farmer, J.D., Des Marais, D.J., 1999. Exploring for a record of ancient Martian life. J. Geophys. Res. 104 (E11), 26977–26995.

Farrand, W.H., Gaddis, L.R., Keszthlyi, L., 2005. Pitted cones and domes on Mars: observations in Acidalia Planitia and Cydonia Mensae using MOC, THEMIS, and TES data. J. Geophys. Res. 110. https://doi.org/10.1029/2004JE002297.

Fawdon, P., Skok, J.R., Balme, M.R., Vye-Brown, C.L., Rothery, D.A., Jordon, C.J., 2015. The geological history of Nili Patera, Mars. J. Geophys. Res. Planets. 120. https://doi.org/10.1002/2015JE004795.

Fournier, R.O., 1989. Geochemistry and dynamics of the Yellowstone National Park Hydrothermal System. Annu. Rev. Earth Planet. Sci. 17, 13–53.

Frydenvang, J., Gasda, P.J., Hurowitz, J.A., Grotzinger, J.P., Wiens, R.C., Newsom, H.E., et al., 2017. Diagenetic silica enrichment and late-stage groundwater activity in Gale crater, Mars. Geophys. Res. Lett. 44(10). 2017GL073323. https://doi.org/10.1002/2017GL073323.

Georgiou, C.D., Deamer, D.W., 2014. Lipids as universal biomarkers of extraterrestrial life. Astrobiology 14 (6), 541–549.

Glansdorff, N., Xu, Y., Labedan, B., 2008. The last universal common ancestor: emergence, constitution and genetic legacy of an elusive forerunner. Biol. Direct 3–29.

Goudge, T.A., Milliken, R.E., Head, J.W., Mustard, J.F., Fassett, C.I., 2017. Sedimentological evidence for a deltaic origin of the western fan deposit in Jezero crater, Mars and implications for future exploration. Earth Planet. Sci. Lett. 458, 357–365.

Goudge, T.A., Mohrig, D., Cardenas, B.T., Hughes, C.M., Fassett, C.I., 2018. Stratigraphy and paleohy-drology of delta channel deposits, Jezero crater, Mars. Icarus 301, 58–75.

Guido, D.M., Campbell, K.A., 2011. Diverse subaerial and sublacustrine hot spring settings of the Cerro Negro epithermal system (Jurassic, Deseado Massif), Patagonia, Argentina. J. Volcanol. Geotherm. Res. 203, 35–47.

Guido, D.M., Campbell, K.A., 2017. Upper Jurassic travertine at El Macanudo, Argentine Patagonia: a fossil geothermal field modified by hydrothermal silicification and acid overprinting. Geol. Mag., 1–19. https://doi.org/10.1017/S0016756817000498.

Guidry, S.A., Chafetz, H.S., 2003. Anatomy of siliceous hot spring: examples from Yellowstone National Park, Wyoming, USA. Sediment. Geol. 157, 71–106.

Gulick, V.C., 1998. Magmatic Intrusions & a hydrothermal origin for fluvial valleys on Mars. J. Geophys. Res. 103, 19365–19388.

Gulick, V.C., 2001. Origin of the valley networks on Mars: a hydrological perspective. Geomorphology 37, 241–268.

Gulick, V.C., Baker, V.R., 1989. Fluvial valleys & Martian Paleoclimates. Nature 341, 514–516.

Gulick, V.C., Baker, V.R., 1990. Origin of Martian volcano valleys. J. Geophys. Res. 95, 14325–14344.

Handley, K.M., Campbell, K.A., Mountain, B.W., Browne, P.R.L., 2005. Abiotic-biotic controls on the origin and development of spicular sinter: in situ growth experiments, Champagne Pool, Waiotapu, New Zealand. Geobiology 3, 93–114.

Handley, K.M., Turner, S.J., Campbell, K.A., Mountain, B.W., 2008. Silicifying biofilm exopolymers on a hot-spring microstromatolite: templating nanometer-thick laminae. Astrobiology 8 (4), 747–770.

Hays, L.M. (Ed.), 2015. NASA Astrobiology Strategy. NASA Headquarters, Washington, DC. NASA/SP-2015-3710.

Henley, R.W., 1996. Chemical and physical context for life in terrestrial hydrothermal systems: chemical reactors for the early development of life and hydrothermal ecosystems. In: Bock, G.R., Goode, J.A. (Eds.), Hydrothermal Ecosystems on Earth (and Mars?), Ciba Foundation Symposium 202. John Wiley and Sons, Chichester, pp. 61–82.

Henley, R.W., Ellis, A.J., 1983. Geothermal systems ancient and modern: a geochemical review. Earth Sci. Rev. 19, 1–50.

Herdianita, N.R., Browne, P.R.L., Rodgers, K.A., Campbell, K.A., 2000. Mineralogical and morpholog-ical changes accompanying aging in silica sinters. Mineral. Deposita 35, 48–62.

Hiesinger, H., Head III, J.W., 2004. The Syrtis Major volcanic province, Mars. J. Geophys. Res. 109E01004. https://doi.org/10.1029/2003JE002143.

Hinman, N.W., Lindstrom, R.F., 1996. Seasonal changes in silica deposition in hot spring systems. Chem. Geol. 132, 237–246.

Hinman, N.W., Walter, M.R., 2005. Textural preservation in siliceous hot spring deposits during early dia-genesis: examples from Yellowstone National Park and Nevada, U.S.A. J. Sediment. Res. 75 (2), 200–215.

Hofmann, B.A., Farmer, J.D., 2000. Filamentous fabrics in low-temperature mineral assemblages: are they fossil biomarkers? Implications for the research for a subsurface fossil record on the early Earth and Mars. Planet. Space Sci. 48, 1077–1086.

Horneck, G., Walter, N., Westall, F., Grenfell, J.L., Martin, W.F., Gomez, F., Leuko, S., Lee, N., Onofri, S., Tsiganis, K., Saladino, R., Pilat-Lohinger, E., Palomba, E., Harrison, J., Rull, F., Muller, C., Strazzulla, G., Brucato, J.R., Rettberg, P., Capria, M.T., 2016. AstRoMap European Astrobiology Roadmap. Astrobiology 16 (3), 201–243.

Hugo, R.C., Cady, S.L., Smythe, W., 2010. The role of extracellular polymeric substances in the silicifi-cation of Calothrix: evidence from microbial mat communities in hot springs at Yellowstone National Park, USA. Geomicrobiol J. 28, 667–675.

Huntington, J.F., 1996. The role of remote sensing in finding hydrothermal mineral deposits on Earth. In: Bock, G.R., Goode, J.A. (Eds.), Hydrothermal Ecosystems on Earth (and Mars?), Ciba Foundation Symposium 202. John Wiley and Sons, Chichester, pp. 214–235.

Hurowitz, J.A., Grotzinger, J.P., Fischer, W.W., McLennan, S.M., Milliken, R.E., Stein, N., Vasavada, R., Blank, D.F., Dehouck, E., Eigenbrode, J.L., Fairén, A.G., Frydengang, J., Gellert, R., Grank, J.A., Gupta, S., Herkenhoff, K.E., Ming, D.W., Rampe, E.B., Schmidt, M.E., Siebach, K.L., Stack-Morgan, K., Sumner, D.Y., Wiens, R.C., 2017. Redox stratification of an ancient lake in Gale crater, Mars. Science 356 (6341), 922. https://doi.org/10.1126/science.aah6849.

Inskeep, W.P., Rusch, D.B., Jay, Z.J., Herrgard, M.J., Kozubal, M.A., Richardson, T.H., Macur, R.E., Hamamura, N., Jennings, R.dM., Fouke, B., Reysenbach, A.-L., Roberto, F., Young, M., Schwartz, A., Boyd, E.S., Badger, J.H., Mathur, E.J., Ortmann, A.C., Bateson, M., Geesey, G., Frazier, M., 2010. Metagenomes from high-temperature chemotrophic systems reveal geochemical controls on microbial community structure and function. PLoS ONE. 5(3) e9773.

Inskeep, W.P., Jay, Z.J., Tringe, S.G., Herrgård, M.J., Rusch, D.B., YNP Metagenome Project Steering Committee and Working Group Members, 2013. The YNP metagenome project: environmental parameters responsible for microbial distribution in the Yellowstone geothermal ecosystem. Front. Microbiol. 4, 8–22.

Jones, B., Renaut, R.W., 1996. Influence of thermophilic bacteria on calcite and silica precipitation in hot springs with water temperatures above 90C: evidence from Kenya and New Zealand. Can. J. Earth Sci. 33 (1), 72–83.

Jones, B., Renaut, R.W., 2003. Petrography and genesis of spicular and columnar geyserite from the Whakarewarewa and Orakeikorako geothermal areas, North Island, New Zealand. Can. J. Earth Sci. 40 (11), 1585–1610.

Jones, B., Renaut, R.W., Rosen, M.R., 1997. Vertical zonation of biota in microstromatolites associated with hot springs, North Island, New Zealand. Palaios 12, 220–236.

Jones, B., Renaut, R.W., Rosen, M.R., 2001. Taphonomy of silicified filamentous microbes in modern geothermal sinters: implications for identification. Palaios 16 (6), 580–592.

Kinsinger, N.M., Wong, A., Li, D., Villalobos, F., Kisailus, D., 2010. Nucleation and crystal growth of nanocrystalline anatase and rutile phase TiO_2 from a water-soluble precursor. Cryst. Growth Des. 10 (12), 5254–5261.

Konhauser, K.O., Phoenix, V.R., Bottrell, S.H., Adam, D.G., Head, I.M., 2001. Microbial silica interactions in Icelandic hot spring sinter: possible analogs for some Precambrian siliceous stromatolites. Sedimentology 48, 415–433.

Langmuir, D., 1997. Aqueous Environmental Geochemistry. Prentice Hall, Upper Saddle River, NJ, USA, 600 pp.

Lowe, D.R., Braunstein, D., 2003. Microstructure of high-temperature (>73°C) siliceous sinter deposited around hot springs and geysers, Yellowstone Park: the role of biological and abiological processes in sedimentation. Can. J. Earth Sci. 40, 1611–1642.

Lynne, B.Y., 2012. Mapping vent to distal-apron hot spring paleo-flow pathways using siliceous sinter architecture. Geothermics 43, 3–24.

Lynne, B.Y., Campbell, K.A., 2003. Diagenetic transformations (opal-A to quartz) of low- and mid-temperature microbial textures in siliceous hot-spring deposits, Taupo Volcanic Zone, New Zealand. Can. J. Earth Sci. 40, 1679–1696.

Lynne, B.Y., Campbell, K.A., 2004. Morphological and mineralogical transformations from opal-A to opal-CT in low-temperature siliceous sinter diagenesis, Taupo Volcanic Zone, New Zealand. J. Sediment. Res. 74 (4), 561–579.

Lynne, B.Y., Campbell, K.A., Moore, J.N., Browne, P.R.L., 2005. Diagenesis of 1900-year-old siliceous sinter (opal-A to quartz) at Opal Mound, Roosevelt Hot Springs, Utah, USA. Sediment. Geol. 179 (3–4), 249–278.

Lynne, B.Y., Campbell, K.A., James, B.J., Browne, P.R.L., Moore, J., 2007. Tracking crystallinity in siliceous hot-spring deposits. Am. J. Sci. 307 (3), 612–641.

McKenzie, E.J., Brown, K.L., Cady, S.L., Campbell, K.A., 2001. Trace metal chemistry and silicification of microorganisms in geothermal sinter, Taupo Volcanic Zone, New Zealand. Geothermics 30 (4), 483–502.

Meyer-Dombard, D.R., Swingley, W., Raymond, J., Havig, J., Shock, E.L., Summons, R.E., 2011. Hydrothermal ecotones and streamer biofilm communities in the Lower Geyser Basin, Yellowstone National Park. Environ. Microbiol 13, 2216–2231. https://doi.org/10.1111/j.1462-2920.2011.02476.x.

Michalski, J.R., Noe Dobrea, E.Z., Niles, P.B., Cuadros, J., 2017. Ancient hydrothermal seafloor deposits in Eridania basin on Mars. Nat. Commun. 8. https://doi.org/10.1038/ncomms15978.

Milliken, R.E., Swayze, G.A., Arvidson, R.E., Bishop, J.L., Clark, R.N., Ehlmann, B.L., Green, R.O., Grotzinger, J.P., Morris, R.V., Murchie, S.L., Mustard, J.F., Weitz, C., 2008. Opaline silica in young deposits on Mars. Geology 36 (11), 847–850.

Ming, D.W., Gellert, R., Morris, R.V., Yen, A.S., Arvidson, E., Brueckner, J., Clark, B.C., Cohen, B.A., Fleischer, I., Klingelhoefer, G., 2008. Geochemical Properties of Rocks and Soils in Gusev Crater, Mars:

APXS Results From Cumberland Ridge to Home Plate. In: Extended Abstract, 39th Lunar and Planetary Science Conference, 10–14 Mar. League City, TX, United States.

Morris, R.V., Vaniman, D.T., Blake, D.F., Gellert, R., Chipera, S.J., Rampe, E.B., et al., 2016. Silicic volcanism on Mars evidenced by tridymite in high-SiO$_2$ sedimentary rock at Gale crater. Proc. Natl. Acad. Sci. 113 (26), 7071–7076. https://doi.org/10.1073/pnas.1607098113.

Mustard, J.F., Adler, M., Allwood, A., Bass, D.S., Beaty, D.W., Bell III, J.F., Brinckerhoff, W.B., Carr, M., Des Marais, D.J., Drake, B., Edgett, K.S., Eigenbrode, J., Elkins-Tanton, L.T., Grant, J.A., Milkovich, S.M., Ming, D., Moore, C., Murchie, S., Onstott, T.C., Ruff, S.W., Sephton, M.A., Steele, A., Treiman, A., 2013. Report of the Mars 2020 Science Definition Team. 154 pp., posted July, 2013, by the Mars Exploration Program Analysis Group (MEPAG) at http://mepag.jpl.nasa.gov/reports/MEP/Mars_2020_SDT_Report_Final.pdf.

Newsom, H.E., 1980. Hydrothermal alteration of impact melt sheets with implications for Mars. Icarus 44, 207–216. https://doi.org/10.1016/0019-1035(80)90066-4.

Nicholson, K., 1993. Exploration techniques. In: Geothermal Fluids. Springer, Berlin, Heidelberg, pp. 141–149.

Nicolau, C., Reich, M., Lynne, B., 2014. Physico-chemical and environmental controls on siliceous sinter formation at the high-altitude El Tatio geothermal field, Chile. J. Volcanol. Geotherm. Res. 282, 60–76.

Nisbet, E.G., 2000. The realms of Archaean life. Nature 405, 625–626.

Nisbet, E.G., Sleep, N.H., 2001. The habitat and nature of early life. Nature 409, 1083–1091.

O'Connell-Cooper, C.D., Spray, J.G., Thompson, L.M., Gellert, R., Berger, J.A., Boyd, N.I., et al., 2017. APXS-derived chemistry of the Bagnold dune sands: comparisons with Gale crater soils and the global Martian average. J. Geophys. Res. Planets. 2017JE005268. https://doi.org/10.1002/2017JE005268.

Orange, F., Lalonde, S.V., Konhauser, K.O., 2013. Experimental silicification of evaporation-driven silica sinter formation and microbial silicification in hot spring systems. Astrobiology 13 (2), 163–176.

Osinski, G.R., Tornabene, L.L., Banerjee, N.R., Cockell, C.S., Flemming, R., Izawa, M.R.M., McCutcheon, J., Parnell, J., Preston, L.J., Pickersgill, A.E., Pontefract, A., Sapers, H.M., Southam, G., 2013. Impact-generated hydrothermal systems on Earth and Mars. Icarus 224, 347–363.

Papike, J.J., Karner, J.M., Shearer, C.K., Burger, P.V., 2009. Silicate mineralogy of martian meteorites. Geochim. Cosmochim. Acta 73 (24), 7443–7485.

Parnell, J., Cullen, D., Sims, M.R., Bowden, S., Cockell, C.S., Court, R., Ehrenfreund, P., Gaubert, F., Grant, W., Parro, V., Rohmer, M., Sephton, M., Stan-Lotter, H., Steele, A., Toporski, J., Vago, J., 2007. Search for life on Mars: selection of molecular targets for ESA's aurora ExoMars mission. Astrobiology 7 (4), 578–604.

Pavlov, A.A., Vasilyev, G., Ostryakov, V.M., Pavlov, A.K., Mahaffy, P., 2012. Degradation of the organic molecules in the shallow subsurface of Mars due to irradiation by cosmic rays. Geophys. Res. Lett. 39, L13202. https://doi.org/10.1029/2012GL052166.

Peary, J., Castenholz, R.W., 1964. Temperature strains of a thermophilic blue-green alga. Nature 202, 720–721.

Price, R., Boyd, E.S., Hoehler, T.M., Wehrmann, L.M., Bogason, E., Valtýsson, H.Þ., Örlygsson, J., Gautason, B., Amend, J.P., 2017. Alkaline vents and steep Na$^+$ gradients from ridge-flank basalts—Implications for the origin and evolution of life. Geology 45, 1135–1138.

Raitala, J., Esestime, P., Korteniemi, J., Kostama, V.-P., Aittola, M., Neukum, G., 2008. Tectonics and water-related episodes on Claritas Fossae, Mars. Lunar Planet. Sci. XXXIX, Abstract #1569.

Rampe, E.B., Ming, D.W., Blake, D.F., Bristow, T.F., Chipera, S.J., Grotzinger, J.P., Morris, R.V., Morrison, S.M., Vaniman, D.T., Yen, A.S., Achilles, C.N., Craig, P.I., Des Marais, D.J., Downs, R.T., Farmer, J.D., Fendrich, K.V., Gellert, R., Hazen, R.M., Kah, L.C., Morookian, J.M., Peretyazhko, T.S., Sarrazin, P., Treiman, A.H., Berger, J.A., Eigenbrode, J., Fairén, A.G., Forni, O., Gupta, S., Hurowitz, J.A., Lanza, N.L., Schmidt, M.E., Siebach, K., Sutter, B., Thompson, L.M., 2017. Mineralogy of an ancient lacustrine mudstone succession from the Murray formation, Gale crater, Mars. Earth Planet. Sci. Lett. 471, 172–185. https://doi.org/10.1016/j.epsl.2017.04.021.

Renaut, R.W., Jones, B., 2011a. Hydrothermal environments, terrestrial. In: Reitner, J., Thiel, V. (Eds.), Encyclopedia of Geobiology. Springer, Dordrecht, Netherland, pp. 467–479.

Renaut, R.W., Jones, B., 2011b. Sinter. In: Reitner, J., Thiel, V. (Eds.), Encyclopedia of Geobiology. Springer, Dordrecht, Netherlands, pp. 808–813.

Reysenbach, A.-L., Cady, S.L., 2001. Microbiology of ancient and modern hydrothermal systems. Trends Microbiol. 9 (2), 79–86.

Reysenbach, A.L., Wickham, G.S., Pace, N.R., 1994. Phylogenetic analysis of the hyperthermophilic pink filament community in Octopus Spring, Yellowstone National Park. Appl. Environ. Microbiol. 60 (6), 2113–2119.

Rice, C.M., Ashcroft, W.A., Batten, D.J., Boyce, A.J., Caulfield, J.B.D., Fallick, A.E., Hole, M.J., Jones, E., Pearson, M.J., Rodgers, G., Saxton, J.M., Stuart, F.M., Trewin, N.H., Turner, G., 1995. A Devonian Auriferous hot spring system, Rhynie, Scotland. J. Geol. Soc. Lond. 152, 229–250.

Rice, C.M., Trewin, N.H., Anderson, L.I., 2002. Geological setting of the Early Devonian Rhynie Cherts, Aberdeenshire, Scotland: an early terrestrial hot spring system. J. Geol. Soc. 159, 203–214.

Rodgers, K.A., Browne, P.R.L., Buddle, T.F., Cook, K.L., Greatrex, R.A., Hamption, W.A., Herdianita, N.R., Holland, G.R., Lynne, B.Y., Martin, R., Newton, Z., Pastars, D., Sannazzaro, K.L., Teece, C.I.A., 2004. Silica phases in sinters and residues from geothermal fields of New Zealand. Earth Sci. Rev. 66, 1–61.

Rossi, A.P., Neukum, G., Pondrelli, M., Zegers, T., Mason, P., Hauber, E., Ori, G.G., Fueten, F., Oosthoek, J., Chicarro, A., Foing, B., 2007. The case for large-scale spring deposits on Mars: light-toned deposits in crater bulges, Valles Marineris and chaos. Lunar Planet. Sci. XXXVIII, Abstract #1549.

Rossi, A.P., Pondrelli, M., Van Gasselt, S., Zegers, T., Hauber, E., Neukum, G., 2008a. Gale Crater Bulge: a candidate multi-stage large spring mound. Lunar Planet. Sci. XXXIX, Abstract #1611.

Rossi, A.P., Neukum, G., Pondrelli, M., Van Gasselt, S., Zegers, T., Hauber, E., Chicarro, A., Foing, B., 2008b. Large-scale spring deposits on Mars? J. Geophys. Res. 113. https://doi.org/10.1029/2007JE003062.

Rothschild, L.J., Mancinelli, R.L., 2001. Life in extreme environments. Nature 409, 1092–1101.

Ruff, S.W., 2015. New observations reveal a former hot spring environment with high habitability and preservation potential in Gusev Crater, Mars. Lunar Planet. Sci. XXXXIV, Abstract #1613.

Ruff, S.W., Farmer, J.D., 2016. Silica deposits on Mars with features resembling hot spring biosignatures at El Tatio in Chile. Nat. Geosci. 713554. https://doi.org/10.1038/ncomms13554.

Ruff, S.W., Farmer, J.D., Calvin, W.M., Herkenhoff, K.E., Johnson, J.R., Morris, R.V., Rice, M.S., Arvidson, R.E., Bell III, J.F., Christensen, P.R., Squyres, R.W., 2011. Characteristics, distribution, origin, and significance of opaline silica observed by the Spirit rover in Gusev crater, Mars. J. Geophys. Res. Planets 116, 2156–2202.

Schultze-Makuch, D., Dohm, J.M., Fan, C., Fairén, A.G., Rodriguez, J.A.P., Baker, V.R., Fink, W., 2007. Exploration of hydrothermal targets on Mars. Icarus 189, 308–324.

Schwartzman, D.W., Lineweaver, C.H., 2004. The hyperthermophilic origin of life revisited. Biochem. Soc. Trans. 32 (2), 168–171.

Schwenzer, S.P., Abramov, O., Allen, C.C., Bridges, J.C., Clifford, S.M., Filiberto, J., Kring, D.A., Lasue, J., McGovern, P.J., Newsom, H.E., Treiman, A.H., Vaniman, D.T., Wiens, R.C., Wittmann, A., 2012. Gale Crater: formation and post-impact hydrous environments. Planet. Space Sci. 70, 84–95.

Shock, E.L., 1996. Hydrothermal systems as environments for the emergence of life. In: Bock, G.R., Goode, J.A. (Eds.), Hydrothermal Ecosystems on Earth (and Mars?), Ciba Foundation Symposium 202. John Wiley and Sons, Chichester, pp. 40–52.

Siljeström, S., Parenteau, M.N., Jahnke, L.L., Cady, S.L., 2017. A comparative ToF-SIMS and GC-MS analysis of phototrophic communities collected from an alkaline silica-depositing hot spring. Org. Geochem. 109, 14–30.

Sillitoe, R.H., 1993. Epithermal models: genetic types, geometrical controls and shallow features. In: Kirkham, R.V., Sinclair, W.D., Thorpe, R.I., Duke, J.M. (Eds.), Mineral Deposit Modeling. Geol. Assoc. Canada Special Paper 40, pp. 403–417.

Sillitoe, R.H., 2015. Epithermal paleosurfaces. Mineral. Deposita 50 (7), 757–793.

Skok, J.R., Mustard, J.F., Ehlmann, B.L., Milliken, R.E., Murchie, S.L., 2010. Silica deposits in the Nili Patera caldera on the Syrtis Major volcanic complex on Mars. Nat. Geosci. 3. https://doi.org/10.1038/NGEO990.

Sojo, V., Herschy, B., Whicher, A., Camprubí, E.Lane, 2016. The origin of life in alkaline hydrothermal vents. Astrobiology 16 (2), 181–197.

Squyres, S.W., Arvidson, R.E., Ruff, S., Gellert, R., Morris, R.V., Ming, D.W., Crumpler, L., Farmer, J.D., Des Marais, D.J., Yen, A., McLennan, S.M., Calvin, W., Bell III, J.F., Clark, B.C., Wang, A., McCoy, T.J., Schmidt, M.E., De Souza Jr., P.A., 2008. Detection of silica-rich deposits on Mars. Science 320 (5879), 1063–1067.

Stetter, K.O., 1996. Hyperthermophiles in the history of life. In: Bock, G.R., Goode, J.A. (Eds.), Hydrothermal Ecosystems on Earth (and Mars?), Ciba Foundation Symposium 202. John Wiley and Sons, Chichester, pp. 1–23.

Stetter, K.O., 2006. Hyperthermophiles in the history of life. Philos. Trans. R. Soc. Lond. B, Biol. Sci. 361 (1474), 1837–1843.

Stolper, E.M., Baker, M.B., Newcombe, M.E., Schmidt, M.E., Treiman, A.H., Cousin, A., et al., 2013. The petrochemistry of Jake_M: a Martian Mugearite. Science 341 (6153), 1239463. https://doi.org/10.1126/science.1239463.

Strazzulli, A., Iacono, R., Giglio, R., Moracci, M., Cobucci-Ponzano, B., 2017. Metagenomics of hyperthermophilic environments: biodiversity and biotechnology. In: Chénard, C., Lauro, F. (Eds.), Microbial Ecology of Extreme Environments. Springer, Cham, pp. 103–135.

Summons, R.E., Amend, J.P., Bish, D., Buick, R., Cody, G.D., Des Marais, D.J., Dromart, G., Eigenbrode, J.L., Knoll, A.H., Sumner, D.Y., 2011. Preservation of martian organic and environmental records: final report of the Mars Biosignature Working Group. Astrobiology 11 (2), 157–181.

Sun, V.Z., Milliken, R.E., Robertson, K.M., 2016a. Hydrated silica on Mars: relating geologic setting to degree of hydration, crystallinity, and maturity through coupled orbital and laboratory studies. Lunar Planet Sci. XXXXVII, Abstract #2416.

Sun, V.Z., Milliken, R.E., Robertson, K.J., Ruff, S.W., Farmer, J.D., 2016b. Spectral characterization and mineralogy/chemistry of opaline silica samples from diverse Mars analog sites. Lunar Planet Sci. XXXXVII, Abstract #2071.

Taylor, S.R., McLennan, S., 2010. Planetary Crusts: Their Composition, Origin and Evolution, first ed. Cambridge University Press, Cambridge, UK, 378 pp.

Thompson, L.M., Schmidt, M.E., Spray, J.G., Berger, J.A., Fairén, A.G., Campbell, J.L., Perrett, G.M., Boyd, N., Gellert, R., Pradler, I., Van Bommel, S.J., 2016. Potassium-rich sandstones within the Gale impact crater, Mars: the APXS perspective. J. Geophys. Res. Planets 121, 1981–2003.

Tosdal, R.M., Dilles, J.H., Cooke, D.H., 2009. From source to sinks in auriferous magmatichydrothermal porphyry and epithermal deposits. Elements 5, 289–295.

Treiman, A.H., Bish, D.L., Vaniman, D.T., Chipera, S.J., Blake, D.F., Ming, D.W., et al., 2016. Mineralogy, provenance, and diagenesis of a potassic basaltic sandstone on Mars: CheMin X-ray diffraction of the Windjana sample (Kimberley area, Gale Crater). J. Geophys. Res. Planets. 121(1)2015JE004932. https://doi.org/10.1002/2015JE004932.

Trewin, H.H., Fayers, S.R., Kelman, R., 2003. Subaqueous silicification of the contents of small ponds in an Early Devonian hot-spring complex, Rhynie, Scotland. Can. J. Earth Sci. 40 (11), 1697–1712.

Vago, J.L., Westall, F., the Pasteur Instrument Teams, Coates, A.J., Lead, Jaumann, R., Korablev, O., Ciarletti, V., Mitrofanov, I., Josset, J.-L., De Sanctis, M.C., Bibring, J.-P., Rull, F., Goesmann, F., Steininger, H., Goetz, W., Brinckerhoff, W., Szopa, C., Raulin, F., Landing Site Selection Working Group, Westall, F., Lead, Edwards, H.G.M., Whyte, L.G., Fairén, A.G., Bibring, J.-P., Bridges, J., Hauber, E., Ori, G.G., Werner, S., Loizeau, D., Kuzmin, R.O., Williams, R.M.E., Flahaut, J., Forget, F., Vago, J.L., Rodionov, D., Korablev, O., Svedhem, H., Sefton-Nash, E., Kminek, G., Lorenzoni, L., Jourdier, L., Mikhailov, V., Zashchirinskiy, A., Alexashkin, S., Calantropio, F., Merlo, A., Poulakis, P., Witasse, O., Bayle, O., Bayón, S., Other Contributors, Carter, J., García-Ruiz, J.M., Baglioni, P., Haldemann, A., Ball, A.J., Debus, A., Lindner, R., Haessig, F., Monteiro, D., Ball, A.J., Debus, A., Lindner, R., Haessig, F., Monteiro, D., Trautner, R., Voland, C., Rebeyre, P., Goulty, D., Didot, F., Durrant, S., Zekri, E., Koschny, D., Toni, A., Visentin, G., Zwick, M., van Winnendael, M., Azkarate, M., Carreau, C., 2017. Habitability on early

Mars and the search for biosignatures with the ExoMars rover. Astrobiology 17 (6–7), 471–510. https://doi.org/10.1089/ast.2016.1533.

Van Kranendonk, M., Baumgartner, R., Boyd, E., Cady, S., Campbell, K., Czaja, A., Damer, B., Deamer, D., Djokic, T., Fiorentini, M., Gangidine, A., Havig, J., Mulkidjanian, A., Ruff, S., Thordarson, P., 2018. Terrestrial hot springs and the origin of life: implications for the search for life beyond Earth. Lunar Planet Sci. XXXXIX, Abstract #2535.

Walter, M.R., 1976a. Geyserites of Yellowstone National Park: an example of abiogenic "stromatolites" In: Walter, M.R. (Ed.), Stromatolites. Elsevier, Amsterdam, pp. 87–112.

Walter, M.R., 1976b. Hot spring sediments in Yellowstone National Park. In: Walter, M.R. (Ed.), Stromatolites. Elsevier, Amsterdam, pp. 489–498.

Walter, M.R., 1996. Ancient hydrothermal ecosystems on earth: a new palaeobiological frontier. In: Bock, G.R., Goode, J.A. (Eds.), Evolution of Hydrothermal Ecosystems on Earth (and Mars?), Ciba Foundation Symposium 202. John Wiley and Sons, Chichester, pp. 112–127.

Walter, M.R., Des Marais, D.J., 1993. Preservation of biological information in thermal spring deposits: developing a strategy for the search for fossil life on Mars. Icarus 101, 129–143.

Walter, M.R., Des Marais, D., Farmer, J.D., 1996. Lithofacies and biofacies of Mid-Paleozoic thermal spring deposits in the Drummond Basin, Queensland, Australia. Palaios 11, 497–518.

Walter, M.R., Mcloughlin, S., Drinnan, A.N., Farmer, J.D., 1998. Paleontology of Devonian thermal spring deposits, Drummond Basin, Australia. Alcheringa 22, 285–314.

Wang, A., Bell III, J.F., Li, R., Johnson, J.R., Farrand, W.H., Cloutis, E.A., Arvidson, R.E., Crumpler, L., Squyres, S.W., McLennan, S.M., Herkenhoff, K.E., Ruff, S.W., Knudson, A.T., Chen, W., Greenberger, R., 2008. Light-toned salty soils and coexisting Si-rich species discovered by the Mars Exploration Rover Spirit at Columbia Hills. J. Geophys. Res. Planets 113, E12S40.

Ward, D.M., Ferris, M.J., Nold, S.C., Bateson, M.M., 1998. A natural view of microbial biodiversity within hot spring cyanobacterial mat communities. Microbiol. Mol. Biol. Rev. 62 (4), 1353–1370.

Weiss, M.C., Sousa, F.L., Mrnjavac, N., Neukirchen, S., Roettger, M., Nelson-Sathi, S., Martin, W.F., 2016. The physiology and habitat of the last universal common ancestor. Nat. Microbiol. https://doi.org/10.1038/NMICROBIOL.2016.116.

Westall, F., Foucher, F., Bost, N., Bertrand, M., Loizeau, D., Vago, J.L., Kminek, G., Gaboyer, F., Campbell, K.A., Bréhéret, J.-G., Gautret, P., Cockell, C.S., 2015a. Biosignatures on Mars: what, where, and how? Implications for the search for Martian life. Astrobiology 15 (11), 998–1029.

Westall, F., Campbell, K.A., Bréhéret, J.G., Foucher, F., Gautret, P., Hubert, A., Sorieul, S., Grassineau, N., Guido, D.M., 2015b. Archean (3.33 Ga) microbe-sediment systems were diverse and flourished in a hydrothermal context. Geology 43 (7), 615–618.

Westall, F., Hickman-Lewis, K., Hinman, N., Gautret, P., Campbell, K.A., Bréhéret, J.G., Foucher, F., Hubert, A., Sorieul, S., Dass, A.V., Kee, T.P., Georgelin, T., Brock, A., 2018. A hydrothermal-sedimentary context for the origin of life. Astrobiology 18 (3), 259–293.

White, D.E., Thompson, G.A., Sandberg, G.H., 1964. Rocks, structure, and geologic history of Steamboat Springs thermal area, Washoe County, Nevada. Geol. Surv. Prof. Paper 458B.

Worms, J.-C., Lammer, H., Barucci, A., Beebe, R., Bibring, J.P., Blamont, J., Blanc, M., Bonnet, R., Brucato, J.R., Chassefière, E., Coradini, A., Crawford, I., Ehrenfreund, P., Falcke, H., Gerzer, R., Grady, M., Grande, M., Haerendel, G., Horneck, G., Koch, B., Lobanov, A., Lopez-Moreno, J.J., Marco, R., Norsk, P., Rothery, D., Swings, J.-P., Tropea, C., Ulamec, S., Westall, F., Zarnecki, J., 2009. ESSC-ESF position paper—science-driven scenario for space exploration: report from the European Space Sciences Committee (ESSC). Astrobiology 9 (1), 23–41.

Yee, N., Phoenix, V.R., Konhauser, K.O., Benning, L.G., Ferris, F.G., 2003. The effect of cyanobacteria on silica precipitation at neutral pH: implications for bacterial silicification in geothermal hot springs. Chem. Geol. 199, 83–90.

Yen, A.S., Morris, R.V., Clark, B.C., Gellert, R., Knudson, A.T., Squyres, S., Mittlefehldt, D.W., Ming, D.W., Arvidson, R., McCoy, T., Schmidt, M., Hurowitz, J., Li, R., Johnson, J.R., 2008. Hydrothermal processes at Gusev Crater: an evaluation of Paso Robles class soils. J. Geophys. Res: Planets 113(E6). https://doi.org/10.1029/2007JE002978.

Yen, A.S., Ming, D.W., Vaniman, D.T., Gellert, R., Blake, D.F., Morris, R.V., et al., 2017. Multiple stages of aqueous alteration along fractures in mudstone and sandstone strata in Gale Crater, Mars. Earth Planet. Sci. Lett. 471, 186–198.

FURTHER READING

Cady, S.L., Farmer, J.D., Grotzinger, J.P., Schopf, J.W., Steele, A., 2003. Morphological biosignatures and the search for life on Mars. Astrobiology 3 (2), 351–368.

Hays, L.E., Graham, H.V., Des Marais, D.J., Hausrath, E.M., Horgan, B., McCollom, T.M., Parenteau, M.N., Potter-McIntyre, S.L., Williams, A.J., Lynch, K.L., 2017. Biosignature preservation and detection in Mars analog environments. Astrobiology 17 (4), 363–400.

Lalonde, S.V., Konhauser, K.O., Reysenbach, A.-L., Ferris, F.G., 2005. The experimental silicification of Aquificales and their role in hot spring sinter formation. Geobiology 3 (1), 41–52.

NASA's Journey to Mars, 2015. Pioneering Next Steps in Space Exploration. NP-2015-008-2018-HQ.

Oehler, D.Z., Cady, S.L., 2014. Biogenicity and syngeneity of organic matter in ancient sedimentary rocks: recent advances in the search for evidence of past life. Challenges 5 (2), 260–283.

Renaut, R.W., Jones, B., Tiercelin, J.J., 1998. Rapid in situ silicification of microbes at Loburu hot springs, Lake Bogoria, Kenya Rift Valley. Sedimentology 45 (6), 1083–1103.

Ruff, S.W., Farmer, J.D., 2017. The case for silica sinter in the Columbia Hills of Mars and why it matters. Lunar Planet. Sci. XLVIII, Abstract #2879.

Toporski, J., Steele, A., Westall, F., Thomas-Keprta, K.L., McKay, D.S., 2002. The simulated silicification of bacteria—new clues to the modes and timing of bacterial precipitation and implications for the search for extraterrestrial microfossils. Astrobiology 2 (1), 1–26.

White, D.E., Brannock, W.W., Murata, K.J., 1956. Silica in hot spring waters. Geochim. Cosmochim. Acta 10, 27–59.

Yen, A.S., Gellert, R., Schröder, C., Morris, R.V., Bell, J.F., Knudson, A.T., et al., 2005. An integrated view of the chemistry and mineralogy of martian soils. Nature 436 (7047), 49–54.

CHAPTER 8

Habitability and Biomarker Preservation in the Martian Near-Surface Radiation Environment

Luis Teodoro, Alfonso Davila, Richard C. Elphic, David Hamilton, Christopher McKay, Richard Quinn

Contents

8.1 INTRODUCTION

Mars has long been a focal point in the search for evidence of life beyond Earth. There is now compelling evidence, accumulated over the past 40 years of Mars exploration, beginning with the NASA Viking Mission through the NASA Mars Science Laboratory, that conditions on the surface of Mars were benign, even habitable, early in the history of the planet (Squyres et al., 2004; Grotzinger et al., 2014). However, the surface of Mars is today an extreme, frozen desert, where liquid water is, for the most part, unstable. Under these conditions, it is unlikely that metabolically active organisms currently exist on the surface or in the near subsurface of the planet (Rummel et al., 2014; Kminek et al., 2010). Yet, transient habitable conditions may occur, when the tilt of the planet's spin axis relative to the sun periodically increases to a point that results in melting of near-surface ground ice at mid- and high latitudes. These periods of high axis tilt, referred to as high–obliquity periods (Costard et al., 2002; Jakosky et al., 2003; Mckay et al., 2013), have a recurrence time of ~125,000 years (Ward, 1973; Laskar et al., 2004). Transient water-ice melting events may also occur after random events such as volcanic eruptions or meteorite impacts (Abramov and Kring, 2005; Cousins and Crawford, 2011). The possibility that viable microorganisms may exist in a dormant state (i.e., freeze-dried) in the regolith on Mars, awaiting these transient more clement conditions, has not been ruled out (Mckay et al., 2013).

From Habitability to Life on Mars
https://doi.org/10.1016/B978-0-12-809935-3.00012-8

In addition to the limited availability of liquid water, the Martian radiation environment also presents major limitations on the possibility of microbial survival on present-day Mars. In contrast to the Earth, due to its thin atmosphere and the lack of a planetary magnetosphere, ionizing radiation in the form of galactic cosmic rays (GCRs) and solar energetic particles (SEPs) is only slightly attenuated before reaching the surface of Mars. GCRs and SEPs can cause two types of potentially detrimental effects on cells and biomolecules: (i) direct radiation damage (e.g., Dartnell et al., 2011; Kminek and Bada, 2006) and (ii) indirect, in which reactive species are formed (e.g., reactive oxygen species (ROS)) and subsequently chemically react with cells and organic matter (Henriksen, 1962). Therefore, ionizing radiation imposes an upper boundary on the amount of time that a microbial organism can remain dormant and viable at the surface or in the near subsurface of Mars. Direct and indirect radiation effects will also adversely impact the preservation of organic biosignature relict of any possible ancient Martian life.

In this chapter, first, we summarize the radiation environment on the surface of Mars from the first principles (Section 8.2). Then, we discuss the known effects of ionizing radiation on living cells (Section 8.3). At the end, we present results from numerical models of the radiation environment near the surface and their implications for habitability and preservation of biosignatures (Section 8.4).

8.2 THE IONIZING RADIATION ENVIRONMENT ON MARS

Our current understanding of the ionizing radiation environment at a given region on the Martian surface is based on numerical algorithms that employ nuclear physics codes (Dartnell et al., 2007) and direct measurements on the surface taken during robotic missions (Hassler et al., 2014). The primary sources of ionizing radiation on Mars are as follows:

(i) Galactic cosmic rays (GCRs), which due to their high energy (with a peak at 1 GeV) penetrate deep into the surface of Mars ($\sim$10 m)

(ii) Solar energetic particles (SEPs), which range in energy from a few kiloelectron volt to many tens of million electron volts per particle (e.g., Dartnell et al., 2007)

(iii) Decay of radioactive elements that are present in the surface of Mars (e.g., thorium, uranium, and potassium)

(iv) Ultraviolet (UV) solar photons that impinge on the top surface of Mars

The UV radiation (i.e., UV photons) environment on Mars has been subject of numerous studies (e.g., Cockell et al., 2000; Dartnell and Patel, 2014; Rummel et al., 2014; and references therein), particularly with respect to its impact on Martian habitability and the preservation of organic molecules that may provide evidence of past life on Mars (i.e., biomarkers). In general, the impact of UV radiation is limited on the surfaces of geologic materials and mechanical mixing, which can occur during Martian dust storms and other physical processes, and is required to move UV radiation-altered materials to

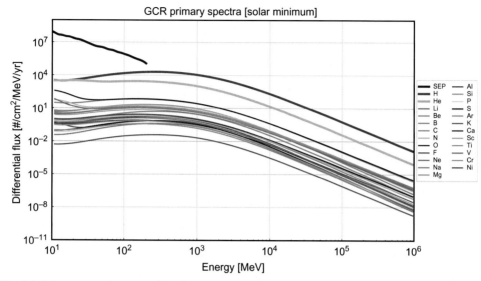

Fig. 8.1 Primary energy spectra for solar energetic particles (SEPs) and galactic cosmic rays (GCRs) at solar minimum. The annual mean SEP flux at Mars orbit as computed by Space ENVironment Information System (SPENVIS) is shown in *blue*. SEPs are made up of H and He. These spectra were also used by Dartnell et al. (2007) and are available as an online database (http://www.spenvis. oma.be). The GCR flux functions of energy bin for ions (H-Ni) are provided by the CREME-06 model.

appreciable depths. In contrast, the other sources of ionizing radiation, GCRs, SEPs, and radioactive decay, are relevant to processes occurring in the top few meters of the Martian surface (Dartnell et al., 2007). Fig. 8.1 shows the primary energy spectra for SEPs and GCRs.

Radiation absorption in geologic materials is a complex process that depends on a number of related parameters including chemical composition and material density. Absorption also depends on the properties of the ionizing radiation (e.g., kinetic energy and speed) and its resulting interactions with the surface materials (i.e., cross section) as it penetrates to depth. To follow the interactions of the overall ionizing radiation distribution throughout an environment as complex as a planetary surface, sophisticated and well-calibrated numerical models have been developed using numerical algorithms or transport codes suitable for both engineering and scientific applications (e.g., Titt et al., 2012). These codes, such as the Monte Carlo N-Particle (extended) (MNCPX, Pelowitz, 2005) and geometry and tracking 4 (Geant4; Agostinelli et al., 2003; Allison et al., 2006), are used to model radiation-particle behavior in three-dimensional space and time. Comparisons between these two codes show that both produce similar results for several models of surface geology, mineral distribution, and water geometry (e.g., Titt et al., 2012). The cross-sectional libraries used in MCNPX and Geant4 have been adapted and used for a range of planetary science applications including the instrument design and calibration and the interpretation of their returned data. Mission instruments that have relied on these codes include the neutron and gamma-ray spectrometers on Mars Odyssey (e.g., Gasnault et al., 1999), the neutron and gamma-ray spectrometers on Lunar Prospector, and the gamma-ray and neutron detector (GRaND) on the Dawn spacecraft (e.g., Prettyman et al., 2006; Lawrence et al., 2013; Teodoro et al., 2015). These codes are also used for astrobiology application, including modeling environmental habitability (e.g., Cai et al., 2017; Ma and Jiang, 1999).

Both GCRs and SEPs are modulated by solar activity. The intensity of GCRs (<30 GeV) is stronger during solar minimum and weaker during solar maximum, while SEP fluxes show the opposite trend due to the increasing number of solar flares during solar maximum (see Potgieter 2013 for a review), and these variations may impact the calculation of ionizing radiation reaching the surface of Mars. The most sophisticated transport codes predict that as GCRs and SEPs impinge on the surface of Mars, interactions with materials in the atmosphere and/or at the shallow surface produce energetic *secondary particles*: mesons (pions and kaons), gamma-ray photons, electrons/positrons, neutrons, and high atomic number and energy (HZE) ions. Secondary mesons, which are not stable, decay over a short timescale producing more muons, gamma-ray photons, and electrons. This leads to a cone-structured particle shower, with a central core of HZE ions (hard component) within spreading cone of electrons, photons, and neutrons (soft component) of an electromagnetic cascade (e.g., Rao and Sreekantan, 1998). The electromagnetic cascade reaches a maximum diameter in the direction transversal to the primary GCR motion, after which it steadily decays when the average particle energy drops below the required threshold for new particle production.

On Mars, the atmosphere and regolith provide radiation shielding, which is defined as the total integrated mass of material in front of a given position per unit area. Radiation penetration distance is calculated by dividing shielding by the average density of the Martian atmosphere and regolith. The location at which the electromagnetic cascade reaches its maximum diameter is known as the Pfotzer maximum (Pfotzer, 1936a,b), and on Mars, due to the planet's thin atmosphere, it takes place approximately between 1.5 and 2 m below the surface, depending on the composition of the regolith. In contrast, on Earth, due to the planet's thicker atmosphere, the Pfotzer maximum takes place at ~15 km above the surface. At the Martian surface (located above the Pfotzer maximum), the average atmospheric shielding is $\sim 1.6 \times 10^2$ kg/m^2 (Simonsen et al., 1991) resulting in an ionizing radiation dose rate of 0.025–0.25 Gy/year (see Simonsen et al., 1991; Pavlov et al., 2002; and references therein). On Earth, at sea level (located below the Pfotzer maximum), atmospheric shielding is $\sim 1 \times 10^4$ kg/m^2, resulting in an ionizing radiation dose rate of 0.0003–0.001 Gy/year (Baumstark-Khan and Facius, 2002). Fig. 8.2 shows the typical cases of secondary ionizing radiation in the shallow Martian subsurface.

As mentioned previously, the interaction of ionizing radiation with matter is usually categorized into two different types: (i) direct effects and (ii) indirect effects. In the case of

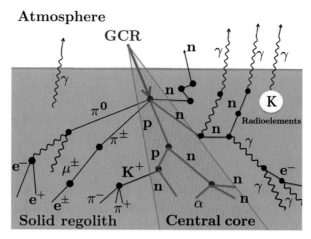

Fig. 8.2 Schematic showing the typical cases of ionizing radiation produced by GCRs on a shallow planetary surface. The secondaries include neutrons (n), gamma rays (γ), protons (p), electrons/positrons (e^-/e^+, respectively), mesons (pions (π) and kaons (K)), and highly charged/high-energy (HZE) ions. For further details, see text. The region in *blue* represents the "central core" of the secondary particle shower. The *red arrow* denotes a galactic cosmic ray (GCR), while the *blue lines* show the central core spreading. Diagram produced by the authors.

direct effects, charged particles (e.g., electrons, positrons, protons, muons, kaons, helium, and HZE ions) ionize matter upon impact. More specifically, high-velocity charged particles (other than electrons) lose energy in matter primarily through ionization and atomic excitation (e.g., Grupen, 2010; Amsler et al., 2008). The mean rate of this energy loss is referred to as the particle's stopping power and is described by the Bethe equation (also known as the Bethe-Bloch equation; see Eq. (A.1) in Appendix A) (e.g., Yao et al., 2006; Amsler et al., 2008). The Bethe equation shows that.

 (i) energy loss is independent of the mass of the incoming particle,
 (ii) energy loss depends quadratically on the charge and velocity of the particle,
(iii) stopping power is relatively independent of the absorbing material.

Further details of these interactions and supporting equations are provided in Appendix A.

The relationships described by the Bethe equation do not include electrons, which Bethe (1930) considered as a particular case of charged-particle energy loss. Interactions between secondary electrons and the matter they pass through occur through a variety of mechanisms. At high energies, the dominant mechanisms of energy loss are collisions that cause ionization and excitation of atoms where energy is attenuated by a multitude of inelastic collisions until it reaches thermal levels and radiation (bremsstrahlung) production occurs due to electron deceleration. The production of X-rays (bremsstrahlung) by the deceleration (stopping) of electrons is not a very efficient process at low energies ($\leq 5\,\mathrm{MeV}$) (Amsler et al., 2008), but it becomes the dominant process at higher energies (see Appendix I for further details). In the second type of effects, neutral particles (e.g., neutrons, neutrinos, mesons, and neutral kaons) or short-wavelength electromagnetic radiations (X-rays and gamma rays) ionize the target only indirectly (e.g., Grupen, 2010). Nuclear breakup by primary particles produces neutrons that subsequently excite elemental nuclei through processes such as neutron scattering and neutron capture. The interactions of the neutrons with the surrounding nuclei and their flux at any location depend on the elemental composition of the material, driven by the different elastic and inelastic cross sections of the elements. Hydrogen, for instance, is especially effective in moderating neutrons (i.e., reducing neutron speed) (e.g., Feldman et al., 1993, 2002). The absorption of neutrons by matter converts elemental nuclei, which are otherwise stable under normal conditions, into unstable, excited states that emit gamma rays as they shed energy and return to their stable, lower-energy states. In particular, excited hydrogen, sulfur, silicon, oxygen, iron, and carbon emit gamma rays, which in turn interact with surrounding matter. The photoelectric effect, in which an atom absorbs a photon and an energetic photoelectron is ejected from one of the atom's bound shells, is the predominant mode of interaction for gamma rays at low energy (typically, $E < 5 \times 10^4$ eV). This interaction takes place with the atom as a whole and cannot take place with free electrons. At intermediate energies (5×10^4 eV $< E < 5 \times 10^7$ eV), Compton scattering in which a photon transmits a fraction of this energy and momentum to an electron loosely bound in the absorbing material is the predominant process. The photoelectric and Compton loss processes differ in several details. Briefly, in Compton loss, the photon energy loss is independent of the atomic number of the absorption material, while in the photoelectric process, energy loss is proportional to atomic number (see Amsler et al., 2008). The flux of higher energy ($>5 \times 10^7$ eV) ionizing radiation particles is low and therefore is not considered in the models.

8.3 RADIATION EFFECTS ON LIVING CELLS

The ionizing radiation field produced by SEPs and GCRs is harmful to life (Henriksen, 1962) through both direct and indirect processes. Severe biological effects occur when ionization takes place at or near some particular radiation-sensitive molecule or cellular component (e.g., Pollard et al., 1955; Steel, 1996); ionization occurring outside these radiation-sensitive volumes is less effective. These processes are described by "radiation target theory" or "radiation target analysis" (e.g., Pollard et al., 1955; Steel, 1996). At the molecular level, direct ionizing radiation effects are based on the interaction of radiation with molecules, with ~60 eV transferred per event (Anchordoquy et al., 2009). These interactions can lead to loss of biological activity via molecular structural rearrangements (Osborne et al., 2000; Kempner, 2017).

The fundamental assumption underpinning radiation target analysis is that each primary ionization is random and has a Poisson distribution. In a single event, the probability of escape (i.e., no interaction) is exponentially dependent on the radiation dose and molecular mass. Thus, number of surviving molecules (or surviving molecular function or property) can be expressed as.

$$N = N_0 \exp(-qDm)$$

where N_0 is the initial number of molecules, D is the radiation dose, m is molecular mass, and $-q$ is a constant (Kempner, 1995). This is the simplest version of the target theory (a single event and single target), and the survival curve is a straight line on a semilogarithmic plot whose slope is $-q$. The dose survival of 0.37 (the D_{37} or D_0 dose) is the dose that will give an average of one hit per target or one inactivating event per cell.

According to this theory, it is possible to assess the efficiency of different types of radiation at producing certain biological effects; some of the model's predictions are as follows:

(i) The survival curve of the irradiated organism is exponential.

(ii) The effect of a given dose is independent of the dose rate of the radiation or of the manner in which it is fractionated.

(iii) There is a difference in the dosage of different forms of radiation, which produce identical biological effects.

Target theory has also been developed for the treatment of more complex situations where radiation effects require multiple primary ionizations. In this case, the inactivation curve displays an initial "shoulder" that may lead to an overestimation of N_0 (e.g., Lea, 1955; Zhao et al., 2015).

8.3.1 Radiation Chemistry of Water

Direct damage of biomolecules occurs when energy losses induced by ionizing radiation excite electrons within the molecules themselves, leading to ionization and radiolysis. However, radiation mainly interacts with water as this makes up to 70% of a cell's volume (20% in bacterial spores). Water in a cell may undergo direct radiolysis to form atomic hydrogen and a hydroxyl radical or may undergo ionization and generate a free electron (e.g., Battista, 1997, 2000; Baumstark-Khan and Facius, 2002):

$$H_2O + radiation \rightarrow H_2O^+ + e^-(aq)$$

The H_2O^+ ion is unstable and will rapidly dissociate (10^{-16} s)

$$H_2O^+ \rightarrow H^+ + \cdot OH$$

while e_{aq}^- can react with water to form a negative ion:

$$H_2O + e^-(aq) \rightarrow H_2O^-$$

The H_2O^- ion is also unstable and dissociates to form atomic hydrogen:

$$H_2O^- \rightarrow H^\cdot + OH^-$$

The net effect of water radiolysis and ionization is the generation of reactive free radicals, including the highly reactive hydroxyl radical that may react with another hydroxyl radical to form hydrogen peroxide:

$$\cdot OH + \cdot OH \rightarrow H_2O_2$$

Hydrogen peroxide may also form via reaction of atomic hydrogen in the presence of oxygen:

$$H^{\cdot} + O_2 \rightarrow HO_2^{\bullet}$$

$$HO_2^{\bullet} + HO_2^{\bullet} \rightarrow H_2O_2$$

The perhydroxyl radical may also dissociate to form superoxide:

$$HO_2^{\bullet} \rightarrow O_2^- + H^+$$

Thus, ionizing radiation can damage biopolymers such as DNA, RNA, and proteins, not only through direct energy deposition but also indirectly by radiochemistry and free-radical diffusion (Baumstark-Khan and Facius, 2002). The reactions shown above represent only a small subset of processes that may occur from the interaction of ionizing radiation with water. Overall, these reactions result in the production of many forms of highly reactive species with unpaired electrons (free radicals). The $\cdot OH$ radical is often the most biologically damaging agent and, as shown above, can also result in the formation of other ROS such as hydrogen peroxide (H_2O_2) (Kiefer and Wiatrowski, 1991; Lehnert, 2007). Hydrogen peroxide and super oxide, as well as other ROS produced via water radiolysis, can act as strong oxidizing agents, and their presence in biological systems can cause chemical damage including the alteration of protein molecules. Cell death from irradiation is believed to be primarily due to DNA damage, and under gamma irradiation, roughly 80% of DNA damage is caused indirectly by irradiation-induced ROS (Ghosal et al., 2005). Of this irradiation-induced damage, 61% of the ionizations take place on the DNA's deoxyribose phosphate backbone, with the remaining 39% distributed among the four nucleotide bases (Purkayastha et al., 2005). The metabolic state of a cell plays an important role in the extent of damage from indirect effects, since metabolically active cells might be able to actively mitigate some of the damage, whereas dormant inactive cells can accumulate damage to lethal levels. Low temperatures can increase resistance against ionizing radiation due to the reduced diffusion of radiation-generated radicals (Baumstark-Khan and Facius, 2002).

8.3.2 Biological Impact of Various Forms of Radiation

Radiation damage effects on biological materials can be described using relative biological effectiveness (RBE), which is the ratio of a standard adsorbed radiation dose to a radiation dose of another type that results in an equal amount of biological damage.

Multiplying the RBE by the adsorbed dose (SI unit gray) yields biological equivalent dose (SI unit sievert):

$$\text{Biological equivalent dose (SI)} = D(\text{Gy}) \times \text{RBE}$$

The International Commission on Radiological Protection (ICRP) publications contain guidelines on the methodology to weight physically absorbed radiation doses. Briefly, the physically absorbed radiation dose depends on the particle type, and in some cases, also on the particle's kinetic energy, the compounded effect of these two physical variables is known as radiological potential. The product of the absorbed dose, D, and the corresponding radiation weighting factor, w_D, is called the equivalent dose, H_T. The standard procedure is to set the X- and γ-ray weightings to 1 and include the other radiation factors via relative biological effectiveness. The factors appropriate for astrobiology research are presented in Table 8.1.

8.4 THE MAXIMUM DORMANCY LIMIT ON MARS

As discussed in the previous section, the radiation environment on the surface of Mars imposes limits on habitability through direct cellular damage and indirectly through the formation of ROS (e.g., Pavlov et al., 2002; Dartnell et al., 2007). Dormant microorganisms, the most likely forms of life that could exist near the surface of Mars today, can only accumulate a certain amount of cellular/molecular damage from exposure to ionizing radiation or ROS before cell death occurs. We call this threshold the maximum dormancy limit (MDL). Therefore, an important habitability parameter is the amount of radiation deposited in the regolith as a function of time and depth. Determination of this parameter requires tracking and calculation of the following:

(i) Propagation of the GCRs and SEPs in the Martian atmosphere and the formation of secondary particle showers due to GCR and SEP interaction with atmospheric molecules.

(ii) Followed by atmospheric secondary impinging on the surface of the planet and those formed in the shallow subsurface. Usually, the outputs of this stage are (a) the flux of the most relevant particles (muons, electrons/positrons, protons, neutrons,

Table 8.1 The radiation biological effectiveness (RBE) of various forms of radiation

Radiation	RBE
X- and γ-rays	1
Electrons/positrons	1
Muons	1
Protons $E \geq 2.0$ MeV	5
α-Particles and HZE	20

Taken from ICRP, 1990. Recommendations of the International Commission on Radiological Protection. Technical report. Pergamon Press, Table 1, p. 6.

photons, and HZE) at discrete locations and (b) energy deposition at finite-size volumes throughout the simulation volume.

(iii) Radiation absorption by materials of astrobiological interest embedded in the subsurface such as cells or organic molecules.

Pavlov et al. (2002) provided the first theoretical assessment of the biological effects of ionizing radiation on Mars using numerical transport codes, but Dartnell et al. (2007) conducted a more detailed study of the propagation of GCRs and SEPs through the Martian atmosphere and regolith to a depth of 20 m. The model used a 70 km atmospheric column reproduced in terms of composition, pressure, density, and temperature with data taken from the Mars Climate Database. The column represented a location on Mars where the surface pressure is 600 Pa (Arabia Terra), at noon on a summer day ($L_s = 270$–300), when atmospheric density is highest. The model predicted that within the top 2 m of regolith, a dormant population of *Deinococcus radiodurans* cells would suffer a million-fold reduction in viable cells in timescales ranging from 10^4 to 10^5 years, with a resulting MDL of ∼650,000 years. However, the radiation assessment detector (RAD) on the Mars Science Laboratory's *Curiosity* rover obtained direct measurements of the GCR and SEP radiation environment on the surface of the planet on 7 August 2012 at Gale Crater (Hassler et al., 2014). The actual biologically weighted absorbed dose (see Section 1.3.3) measured by the RAD was 232 mGy/year at the surface, which is a factor of 3.6 times lower than the value modeled by Dartnell et al. (2007). Based on this surface radiation dose, the MDL at 2 m depth would be ∼2.3 Myr.

For this chapter, we upgraded these previous results with a revised version of the GCR spectra and an updated version of Geant4/PLANETOCOSMICS, as follows:

(i) A 70 km atmospheric column over the Martian polar regions, with composition, pressure, density, and temperature profiles derived from the Martian Ames CGM (http://www-mars.lmd.jussieu.fr/mars/access.html).

(ii) GCR primary spectra from proton to iron nuclei will be obtained from the CREAM09 model (https://creme.isde.vanderbilt.edu/); this is displayed in Fig. 8.1.

(iii) SEP flux over 10–200 MeV obtained from the SPENVIS database and adjusted to the Martian orbit (http://www.spenvis.oma.be/).

Fig. 8.3 presents our modeled particle energy spectra over six orders of magnitude (from 1 MeV to 1 TeV) at the Martian surface. The six most relevant particle species are shown: protons, HZE, neutrons, photons, electrons/positrons, and muons. Their fluxes result from the forward scattering of high-energy particles flowing through the atmosphere, as expected since this hard component of the cascade does not backscatter and some back-scattering is mostly due to the soft component of the shower interacting with the top few decimeters of the Martian surface (Clowdsley et al., 2000). To investigate the role of regolith composition, we considered two end-member scenarios: (i) pure ice (PI) and (ii) dry homogenous (DH) regolith. The former is a 1×10^3 kg/m^3 block of pure ice, while the

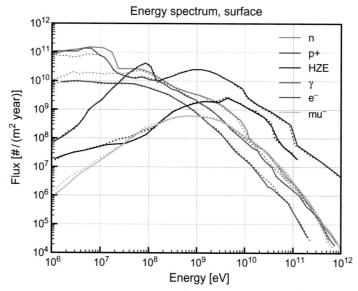

Fig. 8.3 Energy spectra of the photons, neutrons, electrons, protons, HZE, and muons at the Martian surface from SEPs and GCRs (solar minimum) presented in Fig. 8.1. The shallow surface models assume pure ice (PI, dotted) and homogeneous dry (DH, solid); see text for details.

latter is a spatially homogenous $2.8 \times 10^3 \, kg/m^3$ block of completely desiccated regolith with an elemental composition based on average composition values obtained by Pathfinder (Wänke et al., 2001). This approach allowed us to compare our findings with results already in the literature (e.g., Pavlov et al., 2002; Dartnell et al., 2007; Mileikowsky et al., 2000). The feature in the proton spectrum at 100 MeV in Fig. 8.3 is caused by the high-energy cutoff of the SEP spectrum given by SPENVIS (see Fig. 8.1). At energies below 25–40 MeV photons, electrons and neutrons dominate the surface radiation environment, while at energies higher than 30 GeV, protons and HZE have the largest fluxes. It must be noted that the neutron flux is sensitive to soil composition; this is the basis of neutron spectroscopy employed to map volatiles, namely, water, in the shallow subsurface of terrestrial planets (e.g., Feldman et al., 1993). This sensitivity is particularly acute for energies below 20 MeV.

Fig. 8.4 presents the flux of secondary particles below 1 m of Martian regolith. At energies lower than 20 GeV, the flux is mainly dominated by photons, neutrons, and electrons. Like at the surface, at low energies (<40 MeV), there is a strong dependence of the neutron flux on the regolith model composition. This dependency is due the efficient neutron moderation and capture by the hydrogen content of water (e.g., Feldman et al., 1993). For this reason, soils with high water content lead to lower neutron fluxes. At high energies, all the particle fluxes show attenuation when compared with their surface counterparts. This is more evident for HZE particles and is mainly due to ionization

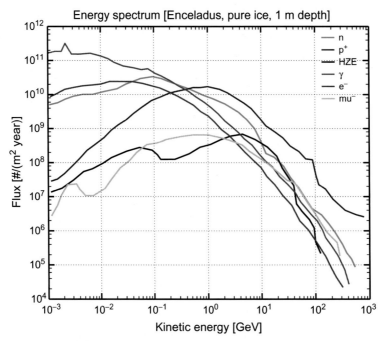

Fig. 8.4 Energy spectra of the secondary particles at 1 m depth by the primary GCRs and SEPs in Fig. 8.1. *Solid and dotted lines* denote DH and PI soils, respectively.

losses to the target and nuclear interactions. The former causes an energy downshift in the spectrum, while the latter leads to fragmentation of the primary HZEs into neutrons and protons. The effects are more obvious in the homogeneous dry soil than in the pure ice counterpart. At high energies, all the particle fluxes show attenuation when compared with their surface counterparts. This is more evident for HZE particles and is mainly due to ionization losses to the target and nuclear interactions.

Fig. 8.5 presents the flux of neutrons, protons, HZE, gamma rays, electrons, and muons as a function of depth. The radiation environment is attenuated least in the pure ice model as its shielding density is lower, but the majority of the considered species present a monotonic attenuation with depth in both regolith models. A Pfotzer maximum is observed for γ-rays at 1 m deeper for the pure ice model than for the denser dry homogenous model. Furthermore, in both regolith scenarios, the gamma flux profile diminishes until it becomes similar to the muon profile at ∼12.5 and ∼6.0 m in pure ice and dry homogenous regolith, respectively. At depths ranging from ∼1 to ∼6.0 m (from ∼2 to ∼12.5), muons become the most dominant species in PI (DH). Electron profiles also present a maximum in the flux profiles within the top 1.5 m of the regolith. The neutron profile in homogeneous dry regolith shows a similar pattern. In pure ice at depths shallower than 1 m, neutrons have the highest secondary particle flux.

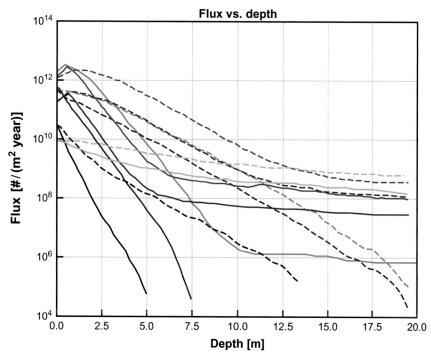

Fig. 8.5 Flux attenuation of electron, muon, neutron, HZE, gamma, and proton within the top 20 m of the Martian subsurface. Solid and dashed lines denote DH and PI soils, respectively.

Fig. 8.6 shows the annual accumulation of physically absorbed dose (left panel) and biologically weighted dose (right panel) as function of depth, for the pure ice and the dry homogeneous regolith models. Briefly, the weighted dose is the sum of the products of each energy deposition event and the factors presented in Table 8.1. Using the assumption of Dartnell et al. (2007) that neutrons are accounted for indirectly, we excluded them from the model.

The radiological effects of these particles are incorporated indirectly via protons, recoil nuclei, and fragments. Our estimated weighted absorbed dose was ~0.84 Gy/year for both regolith models, similar to estimates by Dartnell et al. (2007). The sharp decrease in both models in the top ~0.10 m reflects that SEPs, although dominating the flux of primaries, are not very energetic (see Fig. 8.1). It is worth comparing this dose value with ionizing radiation due to the natural decay of radioisotopes in the regolith (e.g., ^{40}K, ^{226}Ra, ^{238}U, and ^{232}Th). On Earth, the most extreme radiation environment due to natural radioisotope decay is 0.4 Gy/year (Baumstark-Khan and Facius, 2002; Veiga et al., 2006; Anjos et al., 2006), which corresponds to depths of ~0.20 and ~0.60 m in the dry homogeneous and pure ice regolith models, respectively. Mileikowsky et al. (2000) estimated that the intrinsic radiation environment in the Martian regolith

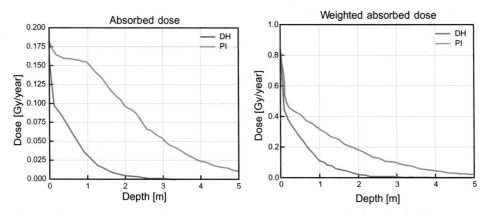

Fig. 8.6 Left and right panel presents deposited radiation and weighted deposited radiation versus depth, respectively. The *blue and orange lines* denote the "pure ice" and "dry homogenous" Martian soils, respectively.

due to natural decay is 4×10^{-4} Gy/year, which corresponds to a depth of $\sim$3.6 m in a dry homogenous regolith model. Therefore, we can assume that below this depth, the effects of GCR's cascade become subdued relative to radioisotope decay.

The weighted absorbed dose is the key parameter required to derive long-term biological effects of the ionizing radiation environment as a function of time and depth. Fig. 8.7 presents the survival time as a function of depth for two microbial populations

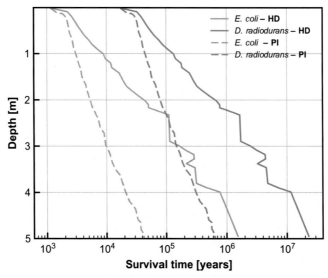

Fig. 8.7 Survival time versus depth, soil type, and microbial species for a 10^6 population reduction. The two microbial species considered are *E. coli* and *D. radiodurans*; these are represented in *gold and green*, respectively. The Martian regolith models are dry homogenous *(solid)* and pure ice *(dashed)*.

Table 8.2 *Deinococcus radiodurans* survival time computed using a 10^6 reduction

Depth	Survival time (years)			
	Mileikowsky et al. (2000)	Pavlov et al. (2002)	Dartnell et al. (2007)	This work
Surface	1.8×10^6	–	1.8×10^4	1.6×10^4
2×10^2 kg/m^2 ($\sim$0.07 m)	–	3.0×10^{4a}	8.7×10^4	8.5×10^4
5×10^3 kg/m^2 ($\sim$1.80 m)	1.4×10^7	–	1.8×10^6	6.0×10^5
1×10^4 kg/m^2 ($\sim$3.60 m)^b	–	4.0×10^7	4.0×10^7	4.0×10^7

All values are computed assuming dry regolith.
[a]Computed using a 10^8 reduction.
[b]Depth at which the GCR influence is subdominant compared with radionucleide decay.

in the two aforementioned regolith models: (i) *D. radiodurans* a radiation-resistant organism and (ii) *Escherichia coli* a radiation-sensitive organism. We assumed complete sterilization would take place after a microbial population reduction of 10^6 (Dartnell et al., 2007), extrapolated to be 600 Gy and 12 kGy for *E. coli* and *D. radiodurans*, respectively (Battista, 1997, 2000). Populations of radiation-sensitive organisms would become extinct within the top 1 m of the Martian surface in <10,000 years in dry regolith and as early as 3000 years for organisms surrounded by pure ice. Populations of radiation-resistant organisms could endure up to 200,000 years within the top 1 m of dry regolith. We emphasize that these predictions assume that organisms are in a dormant state and continuously accumulate radiation damage. Cell repair during spurs of metabolic activity with a frequency higher than these survival times would allow populations to endure longer periods of time. Table 8.2 presents a comparison between some of our *D. radiodurans* models in the DH regolith and some of the published literature. Our models are very similar to estimates by Dartnell et al. (2011) but depart from estimates by Pavlov et al. (2002) and Mileikowsky et al. (2000) mainly at shallow depths. A primary reason for this departure is that these previous studies only considered solar minimum, while Dartnell et al. (2007) used PLANETOCOSMICS/Geant4, which includes all the required interactions to model the ionizing radiation propagation through the regolith. In the code used by Pavlov et al. (2002) for modeling (Blinov and Lazarev, 1999), the contribution of secondary pions, electrons, and gammas was not included, and slow neutron transport was also not included (Pavlov et al., 2002).

8.5 CONCLUSIONS

The main period of Martian habitability occurred >3 billion years ago. Since then, habitability conditions have degraded, perhaps with the exception of relatively short, warm, and wet episodes triggered by volcanic activity, impacts, or orbital fluctuations. Therefore, the potential for extant life near the surface of Mars critically hinges on whether microbial communities could have survived long stretches of dormancy in between punctuated periods of habitable conditions, up to the present. Ionizing radiation on Mars is a "silent killer" that limits the amount of time that microorganisms could remain dormant near the surface of the planet. This maximum dormancy limit (MDL) can be estimated using nuclear physics numerical codes that accurately simulate the interactions of ionizing radiation with geologic materials. In the case of Mars, the MDL could span 1000–200,000 years depending on the radioresistance of dormant cells and the composition of regolith materials. This might be considered an upper limit on the current habitability potential of the planet.

APPENDIX A MATHEMATICAL EXPRESSIONS

Normal Bethe-Bloch Formula

Moderately, relativistic charged particles other than electrons lose energy in matter primarily by ionization and atomic excitation. The mean rate of energy loss referred to as the particle's stopping power, $<dE/dx>$, of a secondary particle with energy E is given by the Bethe equation (also known as the Bethe-Bloch equation; Yao et al., 2006; Amsler et al., 2008; and references therein):

$$-\left\langle \frac{dE}{dx} \right\rangle \bigg|_{Z,B} = K \cdot \frac{z^2}{\beta^2} \cdot \frac{Z}{A} \cdot \left[\frac{1}{2} \ln f(\beta) - \beta^2 - \frac{\delta(\beta\gamma)}{2} \right] \tag{A.1}$$

where Z and A are the charge and atomic number of the traversed material and z the charge number of the projectile nucleus (which corresponds to the total charge for a fully ionized HZE particle). The normalization K is given by $4\pi N_A/(m_e c^2)$, where N_A is Avogadro's constant, m_e the electron rest mass, and c the speed of light. Furthermore, x is the traversed distance and $\beta = v/c$, where c is the light speed and v is the speed of the projectile. Finally, $f(\beta)$ takes into account the details of the effects of the single collisions and the mean exciting energy, while $\delta(\beta\gamma)$ is known as the mean excitation energy. In the current application, we make the following approximation $\delta(\beta\gamma)/2 \approx \delta/2$ (Amsler et al., 2008).

Mixtures of compounds can be thought as thin layers of pure elements in the right proportion. In this case,

$$\left\langle \frac{dE}{dx} \right\rangle \bigg|_{mix.} = \sum_j w_j \left\langle \frac{dE}{dx} \right\rangle \bigg|_j \tag{A.2}$$

where $<dE/dx>|j$ is the mean rate of energy loss in the jth element and w_j denotes the j-layer weight. Eq. (A.1) can be inserted into Eq. (A.2) to find expressions for $\langle Z/A \rangle = \sum w_j Z_j/A_j = n_j Z_j/(\sum n_j A_j)$. For further details, see Amsler et al. (2008).

Electron Bethe-Bloch Formula

The interactions between secondary electrons and the matter they pass through take place through a variety of mechanisms. At high energies, however, the dominant mechanisms of energy loss are collisions that cause ionization and excitation of atoms and radiation production (bremsstrahlung) due to electron negative acceleration. In the former, energy is attenuated by a multitude of inelastic collisions until it reaches thermal levels. Bethe (1930) theorized that the electron energy loss is a particular case of the energy loss by charged particles. Under the assumption that the velocity, v, of the incoming electron exceeds that of the atomic orbital electrons (known as the Born approximation), the mean energy loss per unit length is

$$-\left\langle \frac{dE}{dx} \right\rangle\bigg|_{e^-,coll} = K \cdot \frac{z^2}{\beta^2} \cdot \frac{Z}{A} \cdot \left[\frac{1}{2} \ln g(\beta) + F(\tau) - \frac{\delta(\beta)}{2} \right] \tag{A.3}$$

where $\tau = \beta - 1$ (kinetic energy of the electron in units of $m_e c^2$). The expressions for $g(\beta)$ and $F(\tau)$ can be found in Amsler et al. (2008). The production of X-rays (bremsstrahlung) by the deacceleration (stopping) of electrons is not a very efficient process at low energies ($\lesssim 5\,MeV$; Amsler et al., 2008), but it becomes the dominant process at higher energies. In the context of quantum electrodynamics, using the Thomas-Fermi model, the amount of energy lost by an electron as radiation per unit path length is approximately given by

$$-\left\langle \frac{dE}{dx} \right\rangle\bigg|_{e^-,rad} = 4\alpha \cdot r_e^2 \cdot N_0 \cdot E \cdot Z^2 \cdot \ln \frac{183}{Z^{1/3}} = \frac{1}{x_0} E \tag{A.4}$$

where N_0 is the atomic density of a medium with atomic number Z while α is the fine-structure constant and r_e the classical electron radius. x_0 is the radiation length and is equivalent to the thickness of the target for which the impinging energy is reduced to the eth part of its initial value.

REFERENCES

Abramov, O., Kring, D.A., 2005. Impact-induced hydrothermal activity on early Mars. J. Geophys. Res. 110, E12S09.

Agostinelli, S., Allison, J., Amako, K., Apostolakis, J., Araujo, H., Arce, P., Asai, M., Axen, D., Banerjee, S., Barrand, G., Behner, F., Bellagamba, L., Boudreau, J., Broglia, L., Brunengo, A., Burkhardt, H., Chauvie, S., Chuma, J., Chytracek, R., Cooperman, G., Cosmo, G., Degtyarenko, P., Dell'Acqua, A., Depaola, G., Dietrich, D., Enami, R., Feliciello, A., Ferguson, C., Fesefeldt, H., Folger, G., Foppiano, F., Forti, A., Garelli, S., Giani, S., Giannitrapani, R., Gibin, D., Gomez Cadenas, J.J., Gonzalez, I., Gracia Abril, G., Greeniaus, G., Greiner, W., Grichine, V., Grossheim, A., Guatelli, S., Gumplinger, P., Hamatsu, R., Hashimoto, K., Hasui, H., Heikkinen, A., Howard, A., Ivanchenko, V., Johnson, A., Jones, F.W., Kallenbach, J., Kanaya, N., Kawabata, M., Kawabata, Y., Kawaguti, M., Kelner, S., Kent, P., Kimura, A., Kodama, T., Kokoulin, R., Kossov, M., Kurashige, H., Lamanna, E., Lampen, T., Lara, V., Lefebure, V., Lei, F., Liendl, M., Lockman, W., Longo, F., Magni, S., Maire, M., Medernach, E., Minamimoto, K., Mora de Freitas, P., Morita, Y., Murakami, K., Nagamatu, M., Nartallo, R., Nieminen, P., Nishimura, T.,

Ohtsubo, K., Okamura, M., O'Neale, S., Oohata, Y., Paech, K., Perl, J., Pfeiffer, A., Pia, M.G., Ranjard, F., Rybin, A., Sadilov, S., Di Salvo, E., Santin, G., Sasaki, T., Savvas, N., Sawada, Y., Scherer, S., Sei, S., Sirotenko, V., Smith, D., Starkov, N., Stoecker, H., Sulkimo, J., Takahata, M., Tanaka, S., Tcherniaev, E., Safai Tehrani, E., Tropeano, M., Truscott, P., Uno, H., Urban, L., Urban, P., Verderi, M., Walkden, A., Wander, W., Weber, H., Wellisch, J.P., Wenaus, T., Williams, D.C., Wright, D., Yamada, T., Yoshida, H., Zschiesche, D., GEANT4 Collaboration, July 2003. GEANT4—a simulation toolkit. Nucl. Inst. Methods Phys. Res. A 506, 250–303.

Allison, J., Amako, K., Apostolakis, J., Araujo, H., Arce Dubois, P., Asai, M., Barrand, G., Capra, R., Chauvie, S., Chytracek, R., Cirrone, G.A.P., Cooperman, G., Cosmo, G., Cuttone, G., Daquino, G.G., Donszelmann, M., Dressel, M., Folger, G., Foppiano, F., Generowicz, J., Grichine, V., Guatelli, S., Gumplinger, P., Heikkinen, A., Hrivnacova, I., Howard, A., Incerti, S., Ivanchenko, V., Johnson, T., Jones, F., Koi, T., Kokoulin, R., Kossov, M., Kurashige, H., Lara, V., Larsson, S., Lei, F., Link, O., Longo, F., Maire, M., Mantero, A., Mascialino, B., McLaren, I., Mendez Lorenzo, P., Minamimoto, K., Murakami, K., Nieminen, P., Pandola, L., Parlati, S., Peralta, L., Perl, J., Pfeiffer, A., Pia, M.G., Ribon, A., Rodrigues, P., Russo, G., Sadilov, S., Santin, G., Sasaki, T., Smith, D., Starkov, N., Tanaka, S., Tcherniaev, E., Tome, B., Trindade, A., Truscott, P., Urban, L., Verderi, M., Walkden, A., Wellisch, J.P., Williams, D.C., Wright, D., Yoshida, H., February 2006. Geant4 developments and applications. IEEE Trans. Nucl. Sci. 53, 270–278.

Amsler, C., Doser, M., Antonelli, M., Asner, D.M., Babu, K.S., Baer, H., Band, H.R., Barnett, R.M., Bergren, E., Beringer, J., Bernardi, G., Bertl, W., Bichsel, H., Biebel, O., Bloch, P., Blucher, E., Blusk, S., Cahn, R.N., Carena, M., Caso, C., Ceccucci, A., Chakraborty, D., Chen, M.C., Chivukula, R.S., Cowan, G., Dahl, O., D'Ambrosio, G., Damour, T., de Gouvea, A., DeGrand, T., Dobrescu, B., Drees, M., Edwards, D.A., Eidelman, S., Elvira, V.D., Erler, J., Ezhela, V.V., Feng, J.L., Fetscher, W., Fields, B.D., Foster, B., Gaisser, T.K., Garren, L., Gerber, H.-J., Gerbier, G., Gherghetta, T., Giudice, G.F., Goodman, M., Grab, C., Gritsan, A.V., Grivaz, J.-F., Groom, D.E., Grunewald, M., Gurtu, A., Gutsche, T., Haber, H.E., Hagiwara, K., Hagmann, C., Hayes, K.G., Hernandez-Rey, J.J., Hikasa, K., Hinchliffe, I., Hocker, A., Huston, J., Igo-Kemenes, P., Jackson, J.D., Johnson, K.F., Junk, T., Karlen, D., Kayser, B., Kirkby, D., Klein, S.R., Knowles, I.G., Kolda, C., Kowalewski, R.V., Kreitz, P., Krusche, B., Kuyanov, Y.V., Kwon, Y., Lahav, O., Langacker, P., Liddle, A., Ligeti, Z., Lin, C.-J., Liss, T.M., Littenberg, L., Liu, J.C., Lugovsky, K.S., Lugovsky, S.B., Mahlke, H., Mangano, M.L., Mannel, T., Manohar, A.V., Marciano, W.J., Martin, A.D., Masoni, A., Milstead, D., Miquel, R., Monig, K., Murayama, H., Nakamura, K., Narain, M., Nason, P., Navas, S., Nevski, P., Nir, Y., Olive, K.A., Pape, L., Patrignani, C., Peacock, J.A., Piepke, A., Punzi, G., Quadt, A., Raby, S., Raffelt, G., Ratcliff, B.N., Renk, B., Richardson, P., Roesler, S., Rolli, S., Romaniouk, A., Rosenberg, L.J., Rosner, J.L., Sachrajda, C.T., Sakai, Y., Sarkar, S., Sauli, F., Schneider, O., Scott, D., Seligman, W.G., Shaevitz, M.H., Sjostrand, T., Smith, J.G., Smoot, G.F., Spanier, S., Spieler, H., Stahl, A., Stanev, T., Stone, S.L., Sumiyoshi, T., Tanabashi, M., Terning, J., Titov, M., Tkachenko, N.P., Tornqvist, N.A., Tovey, D., Trilling, G.H., Trippe, T.G., Valencia, G., van Bibber, K., Vincter, M.G., Vogel, P., Ward, D.R., Watari, T., Webber, B.R., Weiglein, G., Wells, J.D., Whalley, M., Wheeler, A., Wohl, C.G., Wolfenstein, L., Womersley, J., Woody, C.L., Workman, R.L., Yamamoto, A., Yao, W.-M., Zenin, O.V., Zhang, J., Zhu, R.-Y., Zyla, P.A., Harper, G., Lugovsky, V.S., Schaffner, P., Particle Data Group, September 2008. Review of particle physics. Phys. Lett. B 667, 1–6.

Anchordoquy, T.J., Molina, M.d.C., Kempner, E.S., February 2009. A radiation target method for size determination of supercoiled plasmid DNA. Anal. Biochem. 385 (2), 229–233.

Anjos, R.M., Veiga, R., Macario, K., Carvalho, C., Sanches, N., Bastos, J., Gomes, P.R.S., May 2006. Radiometric analysis of quaternary deposits from the southeastern Brazilian coast. Mar. Geol. 229, 29–43.

Battista, J.R., 1997. Against all odds: the survival strategies of Deinococcus radiodurans. Annu. Rev. Microbiol. 51 (1), 203–224.

Battista, J.R., 2000. Radiation resistance: the fragments that remain. Curr. Biol. 10 (5), R204–R205. (ISSN 0960-9822).

Baumstark-Khan, C., Facius, R., 2002. Life under conditions of ionizing radiation. In: Horneck, G., Baumstark-Khan, C. (Eds.), Astrobiology. The quest for the conditions of life. In: Physics and astronomy online library. Berlin, Springer. ISBN 3-540-42101-7, pp. 261–284.

Bethe, H., 1930. Zur Theorie des Durchgangs schneller Korpuskularstrahlen durch Materie. Ann. Phys. 397, 325–400.

Blinov, A.V., Lazarev, V.E., 1999. Lateral and temporal changes of nuclear reaction intensity leading to production of cosmogenic nuclides in the Earth atmosphere. Bull. Russ. Acad. Sci. 63 (8), 159–162.

Cai, Z., Kwon, Y.L., Reilly, R.M., 2017. Monte Carlo N-particle (MCNP) modeling of the cellular dosimetry of 64Cu: comparison with MIRDcell S-values and implications for studies of its cytotoxic effects. J. Nucl. Med. 58 (2), 339–345.

Clowdsley, M.S., Wilson, J.W., Kim, M.-H.Y., Singleterry, R.C., Tripathi, R.K., Heinbockel, J.H., Badavi, F.F., Shinn, J.L., 2000. Neutron environments on the Martian surface.1st International Workshop on Space Radiation Research, Arona, Italy, pp. 94–96.

Cockell, C.S., Catling, D.C., Davis, W.L., Snook, K., Kepner, R.L., Lee, P., McKay, C.P., August 2000. The ultraviolet environment of Mars: biological implications past, present, and future. Icarus 146, 343–359.

Costard, F., Forget, F., Mangold, N., et al., 2002. Formation of recent Martian debris flows by melting of near-surface ground ice at high obliquity. Science 295, 110–113.

Cousins, C.R., Crawford, I.A., 2011. Volcano-ice interaction as a microbial habitat on earth and Mars. Astrobiology 11, 695–710.

Dartnell, L.R., Patel, M.R., April 2014. Degradation of microbial fluorescence biosignatures by solar ultraviolet radiation on Mars. Int. J. Astrobiol. 13, 112–123.

Dartnell, L.R., Desorgher, L., Ward, J.M., Coates, A.J., January 2007. Modelling the surface and subsurface Martian radiation environment: implications for astrobiology. Geophys. Res. Lett. 34, L02207.

Dartnell, L.R., Storrie-Lombardi, M.C., Mullineaux, C.W., Ruban, A.V., Wright, G., Griffiths, A.D., Muller, J.-P., Ward, J.M., December 2011. Degradation of cyanobacterial biosignatures by ionizing radiation. Astrobiology 11, 997–1016.

Feldman, W.C., Boynton, W.V., Drake, D.M., 1993. Planetary neutron spectroscopy from orbit. In: Pieters, C.M., Englert, P.A.J. (Eds.), Remote Geochemical Analysis: Elemental and Mineralogical Composition. Cambridge Univ. Press, New York, pp. 213–234

Feldman, W.C., Boynton, W.V., Jakosky, B.M., Mellon, M.T., November 1993. Redistribution of subsurface neutrons caused by ground ice on Mars. J. Geophys. Res. 98, 20855.

Feldman, W.C., Boynton, W.V., Tokar, R.L., Prettyman, T.H., Gasnault, O., Squyres, S.W., Elphic, R.C., Lawrence, D.J., Lawson, S.L., Maurice, S., McKinney, G.W., Moore, K.R., Reedy, R.C., July 2002. Global distribution of neutrons from Mars: results from Mars Odyssey. Science 297, 75–78.

Gasnault, O., D'Uston, C., Barthe, H., Veyan, J.F., July 1999. Simulations and calibrations of the Mars Surveyor 2001 gamma-ray spectrometer performances.The Fifth International Conference on Mars, p. 6079.

Ghosal, D., Omelchenko, M.V., Gaidamakova, E.K., Matrosova, V.Y., Vasilenko, A., Venkateswaran, A., Zhai, M., Kostandarithes, H.M., Brim, H., Makarova, K.S., Wackett, L.P., Fredrickson, J.K., Daly, M.J., 2005. How radiation kills cells: survival of deinococcus radiodurans and shewanella oneidensis under oxidative stress? FEMS Microbiol. Rev. 29 (2), 361–375.

Grotzinger, J.P., Sumner, D.Y., Kah, L.C., et al., 2014. A habitable fluvio-lacustrine environment at Yellowknife Bay, Gale Crater, Mars. Science, 343.

Grupen, C., 2010. Introduction to Radiation Protection: Practical Knowledge for Handling Radioactive Sources, Graduate Texts in Physics. Springer-Verlag, Berlin Heidelberg. ISBN 978-3-642-02585-3. https://doi.org/10.1007/978-3-642-02586-0.

Hassler, D.M., Zeitlin, C., Wimmer-Schweingruber, R.F., Eh- resmann, B., Rafkin, S., Eigenbrode, J.L., Brinza, D.E., Weigle, G., Bottcher, S., Bohm, E., Burmeister, S., Guo, J., Kohler, J., Martin, C., Reitz, G., Cucinotta, F.A., Kim, M.H., Grinspoon, D., Bullock, M.A., Posner, A., Gomez-Elvira, J., Vasavada, A., Grotzinger, J.P., 2014. Mars' surface radiation environment measured with the Mars Science Laboratory's Curiosity rover. Science 343. https://doi.org/10.1126/science.1244797.

Henriksen, T., 1962. Radiation-induced free radicals in frozen aqueous solutions of glycine. Radiat. Res. 17 (2), 158–172.

Jakosky, B.M., Nealson, K.H., Bakermans, C., et al., 2003. Subfreezing activity of microorganisms and the potential habitability of Mars' polar regions. Astrobiology 3, 343–350.

Kempner, E.S., November 1995. The mathematics of radiation target analyses. Bull. Math. Biol. 57 (6), 883–898.

Kempner, E.S., July 2017. Effects of high-energy electrons and gamma rays directly on protein molecules. J. Pharm. Sci. 90 (10), 1637–1646.

Kiefer, J., Wiatrowski, W.A., March 1991. Biological radiation effects. Phys. Today 44, 68.

Kminek, G., Bada, J.L., May 2006. The effect of ionizing radiation on the preservation of amino acids on Mars. Earth Planet. Sci. Lett. 245, 1–5.

Kminek, G., Rummel, J.D.D., Cockell, C.S.S., et al., 2010. Report of the COSPAR mars special regions colloquium. Adv. Space Res. 46, 811–829.

Laskar, J., Correia, A.C.M., Gastineau, M., et al., 2004. Long term evolution and chaotic diffusion of the insolation quantities of Mars. Icarus 170, 343–364.

Lawrence, D.J., Feldman, W.C., Goldsten, J.O., Maurice, S., Peplowski, P.N., Anderson, B.J., Bazell, D., McNutt, R.L., Nittler, L.R., Prettyman, T.H., Rodgers, D.J., Solomon, S.C., Weider, S.Z., January 2013. Evidence for water ice near Mercury's North Pole from MESSENGER neutron spectrometer measurements. Science 339, 292.

Lea, D.E., 1955. Actions of Radiations on Living Cells. Cambridge University Press, New York.

Lehnert, S., 2007. Biomolecular Action of Ionizing Radiation. In: Lehnert, S. (Ed.), Series: Series in Medical Physics and Biomedical Engineering. Taylor & Francis. ISBN 978-0-7503-0824-3.

Ma, C.-M., Jiang, S.B., December 1999. Topical review: Monte Carlo modelling of electron beams from medical accelerators. Phys. Med. Biol. 44, R157–R189.

Mckay, C.P., Stoker, C.R., Glass, B.J., et al., 2013. The icebreaker life mission to Mars: a search for biomolecular evidence for life. Astrobiology 13, 334–354.

Mileikowsky, C., Cucinotta, F.A., Wilson, J.W., Gladman, B., Horneck, G., Lindegren, L., Melosh, J., Rickman, H., Valtonen, M., Zheng, J.Q., June 2000. Natural transfer of viable microbes in space. 1. From Mars to Earth and Earth to Mars. Icarus 145, 391–427.

Osborne, J.C., Miller, J.H., Kempner, E.S., April 2000. Molecular mass and volume in radiation target theory. Biophys. J. 78 (4), 1698–1708.

Pavlov, A.K., Blinov, A.V., Konstantinov, A.N., June 2002. Sterilization of Martian surface by cosmic radiation. Planet. Space Sci. 50, 669–673.

Pelowitz, D.B. (Ed.), 2005. MCNPX User's Manual Version 2.5.0. Technical Report LA-CP-05-0369, Los Alamos National Laboratory, Los Alamos, New Mexico.

Pollard, E.C., Guild, W.R., Hutchinson, F., Setlow, R.B., 1955. The direct action of ionizing radiation on enzymes and antigens. Prog. Biophys. Biophys. Chem. 5, 72.

Potgieter, M.S., 2013. Solar Modulation of Cosmic Rays. Living Reviews in Solar Physics 10. https://doi.org/10.12942/lrsp-2013-3.

Pfotzer, G., 1936a. Dreifachkoinzidenzen der Ultrastrahlung aus vertikaler Richtung in der Stratosphre, Zeitschrift fur Physik 102, 23–40. https://doi.org/10.1007/BF01336829.

Pfotzer, G., 1936b. Dreifachkoinzidenzen der Ultrastrahlung aus vertikaler Richtung in der Stratosphre, Zeitschrift fur Physik 102, 41–58. https://doi.org/10.1007/BF01336830G.

Prettyman, T.H., Hagerty, J.J., Elphic, R.C., Feldman, W.C., Lawrence, D.J., McKinney, G.W., Vaniman, D.T., December 2006. Elemental composition of the lunar surface: analysis of gamma ray spectroscopy data from Lunar Prospector. J. Geophys. Res. (Planets) 111 (E10), E12007.

Purkayastha, S., Milligan, J.R., Bernhard, W.A., 2005. Correlation of free radical yields with strand break yields produced in plasmid DNA by the direct effect of ionizing radiation. J. Phys. Chem. B 109 (35), 16967–16973.

Rao, M.V.S. et al., (Eds.), 1998. Extensive Air Showers. Published by World Scientific Publishing Co. Pte. Ltd. ISBN #9789812817211. https://books.google.com/books?id=t5UZ3mtJVzAC

Rummel, J.D., Beaty, D.W., Jones, M.A., Bakermans, C., Barlow, N.G., Boston, P.J., Chevrier, V.F., Clark, B.C., de Vera, J.-P.P., Gough, R.V., Hallsworth, J.E., Head, J.W., Hipkin, V.J., Kieft, T.L., McEwen, A.S., Mellon, M.T., Mikucki, J.A., Nicholson, W.L., Omelon, C.R., Peterson, R., Roden, E.E., Sherwood Lollar, B., Tanaka, K.L., Viola, D., Wray, J.J., November 2014. A new analysis of Mars "special regions": findings of the second MEPAG special regions science analysis group (SR-SAG2). Astrobiology 14, 887–968.

Simonsen, L.C., Nealy, J.E., Langley Research Center, 1991. Radiation Protection for Human Missions to the Moon and Mars. NASA technical paper. National Aeronautics and Space Administration, Office of Management, Scientific and Technical Information Division. https://books.google.com/books?id=UZ8TAQAAIAAJ.

Squyres, S.W., Grotzinger, J.P., Arvidson, R.E., et al., 2004. In situ evidence for an ancient aqueous environment at Meridiani Planum, Mars. Science 306, 1709–1714.

Steel, G.G., 1996. From targets to genes: a brief history of radiosensitivity. Phys. Med. Biol. 41 (2), 205.

Teodoro, L.F.A., Lawrence, D.J., Eke, V.R., Elphic, R.E., Feldman, W.C., Maurice, S., Siegler, M.A., Paige, D.A., January 2015. The local-time variations of lunar prospector epithermal-neutron data. ArXiv e-prints.

Titt, U., Bednarz, B., Paganetti, H., October 2012. Comparison of MCNPX and Geant4 proton energy deposition predictions for clinical use. Phys. Med. Biol. 57, 6381–6393.

Veiga, R., Sanches, N., Anjos, R.M., Macario, K., Bastos, J., Iguatemy, M., Aguiar, J.G., Santos, A.M.A., Mosquera, B., Carvalho, C., Baptista Filho, M., Umisedo, N.K., 2006. Measurement of natural radioactivity in Brazilian beach sands. Radiat. Meas. 41 (2), 189–196.

Wänke, H., Brückner, J., Dreibus, G., Rieder, R., Ryabchikov, I., April 2001. Chemical composition of rocks and soils at the pathfinder site. Space Sci. Rev. 96, 317–330.

Ward, W.R., 1973. Large-scale variations in the obliquity of Mars. Science 181, 260–262.

Yao, W.M., et al., 2006. Review of particle physics. J. Phys. G: Nucl. Part. Phys. 33 (1), 1.

Zhao, L., Mi, D., Hu, B., Sun, Y., 2015. A generalized target theory and its applications. Nat. Sci. Rep. 5, 14568.

FURTHER READING

Beringer, J., Folger, G., Gianotti, F., Ribon, A., Wellisch, J.P., Barberis, D., Cervetto, M., Osculati, B., 2003. Validation of geant4 hadronic physics. 2003 IEEE Nuclear Science Symposium. Conference Record (IEEE Cat. No. 03CH37515). vol. 1. pp. 494–498.

ICRP, 1990. Recommendations of the International Commission on Radiological Protection. Technical report. Pergamon Press.

Pavlov, A.A., Vasilyev, G., Ostryakov, V.M., Pavlov, A.K., Mahaffy, P., July 2012. Degradation of the organic molecules in the shallow subsurface of Mars due to irradiation by cosmic rays. Geophys. Res. Lett. 39, L13202.

Quinn, R.C., Martucci, H.F.H., Miller, S.R., Bryson, C.E., Grunthaner, F.J., Grunthaner, P.J., June 2013. Perchlorate radiolysis on Mars and the origin of Martian soil reactivity full access. Astrobiology 13, 515–520.

Vago, J., Witasse, O., Svedhem, H., et al., 2015. ESA ExoMars program: the next step in exploring Mars. Sol. Syst. Res. 49, 518–528.

CHAPTER 9

UV and Life Adaptation Potential on Early Mars: Lessons From Extreme Terrestrial Analogs

Donat-Peter Häder, Nathalie A. Cabrol

Contents

9.1 OVERVIEW

The MAVEN mission has provided evidence that most of the early Martian atmosphere was stripped away early by solar wind (e.g., Jakosky et al., 2015, 2017). For the major part of the planet's history, low gravity and weak magnetic field have hardly shielded the surface from the high energetic particles. Yet, thermal differences between day and night and large storms regularly inject substantial amount of dust in the atmosphere, which can preclude solar radiation from significantly penetrating to the surface. The radiation environment of Mars is thus complex, and characterizing its evolution through time is essential to understand (a) the impact of UV radiation on soil geochemistry and its role in the preservation of biosignatures in the geologic record, which were discussed in Chapter 8, and (b) the UV environment and adaptation potential of early subaerial microbial habitats, which are examined here.

9.2 BACKGROUND

The irradiance in watt per square meter from the Sun falls with the square of the distance. The mean extraterrestrial solar irradiance on Earth outside the atmosphere is $1367\,\mathrm{W/m^2}$ (i.e., the solar constant) (Armitage, 1995). In comparison, the irradiance outside the atmosphere of Mars is 0.431 times the solar constant. However, this value varies considerable due to the eccentricity of the Martian orbit. With a very thin atmosphere, the

From Habitability to Life on Mars
https://doi.org/10.1016/B978-0-12-809935-3.00009-8

absence of an ozone layer, and the minor role played by water and carbon dioxide, solar radiation is not strongly attenuated before reaching the surface of Mars (Patel et al., 2002). As a result, neutral attenuation is mainly due to dust aerosol particles. However, although the optical thickness of the Martian atmosphere is nearly 10 times smaller than that of the Earth, data show that dust carried into the atmosphere by strong winds may remain in suspension over extended periods of time due to the planet's low gravity, with particle sizes around 1.5 μm, which increase the atmospheric opacities (τ) between 0.4 and 0.6 at all measured wavelengths (450, 670, 883, and 989 nm) (e.g., Pollack et al., 1977; Colburn et al., 1989; Smith and Lemmon, 1999; Lemmon et al., 2004; Vicente-Retortillo et al., 2017).

A radiative transfer model was used to calculate the transmission of solar UV radiation to the surface of Mars between 190 and 410 nm, which reflects the typical aerosol during quiet periods (McWilliam and Baird, 2002). As the thin Martian atmosphere does not provide significant attenuation by oxygen and ozone, dust aerosol particles are therefore mainly responsible for the neutral attenuation. Thus, at first glance, UV radiation would not seem to pose a substantial hazard. However, the biological efficiency of deleterious UV radiation is calculated by the erythemal action spectrum (Fig. 9.1) defined by the Commission International de L'Eclairage (CIE) (Häder and Erzinger, 2017).

In a low atmospheric dust load scenario, UV-B irradiance reaches $\sim$3.49 W m^{-2}, which is double that of Earth at midlatitude during the summer equinox (e.g., assuming an average total ozone column of 300 DU (Dobson unit), UV-B$=$1.66 W m^{-2} in Erlangen, 49.58° N, 11°) (Häder et al., 2007). However, the record UV-B values measured on Licancabur in the Bolivian Andes at 5920 m a.s.l. (22.83° S, 67.97° W) were 4.1 W m^{-2} during the summer with an average of 250 DU (Cabrol et al., 2014).

In addition to strong UV radiation, the Martian environment is also characterized by high energetic radiation (see Chapter 8). The radiation assessment detector (RAD) onboard the Curiosity rover at Gale crater is allowing detailed measurements (e.g., Hassler et al., 2014). It shows that radiation consists of galactic cosmic rays (GCR) and solar energetic particles (SEP). The GCR dose rate varies between 180 and 225 μGy/day (1 Gy$=$1 J of energy deposited in 1 kg), while the SEP dose rate amounts to about 50 μGy/day. Data indicate that radiation exposure on Mars is much harsher than on Earth, which is explained by a lack of a strong magnetic field and an atmosphere 165 times thinner that cannot deflect high energetic particles. This highly energetic radiation breaks molecular bonds in organic molecules such as DNA and would kill any life form exposed at the surface, including in the top few meters of soil (Granskog et al., 2012; see also Chapter 8). Even the most resilient terrestrial microorganisms, such as the radio-resistant bacterium *Deinococcus radiodurans* (formerly *Micrococcus radiodurans*) would not survive in such an environment (e.g., Rummel et al., 2014).

The present-day radiation environment is thus generally considered deadly for sub-aerial life and habitats. This constitutes a challenge as well for planning and implementing

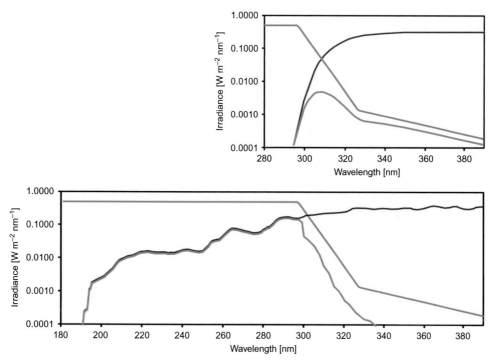

Fig. 9.1 Calculation of the biological efficiency of deleterious UV radiation. The solar spectrum at the Earth's surface (upper panel, *red curve*) is folded by the erythemal action spectrum defined by the Commission International de L'Eclairage *(green curve)* (Häder and Erzinger, 2017) to obtain the effective action spectrum for erythemal damage on Earth *(blue curve)*. The lower panel shows the same procedure using the solar radiation on Mars under the low dust scenario ($\tau = 0.1$) using the same scale as in the upper panel.

the human exploration and colonization of Mars (e.g., Cucinotta et al., 2013). The radiation dose equivalent to the annual cosmic radiation at sea level on our planet is about 0.4 mS (millisieverts). While a 6-month stay exposes astronauts to 75 mS, a 180-day journey to Mars would expose them to 123 mS, with an additional 122 mS for a 500-day stay at the surface.

Geologic, mineralogical, and geochemical evidences support the idea that the current Martian UV radiation environment has been prevalent for the past 3.8 billion years, with some spatiotemporal fluctuations associated with obliquity cycles and atmospheric depth (e.g., Atri et al., 2013). However, over the past decades, the state and evolution of the climate and atmosphere on early Mars has been continuously debated. For some, evidence of abundant surface water activity supports the hypothesis that early Mars had a substantially thicker atmosphere and a warmer and wetter climate than today over extended geologic times (e.g., Baker, 2001; Carr and Head, 2015; Craddock and Lorenz, 2017; Luo et al., 2017). Meanwhile, the results of the MAVEN mission

(Jakosky et al., 2015, 2017) and climate modeling have recently cast doubt upon this scenario (e.g., Wordsworth, 2016; Ehlmann et al., 2016).

$^{38}Ar/^{36}Ar$ data show that Mars rapidly lost over 65% of its atmospheric argon through sputtering by pickup ions (e.g., Jakosky et al., 2017). The inferred loss of CO_2 and O is consistent with an early partial pressure of 1 bar CO_2 and a large early atmospheric loss from intense solar extreme UV radiation during the Sun's T Tauri stage (e.g., Rahmati et al., 2015; Airapetian et al., 2016). MAVEN also suggests that both the thick primordial CO_2 atmosphere inherited from the outgassing period and the magnetosphere were lost by the end of the pre-Noachian (4.1 Ga). While a secondary atmosphere was generated during the heavy bombardment period (4.2–3.8 Ga), large impacts of asteroids that contributed to its formation along with volcanic activity may also have played an important role in its erosion (e.g., Catling and Zahnle, 2009).

Data seem thus to be more aligned with a scenario of continuously unstable atmospheric conditions, including during the Noachian, when this period was previously thought to be the most favorable period for habitability. Meanwhile, studies that focus on the topology of Martian hydrologic networks rather than their abundance emphasize their immature development as an additional argument showing that favorable atmospheric conditions were only episodic and were never stable long enough for the development of systems similar to those of Earth.

Adding to the body of evidence, a recent mineralogical study explains the presence and nature of clays on Mars through three primary modes of formation (i.e., surface, hydrothermal, and crustal). It suggests that none of these modes involve much warmer average temperatures than terrestrial Arctic regions today, with surface formation that can be explained by relatively short wetter climate incursions during favorable obliquities (Bishop et al., 2018). From a geochemical standpoint, the abundance of perchlorates in the Martian soil is also consistent with the hypothesis of long-duration exposure to strong UV radiation (see Chapter 8). Further, the relatively poor abundance of carbonates may indicate that, although the Noachian atmosphere was thicker than today, it may never have reached >0.4–1 bar at most (e.g., Ehlmann et al., 2016).

Overall, the picture of early Mars depicted by recent mission data appears to fit a scenario of an always unstable, relatively thin atmosphere, exposed early to both intense UV radiation and high-energy particles due to the early loss of its magnetosphere. Episodic increase in atmospheric depth during favorable obliquity cycles and dust from impact cratering and storms may have provided temporary shielding to the surface (e.g., Jakosky and Phillips, 2001; Kreslavsky and Head, 2005). However, early subaerial habitats on Mars, if any, may have had to always contend with a challenging radiation environment. With upcoming missions planned to search for biosignatures, it is critical to understand the type of environments this may have created for life and document whether adaptation was ever possible (and what type) to make subaerial microbial habitats possible. In that, unique environmental terrestrial analogs such as the high Andes may provide important clues.

9.3 A POLYEXTREME ENVIRONMENT

Over the past 20 years, the Chilean Atacama Desert, the Altiplano, and the high Andes have provided robust analog study sites for extreme microbial habitats. This region is exposed to the strongest solar irradiance measured to date on our planet (Cabrol et al., 2014; Feister et al., 2015). From sea level up to over 6000 m elevation and from glacial regions to some of the most arid conditions on Earth, unique analog environments allow the reconstruction of a space-for-time substitution experiment that shows plausible scenarios for changing Martian subaerial habitats from the early Noachian to the Hesperian. In the process, they provide critical insights into the evolution of habitability and life potential (see Chapter 6). They also give us pointers on how environmental extremes linked to a thinning unstable atmosphere, increased UV, aridity, and geologic and climate changes, may have impacted biodiversity at local and global scale. More importantly, they show the interplay of environmental polyextremes in a natural lab as it could have unfolded on early Mars (Cabrol et al., 2007; Cabrol, 2018).

The geology, morphology, mineralogy, and geochemistry of the Andes, as well as the rapid impact of climate change on its water resources (e.g., Bradley et al., 2006), make this region a particularly relevant terrestrial analog to early Mars at the transition of the Noachian and Hesperian (e.g., Cabrol et al., 2009). Due to its latitudinal range and elevation, this mountainous tropical region has the highest solar irradiance in the world. With many peaks nearing or over 6000 m elevation, a thin and unstable atmosphere (480–450 mbar on average) generates strong weather and climate variability on a seasonal and interannual base. The column ozone is naturally thinner over the tropics, UV increasing with elevation, clear skies, and low aerosols (Blumthaler et al., 1997). In that context, the tropical Andes represents a particularly valuable test region for quantifying the amount of biologically damaging radiation (UV-A and UV-B) reaching the surface (Piazena, 1996) and its wavelength dependence on ozone in a naturally occurring extreme environment. Field data collected between 18.5 and 23.2 S/68.5–67 W and 4300–5916 m elevation show high levels of UV-B (280–315 nm) well correlated with radiative modeling for exceptionally clear skies (Piazena, 1996; Feister et al., 2015). Winter solstice values are equivalent to those of Antarctica during ozone hole events and double in the Austral summer. High UV-B irradiance occurs in an atmosphere permanently depleted in ozone and the common occurrence of negative ozone anomalies (Figs. 9.2 and 9.3).

The impact of strong UV-B and UV erythemally weighted daily dose on life is compounded by the broad daily temperature variations with sudden and sharp fluctuations brought by clouds, the low relative humidity, and an enhanced yearly negative water balance (Hock, 2008). However, as shown by the field studies in the Andes, the interplay of environmental extremes may not have always necessarily resulted in an amplification of the negative effect of high radiation on early Mars. This interplay was constantly evolving in space and time, factors of fluctuations including latitude; season; obliquity; planetary

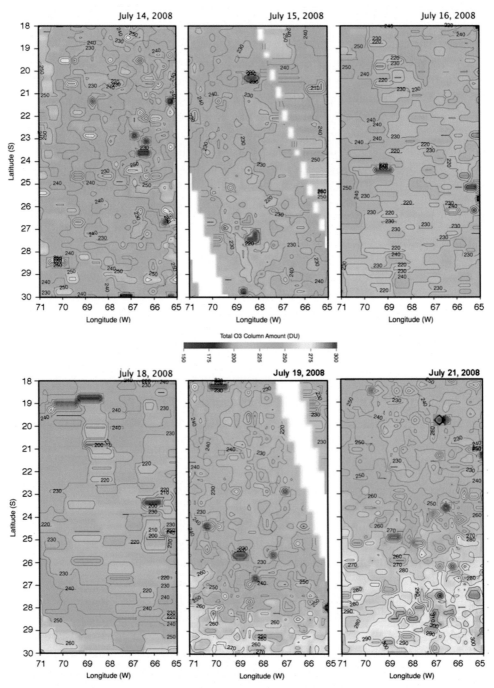

Fig. 9.2 Examples of areas of strong ozone depletion *(dark blue and purple)* compared with the regional standard average for latitude and elevation (220 DU, Dobson units).

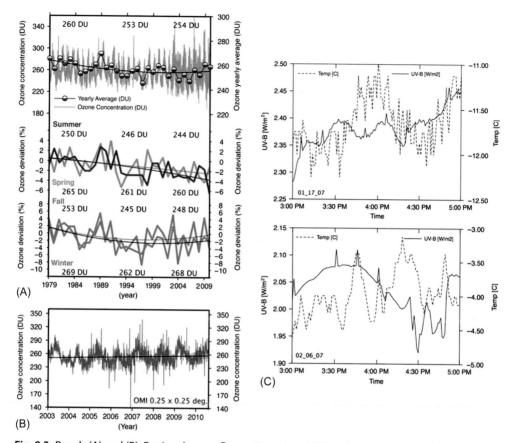

Fig. 9.3 Panels (A) and (B): Regional ozone fluctuations since 1979 in the survey area. Values in panel a indicate decadal means (the 1980s–2000s). *Panel (A)*—Daily ozone variations showing a 2.7% ozone decrease between the 1980s and 1990s (260–253 DU, Dobson units). The 2000 decade was more stable. However, ozone concentration showed greater variability. Middle and bottom panels: Seasonal ozone variations over three decades. Deviation is calculated with respect to the 1980s decadal mean. During the 2000s, winter values almost regained their 1980s average and fall values a third of their 1990s value. Yet, large interannual variations still occurred, for example, 283 DU in 2003 and 247 DU in 2004 (13%) and 281 and 259 DU in 2009 and 2010 (8%), respectively. Both spring and summer values continued to decline throughout the 2000s (−2% and −2.4%, respectively). *Panel (B)*—OMI data showing strong ozone depletion events in the survey area. Data are from Nimbus-7 (from 1979/01/01 to 1993/05/06), Meteor-3 (from 1991/08/22 to 1994/11/24), GOME-1 GDP 4.0 (from 1995/07/01 to 1996/07/21) within 100 km, Earth Probe (from 1996/07/22 to 2005/12/14) at 1.0° × 1.25°, and atmospheric infrared sounder (AIRS) global level-3 database at 1.0° × 1.0° cells. *Panel (C)*—Examples of UV and temperature fluctuations and inversions measured with Eldonet UV dosimeters during the 2007 austral summer at the summit of the Simba volcanoes (5875 m elevation).

changes (loss of atmosphere and magnetic field); and regional to local geochemical, mineralogical, topographic, and sedimentologic characteristics. For instance, as long as water was stable at the surface, abundant total dissolved solids even in a shallow water column would have still provided some shielding to aquatic microbial organisms and so would have an ice cover. On the other hand, the combination of a steady loss of magnetosphere, atmosphere, and water, short UV reaching the surface, high UV/T ratio, and the increasing formation of perchlorates would have been unfavorable not only to the survival of life but also to the preservation of biosignatures (see Cabrol et al., 2007 for more on this aspect of the discussion).

Overall, what the study of the strong radiation and polyextreme environment of the Andes shows us is that in such an environment, the relative spatiotemporal influence of each environmental parameter is what would have been critical to the survival and adaptation of potential microbial habitats. It also shows that this influence may have varied on very short timescales (daily, seasonally, and interannually, e.g., Cabrol et al., 2009; Cabrol and Grin, 2010). Importantly, while climate is the primary force driving changes in planetary habitability (including in the Andes, which is one of the regions of Earth most affected by climate change), extreme microbial habitats are driven by processes at vastly different scales. Elements that dictate their survival or disappearance in the Atacama, Altiplano, and Andes are at the scale of a local slope exposure, water retention, sediment texture, mineralogy, geochemistry, soil pH, local moisture acquisition, temperature, energy, and type of shelter.

Indeed, if the Andes reflects what a changing early Martian environment was for potential microbial habitats, then the keys to their survival and adaptability to an increasing UV environment would ultimately have been found at the meter to micrometer scale (Cabrol, 2018), and this has fundamental implications for the search for biosignatures on Mars because it shows that global trends about the early climate are merely an indication of where to search and provide very little useful information on potential paleohabitat location. As shown by MAVEN, Mars became extreme very early, and in such an environment, microbial habitat distribution and abundance at the surface would have become stochastic. Ultimately, it may be that what we observe in the Andes corresponds to the most favorable conditions for microbial habitat development early Mars ever experienced as within a billion years of its formation; the planet already resembled what it is today. Keys to possible adaptation and survival of life after this short favorable period may be found in the early development of life on Earth and in space and laboratory experiments.

9.4 ADAPTATION AND ITS LIMITS

Before the evolution of oxygenic photosynthetic cyanobacteria 2.4 billion years ago, Earth's atmosphere did not contain oxygen and ozone (Catling and Claire, 2005). Persistent oxygen accumulation developed even much later, about 550 million years ago

(Johnston et al., 2009). Therefore, Earth's early atmosphere resembled the current atmosphere on Mars more than our planet's current atmosphere. Oxygen strongly absorbs UV-C radiation and the stratospheric ozone prevents most of solar UV-B to reach the surface of the Earth. Consequently, the first photosynthetic organisms on our planet had to develop in the presence of strong short-wavelength radiation from the Sun. Both UV-C and UV-B attack relevant biomolecules such as lipids, proteins, and nucleic acids that are modified or destroyed. Since these organisms need to harvest solar radiation, they are simultaneously exposed to detrimental short-wavelength radiation. The photosynthetic apparatus and the nuclear DNA are prime targets (Richa et al., 2015; Gao et al., 2018).

In order to survive in the presence of excessive solar UV-C and UV-B radiation, organisms have developed a number of efficient protective strategies. Probably, the first photosynthetic organisms developed deep down in the water column that is reflected by the fact that some are capable of living with very low irradiances while they are even impaired by high visible light (Macintyre et al., 2002). Another strategy is mat and crust formation where organisms lower in the biofilm are protected by others further up, which are sacrificed at high irradiation. Even though cyanobacteria are not capable of active swimming, they can undergo vertical migrations in the water column changing their buoyancy by producing or reducing gas vacuoles in their cells (Oliver, 1994); by this means, they can adapt their position in the water column in dependence of the incident irradiation.

Early on, cyanobacteria have developed UV-absorbing pigments. Scytonemin is restricted to cyanobacteria and is excreted to the outside of the cell (Garcia-Pichel et al., 1992; Sinha et al., 1999). While this sunscreen pigment absorbs in the UV-A, it also has a strong absorption band in the UV-C that may be regarded as a relic of the early development (Fig. 9.4). The other UV-screening pigments are mycosporine-like amino acids (MAA) that cyanobacteria share with eukaryotic phytoplankton and some macroalgae. MAAs are water-soluble, low-molecular-weight ($< 400\,Da$) molecules absorbing in the UV-B and UV-A. They consist of a cyclohexenone or cyclohexenimine chromophore conjugated to the nitrogen substituent of an amino acid or its imino alcohol (Sinha et al., 1998). It is interesting to note that animals such as zooplankton are not able to synthesize these molecules since they lack the Shikimate pathway, but they can take them up with their diet, store them in their cells, and use them for the same purpose.

In addition to direct damage of essential biomolecules, solar UV can produce reactive oxygen species (ROS) such as singlet oxygen, peroxides, superoxide, and hydroxyl radials (He and Häder, 2002). Cells have developed a number of enzymatic and nonenzymatic reactions to quench ROS and thus mitigate the damage.

Other mitigating strategies include the repair of UV-induced damage. Bacteria are too small to produce and store sufficient concentrations of MAAs to effectively screen out UV radiation before it hits the sensitive DNA and therefore rely on fast repair

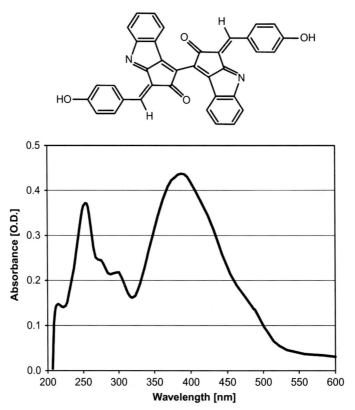

Fig. 9.4 Chemical structure and absorption spectrum of scytonemin. *(Modified from Sinha, R.P., Klisch, M., Vaishampayan, A., Häder, D.-P., 1999. Biochemical and spectroscopic characterization of the cyanobacterium* Lyngbya *sp. inhabiting mango (*Mangifera indica*) trees: presence of an ultraviolet-absorbing pigment, scytonemin. Acta Protozool. 38, 291–298; Rastogi, R.P., Sinha, R.P., Moh, S.H., Lee, T.K., Kottuparambil, S., Kim, Y.-J., Rhee, J.-S., Choi, E.-M., Brown, M.T., Häder, D.-P., 2014. Ultraviolet radiation and cyanobacteria. J. Photochem. Photobiol. B Biol. 141, 154–169.)*

mechanisms. UV-C induces 6,4-photoproducts and their Dewar valence isomers (Sinha and Häder, 2002). In contrast, UV-B induces mainly cyclobutane pyrimidine dimers (CPDs). These are effectively repaired by the enzyme photolyase using the energy of UV-A or blue photons (Sinha and Häder, 2002). In addition, excision repair (base excision repair and nucleotide excision repair) plays important roles in DNA repair. Phytoplankton dwell in the upper mixing layer (UML) of the water column and are moved by the action of wind and waves. When they are close to the surface, they are exposed to excessive solar radiation and encounter UV-induced damage. This is being repaired when the organisms are transported to deeper water with less intense irradiation.

The photosynthetic apparatus is another target of solar UV radiation. The D1 protein in photosystem II is responsible for the electron transport from chlorophyll *a* to the primary acceptor. Excessive radiation kinks the molecule that is subsequently removed and replaced by a newly synthesized protein (Gao et al., 2018).

On Mars, potential organisms would also be exposed to strong ionizing radiation since the planet does not possess a strong magnetic field deflecting this (Kminek and Bada, 2006). In order to analyze the effects of ionizing radiation on phytoplankton, we exposed the green unicellular *Euglena gracilis* to gamma rays and high-energy carbon ion irradiation (Sakashita et al., 2002d,c). Exposure to the ionizing radiation impaired gravitaxis (orientation with respect to the gravitational field of the Earth) and the swimming velocity (Sakashita et al., 2002a,d,c). The damage induced by the ionizing radiation seems to be mediated by induced ROS, since the inhibition could be partially mitigated by adding the ROS quencher Trolox C (Sakashita et al., 2002b).

The current atmosphere on Mars is very thin (6.36 hPa) and equals about 6% of the atmospheric pressure on Earth. It consists of 95.3% carbon dioxide, the reminder are nitrogen (2.7%); argon (1.6%); and traces of oxygen, carbon monoxide, and water (Mahaffy et al., 2013; Montmessin et al., 2017). Thus, the atmospheric pressure on Mars corresponds to a stratospheric air pressure at 35 km above the Earth's surface. Several space experiments have been conducted to evaluate the effect of solar UV and ionizing radiation, vacuum, and widely fluctuating temperatures on various organisms in a scenario closely resembling that on the current Mars surface (Demets et al., 2005). Biopan is a pan-shaped container attached to the outside of a Russian Foton spacecraft that allows exposing biological material to open space. It is closed during launch and reentry and opens when the satellite reaches its orbit. Between 1992 and 1999, four missions have been carried out with 16 experiments. In addition to dormant spores of bacteria, ferns, and mosses, lichens and amino acids were exposed (de La Torre Noetzel, 2007; Cottin et al., 2008). The lichens fully recovered after exposure to 3×10^{-6} hPa for 1 week unattenuated solar UV radiation (>200 nm) with a total fluence of 10.89×10^6 J m^{-2}. Judging from these environments, the authors assume that lichens may be capable of surviving conditions prevailing on Mars (de Vera et al., 2010; De Vera, 2012). The radiation regime during the space flight was recorded using a miniaturized radiometer-dosimeter (R3D) that monitored solar visible and UV radiation in four channels and ionized radiation distributed in 256 channels (Dachev et al., 2005; Häder et al., 2009). One of the objectives of these experiments was to prove that large organic molecules and dormant spores and organisms can survive extended interplanetary space travel posited by the Panspermia hypothesis (Davies, 1996).

Similar experiments with a far longer exposure time have been carried out on several EXPOSE facilities installed outside the ISS (Rabbow et al., 2009). The EXPOSE-E instrument was exposed for 1.1 years before it was returned to Earth for evaluation of the biological samples including bacterial and fungal spores, lichens, and cryptoendolithic microbial communities (Horneck et al., 1999, 2002). Dormant eggs encased in a chitinous capsule of *Daphnia* and *Eucypris* were also found to survive the extended exposure to the hostile environment and showed an 11%–35% hatching success after returning to Earth (Novikova et al., 2011).

Also on the EXPOSE modules, solar visible and UV radiation, as well as ionizing radiation, have been monitored continuously by R3D instruments (Dachev et al., 2012, 2017). One mechanism for protection during extended periods in a hostile environment was also found on Earth in cyanobacteria growing inside rocks under a 5 mm layer of granite in Southern Brazil (Richter et al., 2006). While the light penetration through the overlying rock layer allowed only slow growth, it protected the organisms from high solar irradiation and excessive temperatures.

9.5 CONCLUSION

Conditions on Mars are very harsh for life. Excessive solar UV radiation including UV-C (>200 nm), wide temperature fluctuations, limited water supply, and very thin atmosphere with only traces of oxygen provide a hostile environment. In addition, the lack of a stable magnetic field facilitates high energetic ionizing radiation to hit the surface of the planet. Several life forms ranging from spores of bacteria, fungi, ferns, and mosses have been found to survive similar harsh conditions on Earth and in space, and even dormant lichens and eggs of microcrustaceans tolerate such extreme environments at least for limited periods of time on the order of days to months. Survival strategies include mat and crust formation, vertical migration, UV-screening pigments, endolithic growth inside rocks or meteorites, and efficient repair mechanisms of damage induced by UV and ionizing radiation. However, if long-term survival under such hostile conditions on the order of hundreds to millions of years as posited by the panspermia hypothesis is possible to allow interplanetary travel of early life forms has to be confirmed. However, the main question remains whether or not active growth of primitive organisms is possible under the current hostile conditions on Mars.

ACKNOWLEDGMENTS

The data presented in this chapter were collected as part of the High Lakes Project (HLP), a project supported by the National Aeronautics and Space Administration Astrobiology Institute (NAI) between 2002 and 2009 under grant no. NNA04CC05A. This investigation continues today in the Chilean Altiplano and Andes with the NAI-funded project entitled "Changing planetary environments and the fingerprints of life" (2015–19) that characterizes biosignatures relevant to the Mars 2020 and ExoMars missions under grant no. NNX15BB01A.

REFERENCES

Airapetian, V.S., Glocer, A., Gronoff, G., Hébrard, E., Danchi, W., 2016. Prebiotic chemistry and atmospheric warming of early Earth by an active young Sun. Nat. Geosci. 9, 452–455.
Armitage, P., 1995. Chironomidae as food. In: The Chironomidae. Springer, Netherlands, pp. 423–435.

Atri, D., Hariharan, B., Grießmeier, J.M., 2013. Galactic cosmic ray-induced radiation dose on terrestrial exoplanets. Astrobiology 13, 910–919.

Baker, V.R., 2001. Water and the martian landscape. Nature 412, 228–236.

Bishop, J.L., Fairén, A.G., Michalski, J.R., Gago-Duport, L., Baker, L.L., Velbel, A., Gross, C., Rampe, E.B., 2018. Surface clay formation during short-term warmer and wetter conditions on a largely cold ancient Mars. Nat. Astron. 2, 206–213.

Blumthaler, M., Ambach, W., Ellinger, R., 1997. Increase in solar UV radiation with altitude. J. Photochem. Photobiol. B Biol. 39, 130–134.

Bradley, R., Vuille, M., Diaz, H., Vergara, W., 2006. Threats to water supplies in the tropical Andes. Science 312, 1755.

Cabrol, N.A., 2018. The coevolution of life and environment on Mars: an ecosystem perspective on the robotic exploration of biosignatures. Astrobiology 18 (1), 1–27.

Cabrol, N.A., Grin, E.A. (Eds.), 2010. Lakes on Mars. Elsevier Science, Amsterdam.

Cabrol, N.A., Grin, E.A., Hock, A.N., 2007. Mitigation of environmental extremes as a possible indicator of extended habitat sustainability for lakes on early Mars. Proc. SPIE. 6694. https://doi.org/10.1117/12.731506.

Cabrol, N.A., Grin, E.A., Chong, G., Minkley, E., Hock, A.N., Yu, Y., Bebout, L., Fleming, E., Häder, D.P., Demergasso, C., Gibson, J., Escudero, L., Dorador, C., Lim, D., Woosley, C., Morris, R.L., Tambley, C., Gaete, V., Galvez, M.E., Smith, E., Uskin-Peate, I., Salazar, C., Dawidowicz, G., Majerowicz, J., 2009. The high-lakes project. J. Geophys. Res. Biogeosci. 114. https://doi.org/10.1029/2008JG000818.

Cabrol, N.A., Feister, U., Häder, D.-P., Piazena, H., Grin, E.A., Klein, A., 2014. Record solar UV irradiance in the tropical Andes. Front. Environ. Sci. 2, 1–6.

Carr, M.H., Head, J.W., 2015. Martian surface/near-surface water inventory: sources, sinks, and changes with time. Geophys. Res. Lett. 42, 726–732.

Catling, D.C., Claire, M.W., 2005. How Earth's atmosphere evolved to an oxic state: a status report. Earth Planet. Sci. Lett. 237 (1–2), 1–20.

Catling, D.C., Zahnle, K.J., 2009. The escape of planetary atmospheres. Sci. Am. 300, 36–43.

Colburn, D.S., Pollack, J.B., Haberle, R.M., 1989. Diurnal variations in optical depth at Mars. Icarus 79, 159–189.

Cottin, H., Coll, P., Coscia, D., Fray, N., Guan, Y.Y., Macari, F., Raulin, F., Rivron, C., Stalpor, F., Szopa, C., Chaput, D., Viso, M., Bertrand, M., Chabin, A., Thirkell, L., Westall, F., Brack, A., 2008. Heterogeneous solid/gas organic compounds related to comets, meteorites, Titan and Mars: laboratory and in lower Earth orbit experiments. Adv. Space Res. 42 (12), 2019–2035.

Craddock, R.A., Lorenz, R.D., 2017. The changing nature of rainfall during the early history of Mars. Icarus 293, 172–179.

Cucinotta, F.A., Kim, M.-H.Y., Chappell, L.J., Huff, J.L., 2013. How safe is safe enough? Radiation risk for a human mission to Mars. PLoS ONE. 8(10). e74988. https://doi.org/10.1371/journal.pone.0074988.

Dachev, T.P., Spurny, F., Reitz, G., Tomov, B.T., Dimitrov, P.G., Matviichuk, Y.N., 2005. Simultaneous investigation of galactic cosmic rays on aircrafts and on International Space Station. Adv. Space Res. 36, 1665–1670. https://doi.org/10.1016/j.asr.2005.05.073.

Dachev, T.P., Tomov, B.T., Matviichuk, Y.N., Dimitrov, P.G., Bankov, N.G., Reitz, G., Horneck, G., Häder, D.-P., Lebert, M., Schuster, M., 2012. Relativistic electron fluxes and dose rate variations during April–May 2010 geomagnetic disturbances in the R3DR data on ISS. Adv. Space Res. 50, 282–292.

Dachev, T.P., Bankov, N.G., Tomov, B.T., Matviichuk, Y.N., Dimitrov, P.G., Häder, D.-P., Hornek, G., 2017. Overview of the ISS radiation environment observed during the ESA EXOSE-R2 mission in 2014–2016. Space Weather 15 (11), 1475–1489.

Davies, P.C.W., 1996. Brock, G., Goode, J. (Eds.), The transfer of viable micro-organisms between planets. Evolution of Hydrothermal Ecosystems on Earth (and Mars?): Proceedings of the CIBA Foundation Symposium No. 20. Wiley, New York.

de La Torre Noetzel, R., 2007. BIOPAN experiment LICHENS on the Foton M2 mission: pre-flight verification tests of the *Rhizocarpon geographicum*-granite ecosystem. Adv. Space Res. 40 (11), 1665–1671.

de Vera, J.-P., 2012. Lichens as survivors in space and on Mars. Fungal Ecol. 5, 472–479.

de Vera, J.-P., Möhlmann, D., Butina, F., Lorek, A., Wernecke, R., Ott, S., 2010. Survival potential and photosynthetic activity of lichens under Mars-like conditions: a laboratory study. Astrobiology 10, 215–227.

Demets, R., Shulte, A.W., Baglioni, P., 2005. The past, present and future of BIOPAN. Adv. Space Res. 36, 311–316.

Ehlmann, B.L., Anderson, F.S., Andrews-Hanna, J., Catling, D.C., Christensen, P.R., Cohen, B.A., Dressing, C.D., Edwards, C.S., Elkins-Tanton, L.T., Farley, K.A., Fassett, C.I., Fischer, W.W., Fraeman, A.A., Golombek, M.P., Hamilton, V.E., Hayes, A.G., Herd, C.D.K., Horgan, B., Hu, R., Jakosky, B.M., Johnson, J.R., Kasting, J.F., Kerber, L., Kinch, K.M., Kite, E.S., Knutson, H.A., Lunine, J.I., Mahaffy, P.R., Mangold, N., McCubbin, F.M., Mustard, J.F., Niles, P.B., Quantin-Nataf, C., Rice, M.S., Stack, K.M., Stevenson, D.J., Stewart, S.T., Toplis, M.J., Usui, T., Weiss, B.P., Werner, S.C., Wordsworth, R.D., Wray, J.J., Yingst, R.A., Yung, Y.L., Zahnle, K.J., 2016. The sustainability of habitability on terrestrial planets: insights, questions, and needed measurements from Mars for understanding the evolution of Earth-like worlds. J. Geophys. Res. Planets. 121. https://doi.org/10.1002/2016JE005134.

Feister, U., Cabrol, N.A., Häder, D.-P., 2015. UV irradiance enhancements by scattering of solar radiation from clouds. Atmos. 6 (8), 1211–1228.

Gao, K., Zhang, Y., Häder, D.-P., 2018. Individual and interactive effects of ocean acidification, global warming, and UV radiation on phytoplankton. J. Appl. Phycol. 30, 743–759.

Garcia-Pichel, F., Sherry, N.D., Castenholz, R.W., 1992. Evidence for an ultraviolet sunscreen role of the extracellular pigment scytonemin in the terrestrial cyanobacterium *Chlorogloeopsis* sp. Photochem. Photobiol. 56 (1), 17–23.

Granskog, M.A., Stedmon, C.A., Dodd, P.A., Amon, R.M.W., Pavlov, A.K., de Steur, L., et al., 2012. Characteristics of colored dissolved organic matter (CDOM) in the Arctic outflow in the Fram Strait: assessing the changes and fate of terrigenous CDOM in the Arctic Ocean. J. Geophys. Res. Oceans. 117. WOS:000312821000001. English.

Häder, D.-P., Erzinger, G.S., 2017. Daphniatox—online monitoring of aquatic pollution and toxic substances. Chemosphere 167, 228–235.

Häder, D.-P., Lebert, M., Schuster, M., del Ciampo, L., Helbling, E.W., McKenzie, R., 2007. ELD-ONET—a decade of monitoring solar radiation on five continents. Photochem. Photobiol. 83, 1348–1357.

Häder, D.-P., Richter, P., Schuster, M., Dachev, T.P., Tomov, B., Georgiev, P., Matviichuk, Y., 2009. R3D-B2—measurement of ionizing and solar radiation in open space in the BIOPAN 5 facility outside the FOTON M2 satellite. Adv. Space Res. 43, 1200–1211.

Hassler, D.M., Zeitlin, C., Wimmer-Schweingruber, R.F., Ehresmann, B., Rafkin, S., Eigenbrode, J.L., Brinza, D.E., Weigle, G., Böttcher, S., Böhm, E., Burmeister, S., Guo, J., Köhler, J., Martin, C., Reitz, G., Cucinotta, F.A., Kim, M.H., Grinspoon, D., Bullock, M.A., Posner, A., Gómez-Elvira, J., Vasavada, A., Grotzinger, J.P., MSL Science Team, 2014. Mars' surface radiation environment measured with the Mars Science Laboratory's Curiosity rover. Science. 343, 1244797–1/3, https://doi.org/10.1126/science.1244797.

He, Y.Y., Häder, D.-P., 2002. Reactive oxygen species and UV-B: effect on cyanobacteria. Photochem. Photobiol. Sci. 1 (10), 729–736.

Hock, A.N., 2008. Licancabur Volcano and Life in the Atacama: Environmental Physics and Analogies to Mars. Ph.D. thesis, Univ. of Calif., Los Angeles.

Horneck, G., Wynn-Williams, D.D., Mancinelli, R.L., Cadet, J., Munakata, N., Rontó, G., Edwards, H.G.M., Hock, B., Wänke, H., Reitz, G., Dachev, T., Häder, D.-P., Brillouet, C., 1999. Biological experiments on the EXPOSE facility of the International Space Station.Proc. 2nd European Symposium on the Utilization of the International Space Station, ESTEC, Noordwijk, the Netherlands, 16–18 November, 1998, ESA SP-433, pp. 459–468.

Horneck, G., Hock, B., Wänke, H., Rettberg, P., Häder, D.-P., Dachev, T., Rabbow, E., Reitz, G., Panitz, C., Lux-Endrich, A., Richter, P., Mishev, D., 2002. Lacoste. H. (Ed.), Spores in artificial meteorites, the experiment SPORES on expose. Proceedings of the First European Workshop on Exo-Astrobiology, 16–19 September 2002, Graz, Austria, ESA SP-518. ESA Publications Division, Noordwijk, Netherlands, ISBN 92-9092-828-X, pp. 55–58.

Jakosky, B.M., Phillips, R.J., 2001. Mars' volatile and climate history. Nature 412, 237–244.

Jakosky, B.M., Grebowsky, J.M., Luhmann, J.G., Brain, D.A., 2015. Initial results from the MAVEN mission to Mars. Geophys. Res. Lett. 42 (21), 8791–8802.

Jakosky, B.M., Slipski, M., Benna, M., Mahaffy, P., Elrod, M., Yelle, R., Stone, S., Alsaeed, N., 2017. Mars' atmospheric history derived from upper-atmosphere measurements of $^{38}Ar/^{36}Ar$. Science 355, 1408–1410.

Johnston, D.T., Wolfe-Simon, F., Pearson, A., Knoll, A.H., 2009. Anoxygenic photosynthesis modulated Proterozoic oxygen and sustained Earth's middle age. PNAS 106 (40), 16925–16929.

Kminek, G., Bada, J.L., 2006. The effect of ionizing radiation on the preservation of amino acids on Mars. Earth Planet. Sci. Lett. 245, 1–5.

Kreslavsky, M.A., Head, J.W., 2005. Mars at very low obliquity: atmospheric collapse and the fate of volatiles. Geophys. Res. Lett. 32. https://doi.org/10.1029/2005GL022645.

Lemmon, M.T., Wolff, M.J., Smith, M.D., Clancy, R.T., Banfield, D., Landis, G.A., Ghosh, A., Smith, P.H., Spanovich, N., Whiney, B., Whelley, P., Greeley, R., Thompson, S., Bell III, J.F., Squyes, S.W., 2004. Atmospheric imaging results from the Mars exploration rovers: spirit and opportunity. Science 306 (5702), 1753–1756.

Luo, W., Cang, X., Howard, A.D., 2017. New Martian valley network volume estimate consistent with ancient ocean and warm and wet climate. Nat. Commun. 8. https://doi.org/10.1038/ncomms15766.

Macintyre, H.L., Kana, T.M., Anning, T., Geider, R.J., 2002. Photoacclimation of photosynthesis irradiance response curves and photosynthetic pigments in microalgae and cyanobacteria. J. Phycol. 38, 17–38. https://doi.org/10.1046/j.1529-8817.2002.00094.x.

Mahaffy, P.R., Webster, C.R., Atreya, S.K., Franz, H., Wong, M., Conrad, P.G., Harpold, D., Jones, J.J., Leshin, L.A., Manning, H., Owen, T., Pepin, R.O., Squyres, S., Trainer, M., the MSL Science Team, 2013. Abundance and isotopic composition of gases in the Martian atmosphere from the Curiosity rover. Science Vol. 341. (6143), 263–266.

McWilliam, R.A., Baird, D.J., 2002. Application of postexposure feeding depression bioassays with *Daphnia magna* for assessment of toxic effluents in rivers. Environ. Toxicol. Chem. 21 (7), 1462–1468.

Montmessin, F., Korablev, O., Lefèvre, F., Bertaux, J.-L., Fedorova, A., Trokhimovskiy, A., Chaufray, J.Y., Lacombe, G., Reberac, A., Maltagliati, L., Willame, Y., Guslyakova, S., Gérard, J.-C., Stiepen, A., Fussen, D., Mateshvili, N., Määttänen, A., Forget, F., Witasse, O., Leblanc, F., Vandaele, A.C., Marcq, E., Sandel, B., Gondet, B., Schneider, N., Chaffin, M., Chapron, N., 2017. SPICAM on Mars Express: a 10 year in-depth survey of the Martian atmosphere. Icarus 297 (15), 195–216.

Oliver, R.L., 1994. Floating and sinking in gas-vacuolate cyanobacteria. J. Phycol. 30, 161–173.

Patel, M., Zarnecki, J., Catling, D., 2002. Ultraviolet radiation on the surface of Mars and the Beagle 2 UV sensor. Planet. Space Sci. 50 (9), 915–927.

Piazena, H., 1996. The effect of altitude upon the solar UV-B and UV-A irradiance in the tropical Chilean Andes. Sol. Energy 57, 133–140.

Pollack, J.B., Colburn, D., Kahn, R., Hunter, J., Van Camp, W., Carlston, C.E., Wolf, M.R., 1977. Properties of aerosols in the Martian atmosphere, as inferred from Viking Lander imaging data. J. Geophys. Res. 82 (28), 4479–4496.

Rabbow, E., Horneck, G., Rettberg, P., Schott, J.U., Panitz, C., L'Affitto, A., von Heise-Rotenburg, R., Willnecker, R., Baglioni, P., Hatton, J., Dettman, J., Demets, R., Reitz, G., 2009. EXPOSE, an astrobiological exposure facility on the International Space Station—from proposal to flight. Orig. Life Evol. Biosph. 39 (6), 581–598.

Rahmati, A., Larson, D.E., Cravens, T.E., Lillis, R.J., Dunn, P.A., Halekas, J.S., Connerney, J.E., Eparvier, F.G., Thiemann, E.M.B., Jakosky, B.M., 2015. MAVEN insights into oxygen pickup ions at Mars. Geophys. Res. Lett. 42, 8870–8876.

Richa, R., Sinha, P., Häder, D.-P., 2015. Physiological aspects of UV-excitation of DNA. Top. Curr. Chem. 356, 203–248. https://doi.org/10.1007/128_2014_531.

Richter, P.R., Sinha, R.P., Häder, D.-P., 2006. Scytonemin-rich epilithic cyanobacteria survive acetone treatment. Curr. Trends Microbiol. 2, 13–19.

Rummel, J.D., Beaty, D.W., Jones, M.A., Bakermans, C., Barlow, N.G., Boston, P.J., Chevrier, V.F., Clark, B.C., de Vera, J.P., Gough, R.V., Hallsworth, J.E., Head, J.W., Hipkin, V.J., Kieft, T.L.,

McEwen, A.S., Mellon, M.T., Mikucki, J.A., Nicholson, W.L., Omelon, C.R., Peterson, R., Roden, E.E., Sherwood Lollar, B., Tanaka, K.L., Viola, D., Wray, J.J., 2014. A new analysis of Mars "Special Regions": findings of the Second MEPAG Special Regions Science Analysis Group (SR-SAG2). Astrobiology 14, 887–968.

Sakashita, T., Doi, M., Yasuda, H., Fuma, S., Häder, D.-P., 2002a. Protection of negative gravitaxis in *Euglena gracilis* Z against gamma-ray irradiation by Trolox C. J. Radiat. Res. 43 (Suppl), S257–S259.

Sakashita, T., Doi, M., Yasuda, H., Takeda, H., Fuma, S., Nakamura, Y., Häder, D.-P., 2002b. Comparative study of gamma-ray and high-energy carbon ion irradiation on negative gravitaxis in *Euglena gracilis* Z. J. Plant Physiol. 159, 1355–1360.

Sakashita, T., Doi, M., Yasuda, H., Takeda, H., Fuma, S., Nakamura, Y., Häder, D.-P., 2002c. High-energy carbon ion irradiation and the inhibition of negative gravitaxis in *Euglena gracilis* Z. Int. J. Radiat. Biol. 78, 1055–1060.

Sakashita, T., Doly, M., Nakamura, Y., Fuma, S., Ishii, N., Takeda, H., 2002d. γ-irradiation effect: variation of photosynthetic activity of *Euglena*. Biomed. Environ. Sci. 15, 261–267.

Sinha, R.P., Häder, D.P., 2002. UV-induced DNA damage and repair: a review. Photochem. Photobiol. Sci. 1 (4), 225–236.

Sinha, R.P., Klisch, M., Gröninger, A., Häder, D.-P., 1998. Ultraviolet-absorbing/screening substances in cyanobacteria, phytoplankton and macroalgae. J. Photochem. Photobiol. B 47, 83–94.

Sinha, R.P., Klisch, M., Vaishampayan, A., Häder, D.-P., 1999. Biochemical and spectroscopic characterization of the cyanobacterium *Lyngbya* sp. inhabiting mango (*Mangifera indica*) trees: presence of an ultraviolet-absorbing pigment, scytonemin. Acta Protozool. 38, 291–298.

Smith, P.H., Lemmon, M., 1999. Opacity of the martian atmosphere measured by the imager for Mars Pathfinder. J. Geophys. Res. 104 (E4), 8975–8985.

Vicente-Retortillo, A., Martínez, G.M., Renno, N.O., Lemmon, M.T., de la Torre-Juárez, M., 2017. Determination of dust aerosol particle size at Gale crater using REMS UVS and Mastcam measurements. Geophys. Res. Lett. 44 (8), 3502–3508.

Wordsworth, R.D., 2016. The climate of early Mars. Annu. Rev. Earth Planet. Sci. 44. https://doi.org/10.1146/annurev-earth-060115-012355.

FURTHER READING

Brack, W., Ait-Aissa, S., Burgess, R.M., Busch, W., Creusot, N., Di Paolo, C., Escher, B.I., Hewitt, L.M., Hilscherova, K., Hollender, J., 2016. Effect-directed analysis supporting monitoring of aquatic environments—an in-depth overview. Sci. Total Environ. 544, 1073–1118.

Dachev, T.P., Tomov, B., Matviichuk, Y., Dimitrov, P., Lemaire, J., Gregoire, G., Cyamukungu, M., Schmitz, H., Fujitaka, K., Uchihori, Y., Kitamura, H., Reitz, G., Beaujean, R., Petrov, V., Shurshakov, V., Benghin, V., Spurný, F., 2002. Calibration results obtained with Liulin-4 type dosimeters. Adv. Space Res. 30, 917–925.

Rastogi, R.P., Sinha, R.P., Moh, S.H., Lee, T.K., Kottuparambil, S., Kim, Y.-J., Rhee, J.-S., Choi, E.-M., Brown, M.T., Häder, D.-P., 2014. Ultraviolet radiation and cyanobacteria. J. Photochem. Photobiol. B Biol. 141, 154–169.

CHAPTER 10

Are Recurring Slope Lineae Habitable?

Alfred S. McEwen

Contents

10.1 OVERVIEW

Recurring slope lineae (RSL) are dark flows on steep slopes that extend downhill gradually or incrementally and recur each Mars year. RSL activity often occurs at temperatures exceeding 255 K, meeting one requirement for potential habitability. The temperature-dependent activity is consistent with the stability of brines likely on Mars, and hydrated salts are transiently detected from orbit. These may be the most promising locations for habitable conditions on Mars today, especially in equatorial regions. However, topographic evidence strongly points to the formation of RSL as granular flows, not seepage of water, so the water activity value is probably quite low (<0.5) and habitability remains very challenging. Future spacecraft experiments and laboratory experiments are needed to understand how RSL form and whether they need to be treated like special regions, requiring extra procedures for planetary protection (quarantine).

From Habitability to Life on Mars
https://doi.org/10.1016/B978-0-12-809935-3.00008-6

10.2 INTRODUCTION AND BACKGROUND

Recurring slope lineae (RSL) are relatively dark flows on steep slopes with low albedos (minimal dust cover), typically originating at bedrock outcrops (McEwen et al., 2011, 2014). Individual lineae are up to a few meters wide and up to 1.5 km long. RSL recur annually (by definition) over the same slopes. The lineae grow gradually or incrementally over a period of several months, usually during the warmest time of the year for the particular latitude and slope aspect, and then fade (and often disappear) when inactive. This pattern repeats over multiple years, with varying degrees of interannual variability. They are often associated with pristine small gullies or channels that are otherwise rare on equatorial slopes. Hundreds of individual lineae may be present over a local site and thousands in a single HiRISE image, and there are hundreds of likely RSL sites. Hydrated oxychlorine salts have been detected by Compact Reconnaissance Imaging Spectrometer for Mars (CRISM; Murchie et al., 2007) at five sites (Ojha et al., 2015, 2017a; Ojha, 2016); some RSL may leave relatively bright deposits on the slopes (Chojnacki et al., 2016; see Fig. 10.1).

RSL are common in the southern middle latitudes (SML) where they are most active in southern summer on equator-facing slopes, the equatorial region where activity is usually timed to when the local slope receives the most insolation (McEwen et al., 2014), and in Acidalia/Chryse Planitia with activity in northern spring and summer (Dundas et al., 2015; Stillman et al., 2016). RSL are classified as "fully confirmed" when incremental or gradual growth, fading, and yearly recurrence have all been observed (McEwen et al., 2014). They are called "partially confirmed" when either incremental growth or recurrence has been observed or "candidate" sites when they resemble RSL in single images, but changes have not been observed, or only fading has been seen; often, insufficient repeat coverage has been acquired. The identification of RSL currently depends on ~30 cm/pixel orbital images from the High Resolution Imaging Science Experiment (HiRISE; McEwen et al., 2007) on the Mars Reconnaissance Orbiter (MRO).

There are at least 68 confirmed RSL sites and 406 candidate or partially confirmed sites on Mars, where a "site" corresponds to a HiRISE image that may have multiple slopes with RSL (Stillman et al., 2017). HiRISE has uniquely covered ~2% of the martian surface by mid-2017, and a much smaller fraction has the repeat coverage needed to identify RSL. However, HiRISE has preferentially targeted steep and low-albedo slopes for a variety of science objectives, including the search for RSL. Only ~40% of steep, low-albedo, equator-facing slopes in the SML exhibit candidate RSL in the active summer season (Ojha et al., 2014). They seem to favor west-facing over east-facing slopes (Stillman et al., 2017), but HiRISE preferentially targets the better-illuminated west-facing slopes (MRO observes at ~3 PM), and RSL may be hidden in shadows on east-facing slopes, so an actual east-west preference remains unproven. A global inventory of steep, low-albedo slopes on Mars has not been compiled, but it is likely much less than 1% of Mars' global surface area. A global map of bedrock on slopes based on

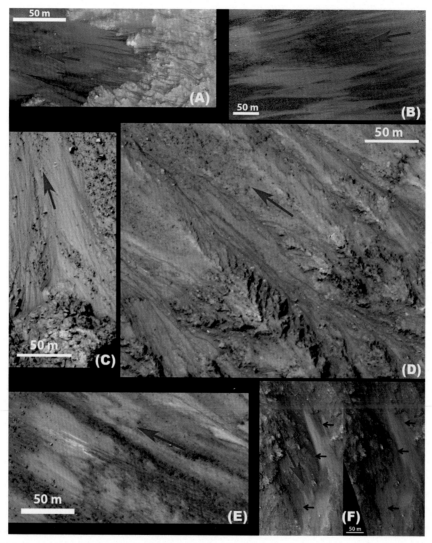

Fig. 10.1 Examples of RSL sites, in IRB color (broad bandpasses centered at ~900, 700, and 500 nm displayed as *red, green,* and *blue,* respectively), with min-max stretches applied to each individual bandpass to increase contrast. North is up and illumination is from the left in all images, and the downhill direction is indicated by *red arrows.* (A) RSL flowing first over bright bedrock and colluvium, then over the darker fan in Coprates Chasma (ESP_050021_1670). The active RSL are dark but leave deposits (seen in places) that appear *greenish* in the IRB color. (B) RSL deposits appear *greenish-reddish* in Ganges Chasm (ESP_047370_1705). (C) These RSL in Asimov crater lengthened unusually late in the SML season, between L_s 11° and 19° when surface temperatures were well below 273 K, but quickly faded (PSP_006926_1320). (D) RSL in Tivat crater (ESP_022973_1335) follow gullies that are much larger than the RSL. The *reddish streak* (lower right) is present all year, but darkens where RSL form over it, suggesting that RSL may darken the surface without necessarily changing its color. (E) RSL in Krupac crater cover bright linear features that appear to be deposits from RSL activity in past years (ESP_045018_1720). (F) RSL fans (*black arrows*) in Coprates Chasm became dark when the air was dusty (left, ESP_030241_1650) but were bright in images acquired both before (not shown) and after (right, ESP_030452_1650) the dusty event.

thermal inertia (Edwards et al., 2009) strongly resembles the global map of RSL occurrences (Stillman et al., 2017), although with some exceptions. They occur over a broad range of elevations, with no preference for regions with greater water stability at the surface (Haberle et al., 2001). RSL are widespread, and the count of confirmed sites will continue to grow as more high-resolution repeat coverage is acquired.

RSL have been observed to recur for up to 5 Mars years (MY; 687 days on Earth), from MY 29 to 33, but there is no upper limit on how long recurrence may continue. (See Piqueux et al. (2015) for description of how Mars years are numbered.) If activity usually ceases after a certain period of time, then that would constrain the quantity of the materials needed (water, salt, or solid grains) or its replenishment rate. An indirect estimate may be possible, assuming that the RSL activity lifetime is random across Mars. There are no fully confirmed RSL sites where activity is known to have ceased, although the level of activity has decreased in places. In some sites with multiple RSL slopes, some slopes appear inactive after RSL were apparent in a prior year, but those particular slopes may not have ever met the definition of fully confirmed. There are ~20 fully confirmed RSL sites that have been well monitored over at least 3 Mars years, and none has ceased activity. If the average lifetime is 100 Mars years, then 3% of the sites should cease in 3 years; $0.03 \times 20 = 0.6$ site turned off versus none observed, so the mean lifetime is probably of order or exceeds 100 Mars years. This logic may be flawed if some of the candidate sites, where repeat imaging did not show recurrence, actually are RSL sites that ceased recurrence.

Most published interpretations of RSL involve water, so it has been recommended that these locations be treated like special regions (SRs) for planetary protection (Rummel et al., 2014; Kminek et al., 2016). An SR is a place where organisms could survive and reproduce; whether or not this is the case at RSL sites is unknown, so they are recommended to be protected like SRs. However, recent results suggest that RSL are granular flows on "angle of repose" slopes (Dundas et al., 2017a), perhaps associated with small quantities of transient brine that may not ever have sufficient water activity (a_w, defined as relative humidity/100) to be habitable. This chapter provides a review of RSL research and a discussion of the growing list of models to explain this activity and implications for habitability.

10.3 RSL IN THE SOUTHERN MIDDLE LATITUDES

The discovery of RSL in the SML and their basic characteristics were described by McEwen et al. (2011). They found RSL only on steep slopes, measured as 27°–38° from the only three HiRISE digital terrain models (DTMs; Kirk et al., 2008) available at that time over known RSL sites. Some RSL appear to terminate in the middle of steep slopes, and McEwen et al. (2011) commented that "…RSL lengths must be controlled by a limited volume of mobile material." They considered multiple hypotheses for the origin of

RSL, including wet and dry flows, but favored some role for briny water in spite of no known source of sufficient water to fill pore spaces as needed for seepage. There has been increasing evidence for widespread hygroscopic salts on Mars from measurements by landers, rovers, orbiters, and analysis of martian meteorites, many of which lower the freezing point of brine (e.g., Hecht et al., 2009; Osterloo et al., 2010; Glavin et al., 2013; Clark and Kounaves, 2016). RSL activity is more consistent with briny than fresh water when surface or shallow subsurface temperatures are above the salt eutectic temperature (as low as ~200 K) and sometimes below 273 K. The peak daily surface temperatures corresponding to seasons of RSL growth in the SML are usually above 273 K and occur between L_s 245° and 314° (late spring to middle summer) (Stillman et al., 2014). (L_s is areocentric longitude of the sun, a measure of season; southern summer begins at L_s 270°.) However, there are exceptions in Asimov and Horowitz craters (McEwen et al., 2011; Ojha et al., 2014) and Hale and Palikir craters (Dundas et al., 2015; Stillman and Grimm, 2018). Only 41% (82 out of 200) of the SML sites with steep, equator-facing rocky slopes with bedrock exposures contain candidate RSL (Ojha et al., 2014), with confirmed RSL present in at least 18 of those locations (Stillman et al., 2017).

10.4 RSL IN EQUATORIAL AND NORTHERN MIDDLE LATITUDES

Evidence for abundant equatorial RSL (McEwen et al., 2014) indicated that preservation of shallow (<1 m) ice or frozen brine was especially unlikely given rapid sublimation rates throughout the year (Mellon et al., 2004; Altheide et al., 2009). Mars Odyssey Neutron Spectrometer data have been interpreted as evidence for patches of shallow equatorial ice (Wilson et al., 2018), but alternate analyses concluded that this interpretation is problematic (Pathare et al., 2018). The equatorial RSL share the same set of distinctive characteristics as in the SML, except that the seasonality varies primarily with slope aspect, or may continue year-round on east- and west-facing slopes. Valles Marineris (VM) has by far the greatest concentration of RSL on Mars (Chojnacki et al., 2016; Stillman et al., 2017) and also has the greatest concentration of steep ~75 m scale slopes (Neumann et al., 2003) over low-albedo regions (Putzig and Mellon, 2007) and with high-rock abundances (Edwards et al., 2009). A dozen confirmed or partially confirmed RSL sites have been documented in other equatorial regions and in Chryse and Acidalia Planitia in the northern middle latitudes (Dundas et al., 2015), where they are clearly active when surface temperatures are below 273 K (Stillman et al., 2016). Confirmed RSL are rare or absent in the northern hemisphere beyond a portion of Chryse and Acidalia Planitia, although there are a few unconfirmed candidate sites near the Jezero crater and northeast Syrtis Major candidate landing sites for the NASA 2020 rover and near the Mawrth Vallis candidate landing site for the ExoMars rover.

10.5 COLOR OBSERVATIONS

CRISM hyperspectral and HiRISE color observations have provided useful constraints on RSL properties, in addition to the CRISM detection of hydrated salts (discussed below). Although water bands have not been detected, Massé et al. (2016) showed via laboratory experiments that wet surfaces could maintain a low albedo when mostly dry, perhaps like RSL in the midafternoon when MRO observes, but that the water absorption bands would be very weak, likely not detectable by CRISM over subpixel areas. CRISM cannot resolve individual flows with a best imaging scale of ~18 m/pixel, but there is evidence for seasonally enhanced ferric and ferrous absorptions over RSL-rich slopes (Ojha et al., 2013). This CRISM result is consistent with the color of RSL fans and some faded RSL as seen by HiRISE, in which these absorptions make the RED (600–800 nm) bandpass relatively bright compared with the BG (400–600 nm) or NIR (800–1000 nm) bands (Fig. 10.1A and B). In other words, these areas appear relatively greenish in HiRISE IRB color with individually contrast-enhanced IR, RED, and BG images displayed as red, green, and blue, respectively (see color images in McEwen et al., 2014; Chojnacki et al., 2016). The spectra could be explained in multiple ways such as coarse grain sizes, precipitation of ferric oxides, and/or wetting of the substrate (Ojha et al., 2013). Darkening and fading due to wetting and drying of soil is well documented (e.g., Pommerol et al., 2013), and some inactive RSL paths are bright or colored, perhaps from deposition of salts and/or ferric oxides (Fig. 10.1A, B, and E). Darkening via larger grain sizes could result from the dynamics of granular flows (Dundas et al., 2017a), but if fading is due to decreasing grain sizes, then how this can happen over a period of a few weeks has not been explained. Fading of RSL due to atmospheric dust fallout would require that the dark lineae be less blue than their surroundings, which is not obvious in the color images. The HiRISE color bands should be effective for distinguishing dust from nondust surface materials on Mars (Delamere et al., 2010) but require atmospheric and slope corrections for quantitative analysis (Fernando et al., 2017). It would also be difficult to explain the fading of inactive lineae adjacent to dark, growing lineae (Stillman et al., 2017) via dust fallout, unless grain flow actively removes the dust. But there are correlations between fading of RSL and fading of dust devil and rock fall tracks in Tivat crater, suggesting that regional seasonal removal of dust may partially explain or complicate RSL fading (Schaefer et al., 2017).

10.6 CRISM DETECTION OF HYDRATED SALTS

A strong constraint on the origin of RSL is the detection of hydrated oxychlorine (perchlorate and chlorate) salts associated with five RSL sites (Ojha et al., 2015, 2017b; Ojha, 2016). The compositional identification is based on comparison to laboratory measurements (Hanley et al., 2014, 2015). Hydrated salts of this composition have been detected from orbit in only two other settings (Ojha et al. 2017a; Ojha, 2016): one associated with ice-exposing new impacts (Byrne et al., 2009; Dundas et al., 2014) and possibly within the north polar layered deposits (Massé et al., 2012). Water ice is exposed and sublimating at the two non-RSL settings, which greatly facilitates deliquescence (Fischer et al., 2016), but what is special about RSL sites? It is probably not just a matter of exposing subsurface materials because hydrated salts have not been seen in new impacts that did not expose ice (Ojha et al., 2017a; Ojha, 2016) or in new gully activity (Núñez et al., 2016). Although the CRISM dataset has not been thoroughly searched at the single-pixel level, the presence of sufficient hydrated oxychlorine salt to be detected by CRISM at ~18 m/pixel seems to be rare, even though anhydrous perchlorates may be ubiquitous in the martian regolith at the level below 1% (Clark and Kounaves, 2016). Perchlorates are highly deliquescent but are expected to form water only in the early morning or late afternoon in nonpolar regions of Mars (Gough et al., 2011). In equatorial regions and in the middle afternoon when CRISM observes, efflorescence is expected to produce anhydrous salts, which lack near-IR absorption bands, rather than hydrated salts (Gough et al., 2011). Furthermore, if perchlorates are distributed in the soil at levels below 1%, then forming enough hydrated salt to be detected from orbit would be difficult at any time of the day, unless a process like flowing water served to concentrate the salt. Cull et al. (2010) found the distribution of perchlorate salts at the Phoenix landing site to be locally concentrated, similar to salt patches that result from aqueous dissolution and redistribution on Earth. The best interpretation is that there must be some water associated with RSL, but not necessarily flowing water.

10.7 RSL ASSOCIATION WITH SMALL GULLIES AND SLUMPS

Although classic martian gullies are concentrated in the middle latitudes, pristine small gullies are often found with RSL in equatorial regions (Fig. 10.2), where fresh-looking gullies are rare. A gully is defined as "A small channel with steep sides caused by erosion and cut in unconsolidated materials by concentrated but intermittent flow of water…and too deep (e.g., >0.5 m) to be obliterated by ordinary tillage" (Neuendorf et al., 2010). The channels (1–20 m wide, sometimes >0.5 m deep) associated with many RSL fit this definition in terms of size. However, martian midlatitude ravines are typically much larger and are called "gullies" in the Mars science literature, so we call the RSL-related landforms "small gullies" even though they are the size of typical gullies on Earth. The largest RSL gullies are in VM and are just barely resolved in images from MRO's Context Camera (Malin et al., 2007), although a recent global CTX-image survey excluded them (Harrison et al., 2015). Auld and Dixon (2016) described a small number of equatorial gullies based on early HiRISE images, before HiRISE began a campaign of imaging steep slopes to look for RSL. HiRISE monitoring has shown that gully formation in the middle latitudes is due to dry debris flows fluidized by seasonal CO_2 (Dundas et al., 2017b).

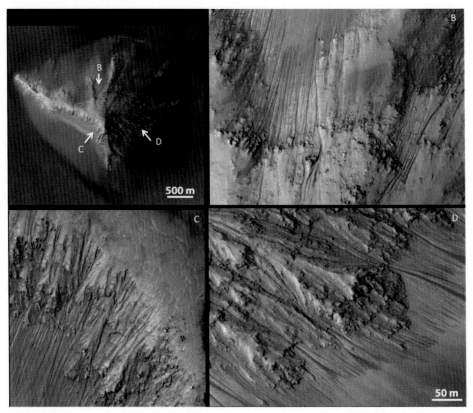

Fig. 10.2 Hill in Juventae Chasma with small gullies and RSL on all sides at L_s 356° in MY 33 (ESP_050390_1755). Locations of full-resolution cutouts B, C, and D indicated in view of the entire hill (upper left). This is also a hill with abundant slumps on the RSL/gully fans.

However, RSL form over the warmest places on Mars, making CO_2 highly unstable throughout the year, so CO_2 fluidization is not a viable mechanism to produce the small gullies at equatorial locations in the present climate.

Were the small channels associated with RSL caused by aqueous erosion, today or in the recent past? If so, it would be odd that they only form over slopes steeper than $\sim$28°. Also, many RSL sites lack apparent gullies, even when lighting conditions are favorable to see subtle topographic shading. Small gullies may not be detected in some HiRISE images, especially on sun-facing slopes where the topographic shading is not accentuated. Many RSL flows are associated with small gullies in their upper reaches, but channels are often not apparent on the smooth fans, which may account for most of the length of lineae and where they are most obvious. There are some equatorial regions where small gullies are present but RSL have not been seen; some of these sites are otherwise identical to RSL slopes in terms of morphology: steep, rocky slopes and channels lead to a smooth fan at ~30° slope and IRB "greenish" fans (e.g., ESP_049071_1630). RSL at such locations may be seen in the future from continued monitoring, or they could have been active in the recent past.

The straightforward interpretation is that RSL activity erodes gullies in the upper slopes and deposits sediment on lower (but still steep) slopes forming fans. We have not seen unambiguous topographic evidence for erosion of these gullies today, so either they erode slowly over many years or the gullies formed in the past and the RSL simply follow these channels. Since we believe that RSL are active for >100 Mars years, slow erosion not detected by HiRISE is plausible. There is often an excellent fit in size between RSL, gullies, and fans, suggesting a genetic relation (Fig. 10.2). In some cases, especially in the middle latitudes, the present-day RSL are much smaller and shorter than the gullies they follow (Fig. 10.1D), so the present-day activity levels did not carve the large gullies. (But there may be smaller gullies in the large-gully headwalls that are well-matched in size to the RSL.) In other places, especially in VM, the RSL precisely fit the small gullies and extend out to the downhill margins of the gully fans. Unless a remarkable coincidence, this suggests that the RSL activity modified the topography. There are a few cases where we see probable topographic changes associated with active RSL (Chojnacki et al., 2016), but in most cases, either the RSL erode and deposit sediment too slowly to detect with HiRISE over several Mars years, or the major erosion is episodic.

HiRISE has detected newly formed topographic slumps associated with RSL fans in at least 12 locations—nine around a hill in Juventae Chasma (Ojha et al., 2017b; see Fig. 10.2), two in Garni crater in Melas Chasma, and one along wall slopes in Coprates Chasma (Chojnacki et al., 2016). Typical lengths are tens of meters with topographic relief of ~1 m (negative on uphill end and higher in the downhill lobe). The slumps are initially relatively dark, similar to RSL, and then fade over a few weeks, also similar to the fading of RSL when they become inactive. Note that slope streaks on dust-mantled slopes, of comparable size, require decades to fade via atmospheric dust fallout (Aharonson et al., 2003). A distinctive landform assemblage is seen at several locations within central and eastern VM: small channels on most slope aspects of isolated hills or crater walls, extending very nearly to the tops of the hills or crater rim, associated with RSL that match the channels in size, and with a set of lobate deposits from slumps at the base of RSL fans (Fig. 10.2). RSL activity in VM generally changes slope aspect with season—N-facing slopes in northern summer and S-facing slopes in southern summer. The slumps begin midway down the RSL fans and have a different seasonality—most active from L_s 0°–120°, the coldest time of the year in VM in general, but especially on south-facing slopes where most of the slumps have been seen. Assuming this association between gullies, RSL, and slumps is not coincidental, an integrated landscape evolution model would be appropriate. Perhaps RSL activity carves the small gullies and deposits sediment near the base of angle-of-repose slopes, slowly oversteepening the slopes, which episodically slump. How this erosion happens and why the slumping happens mostly in the (slightly) colder season are not understood.

10.8 HOW DO RSL FORM?

10.8.1 Melting of Shallow Ice

Chevrier and Rivera-Valentin (2012) proposed that SML RSL formed by melting frozen brines from in a past climate. Frozen brine can remain after pure water ice has sublimated away. This model explained the seasonality and temperature dependence of RSL, including the observation that RSL activity peaks after the warmest surface temperatures are reached, corresponding to peak seasonal temperatures in the shallow subsurface. However, this model did not explain how RSL could recur over many years over such warm slopes and was especially challenged by the discovery of abundant equatorial RSL, where frozen brines should not persist (Altheide et al., 2009).

Stillman et al. (2014) compared RSL activity and surface temperatures in the SML, reporting that RSL are actively lengthening almost only when the peak daily surface temperature exceeds 273 K and concluding that they must be due to freshwater flows. However, the spatial resolution of temperature mapping from orbit exceeds ~200 m, so the temperatures may be dominated by fine-grained materials near RSL rather than the actual lineae. Also, there is no basis for assuming that RSL are active during the warmest time of day. Martian groundwater is expected to be salty (Burt and Knauth, 2003; Tosca et al., 2011), so melting of vapor-deposited ice would seem to be the only viable source of freshwater on Mars today. Grimm et al. (2014) modeled RSL as freshwater flows requiring 2–10 m^3 of H$_2$O per meter of source headwall. While equally favoring origin from shallow groundwater, they suggested that annual recharge of ice by vapor cold trapping might be supplied from the atmosphere or subsurface. However, the martian atmosphere is extraordinarily dry, with <10 precipitable microns of yearly averaged column abundance of water vapor in the southern hemisphere (Smith, 2008), and vapor transport through the subsurface is extremely slow (Hudson et al., 2007; Hudson and Aharonson, 2008). Also, shallow ice should not be present over many of the SML RSL sites, particularly on the equator-facing slopes (Aharonson and Schorghofer, 2006) where SML RSL are found.

10.8.2 Groundwater Models

Stillman et al. (2016, 2017) proposed that the RSL in Acidalia/Chryse and VM are sourced from salty groundwater. However, RSL often initiate near the tops of isolated peaks and ridges, which is difficult to explain via groundwater (McEwen et al., 2011, 2014; Chojnacki et al., 2016; Dundas et al., 2017a). Groundwater will be driven to low elevations by topographic pumping (Showman et al., 2004), but there is no elevation preference for RSL within the >10 km of relief in VM (Chojnacki et al., 2016). Faulting and fractures will complicate where fluids emerge onto the surface, and Watkins et al. (2014) showed that there are many linear structures on slopes with RSL, but the RSL do not appear to initiate along linear structural trends and there should still be a strong preference for lower elevations. Although orbital ground-penetrating radar experiments have not detected subsurface water, that does not rule out its potential existence (Farrell et al., 2009; Stillman and Grimm, 2011). The retention of a global groundwater table to the present day is theoretically difficult but not impossible (Clifford et al., 2010; Grimm et al., 2017). Nevertheless, we can conclude that there is no convincing evidence for widespread shallow or deep groundwater in Mars today, unless the RSL are interpreted as such evidence.

There are many deposits interpreted as chlorides over broad regions of Mars (Osterloo et al., 2010) but little evidence for a direct connection between these deposits and RSL sites (Mitchell et al., 2016). Nevertheless, Mitchell and Christensen (2016) concluded that RSL could be produced by capillary wicking of chloride-rich brines, leaving lag deposits of chlorides below the detection limits of orbiting instruments.

If not groundwater or melting of ice, the only other potential source of water is the atmosphere, but this cannot produce sufficient water to produce saturated flow unless water is somehow extracted from the entire atmosphere and concentrated onto <1% of the surface area of Mars. Seasonal variations in column abundance of atmospheric water over RSL sites do not correlate with RSL activity (McEwen et al., 2014), although the relation between RSL activity and water vapor in the atmosphere could be complex due to exchange with the surface, with subsurface storage in the cold seasons such that only warming is required to initiate deliquescence (Wang et al., 2017). Stillman et al. (2016, 2017) concluded that groundwater is the only recharge mechanism consistent with their interpretation that RSL are due to water flow in a saturated porous medium.

There are abundant equatorial RSL outside of VM, such on the west-facing slopes of Krupac crater (7.8°S, 86.0°E; Figs. 10.1E and 10.3). This site is at a high elevation (+1.8 km) with little surrounding terrain at even higher elevations that might drive groundwater release along the walls of the crater. What is unusual about this site is that the RSL fade near the warmest time of the year. Given the equatorial location and west-facing slopes, seasonal temperatures track subsolar latitude and heliocentric range (perihelion at L_s 251° and aphelion at L_s 71°), so the warmest time of the year for the surface and shallow subsurface is near the equinoxes and perihelion, and that's when we might expect RSL to be active. Instead, RSL are present (dark) most of the year and fade or disappear near perihelion (L_s 232°–264°). Peak activity is near L_s 200° and 10°; they are less active (not changing much, but still relatively dark) at aphelion, when coldest. This pattern has repeated in 2 Mars years, MY 32-33. Clearly, some RSL activity prefers or requires a temperature range below the warmest possible near Mars' equator. The timing of this activity does not match the expected pattern from melting a frozen aquitard at depth, releasing groundwater during the warmest times (Stillman et al., 2016, 2017). What this shows is that RSL activity requires a particular range of temperatures, not necessarily the hottest times at a site.

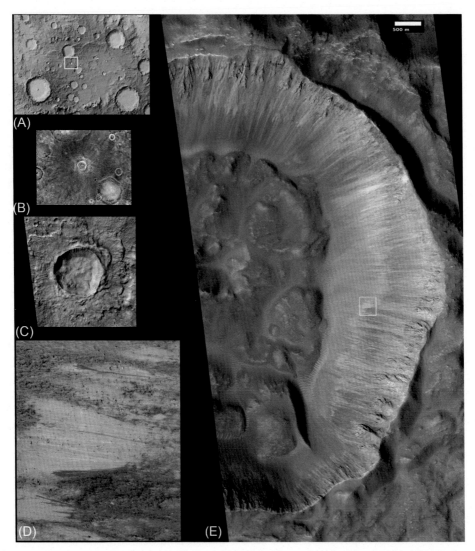

Fig. 10.3 Krupac crater. North is up in all views. (A) MOLA color-coded elevations, with *white box* around Krupac crater. *Red* indicates elevations near +2 km; *yellow* is near +1 km. (B) THEMIS nighttime IR showing bright (warm, rocky) walls of Krupac Crater and extensive dark (cold) radial ejecta, typical of young craters (Tornabene et al., 2006). (C) Image of Krupac Crater by MRO's Context Camera (F17_042407_1722). (D) Full-resolution sample of RSL (ESP_047141_1720); location indicated by *white box* in E. (E) HiRISE browse image covering eastern half of the crater (ESP_047708_1720).

What if groundwater doesn't emerge but rises close to the surface, enhancing the local relative humidity, leading to voluminous deliquescence over nearby slopes? One of the most likely places on Mars for groundwater to reach the surface is deep in VM (Andrews–Hanna et al., 2010). There has been an interpretation of water fogs inside VM (Möhlmann et al., 2009), which would probably require a local source of water

(Leung et al., 2016). Other workers have concluded that the haze in VM is from dust storms that drain into the canyons (Cantor et al., 2001; Inada et al., 2008), but that dust is associated with water (Ojha et al., 2017b). There are also equatorial RSL at high elevations where groundwater upwelling is unlikely and putative fogs have not been seen, such as Krupac crater.

Groundwater and melting of shallow ice do not seem to be viable as general explanations for RSL activity. There remains the possibility that different RSL have different origins. I do not favor this possibility because the RSL have such distinctive and peculiar characteristics, usually consistent across the various locations where they are found. There are also anomalous lineae with nonseasonal growth or fading (e.g., Fig. 7 of Chojnacki et al., 2016) that could require a different explanation. The variable temperatures of RSL activity may indicate that brine compositions vary from place to place, consistent with the known presence of a variety of salt compositions (see next section), if these brines can somehow trigger RSL formation.

10.8.3 Deliquescence

The idea that small amounts of very salty water could be transiently stable on Mars today was mentioned by Leighton and Murray (1966), and hygroscopic salts were detected by the Viking landers (Clark, 1978; Navarro-Gonzalez et al., 2010) and in martian meteorites (Rao et al., 2008; Kounaves et al., 2014). Interest in deliquescence was spurred by observations of perchlorates by the Phoenix lander mission (Smith et al., 2009; Hecht et al., 2009; Möhlmann and Thomsen, 2011), including possible droplets of water on the lander struts (Rennó et al., 2009). There is growing evidence that conditions permitting deliquescence are common on Mars (Gough et al., 2011, 2016; Martín-Torres et al., 2015; Glavin et al., 2013; Fischer et al., 2016; Martinez and Renno, 2013).

Levy (2012) compared RSL to water tracks in Antarctica (Levy et al., 2011; Dickson et al., 2013) and showed that the growth rates of RSL are consistent with liquid flow in the porous upper regolith. Although water tracks share several key characteristics with RSL, such as darkening and downhill lengthening during the warm season and fading in the cold season, they extend onto much lower slope angles than RSL. Water tracks are partially sourced by salt deliquescence (Gough et al., 2017), the mechanism Levy (2012) favored for martian RSL, but the column of air over Antarctica contains about two orders of magnitude more water than martian air (Connolley and King, 1993).

Kossacki and Markiewicz (2014) found theoretical conditions needed for deliquescence of salt on Mars only within a narrow range of parameters they considered, but recent results have widened that activity space. Laboratory work shows that hydrous chlorides and oxychlorine salts, especially in the shallow subsurface, are promising to explain the temperature-dependent activity of RSL, including equatorial latitudes (Nuding et al., 2014; Gough et al., 2011, 2014, 2016; Wang et al., 2017). Once an aqueous solution was formed from calcium chloride, efflorescence did not occur until single-digit RH values were reached (3.9% RH on average; Gough et al., 2016).

Deliquescence and efflorescence can explain the darkening and fading of RSL (Heinz et al., 2016), and these studies proposed that RSL are fossil features—salt deposits from past gully activity—that darken each year due to deliquescence. This model does not explain the incremental or gradual lengthening of RSL, or the observation that they form in different places from year to year, or the fact that many RSL do not follow gullies. But deliquescence along RSL tracks may help explain some observations. For example, there are locations where RSL seem to initially form largely at once, followed by growth only at their tips. HiRISE cannot monitor these sites more often than about twice per month and usually much less often, so it is not possible to distinguish rapid initial growth from instantaneous darkening of previous salt deposits. Rapid initial growth has been described at some well-monitored sites (Schaefer et al., 2015), but we cannot rule out the hypothesis that large lengths of RSL darken all at once as envisioned by Heinz et al. (2016). If so, that does not eliminate the problem of how to explain the incremental or gradual growth and changing RSL paths, but less water may be required.

Does the behavior of RSL in Krupac Crater (Fig. 10.3) make sense in terms of deliquescence? There are some bright lineae (maybe a result of salt deposition or concentration) here that darken incrementally as RSL advance (Fig. 10.1E) and then reappear as the RSL fade. Deliquescence requires particular combinations of temperature and relative humidity (RH). Invariably, RH drops as temperature rises on Mars (Gough et al., 2011; many others). Near perihelion in Krupac crater, the RH may be too low for even metastable water from a particular brine composition, explaining the warm-season fading. With hysteresis, deliquesced liquids may remain metastable through the day to single-digit RH, but if the RH is low enough, it will completely effloresce. Unfortunately, we have very few measurements of RH on Mars and little information of the vertical distribution of water vapor (Tamppari et al., 2010), which might otherwise be used to predict the RH at the surface.

Another observation of potential significance is that RSL activity often correlates with regional dust storm activity (McEwen et al., 2011, 2014; Chojnacki et al., 2016; Stillman et al., 2014, 2017). Large (>100 m scale) and extensive RSL fans in VM become relatively darker (compared to the surroundings) during or shortly after dusty periods, returning to their previous appearance within weeks when the air is back to normal (Fig. 10.1E; Chojnacki et al., 2016, especially supplemental animations 1 and 2). How can we explain this observation? Daytime surface temperatures are lower, and nighttime temperatures are higher, each by ~10°C, when the air is very dusty (Ryan and Henry, 1979). Given the anticorrelation between T and RH, this means that conditions needed for deliquescence, when the air is dusty, may exist for extended time periods of the day and night and the conditions needed for efflorescence should be reduced at midday. However, we are ignorant about actual relative humidity levels at these times and places. A microwave/

submillimeter sounder on a future orbiter could fill this major gap in understanding water on Mars (Forget and Mambo Team, 2004).

In summary, deliquescence may be common on Mars and can explain changing albedos, but it probably cannot produce sufficient water for saturated flow given <20 precipitable microns atmospheric column abundance over most RSL sites (McEwen et al., 2014, Table S4).

10.8.4 Dry RSL Models

Thermal infrared observations could provide a test for wet soils, and Edwards and Piqueux (2016) analyzed data from the Thermal Emission Imaging System (THEMIS: Christensen et al., 2004) over Garni crater, concluding that an upper limit to water abundance is just 3%. Unfortunately, this result is ambiguous due to the temporal variability of RSL in Garni crater (Stillman et al., 2017). There is not a single THEMIS observation of Garni crater in which we can be certain that the RSL were sufficiently active to fill >20% of the THEMIS resolution field of view ($\sim300 \times 300$ m), which Edwards and Piqueux (2016) considered necessary for detection. It is also not clear that another assumption is correct, that of similar thermophysical properties on the floor of Garni crater (where fines accumulate) and the active crater slopes (where fines are removed). A future orbiter designed to study RSL in this manner should have thermal IR resolution better than ~20 m, concurrent submeter resolution imaging, and the ability to target key locations for frequent repeat observations.

Dundas et al. (2017a) measured slope relations of RSL on a set of 10 HiRISE DTMs and found in nearly all cases that the mean slope at the end of a linea is between 28° and 35°, matching the range of slip face slopes for active martian sand dunes (Atwood-Stone and McEwen, 2013). Such slopes are interpreted as the range of critical angles where cohesionless granular flows will stop motion (often called dynamic angle of repose). In some places, there are RSL of widely varying lengths (from tens of meters to 1.5 km), yet all of the lineae terminate where the slope is near 30°. Although flow length could be limited by water supply to seeps and thus terminate before reaching shallow slopes, it does not seem reasonable to argue that the water supply is everywhere just the right amount for the flow length to reach the distance matching certain slope angles. This result seems to be a strong constraint requiring that RSL are fundamentally granular flows, not water seeps. However, the behavior of RSL is extremely unusual for volatile-free flows on Earth, with gradual or incremental growth and hundreds of concurrently active flows over kilometer-scale areas, which recur multiple years in nearly or exactly the same locations. Granular flows should deplete the local slope in fine-grained material, so they should not recur annually in the same places unless resupplied as on active sand dunes. In some places, climbing ripples are seen on RSL fans (Chojnacki et al., 2016; Dundas et al., 2017a), which can then collapse into granular flows, but they have not been seen to climb onto the bedrock where RSL initiate.

How could dry granular flows have the distinctive seasonal behavior of RSL? Schmidt et al. (2017) published a novel idea to explain the RSL by rarefied gas-triggered granular flows. The model requires sudden cooling of warm grains of 100 μm radius or smaller size, which then experience an extremely tiny lift force from the sharp temperature gradients (Knudsen pump) to effectively lower the static angle of repose and initiate movement on slopes already at this angle. However, the static angle of repose is several degrees steeper than the dynamic angle of repose or stopping angle (Cheng and Zhao, 2017), and it is not clear how the RSL slopes are steepened to this static angle during cold seasons. This model requires very low atmospheric pressures and would not work on Earth, which is an attractive way to explain the lack of a good terrestrial analog. However, most RSL slopes are likely to have larger grain sizes than 100 μm radius. The thermal inertias at RSL sites (McEwen et al., 2011; Chojnacki et al., 2016) are consistent with mostly coarse sand and rocks (Putzig and Mellon, 2007). Although thermal inertia is insensitive to the top millimeter, the proposed RSL flow mechanism would remove such a layer of fine grains and preclude the observed yearly recurrence. The Knudsen pump mechanism can provide a tiny lift force on small grains only near rapidly moving shadows, absent along the lengths of many RSL flows, and it is difficult to understand how a small source region near a boulder can generate flows extending hundreds of meters downslope.

There are new dark slope lineae on the Moon that are likely granular flows or boulder tracks (Speyerer et al., 2014), but none reported that behave temporally like RSL, in spite of analysis of 18,516 high-resolution (<2 m/pixel) repeat images with nearly identical photometric angles (Wagner et al., 2017). The Moon does not experience pronounced seasonal temperature changes, lacks a collisional atmosphere, and lacks geologically significant quantities of volatiles, all of which can influence granular flows (Kokelaar et al., 2017). Some combination of martian environmental parameters must be needed for RSL activity.

10.8.5 Hybrid Models

We seem to have contradictory strong constraints on the origin of RSL: the presence of hydrated oxychlorine salts and topographic evidence for cohesionless granular flows. Perhaps small amounts of water serve to trigger granular flows as in the lab experiments of Massé et al. (2016), but those experiments required more water than can reasonably be expected from deliquescence on Mars. Perhaps deliquescence concentrated in the shallow subsurface (Gough et al., 2016; Wang et al., 2017) or expansion of hydroscopic salts can destabilize overlying grains. As in the model of Schmidt et al. (2017), only relatively small forces may be needed to initiate movement on angle-of-repose slopes with a surface layer of low-cohesion grains.

Alternatively, the high temperatures could desiccate the grains of tiny amounts of adsorbed water that provides cohesion (Schorghofer et al., 2002), but it hasn't been explained how that could trigger motion unless the slopes exceeded the static angle of repose, typically several degrees steeper than the stopping angle for granular flows

(Cheng and Zhao, 2017). Also, this prediction does not match the lack of seasonality for martian slope streaks on dust-mantled slopes (Schorghofer and King, 2011).

The annual recurrence of RSL is difficult to explain in all of the models discussed above. RSL activity is depleting something, either water, salt, or small grains, which must be replenished for recurrence. Grains can be produced by eroding the small gullies or by more distributed erosion that doesn't create detectable gullies, but what could cause sufficient erosion? Erosion rates on Mars are 3–30 nm/year in Meridiani Planum (Golombek et al., 2014) and ~100 nm/year for relatively bright sedimentary rocks (Kite and Mayer, 2017). Erosion of 100 nm/year seems far from sufficient to sustain the yearly RSL activity. Although loose grains will be eroded faster than bedrock, RSL activity removes those grains, which must be replenished. Perhaps mechanical salt weathering helps to create sand-sized particles (Malin, 1974; Jagoutz, 2006).

In conclusion, RSL probably extend down slopes as granular flows, and the timing of their activity and albedo changes probably involve salty water, but an exact mechanism is still not understood. This lack of understanding hinders attempts to evaluate habitability.

10.9 IMPLICATIONS FOR HABITABILITY

If RSL are seeps of water from extensive shallow aquifers (Travis et al., 2013; Stillman et al., 2016, 2017; Mitchell and Christensen, 2016) or melting of vapor-transported ice or frozen brine (Chevrier and Rivera-Valentin, 2012), then these features mark the most promising locations on Mars for present-day habitability. However, as discussed above, these models seem highly unlikely. If RSL are largely dry granular flows, perhaps triggered by or otherwise associated with small amounts of brine, then prospects for life as we know it are poor within the RSL because the water activity is low. However, there is evidence consistent with water, and its source is not understood, so we cannot rule out the idea that habitable environments transiently exist in the vicinity of RSL.

For known terrestrial life, key limits are temperature (T) and water activity (a_w); limits recommended to define SRs (with margin) are $T > 245$ K and $a_w > 0.5$, anytime in the past 500 years (Kminek et al., 2016). There is no question that surface temperatures often exceed 245 K when RSL are active, so water activity is the main constraint. Most previous studies have favored the hypothesis that RSL are associated with salty water in some manner, given the temperature-dependent activity and association with transient hydrated salts, but have struggled to explain the origin of sufficient water for downhill seepage. If only a tiny amount of water from deliquescence is present, then these brines should have a low a_w (Tosca et al., 2008; Toner and Catling, 2016) and would not be considered habitable. Ionic strength may be a barrier to water availability in brines even when the water activity was biologically permissive (Fox-Powell et al., 2016). Nevertheless, discussion has continued about possible present-day life on Mars supported by hygroscopic salts (Davila et al., 2010; Ulrich et al., 2010; Parro et al., 2011; Gomez et al., 2012; Oren et al., 2014; Hansen-Goos et al., 2014; Schulze-Makuch et al., 2015; King, 2015; Jänchen et al., 2016; Schuerger et al., 2017), and the RSL mark locations with hygroscopic salts and temperatures permissive of life. Nuding et al. (2017) demonstrated that Ca perchlorate solutions are not sporicidal.

10.10 SHOULD CANDIDATE RSL BE TREATED LIKE SPECIAL REGIONS?

A Mars special region is defined as "A region within which terrestrial organisms are likely to propagate or a region that is interpreted to have a high potential for the existence of extant martian life forms" (Kminek et al., 2010). Confirmed and partially confirmed RSL sites are considered "uncertain" SRs (Rummel et al., 2014), which must be treated like known SRs. Designating only parts of Mars as SRs (or uncertain SRs) has facilitated surface exploration in equatorial regions where no SRs were suspected to exist prior to 2010. The discovery of RSL has altered that paradigm, and there are now locations treated as SRs scattered throughout equatorial low-albedo regions of Mars, the most favorable landing regions, and more are being discovered as more HiRISE images are acquired. However, landing on a slope steeper than 27° has always been avoided, at least for landers whose location has been determined. The current COSPAR study committee recommendation is to evaluate every candidate landing or exploration site with any linear and relatively dark slope feature on a case-by-case basis (Kminek et al., 2016). This includes nearly every slope on Mars. Currently, even weak RSL candidates such as those in Gale crater (Dundas and McEwen, 2015) are being treated like SRs (Jones and Vasavada, 2016). Although features at one location in Gale crater 50 km from the rover met the definition for partially confirmed RSL, all others are candidates for which no clear incremental growth or recurrence has been seen. The Mars Science Laboratory (MSL) rover has imaged small granular flows (Dickson et al., 2016), and the best hypothesis is that the candidate RSL near the rover are dry granular flows that may sometimes overprint or reactivate but lack seasonal recurrence, so they are not RSL. The region around the Curiosity rover is probably the best-monitored spot on Mars (43 overlapping HiRISE images acquired from 2006 to 2017), and there has been no clear detection of growth or recurrence of any slope lineae. Narrow (<0.5 m), barely detected lineae may sometimes not be apparent in a HiRISE image due to a poor signal-noise ratio through dusty air, from highly oblique views of steep slopes, or in blurred images (only since early 2017). NASA directed the MSL project to institute a protocol in daily rover operations to ensure ongoing compliance with its planetary protection categorization, partially in response to concern over the possible role of present-day liquid water in the formation of lineae on local slopes around the rover (Jones and Vasavada, 2016). For further discussion of planetary quarantine on Mars, see Fairén and Schulze-Makuch (2013), Fairén et al. (2017), and Rummel and Conley (2017).

10.11 FUTURE STUDY OF RSL

Laboratory experiments hold great promise for understanding RSL, as they can be performed at only ~2 orders of magnitude smaller scale than in nature. Past experiments have yielded important clues. Gully formation by water has been simulated under martian conditions (e.g., Coleman et al., 2009; Jouannic et al., 2015). Massé et al. (2016) demonstrated that granular flows can be triggered by boiling water or brine, which easily boils at martian atmospheric pressures and surface temperatures. There have been multiple experiments on deliquescence, cited previously. Takagi et al. (2011) demonstrated the behavior of shallow granular flows on steep slopes. Hudson et al. (2007) measured the diffusion of water vapor in martian soil simulants. Wurm et al. (2008) demonstrated that illumination of dust at atmospheric pressure near 6.4 mbar can lift micron-size particles. However, none of these experiments have reproduced all of the environmental conditions of martian RSL over relevant timescales, and none have demonstrated how RSL can form. In particular, for those experiments involving water on slopes, the water or ice is always artificially introduced into the lab setup, and granular flow experiments include some artificial mechanism to initiate flow. A future lab experiment that I recommend would be testing the idea that hygroscopic salts in the shallow subsurface can initiate granular flows on dynamic angle-of-repose slopes under martian conditions.

Unless there is a breakthrough in laboratory experiments or in remote sensing by current assets, future flight projects are needed to understand RSL. The current COSPAR policy and NASA interpretations for Mars SRs means that future surface exploration missions may be required to implement expensive procedures. The solution to this problem is to better understand RSL. New and continued orbital remote sensing could certainly help, especially visible, near-IR, thermal, submillimeter, and radar sounding observations at higher resolutions and sensitivities than past orbital experiments (MEPAG NEX-SAG Report, 2015). Observation at multiple times of the day may also provide key constraints, and the ExoMars Trace Gas Orbiter is expected to achieve a 400 km circular orbit that cycles through all local times of the day several times per season (Zurek et al., 2011). The Color and Stereo Surface Imaging System (CaSSIS; Thomas et al., 2017) can image in four color bands at 4.6 m/pixel, resolving RSL fans and other locations to see if there are color and albedo changes with time of the day, as expected if deliquescence is extensive. For definitive results on habitability, in situ measurements may be required. Innovative concepts to explore RSL sites in situ have been presented in conference abstracts (e.g., Satoh et al., 2013; Tanner et al., 2013; Hatakenaka et al., 2015; Nannen et al., 2016; Mege et al., 2016; Grimm, 2017), but there are not yet any approved missions. If such a mission did discover present-day habitability or detect biosignatures, that would represent a major advance in NASA's top science goal. A negative result for habitability would be less exciting but could ease constraints on future missions and would provide essential insights into a process unknown on Earth.

ACKNOWLEDGMENTS

The author thanks Colin Dundas (USGS), Ashwin Vasavada (JPL), Sarah Sutton (University of Arizona), Ben Clark (Space Science Institute), and Raina Gough (CU) for the reviews and comments.

REFERENCES

Aharonson, O., Schorghofer, N., 2006. Subsurface ice on Mars with rough topography. J. Geophys. Res. 111 (E11), E11007.

Aharonson, O., Schorghofer, N., Gerstell, M.F., 2003. Slope streak formation and dust deposition rates on Mars. J. Geophys. Res. 108 (E12), 12-1. CiteID 5138. https://doi.org/10.1029/2003JE002123.

Altheide, T., Chevrier, V., Nicholson, C., Denson, J., 2009. Experimental investigation of the stability and evaporation of sulfate and chloride brines on Mars. Earth Planet. Sci. Lett. 282, 69–78.

Andrews-Hanna, J.C., Zuber, M.T., Arvidson, R.E., Wiseman, S.M., 2010. Early Mars hydrology: meridiani playa deposits and the sedimentary record of Arabia Terra. J. Geophys. Res. 115. https://doi.org/10.1029/2009JE003485.

Atwood-Stone, C., McEwen, A.S., 2013. Avalanche slope angles in low-gravity environments from active Martian sand dunes. Geophys. Res. Lett. 40, 2929–2934.

Auld, K.S., Dixon, J.C., 2016. A classification of martian gullies from HiRISE imagery. Planet. Space Sci. 131, 88–101.

Burt, D.M., Knauth, L.P., 2003. Electrically conducting, Ca-rich brines, rather than water, expected in the Martian subsurface. J. Geophys. Res. Planets 108 (E4), 7-1. CiteID 8026. https://doi.org/10.1029/2002JE001862.

Byrne, S., et al., 2009. Distribution of mid-latitude ground ice on Mars from new impact craters. Science 325 (5948), 1674–1676. https://doi.org/10.1126/science.1175307.

Cantor, B.A., James, P.B., Caplinger, M., Wolff, M.J., 2001. Martian dust storms: 1999 Mars orbiter camera observations. J. Geophys. Res. 106, 23653–23687.

Cheng, N.-S., Zhao, K., 2017. Difference between static and dynamic angle of repose of uniform sediment grains. Int. J. Sediment Res. 32, 149–154.

Chevrier, V.F., Rivera-Valentin, E.G., 2012. Formation of recurring slope lineae by liquid brines on present-day Mars. Geophys. Res. Lett. 39 (21), 1–5. https://doi.org/10.1029/2012GL054119.

Chojnacki, M., McEwen, A., Dundas, C., Ojha, L., Urso, A., Sutton, S., 2016. Geologic context of recurring slope lineae in Melas and Coprates Chasmata, Mars. J. Geophys. Res. Planets 121, 1204–1231. https://doi.org/10.1002/2015JE004991.

Christensen, P.R., Jakosky, B.M., Kieffer, H.H., Malin, M.C., McSween Jr., H.Y., Nealson, K., Mehall, G.L., Silverman, S.H., Ferry, S., Caplinger, M., Ravine, M., 2004. The thermal emission imaging system (THEMIS) for the Mars 2001 odyssey mission. Space Sci. Rev. 110 (1), 85–130. https://doi.org/10.1023/B:SPAC.0000021008.16305.94.

Clark, B.C., 1978. Implications of abundant hygroscopic minerals in the martian regolith. Icarus 34, 645–665.

Clark, B.C., Kounaves, S.P., 2016. Evidence for the distribution of perchlorates on Mars. Int. J. Astrobiol. 15 (4), 311–318.

Clifford, S.M., Lasue, J., Heggy, E., Boisson, J., McGovern, P., Max, M.D., 2010. Depth of the martian cryosphere: revised estimates and implications for the existence and detection of subpermafrost groundwater. J. Geophys. Res. 115, E07001. https://doi.org/10.1029/2009JE003462.

Coleman, K.A., Dixon, J.C., Howe, K.L., Roe, L.A., Chevrier, V., 2009. Experimental simulation of martian gully forms. Planet. Space Sci. 57 (5–6), 711–716.

Connolley, W.M., King, J.C., 1993. Atmospheric water-vapour transport to Antarctica inferred from radiosonde data. Q. J. R. Meteorol. Soc. 119, 325–342.

Cull, S.C., et al., 2010. Concentrated perchlorate at the Mars Phoenix landing site: evidence for thin film liquid water on Mars. Geophys. Res. Lett. 37, L22203.

Davila, A.F., et al., 2010. Hygroscopic salts and the potential for life on Mars. Astrobiology 10, 617–628.

Delamere, W.A., et al., 2010. Color imaging of Mars by the High Resolution Imaging Science Experiment (HiRISE). Icarus 205, 38–52.

Dickson, J.L., Head, J.W., Levy, J.S., Marchant, D.R., 2013. Don Juan pond, Antarctica: near- surface CaCl₂-brine feeding Earth's most saline lake and implications for Mars. Sci. Rep. 3, 1–7.

Dickson, J.L., Head, J.W., Kulowski, M., 2016. In: Active flows at the Mars Science Laboratory landing site: results from a survey of Mastcam imagery through sol 971. 47th Lunar and Planetary Science Conference, LPI Contribution No. 1903, p. 1726.

Dundas, C.M., McEwen, A.S., 2015. Slope activity in Gale crater, Mars. Icarus 254, 213–218.

Dundas, C.M., Byrne, S., McEwen, A.S., Mellon, M.T., Kennedy, M.R., Daubar, I.J., Saper, L., 2014. HiRISE observations of new impact craters exposing martian ground ice. J. Geophys. Res. Planets 119 (1), 109–127.

Dundas, C.M., McEwen, A.S., Sutton, S., 2015. In: New constraints on the locations, timing and conditions for recurring slope lineae activity on Mars. 46th Lunar and Planetary Science Conference, Held March 16–20, 2015 in The Woodlands, TX, LPI Contribution No. 1832, p. 2327.

Dundas, C.M., McEwen, A.S., Chojnacki, M., Milazzo, M.P., Byrne, S., McElwaine, J.N., Urso, A., 2017a. Granular flows at recurring slope lineae on Mars indicate a limited role for liquid water. Nat. Geosci. https://doi.org/10.1038/s41561-017-0012-5.

Dundas, C.M., McEwen, A.S., Diniega, S., Hansen, C.J., Byrne, S., 2017b. The formation of gullies on Mars today. Geol Soc. Lond. 467. https://doi.org/10.1144/SP467.5.

Edwards, C.S., Piqueux, S., 2016. The water content of recurring slope lineae on Mars. Geophys. Res. Lett. 43, 8912–8919. https://doi.org/10.1002/2016GL070179.

Edwards, C.S., Bandfield, J.L., Christensen, P.R., Furgason, R.L., 2009. Global distribution of bedrock exposures on Mars using THEMIS high-resolution thermal inertia. J. Geophys. Res. Planets 114, E11001. https://doi.org/10.1029/2009JE003363.

Fairén, A.G., Schulze-Makuch, D., 2013. The overprotection of Mars. Nat. Geosci. 6, 510–511.

Fairén, A.G., Parro, V., Schulze-Makuch, D., Whyte, L., 2017. Searching for life on Mars before it is too late. Astrobiology 17. https://doi.org/10.1089/ast.2017.1703.

Farrell, W.M., Plaut, J.J., Cummer, S.A., Gurnett, D.A., Picardi, G., Watters, T.R., Safaeinili, A., 2009. Is the martian water table hidden from radar view? Geophys. Res. Lett. 36, L15206. https://doi.org/10.1029/2009GL038945.

Fernando, J., Doute, S., McEwen, A., Byrne, S., Thomas, N., 2017. In: Mars atmospheric dust contamination of surface albedo and color measurements. 48th Lunar and Planetary Science Conference, Held 20–24 March 2017, at The Woodlands, TX. LPI Contribution No. 1964, id.1635. (also submitted to Planetary and Space Science).

Fischer, E., Martínez, G.M., Rennó, N.O., 2016. Formation and persistence of brine on Mars: experimental simulations throughout the diurnal cycle at the Phoenix landing site. Astrobiology 16 (12), 937–948.

Forget, F., Mambo Team, 2004. In: A Submm Sounder for the Exploration of Mars. 35th COSPAR Scientific Assembly. Held 18–25 July 2004, in Paris, France, p. 4103.

Fox-Powell, M.G., Hallsworth, J.E., Cousins, C.R., Cockell, C.S., 2016. Ionic strength is a barrier to the habitability of Mars. Astrobiology 16, 427–442.

Glavin, D.P., et al., 2013. Evidence for perchlorates and the origin of chlorinated hydrocarbons detected by SAM at the Rocknest aeolian deposit in Gale crater. J. Geophys. Res. Planets 118, 1955–1973. https://doi.org/10.1002/jgre.20144.

Golombek, M.P., Warner, N.H., Ganti, V., Lamb, M.P., Parker, T.J., Fergason, R.L., Sullivan, R., 2014. Small crater modification on Meridiani Planum and implications for erosion rates and climate change on Mars. J. Geophys. Res. Planets 119, 2522–2547.

Gomez, F., et al., 2012. Habitability: where to look for life? Halophilic habitats: earth analogs to study Mars habitability. Planet. Space Sci. 68, 48–55.

Gough, R.V., Chevrier, V.F., Baustian, K.J., Wise, M.E., Tolbert, M.A., 2011. Laboratory studies of perchlorate phase transitions: support for metastable aqueous perchlorate solutions on Mars. Earth Planet. Sci. Lett. 312, 371–377.

Gough, R.V., Chevrier, V.F., Tolbert, M.A., 2014. Formation of aqueous solutions on Mars via deliquescence of chloride-perchlorate binary mixtures. Earth Planet. Sci. Lett. 393, 73–82. https://doi.org/10.1016/j.epsl.2014.02.002.

Gough, R.V., Chevrier, V.F., Tolbert, M.A., 2016. Formation of liquid water at low temperatures via the deliquescence of calcium chloride: implications for Antarctica and Mars. Planet. Space Sci. 131, 79–87. https://doi.org/10.1016/j.pss.2016.07.006.

Gough, R.V., Wong, J., Dickson, J.L., Levy, J.S., Head, J.W., Marchant, D.R., Tolbert, M.A., 2017. Brine formation via deliquescence by salts found near Don Juan pond, Antarctica: laboratory experiments and field observational results. Earth Planet. Sci. Lett. 476, 189–198.

Grimm, R.E., 2017. Mars DartDrop: probing contemporary habitability at recurring slope lineae. Low-Cost Planetary Missions, Pasadena, p. 17. Abstract pdf from http://www.lcpm12.org/.

Grimm, R.E., Harrison, K.P., Stillman, D.E., 2014. Water budgets of martian recurring slope lineae. Icarus 233, 316–327.

Grimm, R.E., Harrison, K.P., Stillman, D.E., Kirchoff, M.R., 2017. On the secular retention of ground water and ice on Mars. J. Geophys. Res. Planets 122, 94–109. https://doi.org/10.1002/2016JE005132.

Haberle, R.M., McKay, C.P., Schaeffer, J., Cabrol, N.A., Grin, E.A., Zent, A.P., Quinn, R., 2001. On the possibility of liquid water on present-day Mars. J. Geophys. Res. 106 (E10), 23317–23326.

Hanley, J., Dalton, J.B., Chevrier, V.F., Jamieson, C.S., Barrows, R.S., 2014. Reflectance spectra of hydrated chlorine salts: the effect of temperature with implications for Europa. J. Geophys. Res. E: Planets 119 (11), 2370–2377. https://doi.org/10.1002/2013JE004565.

Hanley, J., Chevrier, V.F., Barrows, R.S., Swaffer, C., Altheide, T.S., 2015. Near-and mid-infrared reflectance spectra of hydrated oxychlorine salts with implications for Mars. J. Geophys. Res. Planets 120 (8), 1415–1426.

Hansen-Goos, H., Thomson, E.S., Wettlaufer, J.S., 2014. On the edge of habitability and the extremes of liquidity. Planet. Space Sci. 98, 169–181.

Harrison, T.N., Osinski, G.R., Tornabene, L.L., Jones, E., 2015. Global documentation of gullies with the Mars Reconnaissance Orbiter context camera and implications for their formation. Icarus 252, 236–254.

Hatakenaka, R., et al., 2015. In: Preliminary thermal design of the Japanese Mars rover mission. 45th International Conference on Environmental Systems ICES-2015-247 12–16 July 2015, Bellevue, WA. 15 pp.

Hecht, M.H., et al., 2009. Detection of perchlorate and the soluble chemistry of martian soil at the Phoenix lander site. Science 325 (5936), 64–67. https://doi.org/10.1126/science.1172466.

Heinz, J., Schulze-Makuch, D., Kounaves, S.P., 2016. Deliquescence- induced wetting and RSL-like darkening of a Mars analogue soil containing various perchlorate and chloride salts. Geophys. Res. Lett. 43, 4880–4884. https://doi.org/10.1002/2016GL068919.

Hudson, T.L., Aharonson, O., 2008. Diffusion barriers at Mars surface conditions: salt crusts, particle size mixtures, and dust. J. Geophys. Res. 113 (E9), E09008. https://doi.org/10.1029/2007JE003026.

Hudson, T.L., Aharonson, O., Schorghofer, N., Farmer, C.B., Hecht, M.H., Bridges, N.T., 2007. Water vapor diffusion in Mars subsurface environments. J. Geophys. Res. 112 (E5), E05016. https://doi.org/10.1029/2006JE002815.

Inada, A., et al., 2008. Dust haze in Valles Marineris observed by HRSC and OMEGA on board Mars express. J. Geophys. Res. 113, E02004.

Jagoutz, E., 2006. Salt-induced rock fragmentation on Mars: the role of salt in the weathering of Martian rocks. Adv. Space Res. 38 (4), 696–700.

Jänchen, J., Feyh, N., Szewzyk, U., de Vera, J.-P.P., 2016. Provision of water by halite deliquescence for Nostoc Commune biofilms under Mars relevant surface conditions. Int. J. Astrobiol. 15 (2), 107–118.

Jones, M., Vasavada, A., 2016. In: A rover operations protocol for maintaining compliance with planetary protection requirements. 41st COSPAR Scientific Assembly (Cancelled). http://cospar2016.tubitak.gov.tr/en/. Abstract PPP.2–7–16.

Jouannic, G., et al., 2015. Laboratory simulation of debris flows over sand dunes: insights into gully-formation (Mars). Geomorphology 231, 101–115.

King, G.M., 2015. Carbon monoxide as a metabolic energy source for extremely halophilic microbes: implications for microbial activity in Mars regolith. PNAS 112, 4465–4470.

Kirk, R.L., et al., 2008. Ultrahigh resolution topographic mapping of Mars with MRO HiRISE stereo images: meter-scale slopes of candidate Phoenix landing sites. J. Geophys. Res. E: Planets 114. https://doi.org/10.1029/2007JE003000.

Kite, E.S., Mayer, D.P., 2017. Mars sedimentary rock erosion rates constrained using crater counts, with applications to organic-matter preservation and to the global dust cycle. Icarus 286, 212–222.

Kminek, G., et al., 2010. Report of the COSPAR mars special regions colloquium. Adv. Space Res. 46, 811–829.

Kminek, G., et al., 2016. COSPAR panel on planetary protection colloquium, Bern, Switzerland 2015 (Meeting Reports). Space Res. Today 195, 42–51.

Kokelaar, B.P., Bahia, R.S., Joy, K.H., Viroulet, S., Gray, J.M.N.T., 2017. Granular avalanches on the moon: mass-wasting conditions, processes, and features. J. Geophys. Res. Planets 122, 1893–1925. https://doi.org/10.1002/2017JE005320.

Kossacki, K.J., Markiewicz, W.J., 2014. Seasonal flows on dark Martian slopes, thermal condition for liquescence of salts. Icarus 233, 126–130.

Kounaves, S.P., Carrier, B.L., O'Neil, G.D., Stroble, S.T., Claire, M.W., 2014. Evidence of martian perchlorate, chlorate, and nitrate in Mars meteorite EETA79001: implications for oxidants and organics. Icarus 229, 206–213.

Leighton, R.B., Murray, B.C., 1966. Behavior of carbon dioxide and other volatiles on Mars. Science 153, 136–144.

Leung, C.W.S., Rafkin, S.C.R., Stillman, D.E., Chojnacki, M., McEwen, A.S., 2016. Fogs and clouds are a potential indicator of a local water source in Valles Marineris. In: 47th Lunar and Planetary Science Conference, 21–25 March, The Woodlands, TX. 2878.pdf.

Levy, J.S., 2012. Hydrological characteristics of recurrent slope lineae on Mars: evidence for liquid flow through regolith and comparisons with Antarctic terrestrial analogs. Icarus 219, 1–4.

Levy, J.S., Fountain, A.G., Gooseff, M.N., Welch, K.A., Lyons, W.B., 2011. Water tracks and permafrost in Taylor Valley, Antarctica: extensive and shallow groundwater connectivity in a cold desert ecosystem. Geol. Soc. Am. Bull. 123. https://doi.org/10.1130/B30436.1.

Malin, M.C., 1974. Salt weathering on Mars. J. Geophys. Res. 79, 3888–3894.

Malin, M.C., et al., 2007. Context camera investigation on board the Mars Reconnaissance Orbiter. J. Geophys. Res. 112 (E5). https://doi.org/10.1029/2006JE002808.

Martín-Torres, F.J., et al., 2015. Transient liquid water and water activity at Gale crater on Mars. Nat. Geosci. 8, 1–5. https://doi.org/10.1038/ngeo2412.

Martinez, G.M., Renno, N.O., 2013. Water and brines on Mars: current evidence and implications for MSL. Space Sci. Rev. 175, 29–51.

Massé, M., Bourgeois, O., Le Mouélic, S., Verpoorter, C., Spiga, A., Le Deit, L., 2012. Wide distribution and glacial origin of polar gypsum on Mars, Earth Planet. Sci. Lett. 317–318, 44–55. https://doi.org/10.1016/j.epsl.2011.11.035.

Massé, M., et al., 2016. Transport process induced by metastable boiling water under martian surface conditions. Nat. Geosci. 9, 425–428.

McEwen, A.S., et al., 2007. Mars Reconnaissance Orbiter's High Resolution Imaging Science Experiment (HiRISE). J. Geophys. Res. 112, E05S02.

McEwen, A.S., Ojha, L., Dundas, C.M., Mattson, S.S., Byrne, S., Wray, J.J., Cull, S.C., Murchie, S.L., Thomas, N., Gulick, V.C., 2011. Seasonal flows on warm Martian slopes. Science 333 (6043), 740–743. https://doi.org/10.1126/science.1204816.

McEwen, A.S., Dundas, C.M., Mattson, S.S., Toigo, A.D., Ojha, L., Wray, J.J., Chojnacki, M., Byrne, S., Murchie, S.L., Thomas, N., 2014. Recurring slope lineae in equatorial regions of Mars. Nat. Geosci. 7 (1), 53–58. https://doi.org/10.1038/ngeo2014.

Mege, D., et al., 2016. The Highland Terrain Hopper (HOPTER): concept and use cases of a new locomotion system for the exploration of low gravity solar system bodies. Acta Astronaut. 121, 200–220.

Mellon, M.T., Feldman, W.C., Prettyman, T.H., 2004. The presence and stability of ground ice in the southern hemisphere of Mars. Icarus 169, 324–340.

MEPAG NEX-SAG Report, 2015. Report from the Next Orbiter Science Analysis Group (NEX-SAG), Chaired by Campbell, B., Zurek, R., 77 pp., December, 2015. Available at: http://mepag.nasa.gov/reports.cfm.

Mitchell, J.L., Christensen, P.R., 2016. Recurring slope lineae and chlorides on the surface of Mars. J. Geophys. Res. Planets 121, 1411–1428. https://doi.org/10.1002/2016JE005012.

Möhlmann, D.T.F., Thomsen, K., 2011. Properties of cryobrines on Mars. Icarus 212, 123–130.

Möhlmann, D.T.F., Nieman, M., Formisano, V., Savijarvi, H., Wolkenberg, P., 2009. Fog phenomena on Mars. Planet. Space Sci. 57, 1987–1992.

Murchie, S., et al., 2007. Compact reconnaissance imaging spectrometer for Mars (CRISM) on Mars Reconnaissance Orbiter (MRO). J. Geophys. Res. 112, E05S03. https://doi.org/10.1029/2006JE002682.

Nannen, V., et al., 2016. UTOPUS traction technology: a new method for planetary exploration of steep and difficult terrain. In: Proceedings of the ISTVS 8th Americas Regional Conference, Detroit, MI, September 12–14, pp. 1–12.

Navarro-Gonzalez, R., Vargas, E., Rosa, J., Raga, A.C., McKay, C.P., 2010. Reanalysis of the Viking results suggests perchlorate and organics at midlatitudes on Mars. J. Geophys. Res. 115, E12010.

Neuendorf, K.K.E., Mehl Jr., J.P., Jackson, J.A. (Eds.), 2010. The Glossary of Geology, fifth ed. American Geological Institute, Alexandria VA, 800 pp.

Neumann, G.A., Abshire, J.B., Aharonson, O., Garvin, J.B., Sun, X., Zuber, M.T., 2003. Mars Orbiter Laser Altimeter pulse width measurements and footprint-scale roughness. Geophys. Res. Lett. 30, 1561. https://doi.org/10.1029/2003GL017048.

Nuding, D.L., Rivera-Valentin, E.G., Davis, R.D., Gough, R.V., Chevrier, V.F., Tolbert, M.A., 2014. Deliquescence and efflorescence of calcium perchlorate: an investigation of stable aqueous solutions relevant to Mars. Icarus 243, 420–428.

Nuding, D.L., Gough, R.V., Venkateswaran, K.J., Spry, J.A., Tolbert, M.A., 2017. Laboratory investigations on the survival of Bacillus Subtilis spores in deliquescent salt Mars analog environments. Astrobiology 17. https://doi.org/10.1089/ast.2016.1545.

Núñez, J.I., Barnouin, O.S., Murchie, S.L., Seelos, F.P., McGovern, J.A., Seelos, K.D., Buczkowski, D.L., 2016. New insights into gully formation on Mars: constraints from composition as seen by MRO/CRISM. Geophys. Res. Lett. 43 (17), 8893–8902.

Ojha, L., 2016. Geophysical and Remote Sensing Study Of Terrestrial Planets (PhD Dissertation). Georgia Institute of Technology. August, 2016.

Ojha, L., Wray, J.J., Murchie, S.L., McEwen, A.S., Wolff, M.J., Karunatillake, S.S., 2013. Spectral constraints on the formation mechanism of recurring slope lineae. Geophys. Res. Lett. 40. https://doi.org/10.1002/2013GL057893.

Ojha, L., McEwen, A., Dundas, C., Byrne, S., Mattson, S., Wray, J., Masse, M., Schaefer, E., 2014. HiRISE observations of recurring slope lineae (RSL) during southern summer on Mars. Icarus 231, 365–376. https://doi.org/10.1016/j.icarus.2013.12.021.

Ojha, L., Wilhelm, M.B., Murchie, S.L., McEwen, A.S., Wray, J.J., Hanley, J., Massé, M., Chojnacki, M., 2015. Spectral evidence for hydrated salts in recurring slope lineae on Mars. Nat. Geosci. 8, 829–832. https://doi.org/10.1038/NGEO2546.

Ojha, L., Murchie, S.L., Massé, M., McEwen, A.S., Dundas, C., Wray, J.J., Wilhelm, M.B., Hanley, J., 2017a. Oxychlorine Salts in the Northern Mid-Latitudes of Mars: Implications for Ice Stability and Habitability. Icarus (Submitted).

Ojha, L., et al., 2017b. Seasonal slumps in Juventae Chasma, Mars. J. Geophys. Res. Planets 122 (10), 2193–2214.

Oren, A., Bardavid, R.E., Mana, L., 2014. Perchlorate and halophilic prokaryotes: implications for possible halophilic life on Mars. Extremophiles 18, 75–80.

Osterloo, M.M., Anderson, F.S., Hamilton, V.E., Hynek, B.M., 2010. Geologic context of proposed chloride-bearing materials on Mars. J. Geophys. Res. Planets 115, E10012.

Parro, V., et al., 2011. A microbial oasis in the Hypersaline Atacama subsurface discovered by a life detector chip: implications for the search for life on Mars. Astrobiology (10), 969–996.

Pathare, A.V., Feldman, W.C., Prettyman, T.H., Maurice, S., 2018. Driven by excess? Climatic implications of new global mapping of near-surface water-equivalent hydrogen on Mars. Icarus 301, 97–116.

Piqueux, S., Byrne, S., Kieffer, H.H., Titus, T.N., Hansen, C.J., 2015. Enumeration of Mars years and seasons since the beginning of telescopic exploration. Icarus 251, 332–338.

Pommerol, A., Thomas, N., Jost, B., Beck, P., Okubo, C., McEwen, A.S., 2013. Photometric properties of Mars soils analogs. J. Geophys. Res. Planets 118, 2045–2072. https://doi.org/10.1002/jgre.20158.

Putzig, N.E., Mellon, M.T., 2007. Apparent thermal inertia and the surface heterogeneity of Mars. Icarus 191, 68–94.

Rao, M.N., Nyquist, L.E., Wentworth, S.J., Sutton, S.R., Garrison, D.H., 2008. The nature of Martian fluids based on mobile element studies in salt-assemblages from Martian meteorites. J. Geophys. Res. 113 (E6), E06002.

Rennó, N.O., Bos, B.J., Catling, D., Clark, B.C., Drube, L., Fisher, D., Goetz, W., Hviid, S.F., Keller, H.U., Kok, J.F., Kounaves, S.P., Leer, K., Lemmon, M., Madsen, M.B., Markiewicz, W.J., Marshall, J., McKay, C., Mehta, M., Smith, M., Zorzano, M.P., Smith, P.H., Stoker, C., Young, S.M.M., 2009. Possible physical and thermodynamical evidence for liquid water at the Phoenix landing site. J. Geophys. Res. 114 (C10), E00E03.

Rummel, J.D., Conley, C.A., 2017. Four fallacies and an oversight: searching for Martian life. Astrobiology 17. https://doi.org/10.1089/ast.2017.1749.

Rummel, J.D., et al., 2014. A new analysis of Mars "Special Regions": findings of the second MEPAG Special Regions Science Analysis Group (SR-SAG2). Astrobiology 14, 887–968.

Ryan, J.A., Henry, R.M., 1979. Mars atmospheric phenomena during major dust storms, as measured at surface. J. Geophys. Res. 84, 2821–2829.

Satoh, T., et al., 2013. In: Overview of Japan's MELOS1 mission: Mars exploration for life/organism search. International Astrobiology Workshop 2013, Proceedings of the conference Held 28–30 November, 2013 in Sagamihara, Kanagawa, Japan, LPI Contribution No. 1766, p. 1049.

Schaefer, E.I., McEwen, A.S., Mattson, S., Ojha, L., 2015. In: Quantifying the behavior of recurring slope lineae (RSL). 46th Lunar and Planetary Science Conference, Held March 16–20, 2015 in The Woodlands, TX, LPI Contribution No. 1832, p. 293.

Schaefer, E.I., McEwen, A.S., Sutton, S., 2017. In: Recurring slope lineae (RSL) at Tivat crater: part of an assemblage of darkening features? 48th Lunar and Planetary Science Conference. LPI Contribution No. 1964, id.2770.

Schmidt, F., Andrieu, F., Costard, F., Kocifaj, M., Merescescu, A.G., 2017. Formation of recurring slope lineae on Mars by rarefied gas-triggered granular flows. Nat. Geosci. 10, 270–274. https://doi.org/10.1038/ngeo2917.

Schorghofer, N., King, C.M., 2011. Sporadic formation of slope streaks on Mars. Icarus 216 (1), 159–168.

Schorghofer, N., Aharonson, O., Khatiwala, S., 2002. Slope streaks on Mars: correlations with surface properties and the potential role of water. Geophys. Res. Lett. 29 (23), 41-1. CiteID 2126. https://doi.org/10.1029/2002GL015889.

Schuerger, A.C., Ming, D.W., Golden, D.C., 2017. Biotoxicity of Mars soils: 2. Survival of Bacillus subtilis and Enterococcus faecalis in aqueous extracts derived from six Mars analog soils. Icarus 290, 215–223.

Schulze-Makuch, D., Rummel, J.D., Benner, S.A., Levin, G., Parro, V., Kounaves, S., 2015. Nearly forty years after Viking: are we ready for a new life-detection mission? Astrobiology 15, 413–419.

Showman, A.P., Mosqueira, I., Head, J.W., 2004. On the resurfacing of Ganymede by liquid-water volcanism. Icarus 172, 625–640. https://doi.org/10.1016/j.icarus.2004.07.011.

Smith, M.D., 2008. Spacecraft observations of the martian atmosphere. Annu. Rev. Earth Planet. Sci. 36, 191–219.

Smith, P.H., et al., 2009. H_2O at the Phoenix landing site. Science 58, 58–61. https://doi.org/10.1126/science.1172339.

Speyerer, E., Wagner, R., Robinson, M., 2014. In: Automatic identification of changes on the lunar surface. EGU General Assembly 2014, Held 27 April–2 May, 2014 in Vienna, Austria. id.12252.

Stillman, D.E., Grimm, R.E., 2011. Radar penetrates only the youngest geological units on Mars. J. Geophys. Res. 116, E03001. https://doi.org/10.1029/2010JE003661.

Stillman, D.E., Grimm, R.E., 2018. Two pulses of seasonal activity in martian southern mid-latitude recurring slope lineae (RSL). Icarus 302, 126–133.

Stillman, D.E., Michaels, T.I., Grimm, R.E., Harrison, K.P., 2014. New observations of martian southern mid-latitude recurring slope lineae (RSL) imply formation by freshwater subsurface flows. Icarus 233, 328–341.

Stillman, D.E., Michaels, T.I., Grimm, R.E., Hanley, J., 2016. Observations and modeling of northern mid-latitude recurring slope lineae (RSL) suggest recharge by a present-day martian briny aquifer. Icarus 265, 125–138. https://doi.org/10.1016/j.icarus. 2015.10.007.

Stillman, D.E., Michaels, T.I., Grimm, R.E., 2017. Characteristics of the numerous and widespread recurring slope lineae (RSL) in Valles Marineris, Mars. Icarus 285, 195–210.

Takagi, D., McElwaine, J.N., Huppert, H.E., 2011. Shallow granular flows. Phys. Rev. E 83, 031306-1–031306-10.

Tamppari, L.K., et al., 2010. Phoenix and MRO coordinated atmospheric measurements. J. Geophys. Res. 115, E00E17.

Tanner, M.M., Burdick, J.W., Nesnas, I.A.D., 2013. Online motion planning for tethered robots in extreme terrain. In: 2013 IEEE International Conference on Robotics and Automation (ICRA) Karlsruhe, Germany, May 6–10, 2013, pp. 5557–5564.

Thomas, N., et al., 2017. The colour and stereo surface imaging system (CaSSIS) for the ExoMars Trace Gas Orbiter. Space Sci. Rev. 212 (3–4), 1897–1944.

Toner, J.D., Catling, D.C., 2016. Water activities of $NaClO_4$, $Ca(ClO_4)_2$, and $Mg(ClO_4)_2$ brines from experimental heat capacities: water activity >0.6 below 200 K. Geochim. Cosmochim. Acta 181, 164–174.

Tornabene, L.L., Moersch, J.E., McSween, H.Y., McEwen, A.S., Piatek, J.L., Milam, K.A., Christensen, P.R., 2006. Identification of large (2–10 km) rayed craters on Mars in THEMIS thermal infrared images: implications for possible Martian meteorite source regions. J. Geophys. Res. 111 (E10), E10006.

Tosca, N.J., Knoll, A.H., McLennan, S.M., 2008. Water activity and the challenge for life on early Mars. Science 320, 1204–1207.

Tosca, N.J., McLennan, S.M., Lamb, M.P., Grotzinger, J.P., 2011. Physicochemical properties of concentrated Martian surface waters. J. Geophys. Res. 116, E05004.

Travis, B.J., Feldman, W.C., Maurice, S., 2013. A mechanism for bringing ice and brines to the near surface of Mars. J. Geophys. Res. Planets 118, 877–890.

Ulrich, R., Kral, T., Chevrier, V., Pilgrim, R., Roe, L., 2010. Dynamic temperature fields under Mars landing sites and implications for supporting microbial life. Astrobiology 10, 643–650.

Wagner, R.V., Speyerer, E.J., Robinson, M.S., 2017. Exploring the moon with automated feature detection. In: Third Planetary Data Workshop and The Planetary Geologic Mappers Annual Meeting, Held June 12–15, 2017 in Flagstaff, AZ. LPI Contribution No. 1986, id.7074.

Wang, A., Ling, Z.C., Yan, Y.C., McEwen, A.S., Mellon, M.T., Smith, M.D., Jolliff, B.L., Head, J., 2017. Atmosphere-surface H_2O exchange to sustain the recurring slope lineae (RSL) on Mars. In: 48th Lunar and Planetary Science Conference, Held 20–24 March 2017, at The Woodlands, TX. LPI Contribution No. 1964, id.2351.

Watkins, J., Ojha, L., Chojnacki, M., Reith, R., Yin, A., 2014. Structurally controlled subsurface fluid flow as a mechanism for the formation of recurring slope lineae. In: 45th Lunar and Planetary Science Conference, LPI Contribution No. 1777, p. 2911.

Wilson, J.T., Eke, V.R., Massey, R.J., Elphic, R.C., Feldman, W.C., Maurice, S., Teodoro, L.F.A., 2018. Equatorial locations of water on Mars: improved resolution maps based on Mars Odyssey neutron spectrometer data. Icarus 299, 148–160.

Wurm, G., Teiser, J., Reiss, D., 2008. Greenhouse and thermophoretic effects in dust layers: the missing link for lifting of dust on Mars. Geophys. Res. Lett. 35, L10201.

Zurek, R.W., Chicarro, A., Allen, M.A., Bertaux, J.-L., Clancy, R.T., Daerden, F., Formisano, V., Garvin, J.B., Neukum, G., Smith, M.D., 2011. Assessment of a 2016 mission concept: the search for trace gases in the atmosphere of Mars. Planet. Space Sci. 59 (2–3), 284–291.

CHAPTER 11

The NASA Mars 2020 Rover Mission and the Search for Extraterrestrial Life

Kenneth H. Williford, Kenneth A. Farley, Kathryn M. Stack, Abigail C. Allwood, David Beaty, Luther W. Beegle, Rohit Bhartia, Adrian J. Brown, Manuel de la Torre Juarez, Svein-Erik Hamran, Michael H. Hecht, Joel A. Hurowitz, Jose A. Rodriguez-Manfredi, Sylvestre Maurice, Sarah Milkovich, Roger C. Wiens

Contents

From Habitability to Life on Mars
https://doi.org/10.1016/B978-0-12-809935-3.00010-4

11.1 INTRODUCTION

Beginning over 40 years ago with the Viking missions, the National Aeronautics and Space Administration (NASA) strategy for the exploration of Mars has centered on the question of whether that planet ever hosted life. The lack of compelling positive results from the Viking biology experiments (Klein et al., 1976; Klein, 1998), together with an emerging understanding of past and present Mars surface conditions, has shifted the focus from extant to ancient life. The current surface of Mars is extremely inhospitable, whereas habitable conditions were apparently widespread in the distant past when Mars had a denser atmosphere and an abundant surface water (Grotzinger et al., 2014; Arvidson et al., 2014; Arvidson, 2016 and references therein; Wordsworth, 2016). NASA's strategy for Mars exploration has evolved in recent years from a "follow the water" approach by the Mars Exploration Rovers (MER; Squyres et al., 2004), to a search for ancient habitable environments by the Mars Science Laboratory (MSL; Grotzinger et al., 2012), to a search for signs of ancient life by Mars 2020 (MEPAG, 2015). The analogous search for the earliest evidence of life on Earth is fraught with ambiguity (e.g., French et al., 2015; Schopf and Packer, 1987; Schopf, 1993; Brasier et al., 2002, 2015; Schopf and Kudryavstev, 2012), in part due to active geologic processes that—over eons—destroy rocks or alter their records beyond recognition. Although preservation processes have acted differently on Mars where plate tectonics, metamorphism, and modern-day weathering are reduced or absent, the burden of proof for the confirmation of extraterrestrial life will be high. Compelling confirmation of a past martian biosphere may not be possible with in situ (remote robotic) science alone, which motivates collection of samples by Mars 2020 for possible Earth return. Beyond possible paradigm-shifting advances in the field of astrobiology, sample return could revolutionize our understanding of Mars' early geologic, geochemical, and climatic evolution relative to other terrestrial planets and the solar system as a whole. A sample return campaign would also offer compelling demonstrations of key technical capabilities required for eventual human exploration.

This chapter was written and published before the mission's planned launch in 2020, during a time of rapid development and against a backdrop of ongoing science and engineering trades. Although accurate at the time of writing, details herein must be considered in this context—this is a snapshot in time of a mission in development by NASA, the Jet Propulsion Laboratory, and the Mars 2020 project to explore the surface of Mars, seek evidence of ancient life, collect a returnable cache of samples, and prepare for future exploration. Collection of returnable samples by Mars 2020 represents a critical first step toward a potential multimission Mars sample return (MSR) effort. Mars 2020 is the only mission related to MSR that is currently approved or funded, and as such, all references in this document to the overarching sample return effort must be considered conceptual. The National Research Council recommended that the highest priority for large planetary science missions in the decade 2013–22 should be to "initiate the Mars sample return campaign" (NRC, 2011, e.g., p. 258). With this in mind, the Mars 2020 team is developing and will conduct the mission in a way that maximizes the prospects for eventual sample return.

11.1.1 Background and Previous Missions

The possibility of life on Mars has captivated the human imagination for centuries and has been a key scientific driver for NASA's exploration of that planet since the early Mariner missions. The search for evidence of life on the martian surface was a primary mission objective of the twin Viking landers in 1976. The Viking biological investigation was a groundbreaking scientific and technical achievement that continues to inform planetary missions, but the result is generally considered to have been negative (Klein et al., 1976; Klein, 1998). The observation of $^{14}CO_2$ that evolved during the labeled release experiment remained an anomalous and equivocal element of the Viking biological investigation until the detection of perchlorate in martian regolith by the Phoenix lander (Hecht et al., 2009) provided direct evidence supporting an abiotic explanation (Zent and McKay, 1994; Yen et al., 2000; Navarro-González et al., 2010; Quinn et al., 2013) consistent with other Viking results.

With the widely publicized—and highly debated—claim of potential ancient biosignatures in martian meteorite ALH84001 (McKay et al., 1996) and the successful demonstration of the first Mars rover Sojourner by the NASA Pathfinder mission a year later, interest in life on Mars was powerfully reinvigorated. Distinct from that of the Viking era, the emphasis of the new age of Mars surface exploration is a search for evidence of *ancient* life. While it is not impossible that Mars is currently inhabited in the relatively deep subsurface where stable liquid water is possible, Mars was a far more habitable planet in the distant past prior to the loss of its atmosphere and, along with it, widely clement surface conditions. Furthermore, the search for extant life on Mars introduces a number of planetary protection considerations that make the delivery of highly capable scientific payloads (and the engineering systems required to deploy and operate them) extraordinarily challenging. By contrast, the search for geologic evidence of ancient life in areas of Mars currently uninhabitable by known Earth life relaxes planetary protection concerns. As an additional benefit, exploration of ancient environments in search of ancient biosignatures yields valuable scientific knowledge about the long-term evolution of Mars as a planetary system.

The exploration strategy for the new wave of Mars surface missions beginning in 2004—the MER Spirit and Opportunity—was to "follow the water." The requirement for liquid water unites all known life forms, and it was thus reasoned that past liquid water was the primary prerequisite for the past life on Mars. Indeed, MER found multiple lines of evidence confirming the past presence of liquid water including trough cross stratification and diagenetic hematite concretions termed "blueberries" (Squyres et al., 2004) and the iron sulfate mineral jarosite (Klingelhöfer et al., 2004).

MSL *Curiosity* followed 8 years later with a new and larger rover platform carrying a more complex and capable scientific payload. With the intention to explore a Noachian-Hesperian sedimentary sequence at Mount Sharp in Gale Crater, Curiosity sought evidence of habitability beyond the presence of liquid water and continues to explore an environment that presumably records the *loss* of widespread surface habitability that occurred as a result of atmospheric loss and the "great drying" of Mars. Early in the mission, *Curiosity* explored an ancient lacustrine mudstone at Yellowknife Bay and uncovered the most robust and comprehensive evidence yet assembled for a habitable extraterrestrial environment. Work at Yellowknife Bay revealed lithologic and textural evidence for abundant surface water with geochemical and mineralogical evidence of its circumneutral pH and available redox couples to support chemoautotrophic microbial metabolism at the time of deposition (Grotzinger et al., 2014). Among its many other scientific and technical achievements, MSL further advanced the understanding of the evolution of habitability on ancient Mars with the discovery of a record of sustained deposition in a fluvio-deltaic-lacustrine setting (Grotzinger et al., 2015), a deuterium/hydrogen measurement of Hesperian clay hydroxyls in Yellowknife Bay sediments demonstrating deposition prior to complete atmospheric loss (Mahaffy et al., 2015) and direct measurements of the depositional and exposure ages of Yellowknife Bay via the first radiometric dates from the surface of another planet (Farley et al., 2013). MSL has also made the first detections of organic molecules on the surface of Mars in the form of chlorobenzene (150–300 ppb) and C_2-C_4 dichloroalkanes (up to 70 ppb). These molecules are interpreted as reaction products of martian chlorine (e.g., as oxychlorine) with organic matter synthesized on Mars or delivered via meteoritic infall (Freissinet et al., 2015). MSL has made major advances in the study of extraterrestrial habitability, including an exploration model for organic molecules that focuses the search for instances of "scarp retreat" in order to access the most recently exhumed rock that has been protected from destructive, ionizing radiation (Farley et al., 2013). However, the lack of any clear biosignatures observed by *Curiosity*'s extremely capable scientific payload in an environment known to have been habitable underscores the challenges associated with the key objective of Mars 2020 to seek the signs of ancient life.

11.2 MISSION OBJECTIVES

Mars 2020 has an ambitious set of objectives that are derived from the recommendations of the Planetary Decadal Survey (NRC, 2011) and the Mars 2020 Science Definition Team report (Mustard et al., 2013). These objectives build upon the successes of MSL and the earlier Mars surface missions, though Mars 2020 is distinguished by its objective to make progress toward MSR by assembling a cache of scientifically selected samples that could be retrieved and returned to Earth by future missions. Mars 2020 brings an extremely capable payload to the surface that will facilitate sample selection by documenting the geologic and astrobiological context of the landing site, effectively assembling the "field notes" for the sample set. The exploration process required to document field context for a returnable sample set will generate important new scientific discoveries about Mars, past and present. These discoveries will advance planetary science, of course, but they also lay important new groundwork for future human exploration of the surface of Mars.

The Mars 2020 mission objectives are as follows, taken from the Program Level Requirements Appendix (PLRA), the agreement between NASA and JPL establishing the level 1 requirements for the mission:

(A) Characterize the processes that formed and modified the geologic record within a field exploration area on Mars selected for evidence of an astrobiologically relevant ancient environment and geologic diversity.

(B) Perform the following astrobiologically relevant investigations on the geologic materials at the landing site:

    (1) Determine the habitability of an ancient environment.

    (2) For ancient environments interpreted to have been habitable, search for materials with high biosignature preservation potential.

    (3) Search for potential evidence of past life using the observations regarding habitability and preservation as a guide.

(C) Assemble rigorously documented and returnable cached samples for possible future return to Earth:

    (1) Obtain samples that are scientifically selected, for which the field context is documented, that contain the most promising samples identified in objective B and that represent the geologic diversity of the field site.

    (2) Plan for compliance with expected future needs in the areas of planetary protection and engineering so that the cached samples could be returned in the future if NASA chooses to do so.

(D) Contribute to the preparation for human exploration of Mars by making significant progress toward filling at least one major Strategic Knowledge Gap (SKG). The highest priority SKG measurements that are synergistic with Mars 2020 science objectives and compatible with the mission concept are as follows:

    (1) Demonstration of in-situ resource utilization (ISRU) technologies to enable propellant and consumable oxygen production from the martian atmosphere for future exploration missions.

    (2) Characterization of atmospheric dust size and morphology to understand its effects on the operation of surface systems and human health.

    (3) Surface weather measurements to validate global atmospheric models.

(4) A set of engineering sensors embedded in the M2020 heat shield and backshell to gather data on the aerothermal conditions, thermal protection system, and aerodynamic performance characteristics of the M2020 entry vehicle during its entry and descent to the Mars surface.

The primary scientific question that integrates science objectives A–C is whether Mars was ever inhabited. In situ exploration using the scientific payload immediately provides new insights for Mars geology and planetary science (objectives A and B) while guiding the selection of and providing the critical scientific context for a cache of samples (objective C) that has the potential to revolutionize our understanding of life, terrestrial planets, and evolution of the solar system. The mission also supports NASA's goal to send humans to Mars in the coming decades: the more that is known about the martian environment, the better prepared humankind will be to send people to the surface and return them safely to Earth.

11.3 MISSION OVERVIEW AND COMPARISON TO MSL

The Mars 2020 mission began development in early 2013, just a few months after the MSL *Curiosity* rover's spectacular landing success at Gale Crater. Mars 2020 builds directly on MSL, relying heavily on its design and verification data and in many cases using flight-spare hardware. This "heritage" approach reduces new engineering development and associated uncertainty and is expected to yield as a dividend a substantially lower overall cost and shorter development cycle.

The major technical advances required for Mars 2020 to accomplish its goals of seeking the signs of ancient life and to prepare samples for possible Earth return include a new suite of seven scientific instruments and a sophisticated robotic system to collect, seal, and cache samples. Both are described more fully in the following sections. Mars 2020 will also be outfitted with strengthened wheels to eliminate the damage experienced by *Curiosity* as it drove across rock-strewn martian surfaces (Arvidson et al., 2017).

Mars 2020 will launch from Cape Canaveral sometime within a ~30-day window in July and August of 2020 (hence the current mission name; the rover will likely be renamed before launch). It will arrive at Mars after a 7.5-month cruise in February of 2021. The fundamental design of the cruise and the entry, descent, and landing (EDL) systems remain unchanged from MSL, but a microphone and new high-definition cameras on the rover and the descent stage will acquire unique sound and video to document the EDL process.

Mars 2020 will have a prime mission duration of at least 1 Mars year or approximately 2 Earth years. Due to the expected challenges of accomplishing a robust program of in situ exploration comparable with MSL while also assembling a cache of samples during the prime mission, the project has elected to qualify flight hardware for an additional half Mars year (i.e., beyond the qualification for MSL). Like MSL, the Mars 2020 project includes a multimission radioisotope thermoelectric generator (MMRTG) that is expected to yield sufficient power for rover operations substantially exceeding the length of the primary mission.

11.4 SCIENCE PAYLOAD AND IN SITU INVESTIGATIONS

The science payload was carefully selected to support the mission's science goals and represents a mixture of fundamentally new instruments and enhanced versions of instruments on board *Curiosity*. Sensing units for two of the instruments are mounted on the Remote Sensing Mast (Mastcam-Z and SuperCam), and two are mounted to the turret located at the end of the robotic arm [Planetary Instrument for X-ray Lithochemistry (PIXL) and scanning habitable environments with raman and luminescence for organics and chemicals (SHERLOC)]. The Radar Imager for Mars' Subsurface Experiment (RIMFAX) antenna is mounted on the underside of the rover, while Mars Environmental Dynamics Analyzer (MEDA) has sensors distributed around the rover. The Mars Oxygen In-Situ Resource Utilization Experiment (MOXIE) is located inside the rover, as are the most temperature-sensitive electronics components of the other instruments. Configuration of instruments and other key components on the rover is shown in Fig. 11.1.

11.4.1 Mastcam-Z

Mastcam-Z features a stereo pair of multispectral, color zoom cameras. Maximum image size is 1600×1200 pixels ($\sim$2 megapixels), and maximum pixel scale at 2 m is about

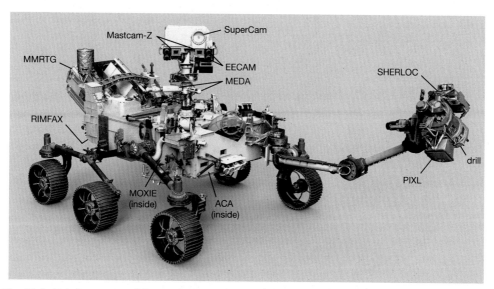

Fig. 11.1 Artist's concept of the Mars 2020 rover with key elements indicated as discussed in the text.

150 μm per pixel (enabling features of ~0.5 mm to be distinguished). Mastcam-Z shares strong heritage with the MSL Mastcam instrument, but includes new multispectral filters and a new zoom capability that significantly enhances stereo imaging flexibility. "Wide-angle" Mastcam-Z stereo panoramic images are crucial to understand geologic structures in the exploration environment, and "telephoto" images of individual targets and the robotic arm workspace immediately in front of the rover are key data products to evaluate lithology and support scientific targeting by other instruments.

11.4.2 RIMFAX

The RIMFAX, contributed by Norway, is a ground-penetrating radar with a frequency range of 150–1200 MHz and a penetration depth >10 m depending on surface materials. Although ground-penetrating radars have orbited Mars (MARSIS on Mars Express and SHARAD on Mars Reconnaissance Orbiter), RIMFAX is a completely new technology for Mars surface missions. Ground-penetrating radar will enable the science team to trace geologic structures observed on the surface into the subsurface, enhancing reconstructions of local and regional stratigraphy. RIMFAX also has the capability to detect water and ice in the subsurface, although landing sites with orbital evidence for ice or liquid water within 5 m of the surface will be avoided for planetary protection reasons.

11.4.3 SuperCam

SuperCam builds upon the strong heritage and scientific success of the MSL ChemCam instrument and is developed primarily in partnership between the United States and France. SuperCam is a remote sensing instrument comprising a remote micro imager (RMI) and a laser-induced breakdown spectroscopy (LIBS) similar to MSL ChemCam (Wiens et al., 2012, 2015, 2017; Maurice et al., 2012, 2016) but now with color RMI. New capabilities for SuperCam include visible-infrared (VISIR) and remote Raman spectroscopy. The latter technique uses a pulsed laser at 532 nm (green Raman) to interrogate samples over a very short period of time (~5 ns) and uses a time-gated intensified detector to minimize interference from ambient light and fluorescence. SuperCam's VISIR spectroscopy consists of an infrared (IR) spectrometer that covers the 1.3–2.6 μm range and the 0.4–0.85 μm range with the spectrometers used for LIBS and Raman spectroscopy. SuperCam includes a scientific microphone that will remotely determine physical properties of the targets from the acoustic signal of the (LIBS) laser interaction with the surface and can also be used for characterization of atmospheric turbulence (wind gusts and dust devils).

Coboresighted VISIR and remote Raman spectroscopies enable a more balanced mineral characterization and more confident identification than either one technique alone. For example, Raman spectroscopy easily identifies feldspar minerals, which are difficult (at best) to identify with VISIR, while VISIR spectroscopy may offer better sensitivity than Raman for certain minerals such as hydrated silicates. These two techniques together can observe silicates, carbonates, sulfates, phosphates, sulfides, and organic molecules. The combination of both elemental and mineral signatures provides strong synergy in characterizing any target. Like ChemCam, elemental compositions are obtained for all major elements to high precision, and minor and trace elements including Li, B, C, N, Cr, V, Mn, Ni, Cu, Rb, Sr, and Ba will be detected and quantified (Wiens et al., 2015; Maurice et al., 2016). An added advantage to all techniques is that the LIBS analysis removes surface dust, enabling a clear view of the surfaces.

The SuperCam mast unit contains a Nd:YAG laser that provides both LIBS (1064 nm) and Raman (532 nm) interrogation, a telescope to project the laser light and collect the light for all techniques, the RMI camera, and the IR spectrometer, microphone, and electronics. Light for the LIBS, the Raman, and the VIS portion of the VISIR spectra is transferred to the rover body by a fiber optic cable. The body unit contains three spectrometers to cover these observations, along with electronics to control the instrument and communicate with the rover. The mast unit is provided by the French Space Agency (CNES), while the body unit is built by Los Alamos National Laboratory under contract with NASA. An extensive calibration target assembly with some 28 targets— including mineral separates, rock, and glasses that have been crushed and flash sintered for homogeneity (Cousin et al., 2017)—is mounted on the back of the rover. The assembly is a Spanish contribution, while the targets themselves are a multinational effort involving France, Denmark, Canada, Spain, and the United States.

With LIBS capability extending to a distance of ~7 m, Raman to ~12 m, VISIR up to 10 km, and RMI to infinity, SuperCam offers a diverse toolkit to guide exploration at a distance. Analytic spot sizes vary by technique and with distance but at 2 m are expected to be ~400 μm for LIBS, 1.4 mm for Raman, and 2.2 mm for IR. An IR "long raster" mode, with SuperCam rapidly analyzing ~100 locations across a transect, will reveal mineralogical distinctions at the outcrop scale. A single-spot, submillimeter-scale analytic resolution at a distance of ~2 m in the rover workspace enables rapid interrogation of individual grains or clasts, matrices, veins, and alteration rinds—key components to reconstruct formation and alteration processes in the exploration environment. Using fine-scale targeting, SuperCam will be able to analyze compositions within the drill hole itself, which will be an important feature for understanding the nature of the samples that could later be returned to Earth.

11.4.4 PIXL

The PIXL is a robotic arm-mounted X-ray fluorescence (XRF) spectrometer and microcontext camera system that will be used to measure the elemental chemistry of rocks and soils and to map chemical variations in relation to visible fine-scale textures and microstructures. PIXL follows the highly successful alpha particle X-ray spectrometer (APXS) instruments on previous Mars rovers that measured the average elemental composition of rocks and soils over an area of several square centimeters. With the capability to analyze elemental lithochemistry with spatial sampling of $\sim$100 μm, PIXL will allow scientists to accurately link chemical variations to small-scale features such as individual grains, laminae, and interstitial cements.

The turret-mounted PIXL sensor head uses a 28 kV X-ray tube and polycapillary X-ray focusing optic to produce a $\sim$100 μm–diameter, high-flux X-ray beam on a target 25 mm away. The high-intensity beam causes X-ray fluorescence that is measured with dual silicon drift detectors, enabling measurement of major and minor elements in as little as 10 s. The sensor head is mounted on a hexapod motion system that can be used to precisely scan the instrument in three dimensions. Using the hexapod to raster the beam across the surface of a rock, PIXL can acquire square-centimeter-scale elemental maps with submillimeter-scale spatial resolution in a matter of hours. For rapid analyses, line and grid scans can be performed. Scans can be performed on both flat surfaces and uneven surfaces with several millimeters or more of surface topography. However, maps are best acquired on flat surfaces. A $\sim$45 mm–diameter abrasion tool and dust removal tool on the rover turret will create a flat surface where desired and provide access to less weathered rock.

PIXL's element data can be used in multiple ways to gain detailed insights to the nature of geologic materials encountered. Examples include the following:

(1) Use spatial covariations of elements to constrain mineralogy.
(2) Use element maps to recognize textures or microstructures that aren't apparent in visual images, for example, due to tool markings from the interaction of the abrading bit with the rock surface.
(3) Determine the chemical makeup of individual features such as sedimentary grains, igneous crystals, laminae, veinlets, and interstitial cements.
(4) Identify spatial variations in the relative abundance of elements to recognize and characterize alteration gradients, alteration rims on grains, zoned cements, and crystals.
(5) Sum spectra together to determine the bulk rock chemistry of a scanned area.

Thus, in addition to providing the bulk rock chemical analyses to which Mars scientists have become accustomed on past rover missions, PIXL will also enable true petrologic analysis of martian rocks.

11.4.5 SHERLOC

The SHERLOC instrument is a noncontact robotic arm–mounted spectrometer with two imaging systems (Beegle et al., 2015). SHERLOC provides high-spatial-resolution (15 μm/pixel) imaging coregistered with hyperspectral maps of mineral and organic composition acquired over approximately square centimeter areas using a $\sim$100 μm beam diameter.

SHERLOC's 248.6 nm deep ultraviolet (DUV) scanning laser generates fluorescence emission from aromatic organics and DUV resonance Raman scattering from aliphatic and aromatic organics and astrobiologically relevant minerals. When using a DUV light source, the weaker Raman scattering takes place in a wavelength range (253–274 nm) that is outside the otherwise obscuring fluorescence range that starts beyond 270 nm and extends into the visible. This enables detection of both Raman scattered photons and stimulated fluorescence emission on the same CCD. The SHERLOC CCD is identical to those used for MSL ChemCam and Mars 2020 SuperCam (Wiens et al., 2012). Detectable organic functional groups include the C—H, CN, C=O, and C=C bonds, and detectable mineral species include carbonates, perchlorates, sulfates, and phyllosilicates. Expected detection limits are as low as 10^{-6} (w/w) for aromatic organics and 10^{-3} for aliphatic organics, and mineral spectra from grains as small as 50 μm can feasibly be recognized.

SHERLOC employs two imaging systems: the Wide Angle Topographic Sensor for Operations and Engineering (WATSON) and the autofocus and context imager (ACI). WATSON is a reflight of the MSL Mars Hand Lens Imager (MAHLI) camera head (Edgett et al., 2012). WATSON will provide color images with a spatial resolution sufficient (at closest approach) to distinguish sand from silt and smaller-sized grains—an important sedimentologic distinction in the evaluation of habitable environments and potential biosignature preservation. Like MAHLI, WATSON also images rover hardware for engineering capabilities including monitoring of wheel wear and produces rover "selfies" for education and public engagement. The ACI enables focusing of the DUV laser for standoff analysis from a distance of 48 ± 7 mm above a target and is coboresighted with the laser scanning system to document the visible context for fluorescence and Raman maps.

SHERLOC DUV fluorescence and Raman maps will be coregistered with WATSON images and PIXL maps (see above), providing a depth of petrographic information about martian materials previously only available from analyses of meteorites in Earth-based laboratories. Such coordinated, grain-scale observations of elemental, mineral, and organic compositions in their textural and stratigraphic context powerfully support mission objectives to document rock formation and alteration processes, evaluate habitability, seek signs of ancient life, and select sampling targets with high potential to preserve signs of life and planetary evolution.

11.4.6 MEDA

The MEDA is contributed by Spain, led by the Centro de Astrobiologia (CAB), with support for the US investigation team members from the NASA Science Mission and Human Exploration and Operations Mission Directorates. MEDA is an integrated suite of sensors providing in situ near-surface weather measurements and dust characterization.

MEDA is an evolution of the MSL Rover Environmental Monitoring Station (REMS; Gómez-Elvira et al., 2012) and PanCam/HazCam (Bell et al., 2003). It will measure wind speed and direction, atmospheric pressure, air and ground temperature, relative humidity, radiation fluxes near the surface of Mars, and optical properties of dust. These environmental measurements will inform Mars climate models and support eventual human exploration.

MEDA includes a pressure sensor that shares direct heritage with REMS. The relative humidity sensor, like the pressure sensor, is contributed by the Finnish Meteorological Institute and incorporates a new sensor membrane for increased dynamic range. The wind sensor emerges from a collaboration between the CAB; the Universitat Politècnica de Catalunya (UPC); the Seville Institute for Microelectronics (IMSE); and the Computadoras, Redes e Ingeniería (CRISA). Details of the sensor design have been changed to increase mechanical robustness and consume less power, and the number of hot plates has been doubled to increase functional redundancy.

The air temperature sensor (ATS), developed by CAB, consists of five sets of triple thin wire thermocouple sensors based on the concept used by Mars Pathfinder and Viking. Three ATS sensors are located around the rover mast at around 278 mm height from the rover deck (1458 mm from the ground) and 120° from each other to ensure that one sensor is always upwind of the mast. The other two sets of thermocouples are positioned at the sides of the rover and about 880 mm above the ground. This vertical distribution of the ATS will help characterize the temperature gradient on the surface layer above the regolith.

The Thermal Infrared Sensor (TIRS), by CAB and CRISA, has upward and downward pointing thermopiles using similar detectors as on REMS but sampling different wavelength ranges. In addition to surface temperature, the TIRS will measure downward and upward radiation in the 6.5–30 μm range to document net radiative TIR fluxes forcing the atmosphere at the surface of Mars. It also provides atmospheric temperature of the low layer above the rover location averaged over a few tens of meters, constraining vertical temperature gradients that drive convection.

The radiation and dust sensor (RDS), developed by the Instituto Nacional de Tecnica Aeroespacial (INTA), represents an evolution of two sensors previously flown to Mars. The first sensor, Skycam, is a JPL-provided CCD camera that inherits the electronics of the MER HazCams and incorporates a new lens to minimize internal reflections and stray light. Skycam also includes a neutral density filter to enable direct observation of the sun with minimal blooming and a ring mask to compare direct with scattered sunlight. Skycam will measure the rate of irradiance decay as a function of the distance to the sun disc, and its comparison to radiative transfer models will give a constraint on atmospheric particle-size distribution. Side-looking photodiodes on the RDS will measure scattering properties related to particle shape. Uplooking photodiodes will characterize optical opacity of the atmosphere as a function of wavelength from the UV to the near IR.

11.4.7 MOXIE

The MOXIE is a ~1% scale model of an oxygen processing plant that is intended to support a human expedition sometime in the 2030s. A technology demonstrator, MOXIE ingests the thin CO_2 that comprises 96% of the martian air and produces O_2 as a primary product. On a future mission, such a process could produce ~30t of liquid oxygen for ascent vehicle propellant in the 16months preceding launch of a human crew to Mars, representing ~78% of the propellant mass needed for a CH_4/O_2 propulsion system. To bring this amount of oxygen from Earth would otherwise require four to five heavy lift launches (Drake, 2007).

MOXIE's solid oxide electrolysis (SOXE) stack for converting CO_2 to O_2 is designed and built by Ceramatec, Inc. (Hartvigsen et al., 2015). Its working elements are stacked, scandia-stabilized zirconia electrolyte-supported cells with thin screen-printed electrodes, coated with a catalytic cathode on one side and an anode on the other. When CO_2 flows over the catalyzed cathode surface at ~800°C under an applied electric potential, it is electrolyzed according to the reaction $CO_2 + 2e^- \Rightarrow CO + O^-$. The resulting oxygen ions are electrochemically drawn through the solid oxide electrolyte to the anode, where they are oxidized ($O^- \Rightarrow O + 2e^-$) to produce gaseous O_2. A scroll pump developed by Air Squared, Inc. collects and compresses the CO_2 for the reaction.

In development at the Jet Propulsion Laboratory, MOXIE is expected to produce ~10 g/h of O_2 on Mars with >99.6% purity (Hoffman et al., 2015). Oxygen production is expected to be limited both by the compressor capacity and by the external conditions that determine the density and quantity of air that can be drawn in. On Mars, autonomous MOXIE operations will be optimized, and degradation mechanisms will be studied.

11.4.8 Returned Sample Science

A unique aspect of Mars 2020 is the inclusion of returned sample science (RSS) as a distinct investigation. RSS is not associated with any one instrument, but rather is concerned with maximizing the scientific value of samples to be collected during the surface mission. The reasons why a set of martian samples would be valuable if returned to Earth have been described in multiple reports in the literature, and the interested reader is referred to Mustard et al. (2013), E2E-iSAG (McLennan et al., 2011), and NRC (2011) and references therein. Of particular significance is E2E-iSAG (McLennan et al., 2011), which describes specific scientific objectives for a sample return enterprise and translates that into terms that are useful for a project development team, such as how many samples are needed, the physical attributes (e.g., mass, volume, and mechanical integrity) needed in order to support the requisite measurements as currently envisioned, and how the collection could be organized into suites.

To represent the needs and interests of the future sample analysis community during the development phase of the Mars 2020 project, a "Returned Sample Science Board" (RSSB) comprising 14 scientists was selected by NASA from a pool of applicants. As part of the Mars 2020 science team, the RSSB has participated in landing site selection discussions, consulted on key science-engineering trades (e.g., RSSB, 2016) at the request of Mars 2020 project science, and is represented along with the selected payloads on the Project Science Group (PSG). Many of the questions put to the RSSB during the period 2015–17 related to the degree of importance of certain factors that relate to sample quality, such as temperature, contamination, and fracturing. In addition, the RSSB contributed to the development of the project's contamination knowledge strategy (Farley et al., 2017).

The term of the RSSB extends until the participating scientist selection process, projected to complete prior to launch. Some participating scientists will be selected specifically for their expertise in returned sample science, and these individuals will join the Mars 2020 science team in an integrated approach to scientific decision making in support of in situ exploration and returned sample science during the surface mission.

11.5 SAMPLING AND CACHING SYSTEM

The major new development for the Mars 2020 flight system is the Sampling and Caching System (SCS). The Mars 2020 SCS represents an unprecedented set of technical challenges and is critical to the success of the Mars 2020 mission and to the progress toward MSR. Although a number of previous missions have had surface preparation and sample collection capabilities—for example, MSL *Curiosity* has an advanced drilling and sample handling system (Anderson et al., 2012)—the Mars 2020 SCS is a fundamentally new design that enables surface preparation and core acquisition, as well as sealing, onboard storage, and release of sample tubes to the surface, all bound by a set of organic, inorganic, and biological cleanliness requirements more stringent than those for any prior planetary mission. Key elements of the Mars 2020 sampling and caching system are shown in Fig. 11.2.

11.5.1 Sample-Related Requirements

Mars 2020 is designed to meet an ambitious and stringent set of scientific requirements on the samples to be cached (Table 11.1). The mission will be capable of acquiring at least 31 samples, each consisting of about 15 g of rock or regolith. The desired mass of each sample was derived from estimates of how much mass is likely required for analysis and archiving of returned samples (McLennan et al., 2011), while the total number of samples is dictated by the duration of Mars 2020's surface investigation and notional lift capabilities of a return mission. Each rock sample will consist of a cylindrical core about 1 cm in diameter from the uppermost ~5 cm of the rock and will be drilled directly into an individual ultraclean sample tube. Unconsolidated material such as regolith will be collected through a specialized sampling bit that allows particles to flow by gravity into the sample tube.

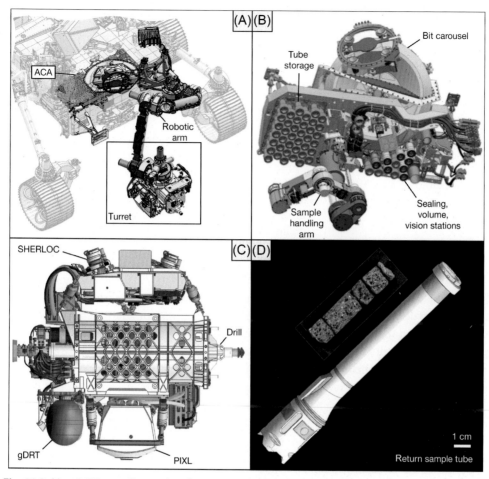

Fig. 11.2 Mars 2020 sampling and caching system, including a front view of the rover (A) with turret, robotic arm, and adaptive caching assembly (ACA) indicated; a view of the ACA from below (B); another view of the turret (C) showing PIXL, SHERLOC, and drill (one stabilizer shown; bit is retracted); and a drawing of a sample tube (D) with a representative test core shown to scale.

Scientific integrity requirements exist to minimize rock fragmentation that would preclude identification of structures (especially biosignatures), to restrict sample temperature to <60°C during drilling and storage on Mars, and to minimize volatile loss via hermetic sealing of each tube immediately after acquisition. The most demanding requirements on sample integrity are those that limit terrestrial contamination. These include a requirement of absolute sterility (less than one viable organism per sample tube), restrictions on the abundance of many elements critical for geochemical study, and extraordinarily stringent requirements on organic contamination.

Table 11.1 Requirements on the samples to be prepared for caching by the Mars 2020 mission

Category	Requirement
Number of samples	Capable of at least 31 in total, with 20 in the primary mission
Sample mass (each)	10–15 g cylindrical cores
Contamination limits	
Inorganic	Limits on 21 key elements based on typical concentrations in martian meteorites
Organic	<10 ppb total organic carbon
	<1 ppb of 10 critical marker compounds
Biologic	<1 viable terrestrial organism per sample
Drilling and storage temperature	<60°C at all times including during depot on Mars surface
Individual sample tube sealing	Hermetic (to prevent volatile loss and contamination)
Sample disaggregation	Maintain large pieces to allow structure investigations during drilling, storage, and possible Earth return

Separate from sample cleanliness requirements, the Mars 2020 mission will acquire detailed contamination knowledge covering all relevant stages of the mission. This knowledge will be critical for confirming the martian origin of any potential biosignatures detected in the returned samples. Prior to spacecraft assembly, witness plates and spacecraft swabs will be used to characterize vapor-deposited and particulate organic and biological contamination. This includes a thorough genetic inventory of potential microbial contaminants in relevant development environments. Also during this phase, lot-identical spacecraft components, including a planned "contamination model" of the drilling system, will be archived for future analysis.

Once the spacecraft is assembled, it is no longer possible to directly measure Earth-sourced contamination until samples are returned, so Mars 2020 will feature a system designed to accumulate contaminants during Mars surface operations and the roundtrip flight. Six sample tubes ("witness tubes") will be modified to carry a witness plate assembly (WPA) featuring ultraclean surfaces and meshes that will trap molecular and particulate contaminants. The witness tubes will be stored and processed identically to sample-containing tubes, with the exception of bit-on-rock contact. By carrying six such tubes, it will be possible for the science team to devise appropriate strategies for when to expose and seal these tubes as the contamination environment aboard the rover evolves.

If specific organic compounds or organisms are someday detected in returned samples, a comparison between the preflight contaminant analyses and the witness tube measurements can be made to assess whether there is evidence for an Earth source. If necessary, the contamination model of the drill can be used to assess the role of bit-on-rock contact in sourcing and possibly modifying organic compounds. In the end, it may be necessary to make spatially resolved measurements (i.e., compare interior portions of sample with exterior portions) to compellingly confirm a martian source of potential biosignatures.

11.5.2 Design of the Sampling and Caching System

The Mars 2020 sampling and caching system (SCS) is a dual-purpose subsystem supporting both in situ exploration and sample acquisition and storage. Features of the SCS that support sampling include a rotary-percussive drill similar to MSL but modified to generate cores rather than cuttings, a rotating bit carousel containing multiple bits for rock coring and regolith collection, and an adaptive caching assembly (ACA) with its own robotic sample handling arm (SHA) enabling filled sample or witness tubes to be photographed, sealed, stored, and eventually placed on the surface of Mars. The SCS supports in situ science investigations with a gas-based dust removal tool (gDRT) and drill bits designed for surface abrasion. The abrading bit is used to prepare a ∼4 cm–diameter patch of rock to facilitate proximity science including imaging and mapping by the PIXL and SHERLOC instruments. The gDRT uses a high-velocity jet of nitrogen gas to remove dust and abrasion fines that might otherwise obscure surface targets.

To collect a rock sample, the SHA inserts an empty sample tube into a coring bit, and this bit is rotated into docking position for insertion into the drill. The drill containing the coring bit is placed onto the desired target and preloaded, a "hole start" routine is run, and coring continues in either rotary-percussive or rotary-only mode, depending upon the mechanical characteristics of the rock. An eccentric race at the end of the sample tube facilitates core breakoff, the drill is removed from the borehole, and the robotic arm docks again to the bit carousel, which removes the bit and rotates, presenting the rear portion of the bit to the ACA. The filled sample tube is removed from the bit by the SHA and taken through a number of stations including imaging, volume estimation, and sealing prior to storage in its original location.

11.5.3 Sample Caching and Potential Return

In order to maximize flexibility and minimize risk, Mars 2020 has adopted an approach called "adaptive" caching by which multiple sealed sample tubes will be dropped by the ACA onto the surface of Mars (e.g., Beaty et al., 2015). This is in contrast to an approach in which all samples are stored on board the rover in a single container until a decision is made to transfer the container—the monolithic cache—on to the martian surface. Evaluation of the relative benefits of these two basic caching architectures dates back at least as far as early MSR concepts emerging in the years following Mars Pathfinder. The main advantage of the adaptive approach adopted by Mars 2020 is that it allows the mission to offload its sample cargo at opportune times, thereby eliminating the possibility that all of the samples (e.g., in a monolithic cache) get stranded in a disabled rover. Adaptive caching also permits the potential return mission to select the highest value samples for return rather than to take all samples in the monolithic cache.

The current notional strategy is that multiple samples will be deposited at a "depot" site selected to optimize relocation by the retrieval mission and where dust accumulation is expected to be minimal. A likely scenario for depot caching is as follows. A first region of interest (ROI 1) is explored, approximately half (~10) of prime mission samples are collected, and the rover drives to ROI 2 with samples stored on board. Prior to exploring ROI 2, a suitable depot location is identified nearby, and the samples from ROI 1 are dropped in a collection. This offloads the risk associated with further transport of the sample cargo. ROI 2 is explored, ~10 more samples are collected, the rover returns to the depot, and the samples from ROI 2 are added to the initial collection. At this point, the team is free to pursue higher risk exploration near ROI 2 or at a third ROI some distance away. Once more, samples are collected; the team will face the decision of whether to return to the first depot or to establish a new depot. An obvious risk of establishing multiple depots is that the retrieval mission may not be designed to traverse among far-flung depots.

Because follow-on sample return missions are not yet confirmed, the length of time the samples collected by Mars 2020 must retain their integrity on the surface is unknown. Mars 2020 is verifying that sample tubes and seals will last at least 10 years on the surface of Mars and 10 years in Mars orbit.

11.6 EXTANT LIFE AND PLANETARY PROTECTION

The combination of low and widely varying temperature, low atmospheric pressure, low water activity, and high ultraviolet and ionizing radiation environment at the surface of modern Mars is extremely inhospitable for life. Geologic evidence and models of planetary evolution suggest that similarly inhospitable conditions have prevailed for most of Mars history—probably for >3 billion years (3 Ga; Carr and Head, 2010)—and martian life has not been detected by any previous mission. It thus seems unlikely, but not impossible, that present-day Mars supports a biosphere. If martian life does exist today, it would likely be confined to subsurface environments that are protected from deleterious radiation and where liquid water is stable. These caveats notwithstanding, the precautionary principle has led to policies for planetary protection designed to limit "forward" contamination of extraterrestrial environments by Earth organisms and "backward" contamination of Earth by extraterrestrial organisms.

Samples collected by Mars 2020 will be sealed in tubes and placed on the surface of Mars where their exteriors will be in contact with the martian environment (and possible martian organisms). Further, containment of samples in such a way that Earth is protected from accidental release of a possible martian organism would be the responsibility of follow-on missions. The Mars 2020 project is working with future mission planners to design a system that would enable future missions to meet planetary protection responsibilities and will be archiving information about Mars 2020 that future missions would need to demonstrate that samples are safe to return.

Surface environments on Mars "within which terrestrial organisms are likely to replicate" (e.g., with water activity between 0.5 and 1.0 and temperature above − 25°C) have been designated "special" or "uncertain regions" (Rummel et al., 2002, 2014; Committee to Review the MEPAG Report on Mars Special Regions et al., 2015; COSPAR, 2015). Mars 2020 will not target special regions because its goal is to look instead at the potentially far more habitable environments of pre-3 Ga Mars. As was the case for MSL, the Mars 2020 strategy to prevent forward contamination responds to NASA policy for Mars surface missions that do not access special regions. MSL met the biological cleanliness (or "bioburden") requirements to prevent forward contamination with significant margin (Benardini et al., 2014). Mars 2020 will use a heritage approach with limited improvements to ensure compliance with high confidence.

11.7 LANDING SITE SELECTION

Following previous practice (e.g., Golombek et al., 2012), the selection of a landing site for Mars 2020 will occur late in mission development to ensure that the greatest amount of orbital data has been acquired before a decision is made. Previous landing site selection efforts, additional orbital data acquired since the MSL site selection process, and heritage engineering approach used for Mars 2020 have enabled more thorough and rapid scientific and engineering safety analyses of landing sites than ever before. Landing site selection criteria are different for Mars 2020 than for previous rover missions. With objectives to seek evidence of ancient life and prepare a cache of samples, strong evidence for habitability and conditions conducive to biosignature preservation in the depositional environment and lithologic diversity assumes new importance.

The bulk of the record of life on Earth derives from sedimentary rocks deposited in relatively shallow, subaqueous settings. Experience studying the Earth's sedimentary rock record also provides a clear and robust model to guide the search for biosignatures in fluvio-deltaic or lacustrine environments on Mars. For these reasons, many in the Mars science community favor a landing site with clear evidence for large, standing bodies of water, for example, geomorphic evidence for a delta, often accompanied by mineralogical evidence for hydrous minerals.

However, questions remain whether habitable environments present at the Mars surface subsisted long enough to allow life to gain a recognizable foothold. Current thinking suggests that life on Earth emerged among the diverse and abundant chemical disequilibria (including abiotic organic synthesis) that result from energetic water-rock interaction in subsurface hydrothermal systems (e.g., Russell et al., 2014). The great significance of Earth's "deep biosphere" (organisms living >1 m below the surface; Edwards et al., 2012) has been appreciated only relatively recently (e.g., Kallmeyer et al., 2012). While there is nothing to suggest that the deep biosphere is a recent phenomenon, very little is known about its distribution and extent throughout Earth's history—and of particular relevance to the Mars 2020 mission—and how it may be preserved in the rock record. The influence of a magnetic field on the habitability of surface environments on Mars is also unknown. Most sites considered for Mars 2020 appear to have been deposited after the magnetic field was lost. For these reasons, the most ancient (early Noachian) landing sites featuring evidence for hydrothermal activity enjoy strong community support.

In addition to past habitability, the ideal Mars 2020 landing site would feature geologic diversity adequate to satisfy the needs and desires of the community of scientists who would someday analyze returned samples. There is a great diversity of investigations to which returned samples are likely to be subjected, and many investigations have specific target lithologies. For example, there is strong and obvious interest in analyzing igneous rocks to better understand the nature and timing of early planetary differentiation and in sedimentary rocks that may carry a record of ancient climatic evolution. Also of interest are rocks that may document an early martian magnetic field and those that would provide a radiometrically determined age of a laterally extensive cratered surface to test crater chronology models. Locating a single site that meets *all* of these objectives is unlikely, and prioritization will almost certainly be required.

As of this writing, three Mars 2020 landing site workshops have been held, and the list of possible landing sites has been narrowed to three: Columbia Hills, Jezero Crater, and Northeast Syrtis. Leading hypotheses regarding the geologic origin and potential habitability have been developed for each site. A delta at Jezero Crater samples a lithologically diverse watershed and provides evidence for subaqueous sediment deposition in a crater-filling lake (Fassett and Head, 2005; Ehlmann et al., 2008; Schon et al., 2012; Goudge et al., 2015, 2017). Among a diversity of igneous rocks observed by the Spirit rover near the Columbia Hills in Gusev Crater is a silica deposit interpreted to be the preserved remnants of a surface hydrothermal system (Ruff et al., 2011; Schmidt et al., 2008; Squyres et al., 2008) that could contain biosignatures (Ruff and Farmer, 2016). Northeast Syrtis exposes extremely ancient, early Noachian crust without clear geomorphic evidence for persistent surface water but with mineralogical indicators consistent with past water-rock interaction across a range of temperatures, including phyllosilicates and alteration of olivine to Mg carbonate and serpentine (Bramble et al., 2017; Ehlmann et al., 2008, 2009; Ehlmann and Mustard, 2012) or talc carbonate (Brown et al., 2010).

11.8 MARS 2020 AND THE SEARCH FOR LIFE BEYOND EARTH

11.8.1 Mars 2020 and In Situ Astrobiology

Mars 2020 builds upon previous rover missions' approaches for documenting geologic context and assessing ancient habitability. Upon landing and system checkout, the science team will select an initial exploration area or "region of interest" (ROI), roughly 1 km^2 in area. The first ROI will be selected on the basis of perceived scientific potential and proximity to the rover landing site from a number of previously identified ROIs that will have been mapped and prioritized by the science team based on orbital data (e.g., MRO HiRISE imagery and CRISM spectra). As the rover approaches and arrives at the first ROI, stereo color images from Mastcam-Z and enhanced engineering cameras (EECAMs) at increasing spatial resolution will begin to reveal information about lithology, structure, and stratigraphic relations of geologic units present within the ROI. Radar soundings from RIMFAX will be correlated with image data to extend the structural and stratigraphic investigation into the subsurface. Multispectral imagery from Mastcam-Z, together with LIBS, VNIR, and Raman spectroscopy from SuperCam, will provide information about mineralogy at a distance of a few to tens of meters from the rover, enabling further development of a model for deposition and subsequent diagenesis of rocks in the ROI. As the contextual, ground-based "remote science" dataset is acquired, new understanding of the local geology will be used to refine ROI maps based previously on orbital data alone. Remote science data will lead the team to select scientifically compelling and safely accessible targets for surface preparation and "proximity science" with the turret-mounted instruments PIXL and SHERLOC, whose data will ultimately inform the choice of sampling locations based upon the likelihood of a target to preserve records of ancient life and planetary evolution that could someday be detected in Earth-based laboratories.

Certain elements of Mars 2020 astrobiology exploration strategy are landing site independent. Evidence of past liquid water and its persistent interaction with rock to yield metabolic substrates is the dominant exploration target regardless of habitat type. Potential for and diversity of disequilibria are important components of habitability, so locations with apparently primary mineralogical and morphologic complexity may be preferable to relatively homogeneous locations. Evidence of diagenetic alteration can be favorable or unfavorable, depending upon the circumstance. Oxidizing or acidic fluids can destroy primary elemental or molecular biosignatures (e.g., Sumner, 2004), and recrystallization or impact processes can destroy morphologic biosignatures. From that point of view, rocks that are as mechanically and chemically "primary" are considered to be higher priority targets for potential biosignature preservation. On the other hand, the aforementioned fracture networks can represent important subsurface habitable environments and must not be overlooked. Similarly, impact-generated hydrothermal systems may destroy evidence of earlier life in the host rock while forming their own postimpact habitats (Osinski et al., 2013) and potentially preserving records of subsurface life (Parnell et al., 2010; Sapers et al., 2014, 2015).

As the shift in focus of the MSL mission from habitability to taphonomy demonstrates (e.g. Grotzinger, 2014), the element of time is critical in both directions. Long times are important—for a habitat to generate sufficient biosignatures to enable detection, the system must persist long enough for a biome to take hold and produce a relatively robust and recalcitrant environmental expression. Short times are also important—rapid burial and/or rapid crystallization are key to preservation of biosignatures before they can be degraded by ambient environmental processes, and short surface exposure time is important to protect any complex molecular biosignatures from being destroyed (or rendered ambiguous) by ionizing radiation. Each of these requirements for biosignature formation and preservation has implications for the Mars 2020 mission's exploration strategy. Mars 2020 will seek local environments with evidence for relatively persistent water (e.g., lacustrine sediments and nodes in fracture networks), maximal primary chemical diversity, and favorable taphonomic conditions (e.g. rapid burial or entombment by evaporitic minerals and impact glass).

Certain details of the mission's surface exploration strategy are heavily site-dependent. For example, exploration of fluvio-deltaic and lacustrine environments in Jezero Crater might prioritize a search for low-energy distal facies with a high capacity to preserve organic matter and any carbonate-bearing facies nearer to ancient shorelines that could have offered a physical and chemical environment particularly conducive to the formation and preservation of biosignatures in forms recognizable to rover instrumentation (e.g., fossilized microbial mats or stromatolites). By contrast, habitable environments at NE Syrtis were likely in the subsurface where water interacted with rock to generate chemical disequilibria that could have been harnessed by microbial life. The exploration strategy at NE Syrtis might focus on identifying fluid flow networks, perhaps seeking nodes in ancient fracture/vein systems where fluids of varying compositions could have interacted to generate chemically fertile microenvironments. Exploration targets at Columbia Hills have already been identified using ground-based data from the MER Spirit rover. High-silica materials with digitate morphology observed near Home Plate have been interpreted to represent potential biosignatures (Ruff and Farmer, 2016), and a spatially resolved investigation of the elemental and molecular chemistry of these features with PIXL and SHERLOC would be a priority in order to determine whether there is more compelling evidence of ancient life than what MER *Spirit* observed.

11.8.2 Martian Biosignatures

What is a martian biosignature? Will we know it when we see it? Mars rover scientists and engineers often joke about finding "the dinosaur bone," but the modern concept of astrobiology holds that the conditions enabling complex, multicellular life on Earth are probably unique in the solar system, and thus any extraterrestrial life is likely to have been microbial. Rather than the dinosaur bone, then, we may consider the astrobiological "holy grail" of a Mars rover mission to be something like a stromatolite—a finely layered sedimentary rock that may represent a fossil microbial mat. Many stromatolites as we know them from ancient Earth rocks would be readily detectable by Mars 2020 instrumentation, with submillimeter- to centimeter-scale lamination that sometimes exhibits morphologically correlated elemental and molecular heterogeneity (including biotic organic matter preserved on billion year timescales as kerogen and distinctly confined to certain laminae). For example, variably silicified, carbonate stromatolites of the ~3.4 Ga Strelley Pool formation in Western Australia represent some of the oldest widely accepted evidence for life on Earth and show morphologic, elemental, molecular, and isotopic signs of life within a geologic context clearly indicative of deposition in a habitable environment (Allwood et al., 2006, 2009; Bontognali et al., 2012; Lepot et al., 2013; Flannery et al., 2018).

Recently, the Mars 2020 PIXL and SHERLOC teams conducted a coordinated investigation of a particularly well-preserved kerogen- and carbonate-bearing sample of a Strelley Pool formation stromatolite using laboratory development models of the Mars 2020 instruments (Fig. 11.3). PIXL used microfocus X-ray fluorescence to observe features consistent with a variably silicified matrix of original dolomite having spatially

Fig. 11.3 Coregistered X-ray fluorescence (PIXL) and UV Raman (SHERLOC) maps over a ~2.5 × 8 mm laminated region of a stromatolite from the ~3.4 Ga Strelley Pool formation, Western Australia. Leftmost panel is similar to an image acquired from the WATSON camera on SHERLOC; center panel is a PIXL map showing distribution of Si, Ca, and Fe; and rightmost panel is a SHERLOC UV Raman map showing intensity of peaks corresponding to dolomite, kerogen, and quartz.

variable Fe concentrations consistent with earlier observations by Allwood et al. (2009). Coregistered SHERLOC UV fluorescence and Raman measurements confirmed the carbonate mineralogy variably altered to quartz, with kerogen preferentially distributed in zones of silicification. This type of elemental and molecular heterogeneity in correlation with primary morphology (in this case, millimeter-scale laminae) represents an important class of biosignature clearly detectable by the Mars 2020 payload.

In general, we suggest that potential biosignatures can be defined as (1) co-occurring concentrations of biologically important elements, molecules, and/or minerals; (2) exhibiting heterogeneity correlated in space with complex or otherwise biologically suggestive morphologies; (3) observed within a geologic context consistent with habitability; and (4) the likelihood of biogenicity dependent upon the relative parsimony of any abiotic explanations for the sum of the observed phenomena.

11.8.3 Astrobiologic Considerations for MSR

Mars 2020 has the tremendous opportunity and profound responsibility to provide the in situ geologic and environmental context for what would be among the most precious samples in the history of science if successfully returned to Earth. The ability to achieve scientific consensus about any evidence for ancient life observed in samples returned from Mars strongly depends on the quality of science conducted during the Mars 2020 mission. To this end, the science team is developing a three-part approach to contextualization of samples.

The first component of sample contextualization will include a set of data and interpretations largely independent of the decision to collect any particular sample. Using an approach similar to previous rover missions, Mars 2020 will explore the landing site, conducting an analysis of representative materials that is as comprehensive as possible in order to generate interpretive models explaining the emplacement and alteration of rocks exposed at the surface. This work will be in service of planetary science generally—the science team will pursue and communicate discoveries related to the evolution of Mars as a system, including its capacity to support life. This process will also guide the eventual selection of samples that are themselves dominantly contextual. In addition to any primarily "astrobiology" samples with high potential to preserve biosignatures, future scientists will need access to a representative set of "context" samples collected primarily for this purpose. These samples would be of high scientific value in their own right, selected as faithful recorders of the conditions of deposition/emplacement, the nature of the most significant alteration events, and a time series of environmental change encompassing the interval represented by astrobiology-oriented samples.

The second component will include a minimum set of systematic observations to be performed as consistently as possible on or in association with every sampling target. Analytic consistency will support the rapid construction and dissemination of "dossiers" for each sample, the compilation of which will represent an early guidebook to the samples available for return. The third component will include detailed observations specifically tailored to individual samples or sampling locations. These may represent the richest, but most complex and difficult to interpret data from the mission, and their full fruition may not come until sample analysis on Earth. It is in this third realm that any potential biosignatures are likely to emerge, and it is these data that may provide the most compelling rationale for sample return.

Mars 2020 will carry more sample tubes (~40) than the currently envisioned carrying capacity for a retrieval mission (31). Planning for success, scientists of the future will be able to debate the merits of the Mars 2020 samples after the sampling phase of the mission concludes in order to determine the highest value subset for return to Earth. The Mars 2020 science team is designing the sample documentation program with these future deliberations in mind.

11.8.4 Some Thoughts on Returned Sample Science

Assuming that samples collected by Mars 2020 are successfully returned, the Earth-based phase of returned sample science would begin in earnest with documentation of any conditions associated with touchdown and field recovery of the Earth entry capsule that could impact sample properties. Planetary protection concerns related to returned extraterrestrial samples that could feasibly contain viable extraterrestrial organisms require that extraordinary measures be taken to ensure containment. For example, the return capsule must be designed to survive impact without the aid of a parachute to avoid the unintended and uncontrolled exposure of martian materials to the Earth environment. Upon landing, the return capsule would be removed to a receiving facility with a degree of biological security that may exceed any that exists today. After an initial phase of analysis, it is possible that the samples would be transferred to a separate, lower biosecurity and longer-term curation facility from which subsamples could be distributed to external laboratories for further analysis.

The search for evidence of ancient life in returned martian samples would likely use an approach fundamentally similar to that described in Section 11.8.2, with the addition of a wide variety of analytic techniques not featured by Mars 2020. The analytic possibilities are numerous, and it is likely that approaches not yet imagined would be employed. Here, we offer several examples of existing techniques that would almost certainly play an important role in returned sample astrobiology. Solvent extraction followed by gas chromatography and mass spectrometry is among the most powerful methods available to study the ancient record of life on Earth, and this approach applied to returned martian samples would enable the detection of any organic molecules of sufficient complexity or distributions of molecules in patterns that cannot be explained abiotically (e.g., Summons et al., 2011). Clearly, nondestructive or minimally destructive techniques would be favored whenever feasible for returned sample analysis as they minimize sample consumption and provide data in petrographic context. Careful sample preparation followed by light and electron microscopy would enable a search for microbial fossils and spatially resolved elemental, molecular, and isotopic analysis. Biogenicity assessment of putative microfossils in ancient Earth rocks has a long and controversial history (e.g., Schopf, 1993; Brasier et al., 2002), but criteria for such assessments have evolved considerably over the last several decades (e.g., Schopf and Walter, 1983; Buick, 1990; Sugitani et al., 2007; Wacey, 2009), and similar approaches have been applied to nonmicrofossil biosignatures including putative microbial microalteration textures in igneous rocks (McLoughlin and Grosch, 2015). Recent developments in spatially resolved analysis of organic and mineral matter distinguish morphologically correlated compositional heterogeneities and represent important new capabilities for the confident detection of biosignatures at or below the scale of individual microorganisms (e.g., Williford et al., 2013, 2016; Wacey et al., 2016). In particular, the evolution of techniques such as atom probe tomography (e.g., Miller et al., 2012; Valley et al., 2014) will no doubt be important to the future analysis of returned extraterrestrial samples.

Although the primary astrobiological motivation for Mars 2020 and MSR as currently envisioned is to investigate the possibility of ancient life on Mars, a pristine and scientifically selected set of samples from the surface of that planet would also offer an excellent opportunity to seek signs of extant (living or recently deceased) extraterrestrial life. This search would be complementary to the planetary protection goal to determine whether the samples contain any martian organisms that are hazardous to Earth life. As such and as was the case for the lunar samples returned by the Apollo program, this biological analysis would likely occur early (in part to assess any risks to human workers) and certainly within the high biosecurity receiving facility. Because it would be impossible to conclusively prove the absence of a dangerous martian organism without complete consumption of the sample set (an absurd proposition!), future scientists and policy makers will likely face a decision to either sterilize the samples or contain them in the high biosecurity receiving facility indefinitely (or until human exploration of Mars matures to a degree that existential threats to the biosphere of Earth can be confidently rejected). Removal to a lower security, longer-term curation facility, and distribution to external laboratories may thus require sterilization of the samples and acceptance of any accompanying alterations to sample properties that might result. Workers of the future may have to balance a desire to conduct specific, idiosyncratic analyses in unique individual laboratories against the potential scientific cost of sample sterilization.

How then might extant life be detected in martian samples, and if detected, how could scientists determine conclusively whether the organisms were Earth-sourced contaminants (the obvious null hypothesis) or indigenous martian life? If the contamination could be confidently rejected and the recovery of indigenous martian life confirmed, the implications would be profound. Such a discovery would immediately raise an important second-order question: are the martian organisms the result of an independent origin of life on Mars (independent genesis) or are Earth and Mars life related through impact exchange (panspermia)?

One might imagine (based on the current state of the art for life detection in low biomass terrestrial samples) that the first detection of a possible extant microorganism in a returned martian sample would result from fluorescence or electron microscopy (e.g., Kallmeyer et al., 2008; Morono et al., 2009). If a cell-like entity (CLE) were detected, researchers would face difficult decisions about downstream destructive analysis. Possible choices with the potential to yield a confident determination of biogenicity would include lysis and sequencing (e.g., Gawad et al., 2016) or cryo–electron tomography (e.g., Dobro et al., 2017). Unsuccessful sequencing after lysis would be uninformative unless sufficient unprocessed CLE material remained to permit further analysis. Successful amplification and sequencing showing a match to any known Earth life would suggest contamination. A sequence that did not match any previously known and showed extraordinarily deep divergence (e.g., such that the sequence was unassignable to any of the three domains of Earth life) could be explained by either (1) panspermia or (2) extraordinary convergent evolution to DNA after independent genesis. Successful cryo-EM could show cellular anatomy distinct from any known form of Earth life, but this result would also likely be inconclusive given the current pace of discovery in this field (Dobro et al., 2017). It is possible that a combination of mass spectrometry and crystallography could reveal a system of pseudo/xenoproteins and/or xenonucleic acids that would represent strong evidence for independent genesis on Mars. Regardless of the analytic choice and the result, information from a single CLE would almost certainly be insufficient to disprove the null hypothesis of terrestrial contamination, and a variety of analytic techniques applied on a larger number of CLEs would be required. Therefore, the probable approach prior to any destructive analysis would be to repeat the procedure that yielded the first CLE detection, consuming additional material from the sample in question and/or from other returned samples until either (1) more CLEs were detected or (2) the willingness to expend sample material on the search for extant life was overwhelmed by the desire to pursue the more central goals to understand the evolution of Mars as a system, including the search for evidence of ancient life.

Return and analysis of samples from Mars would represent an important turning point for astrobiology. For a field that has so far been restricted to analysis of Earth-based analogs, models, and remote observations, the opportunity to bring scientifically selected samples of a once habitable planet into the laboratory for interrogation by any technique available on Earth would be transformative. Perhaps the most exciting, but least likely, result (given the current inhospitability of surface environments) of MSR science would be the determination that Mars is currently inhabited by organisms from an independent genesis. More likely and similarly meaningful would be a determination that Mars was inhabited in the ancient past—answering a central motivating question for Mars 2020 and MSR. It may be difficult or impossible to determine whether any evidence of ancient martian life in samples collected by Mars 2020 resulted from independent genesis or panspermia. It is also possible, of course, that Mars was never inhabited or that samples and in situ data collected by Mars 2020 reveal no convincing evidence of life. A complete lack of biosignatures in a diverse and carefully selected set of samples from a clearly habitable ancient environment on Mars could place important constraints on the spatiotemporal ubiquity of life on habitable planets. The explosion of scientific knowledge sparked by the return and analysis of lunar samples demonstrates that, even in the event of this third, astrobiologically negative result, the scientific rewards for MSR would be tremendous.

Regardless of the astrobiology results, successful return and analysis of samples scientifically selected and with geologic context exhaustively documented by Mars 2020 fairly guarantees a revolution in human understanding of our planetary neighborhood. If conclusive martian biosignatures *are* detected, however, a new world opens before us, and future generations can finally leap from the Earthly tree of life to begin exploring the forest.

ACKNOWLEDGMENTS

David Flannery and the rest of the PIXL team is acknowledged for providing the PIXL map of the Strelley Pool stromatolite, and the SHERLOC team is acknowledged for providing the Raman map. We thank reviewer Bob Craddock for comments that led to an improved manuscript. Part of this research was done at the Jet Propulsion Laboratory, California Institute of Technology, under a grant from the National Aeronautics and Space Administration.

REFERENCES

Allwood, A.C., Walter, M.R., Kamber, B.S., Marshall, C.P., Burch, I.W., 2006. Stromatolite reef from the Early Archean era of Australia. Nature 441, 714–718.

Allwood, A.C., Grotzinger, J.P., Knoll, A.H., Burch, I.W., Anderson, M.S., Coleman, M.L., Kanik, I., 2009. Controls on development and diversity of Early Archean stromatolites. Proc. Natl. Acad. Sci. USA 106, 9548–9555.

Anderson, R.C., Jandura, L., Okon, A.B., Sunshine, D., Roumeliotis, C., Beegle, L., Hurowitz, J., Kennedy, B., Limonadi, D., McCloskey, S., Robinson, M., Seybold, C., Brown, K., 2012. Collecting samples in Gale Crater, Mars: an overview of the Mars Science Laboratory sample acquisition, sample processing and handling system. Space Sci. Rev. 170, 57–75.

Arvidson, R.E., 2016. Aqueous history of Mars as inferred from landed mission measurements of rocks, soils, and water ice. J. Geophys. Res. Planets 121, 1602–1626.

Arvidson, R.E., Squyres, S.W., Bell III, J.F., Catalano, J.G., Clark, B.C., Crumpler, L.S., de Souza Jr., P.A., Fairén, A.G., Farrand, W.H., Fox, V.K., Gellert, R., Ghosh, A., Golombek, M.P., Grotzinger, J.P., Guinness, E.A., Herkenhoff, K.E., Jolliff, B.L., Knoll, A.H., Li, R., McLennan, S.M., Ming, D.M., Mittlefehldt, D.W., Moore, J.M., Morris, R.V., Murchie, S.L., Parker, T.J., Paulsen, G., Rice, J.W., Ruff, S.W., Smith, M.D., Wolff, M.J., 2014. Ancient aqueous environments at Endeavour Crater, Mars. Science 343, 1248097. https://doi.org/10.1126/science.1248097.

Arvidson, R.E., DeGrosse Jr., P., Grotzinger, J.P., Heverly, M.C., Shechet, J., Moreland, S.J., Newby, M.A., Stein, N., Steffy, A.C., Zhou, F., Zastrow, A.M., Vasavada, A.R., Fraeman, A.A., Stilly, E.K., 2017. Relating geologic units and mobility system kinematics contributing to Curiosity wheel damage at Gale Crater, Mars. J. Terrramech. 73, 73–93. https://doi.org/10.1016/j.jterra.2017.03.001.

Beaty, D.W., Hays, L.E., Parrish, J., Whetsel, C., 2015. In: Caching scenarios for the Mars 2020 rover, and possible implications for the science of potential Mars sample return. (abs.).46th Lun. and Plan. Sci. Conf. abs. #1672. http://www.hou.usra.edu/meetings/lpsc2015/pdf/1672.pdf.

Beegle, L., Bhartia, R., White, M., DeFlores, L., Abbey, W., Wu, Y.-H., Cameron, B., Moore, J., Fries, M., Burton, A., Edgett, K.S., Ravine, M.A., Hug, W., Reid, R., Nelson, T., Clegg, S., Wiens, R., Asher, S., Sobron, P., 2015. In: SHERLOC: scanning habitable environments with Raman & luminescence for organics & chemicals. Aerospace Conference. IEEE, Big Sky, MT.

Bell III, J.F., Squyres, S.W., Herkenhoff, K.E., Maki, J.N., Arneson, H.M., Brown, D., Collins, S.A., Dingizian, A., Elliot, S.T., Hagerott, E.C., Hayes, A.G., Johnson, M.J., Johnson, J.R., Joseph, J., Kinch, K., Lemmon, M.T., Morris, R.V., Scherr, L., Schwochert, M., Shepard, M.K., Smith, G.H., Sohl-Dickstein, J.N., Sullivan, R.J., Sullivan, W.T., Wadsworth, M., 2003. The Mars Exploration Rover Athena Panoramic Camera (PanCam) investigation. J. Geophys. Res 108 (E12), 8063–8089. https://doi.org/10.1029/2003JE002070.

Benardini, J.N., La Duc, M.T., Beaudet, R.A., Koukol, R., 2014. Implementing planetary protection measures on the Mars Science Laboratory. Astrobiology 14, 27–32.

Bontognali, T.R.R., Sessions, A.L., Allwood, A.C., Fischer, W.W., Grotzinger, J.P., Summons, R.E., Eiler, J.M., 2012. Sulfur isotopes of organic matter preserved in 3.45-billion-year-old stromatolites reveal microbial metabolism. Proc. Natl. Acad. Sci. USA 109, 15146–15151.

Bramble, M.S., Mustard, J.F., Salvatore, M.R., 2017. The geological history of Northeast Syrtis Major, Mars. Icarus 2017, 66–93.

Brasier, M.D., Green, O.R., Jephcoat, A.P., Kleppe, A.K., van Kranendonk, M.J., Lindsay, J.F., Steele, A., Grassineau, N., 2002. Questioning the evidence for Earth's oldest fossils. Nature 416, 76–81.

Brasier, M.D., Antcliffe, J., Saunders, M., Wacey, D., 2015. Changing the picture of Earth's earliest fossils (3.5–1.9 Ga) with new approaches and new discoveries. Proc. Natl. Acad. Sci. USA 112, 4859–4864.

Brown, A.J., Hook, S.J., Baldridge, A.M., Crowley, J.K., Bridges, N.T., Thompson, B.J., Marion, G.M., de Souza Filho, C.R., Bishop, J.L., 2010. Hydrothermal formation of clay-carbonate alteration assemblages in the Nili Fossae region of Mars. Earth Planet. Sci. Lett. 297, 174–182.

Buick, R., 1990. Microfossil recognition in Archean rocks: an appraisal of spheroids and filaments from a 3500 M.Y. old chert-barite unit at North Pole, Western Australia. PALAIOS 5, 441–459.

Carr, M.H., Head III, J.W., 2010. Geologic history of Mars. Earth Planet. Sci. Lett. 294, 185–203.

Committee to Review the MEPAG Report on Mars Special Regions, Space Studies Board, Division on Engineering and Physical Sciences, The National Academies of Sciences, Engineering, and Medicine, European Space Sciences Committee, European Science Foundation, 2015. Review of the MEPAG Report on Mars Special Regions. The National Academies Press, Washington, DC.

COSPAR (Committee on Space Research), 2015. COSPAR's planetary protection policy. Space Res. Today 193, 7–19.

Cousin, A., Bernard, S., Dromart, G., Drouet, C., Fabre, C., Fouchet, T., Gasnault, O., Madsen, M., Manrique, J.A., Maurice, S., Meslin, P.Y., Montagnac, G., Rull, F., Sautter, V., Thiebaut, C., Virmontois, C., Wiens, R., 2017. In: Development of onboard calibration targets for the Mars 2020/SuperCam remote sensing suite. (abs.).48th Lun. and Plan. Sci. Conf. abs. #2082.

Dobro, M.J., Oikonomou, C.M., Piper, A., Cohen, J., Guo, K., Jensen, T., Tadayon, J., Donermeyer, J., Park, Y., Solis, B.A., Kjaer, A., Jewett, A.I., McDowall, A.W., Chen, S., Chang, Y.-W., Shi, J., Subramanian, P., Iancu, C.V., Li, Z., Briegel, A., Tocheva, E.I., Pilhofer, M., Jensen, G.J., 2017. Uncharacterized bacterial structures revealed by electron cryotomography. J. Bacteriol. 199, e00100–e00117.

Drake, B. (Ed.), 2007. Mars design reference architecture 5.0. NASA/SP–2009566-ADD. Available at: https://www.nasa.gov/pdf/373667main_NASA-SP-2009-566-ADD.pdf.

Edgett, K.S., et al., 2012. Curiosity's Mars hand lens imager (MAHLI) investigation. Space Sci. Rev. 170, 259–317.

Edwards, K.J., Becker, K., Colwell, F., 2012. The deep, dark energy biosphere: intraterrestrial life on Earth. Annu. Rev. Earth Planet. Sci. 40, 551–568.

Ehlmann, B.L., Mustard, J.F., 2012. An in-situ record of major environmental transitions on early Mars at Northeast Syrtis Major. Geophys. Res. Lett. 39, L11202.

Ehlmann, B.L., Mustard, J.F., Fassett, C.I., Schon, S.C., Head III, J.W., Des Marais, D.J., Grant, J.A., Murchie, S.L., 2008. Clay minerals in delta deposits and organic preservation potential on Mars. Nat. Geosci. 1, 355–358.

Ehlmann, B.L., Mustary, J.F., Swayze, G.A., Clark, R.N., Bishop, J.L., Poulet, F., Des Marais, D.J., Roach, L.H., Milliken, R.E., Wray, J.J., Banouin-Jha, O., Murchie, S.L., 2009. Identification of hydrated silicate minerals on Mars using MRO-CRISM: geologic context near Nili Fossae and implications for aqueous alteration. J. Geophys. Res. 114, E00D08.

Farley, K.A., Malespin, C., Mahaffy, P., Grotzinger, J.P., Vasconcelos, P.M., Milliken, R.E., Malin, M., Edgett, K.S., Pavlov, A.A., Hurowitz, J.A., Grant, J.A., Miller, H.B., Arvidson, R., Beegle, L., Calef, F., Conrad, P.G., Dietrich, W.E., Eigenbrode, J., Gellert, R., Gupta, S., Hamilton, V., Hassler, D.M., Lewis, K.W., McLennan, S.M., Ming, D., Navarro-González, R., Schwenzer, S.P., Steele, A., Stolper, E.M., Sumner, D.Y., Vaniman, D., Vasavada, A., Williford, K., Wimmer-Schweingruber, R.F., The MSL Science Team, 2013. In situ radiometric and exposure age dating of the martian surface. Science 343 (6169), 1247166. https://doi.org/10.1126/science.1247166.

Farley, K.A., Williford, K., Beaty, D.W., McSween, H.Y., Czaja, A.D., Goreva, Y.S., Hausrath, E.M., Hays, L.E., Herd, C.D.K., Humayun, M., McCubbin, F.M., McLennan, S.M., Pratt, L.M., Sephton, M.A., Steele, A., Weiss, B.P., 2017. In: Contamination knowledge strategy for the Mars 2020 sample-collecting rover (abs.).48th Lun. and Plan. Sci. Conf., Houston, TX. abs. #2251.

Fassett, C.I., Head III, J.W., 2005. Fluvial sedimentary deposits on Mars: ancient deltas in a crater lake in the Nili Fossae region. Geophys. Res. Lett. 32, L14201.

Flannery, D., Allwood, A.C., Summons, R.E., Williford, K.H., Abbey, W., Matys, E.D., Ferralis, N., 2018. Spatially-resolved isotopic study of carbon trapped in ~3.43 Ga Strelley Pool Formation stromatolites. Geochim. Cosmochim. Acta 223, 21–35.

Freissinet, C., Glavin, D.P., Mahaffy, P.R., Miller, K.E., Eigenbrode, J.L., Summons, R.E., Brunner, A.E., Buch, A., Szopa, C., Archer Jr., P.D., Franz, H.B., Atreya, S.K., Brinkerhoff, W.B., Cabane, M., Coll, P., Conrad, P.G., Des Marais, D.J., Dworkin, J.P., Fairén, A.G., François, P., Grotzinger, J.P., Kashyap, S., ten Kate, I.L., Leshin, L.A., Malespin, C.A., Martin, M.G., Martin-Torres, F.J., McAdam, A.C., Ming, D.W., Navarro-González, R., Pavlov, A.A., Prats, B.D., Squyres, S.W., Steele, A., Stern, J.C., Sumner, D.Y., Sutter, B., Zorzano, M.-P. the MSL Science Team, 2015. Organic molecules in the Sheepbed Mudstone, Gale Crater, Mars. J. Geophys. Res. Planets 120, 495–514.

French, K.L., Hallmann, C., Hope, J.M., Schoon, P.L., Zumberge, J.A., Hoshino, Y., Peters, C.A., George, S.C., Love, G.D., Brocks, J.J., Buick, R., Summons, R.E., 2015. Reappraisal of hydrocarbon biomarkers in Archean rocks. Proc. Natl. Acad. Sci. USA 112, 5915–5920.

Gawad, C., Koh, W., Quake, S.R., 2016. Single-cell genome sequencing: current state of the science. Nat. Rev. Genet. 17, 175–188.

Golombek, M., Grant, J., Kipp, D., Vasavada, A., Kirk, R., Fergason, R., Bellutta, P., Calef, F., Larsen, K., Katayama, Y., Huertas, A., Beyer, R., Chen, A., Parker, T., Pollard, B., Lee, S., Sun, Y., Hoover, R., Sladek, H., Grotzinger, J., Welch, R., Noe Dobrea, E., Michalski, J., Watkins, M., 2012. Selection of the Mars Science Laboratory landing site. Space Sci. Rev. 170, 641–737.

Gómez-Elvira, J., Armiens, C., Castañer, L., Domínguez, M., Genzer, M., Gómez, F., Haberle, R., Harri, A.-M., Jiménez, V., Kahanpää, H., Kowalski, L., Lepinette, A., Martín, J., Martínez-Frías, J., McEwan, I., Mora, L., Moreno, J., Navarro, S., de Pablo, M.A., Peinado, V., Peña, A., Polkko, J., Ramos, M., Renno, N.O., Ricart, J., Richardson, M., Rodríguez-Mangredi, J., Romeral, J., Sebastián, E., Serrano, J., de la Torre Juárez, M., Torres, J., Torrero, F., Urquí, R., Vázquez, L., Velasco, T., Verdasca, J., Zorzano, M.-P., Martín-Torres, J., 2012. REMS: the environmental sensor suite for the Mars Science Laboratory Rover. Space Sci. Rev. 170, 583–640.

Goudge, T.A., Mustard, J.F., Head, J.W., Fassett, C.I., Wiseman, S.M., 2015. Assessing the mineralogy of the watershed and fan deposits of the Jezero crater paleolake system, Mars. J. Geophys. Res. Planets 120, 775–808.

Goudge, T.A., Milliken, R.E., Head III, J.W., Mustard, J.F., Fassett, C.I., 2017. Sedimentological evidence for a deltaic origin of the western fan deposit in Jezero crater, Mars and implications for future exploration. Earth Planet. Sci. Lett. 458, 357–365.

Grotzinger, J.P., 2014. Habitability, taphonomy, and the search for organic carbon on Mars. Science 343, 386–387.

Grotzinger, J.P., Crisp, J., Vasavada, A.R., Anderson, R.C., Baker, C.J., Barry, R., Blake, D.F., Conrad, P., Edgett, K.S., Ferdowski, B., Gellert, R., Gilbert, J.B., Golombek, M., Gómez-Elvira, J., Hassler, D.M., Jandura, L., Litvak, M., Mahaffy, P., Maki, J., Meyer, M., Malin, M.C., Mitrofanov, I., Simmonds, J.J., Vaniman, D., Welch, R.V., Wiens, R.C., 2012. Mars Science Laboratory Mission and science investigation. Space Sci. Rev. 170, 5–56.

Grotzinger, J.P., Sumner, D.Y., Kah, L.C., Stack, K., Gupta, S., Edgar, L., Rubin, D., Lewis, K., Schieber, J., Mangold, N., Milliken, R., Conrad, P.G., Desmarais, D., Farmer, J., Siebach, K., Calef, F., Hurowitz, J., Mclennan, S.M., Ming, D., Vaniman, D., Crisp, J., Vasavada, A., Edgett, K.S., Malin, M., Blake, D., Gellert, R., Mahaffy, P., Wiens, R.C., Maurice, S., Grant, J.A., Wilson, S., Anderson, R.C., Beegle, L., Arvidson, R., Hallet, B., Sletten, R.S., Rice, M., Bell, J., Griffes, J., Ehlmann, B., Anderson, R.B., Bristow, T.F., Dietrich, W.E., Dromart, G., Eigenbrode, J., Fraeman, A., Hardgrove, C., Herkenhoff, K., Jandura, L., Kocurek, G., Lee, S., Leshin, L.A., Leveille, R., Limonadi, D., Maki, J., Mccloskey, S., Meyer, M., Minitti, M., Newsom, H., Oehler, D., Okon, A., Palucis, M., Parker, T., Rowland, S., Schmidt, M., Squyres, S., Steele, A., Stolper, E., Summons, R., Treiman, A., Williams, R., Yingst, A., MSL Science Team, 2014. A habitable fluvio-lacustrine environment at Yellowknife Bay, Gale Crater, Mars. Science 343, 1242777.

Grotzinger, J.P., Gupta, S., Malin, M.C., Rubin, D.M., Schieber, J., Siebach, K., Sumner, D.Y., Stack, K.M., Vasavada, A.R., Arvidson, R.E., Calef III, F., Edgar, L., Fischer, W.F., Grant, J.A., Griffes, J., Kah, L.C., Lamb, M.P., Lewis, K.W., Mangold, N., Minitti, M.E., Palucis, M., Rice, M., Williams, R.M.E., Yingst, R.A., Blake, D., Blaney, D., Conrad, P., Crisp, J., Dietrich, W.E., Dromart, G., Edgett, K.S., Ewing, R.C., Gellert, R., Hurowitz, J.A., Kocurek, G., Mahaffy, P., McBride, M.J., McLennan, S.M., Mischna, M., Ming, D., Milliken, R., Newsom, H., Oehler, D., Parker, T.J., Vaniman, D., Wiens, R.C., Wilson, S.A., 2015. Deposition, exhumation, and paleoclimate of an ancient lake deposit, Gale crater, Mars. Science 350, 177–189.

Hartvigsen, J.J., Elangovan, S., Larsen, D., Elwell, J., Bokil, M., Frost, L., Clark, L.M., 2015. In: Challenges of solid oxide electrolysis for production of fuel and oxygen from Mars atmospheric CO_2.Proc. ECS Conf. on Electrochemical Energy Conversion & Storage with SOFC-XIV, Glasgow, Scotland.

Hecht, M.H., Kounaves, S.P., Quinn, R.C., West, S.J., Young, S.M.M., Ming, D.W., Catling, D.C., Clark, B.C., Boynton, W.V., Hoffman, J., DeFlores, L.P., Gospodinova, K., Kapit, J., Smith, P.H., 2009. Detection of perchlorate and the soluble chemistry of martian soil at the Phoenix lander site. Science 325, 64–67.

Hoffman, J.A., Rapp, D., Hecht, M.H., 2015. In: The Mars oxygen ISRU experiment (MOXIE) on the Mars 2020 Rover.AIAA SPACE 2015 Conference and Exposition, AIAA SPACE Forum, Pasadena, CA (AIAA 2015-4561).

Kallmeyer, J., Smith, D.C., Spivack, A.J., D'Hondt, S., 2008. New cell extraction procedure applied to deep subsurface sediments. Limnol. Oceanogr. Methods 6, 236–245.

Kallmeyer, J., Pockalny, R., Ram Adhikari, R., Smith, D.C., D'Hondt, S., 2012. Global distribution of microbial abundance and biomass in subseafloor sediment. Proc. Natl. Acad. Sci. USA 109, 16213–16216.

Klein, H.P., 1998. The search for life on Mars: what we learned from viking. J. Geophys. Res. 103, 28463–28466.

Klein, H.P., Horowitz, N.H., Levin, G.V., Oyama, V.I., Lederberg, J., Rich, A., Hubbard, J.S., Hobby, G.L., Straat, P.A., Berdahl, B.J., Carle, G.C., Brown, F.S., Johnson, R.D., 1976. The viking biological investigation: preliminary results. Science 194, 99–105.

Klingelhöfer, G., Morris, R.V., Bernhardt, B., Schroder, C., Rodionov, D.S., de Souza, P.A., Yen, A., Gellert, R., Evlanov, E.N., Zubkov, B., Foh, J., Bonnes, U., Kankeleit, E., Gutlich, P., Ming, D.W., Renz, F., Wdowiak, T., Squyres, S.W., Arvidson, R.E., 2004. Jarosite and hematite at Meridiani Planum from opportunity's Mössbauer spectrometer. Science 306, 1740–1745.

Lepot, K., Williford, K.H., Ushikubo, T., Sugitani, K., Mimura, M., Spicuzza, M.J., Valley, J.W., 2013. Texture-specific isotopic compositions in 3.4 Gyr old organic matter support selective preservation in cell-like structures. Geochim. Cosmochim. Acta 112, 66–86.

Mahaffy, P.R., Webster, C.R., Stern, J.C., Brunner, A.E., Atreya, S.K., Conrad, P.G., Domagal-Goldman, S., Eigenbrode, J.L., Flesch, G.J., Christensen, L.E., Franz, H.B., Freissinet, C., Glavin, D.P., Grotzinger, J.P., Jones, J.H., Leshin, L.A., Malespin, C., McAdam, A.C., Ming, D.W., Navarro-Gonzalez, R., Niles, P.B., Owen, T., Pavlov, A.A., Steele, A., Trainer, M.G., Williford, K.H., Wray, J.J., The MSL Science Team, 2015. The imprint of atmospheric evolution on the D/H of Hesperian clay minerals on Mars. Science 347, 412–414.

Maurice, S., Wiens, R.C., Saccoccio, M., Barraclough, B., Gasnault, O., Forni, O., Mangold, N., Baratoux, D., Bender, S., Berger, G., Bernardin, J., Berthé, M., Bridges, N., Blaney, D., Bouyé, M., Cais, P., Clark, B., Clegg, S., Cousin, A., Cremers, D., Cros, A., DeFlores, L., Derycke, C., Dingler, B., Dromart, G., Dubois, B., Dupieux, M., Durand, E., d'Uston, L., Fabre, C., Faure, B., Gaboriaud, A., Gharsa, T., Herkenhoff, K., Kan, E., Kirkland, L., Kouach, D., Lacour, J.-L., Langevin, Y., Lasue, J., Le Mouélic, S., Lescure, M., Lewin, E., Limonadi, D., Manhes, G., Mauchien, P., McKay, C., Meslin, P.-Y., Michel, Y., Miller, E., Newsom, H.E., Orttner, G., Paillet, A., Pares, L., Parot, Y., Perez, R., Pinet, P., Poitrasson, F., Quertier, B., Sallé, B., Sotin, C., Sautter, V., Seran, H., Simmonds, J.J., Sirven, J.-B., Stiglich, R., Streibig, N., Thocaven, J.-J., Toplis, M., Vaniman, D., 2012. The ChemCam instruments on the Mars Science Laboratory (MSL) Rover: science objectives and mast nit. Space Sci. Rev. 170, 95–166. https://doi.org/10.1007/s11214-012-9912-2.

Maurice, S., Clegg, S., Wiens, R.C., Gasnult, O., Rapin, W., Forni, O., Cousin, A., Sautter, V., Mangold, N., Le Deit, L., Nachon, M., Anderson, R., Lanza, N., Fabre, C., Payré, V., Lasue, J., Meslin, P.-Y., Sirven, J.-B., Melikechi, N., Le Mouelic, S., Frydenvang, J., Vasavada, A., Bridges, J.C., Bender, S.C., Schroeder, S., Francis, R., Pinet, P., Newsom, H., Ollila, A., Herkenhoff, K., Madsen, M.B., Dromart, G., Beck, P., Lewin, E., Lacour, J.-L., Langevin, Y., Gondet, B., d'Uston, L., Berger, G., Toplis, M., Johnson, J.R., Dyar, M.D., Bridges, N., Vaniman, D., Barraclough, B., 2016. ChemCam activities and discoveries during the Mars Science Laboratory nominal mission in Gale crater, Mars. J. Anal. At. Spectrom. 31, 863–889. https://doi.org/10.1039/c5ja00417a.

McKay, D.S., Gibson Jr., E.K., Thomas-Keprta, K.L., Vali, H., Romanek, C.S., Clemett, S.J., Chillier, X.D.F., Maechling, C.R., Zare, R.N., 1996. Search for past life on Mars: possible relic biogenic activity in Martian meteorite ALH84001. Science 273, 924–930.

McLennan, S.M., Sephton, M.A., Allen, C., Allwood, A.C., Barbieri, R., Beaty, D.W., Boston, P., Carr, M., Grady, M., Grant, J., Heber, V.S., Herd, C.D.K., Hofmann, B., King, P., Mangold, N., Ori, G.G., Rossi, A.P., Raulin, F., Ruff, S.W., Sherwood Lollar, B., Symes, S., Wilson, M.G., 2011. Planning for Mars Returned Sample Science: final report of the MSR End-to-End International Science Analysis Group (E2E-iSAG), by The Mars Exploration Program Analysis Group (MEPAG) December, 101 pp. http://mepag.jpl.nasa.gov/reports/.

McLoughlin, N., Grosch, E.G., 2015. A hierarchical system for evaluating the biogenicity of metavolcanic- and ultramafic-hosted microalteration textures in the search for extraterrestrial life. Astrobiology 15, 901–921.

MEPAG, 2015. Mars Scientific Goals, Objectives, Investigations, and Priorities: 2015. V. Hamilton, ed., by The Mars Exploration Program Analysis Group (MEPAG) white paper posted June, 74 p. http://mepag.nasa.gov/reports.cfm.

Miller, M.K., Kelly, T.F., Rajan, K., Ringer, S.P., 2012. The future of atom probe tomography. Mater. Today 15, 158–165.

Morono, Y., Terada, T., Masui, N., Inagaki, F., 2009. Discriminative detection and enumeration of micro-bial life in marine subsurface sediments. ISME J. 3, 503–511.

Mustard, J.F., Adler, M., Allwood, A., Bass, D.S., Beaty, D.W., Bell III, J.F., Brinkerhoff, W.B., Carr, M., Des Marais, D.J., Drake, B., Edgett, K.S., Eigenbrode, J., Elkins-Tanton, L.T., Grant, J.A., Milkovich, S. M., Ming, D., Moore, C., Murchie, S., Onstott, T.C., Ruff, S.W., Sephton, M.A., Steele, A., Treiman, A., 2013. Report of the Mars 2020 Science Definition Team, by The Mars Exploration Program Analysis Group (MEPAG) July, 154 pp. http://mepag.jpl.nasa.gov/reports/MEP/Mars_2020_SDT_Report_Final.pdf.

National Research Council (US), Committee on the Planetary Science Decadal Survey, Space Studies Board, 2011. Vision and Voyages for Planetary Science in the Decade 2013-2022. National Academies Press, Washington, DC.

Navarro-González, R., Vargas, E., de la Rosa, J., Raga, A.C., McKay, C.P., 2010. Reanalysis of the Viking results suggests perchlorate and organics at midlatitudes on Mars. J. Geophys. Res. 115, E122010.

Osinski, G.R., Tornabene, L.L., Banerjee, N.R., Cockell, C.S., Flemming, R., Izawa, M.R.M., McCutcheon, J., Parnell, J., Preston, L.J., Pickersgill, A.E., Pontrefact, A., Sapers, H.M., Southam, G., 2013. Impact-generated hydrothermal systems on Earth and Mars. Icarus 224, 347–363.

Parnell, J., Boyce, A., Thackrey, S., Muirhead, D., Lindgren, P., Mason, C., Taylor, C., Still, J., Bowden, S., Osinski, G.R., Lee, P., 2010. Sulfur isotope signatures for rapid colonization of an impact crater by thermophilic microbes. Geology 38, 271–274.

Quinn, R.C., Martucci, H.F.H., Miller, S.R., Bryson, C.E., Grunthaner, F.J., Grunthaner, P.J., 2013. Perchlorate radiolysis on Mars and the origin of martian soil reactivity. Astrobiology 13, 515–520.

Returned Sample Science Board (RSSB), 2016. In: Planning for the collection of a compelling set of Mars samples in support of a potential future Mars sample return (abs.).2016 GSA Annual Meeting, Denver, CO. #141-15. https://gsa.confex.com/gsa/2016AM/webprogram/Paper279910.html.

Ruff, S.W., Farmer, J.D., 2016. Silica deposits on Mars with features resembling hot spring biosignatures at El Tatio in Chile. Nat. Commun. 7, 13554.

Ruff, S.W., Farmer, J.D., Calvin, W.M., Herkenhoff, K.E., Johnson, J.R., Morris, R.V., Rice, M.S., Arvidson, R.E., Bell III, J.F., Christensen, P.R., Squyres, S.W., 2011. Characteristics, distribution, origin, and significance of opaline silica observed by the Spirit rover in Gusev crater, Mars. J. Geophys. Res. 116, E00F23.

Rummel, J.D., COSPAR Planetary Protection Panel, COSPAR, International Astronomical Union, and International Council for Science. 2002. Report of the Workshop on Planetary Protection held under the auspices of the Committee on Space Research and the International Astronomical Union of the International Council for Science at Williamsburg, VA, USA on 2–4 April. COSPAR, Paris.

Rummel, J.D., Beaty, D.W., Jones, M.A., Bakermans, C., Barlow, N.G., Boston, P.J., Chevrier, V.F., 2014. A new analysis of Mars "special regions": Findings of the second MEPAG special regions science analysis group (SR-SAG2). Astrobiology 14, 887–968.

Russell, M.J., Barge, L.M., Bhartia, R., Bocanegra, D., Bracher, P.J., Branscomb, E., Kidd, R., McGlynn, S., Meier, D.H., Nitschke, W., Shibuya, T., Vance, S., White, L., Kanik, I., 2014. The drive to life on wet and icy worlds. Astrobiology 14, 308–343.

Sapers, H.M., Osinski, G.R., Banerjee, N.R., Preston, L.J., 2014. Enigmatic tubular features in impact glass. Geology 42, 471–474.

Sapers, H.M., Banerjee, N.R., Osinski, G.R., 2015. Potential for impact glass to preserve microbial metabolism. Earth Planet. Sci. Lett. 430, 95–104.

Schmidt, M.E., Ruff, S.W., McCoy, T.J., Farrand, W.H., Johnson, J.R., Gellert, R., Ming, D.W., Morris, D.V., Cabrol, N., Lewis, K.W., Schroder, C., 2008. Hydrothermal origin of halogens at Home Plate, Gusev Crater. J. Geophys. Res. 113, E06S12.

Schon, S.C., Head, J.W., Fassett, C.I., 2012. An overfilled lacustrine system and progradational delta in Jezero crater, Mars: implications for Noachian climate. Planet. Space Sci. 67, 28–45.

Schopf, J.W., 1993. Microfossils of the Early Archean Apex chert: new evidence of the antiquity of life. Science 260, 640–646.

Schopf, J.W., Kudryavstev, A.B., 2012. Biogenicity of Earth's earliest fossils: a resolution of the controversy. Gondwana Res. 22, 761–771.

Schopf, J.W., Packer, B.M., 1987. Early Archean (3.3-billion to 3.5-billion-year-old) microfossils from Warrawoona Group, Australia. Science 237, 70–73.

Schopf, J.W., Walter, M.R., 1983. Archean microfossils: new evidence of ancient microbes. In: Schopf, J.W. (Ed.), Earth's Earliest Biosphere, Its Origin and Evolution. Princeton University Press, Princeton, pp. 214–239.

Squyres, S.W., Arvidson, R.E., Bell, J.F.I.I.I., Brückner, J., Cabrol, N.A., Calvin, W., Carr, M.H., Christensen, P.R., Clark, B.C., Crumpler, L., Des Marais, D.J., d'Uston, C., Economou, T., Farmer, J., Farrand, W., Folkner, W., Golombek, M., Gorevan, S., Grant, J.A., Greeley, R., Grotzinger, J., Haskin, L., Herkenhoff, K.E., Hviid, S., Johnson, J., Klingelhöfer, G., Knoll, A.H., Landis, G., Lemmon, M., Li, R., Madsen, M.B., Malin, M.C., McLennan, S.M., McSween, H.Y.,

Ming, D.W., Moersch, J., Morris, R.V., Parker, T., Rice Jr., J.W., Richter, L., Rieder, R., Sims, M., Smith, M., Smith, P., Soderblom, L.A., Sullivan, R., Wänke, H., Wdowiak, T., Wolff, M., Yen, A., 2004. The opportunity Rover's Athena science investigation at Meridiani Planum, Mars. Science 306, 1698–1703.

Squyres, S.W., Arvidson, R.E., Ruff, S., Gellert, R., Morris, R.V., Ming, D.W., Crumpler, L., Farmer, J.D., Des Marais, D.J., Yen, A., McLennan, S.M., Calvin, W., Bell III, J.F., Clark, B.C., Wang, A., McCoy, T.J., Schmidt, M.E., de Souza Jr., P.A., 2008. Detection of silica-rich deposits on Mars. Science 230, 1063–1067.

Sugitani, K., Grey, K., Allwood, A., Nagaoka, T., Mimura, K., Minami, M., Marshall, C.P., Van Kranendonk, M.J., Walter, M.R., 2007. Diverse microstructures from Archaean chert from the Mount Goldsworthy–Mount Grant area, Pilbara Craton, Western Australia: microfossils, dubiofossils, or pseudofossils? Precambrian Res. 158, 228–262.

Summons, R.E., Amend, J.P., Bish, D., Buick, R., Cody, G.R., Des Marais, D.J., Dromart, G., Eigenbrode, J.L., Knoll, A.H., Sumner, D.Y., 2011. Preservation of martian organic and environmental records: Final report of the Mars Biosignature Working Group. Astrobiology 11, 157–181.

Sumner, D.Y., 2004. Poor preservation potential of organics in Meridiani Planum hematite-bearing sedimentary rocks. J. Geophys. Res. Planets 109, E12007.

Valley, J.W., Cavosie, A.J., Ushikubo, T., Reinhard, D.A., Lawrence, D.F., Larson, D.J., Clifton, P.H., Kelly, T.F., Wilde, S.A., Moser, D.A., Spicuzza, M.J., 2014. Hadean age for a post-magma-ocean zircon confirmed by atom-probe tomography. Nat. Geosci. 7, 219–223.

Wacey, D., 2009. Early Life on Earth: A Practical Uide. Springer, New York.

Wacey, D., Battison, L., Garwood, R.J., Hickman-Lewis, K., Brasier, M.D., 2016. Advanced analytical techniques for studying the morphology and chemistry of Proterozoic microfossils. Geol. Soc. Lond. Spec. Publ. 448, 81–104.

Wiens, R.C., Maurice, S., Barraclough, B., Saccoccio, M., Barkley, W.C., Bell III, J.F., Bender, S., Bernardin, J., Blaney, D., Blank, J., Bouye, M., Bridges, N., Cais, P., Clanton, R.C., Clark, B., Clegg, S., Cousin, A., Cremers, D., Cros, A., DeFlores, L., Delapp, D., Dingler, R., D'Uston, C., Dyar, M.D., Elliott, T., Enemark, D., Fabre, C., Flores, M., Forni, O., Gasnault, O., Hale, T., Hays, C., Herkenhoff, K., Kan, E., Kirkland, L., Kouach, D., Landis, D., Langevin, Y., Lanza, N., LaRocca, F., Lasue, J., Latino, J., Limonadi, D., Lindensmith, C., Little, C., Mangold, N., Manhes, G., Mauchien, P., McKay, C., Miller, E., Mooney, J., Morris, R.V., Morrison, L., Nelson, T., Newsom, H., Ollila, A., Ott, M., Pares, L., Perez, R., Poitrasson, F., Provost, C., Reiter, J.W., Roberts, T., Romero, F., Sautter, V., Salazar, S., Simmonds, J.J., Stiglich, R., Storms, S., Streibig, N., Thocaven, J.-J., Trujillo, T., Ulibarri, M., Vaniman, D., Warner, N., Waterbury, R., Whitaker, R., Witt, J., Wong-Swanson, B., 2012. The ChemCam instruments on the Mars Science Laboratory (MSL) Rover: body unit and combined system performance. Space Sci. Rev. 170, 167–227. https://doi.org/10.1007/S11214-012-9902-4. LAUR 11-05584.

Wiens, R.C., Maurice, S., MSL Science Teams, 2015. ChemCam: chemostratigraphy by the first Mars microprobe. Elements 11, 33–38.

Wiens, R.C., Maurice, S., Rull Perez, F., 2017. The SuperCam remote sensing instrument suite for the Mars 2020 rover mission: a preview. Spectroscopy 32, 50–55.

Williford, K.H., Ushikubo, T., Schopf, J.W., Lepot, K., Kitajima, K., Valley, J.W., 2013. Preservation and detection of microstructural and taxonomic correlations in the carbon isotopic compositions of individual Precambrian microfossils. Geochim. Cosmochim. Acta 104, 165–182.

Williford, K.H., Ushikubo, T., Lepot, K., Kitajima, K., Hallmann, C., Spicuzza, M.J., Kozdon, R., Eigenbrode, J.L., Summons, R.E., Valley, J.W., 2016. Carbon and sulfur isotopic signatures of ancient life and environment at the microbial scale: Neoarchean shales and carbonates. Geobiology 14, 105–128.

Wordsworth, R.D., 2016. The climate of early Mars. Annu. Rev. Earth Planet. Sci. 44, 381–408.

Yen, A.S., Kim, S.S., Hecht, M.H., Frant, M.S., Murray, B., 2000. Evidence that the reactivity of the Martian soil is due to superoxide ions. Science 289, 1909–1912.

Zent, A.P., McKay, C.P., 1994. The chemical reactivity of the martian soil and implications for future missions. Icarus 108, 146–157.

CHAPTER 12

Searching for Traces of Life With the ExoMars Rover

Jorge L. Vago, Andrew J. Coates, Ralf Jaumann, Oleg Korablev, Valérie Ciarletti, Igor Mitrofanov, Jean-Luc Josset, Frances Westall, M. Cristina De Sanctis, Jean-Pierre Bibring, Fernando Rull, Fred Goesmann, William Brinckerhoff, François Raulin, Elliot Sefton-Nash, Håkan Svedhem, Gerhard Kminek, Daniel Rodionov, Pietro Baglioni, The ExoMars Team

Contents

Abbreviations

ADRON	active detector for gamma rays and neutrons
ALD	analytic laboratory drawer
CLUPI	close-up imager, accommodated on the drill box's external surface

From Habitability to Life on Mars
https://doi.org/10.1016/B978-0-12-809935-3.00011-6

309

CS	crushing station
CSTM	core sample transport mechanism
Der	derivatization, an agent added to sample material that reacts with indigenous organic molecules to reduce their polarity, rendering the resulting compounds more volatile
DM	descent module, the capsule that enters the atmosphere for landing
DMF-DMA	*N,N*-dimethylformamide dimethyl acetal, one of the MOMA derivatization agents
EC	experiment cycle
EDM	entry, descent, and landing demonstrator module, part of ExoMars 2016
FOV	field of view
GCMS	gas chromatograph-mass spectrometer, one of the MOMA operation modes
GPR	ground-penetrating radar
HRC	high-resolution camera, part of PanCam
IR	infrared
ISEM	infrared spectrometer for ExoMars, on the rover mast
LDMS	laser-desorption mass spectrometry, one of the MOMA operation modes
LSSWG	Landing Site Selection Working Group
Ma_MISS	Mars multispectral imager for subsurface studies, the IR spectrometer integrated in the drill
MER	Mars Exploration Rovers, Spirit and Opportunity
MicrOmega	micro observatoire pour la mineralogie, l'eau, les glaces et l'activité
MOLA	Mars orbiter laser altimeter, an instrument on NASA's 1996 Mars Global Surveyor. 0-MOLA altitude is considered the reference elevation for Mars
MOMA	Mars organic molecule analyzer, in the rover's analytic laboratory
MSL	Mars Science Laboratory, NASA's Curiosity rover
MSR	Mars sample return
MTBSTFA/ DMF	*N*-tert-butyldimethylsilyl-*N*-methyltrifluoroacetamide/dimethylformamide as a 3:1 mixture, one of the MOMA derivatization agents
PAHs	polycyclic aromatic hydrocarbons
PanCam	panoramic camera, on the rover mast
Pyr	pyrolysis, heating a sample to release volatile (fragments) organic molecules
RLS	Raman laser spectrometer, accommodated in the rover's analytic laboratory
ROCC	Rover Operations and Control Center
RSM	reference surface mission
SAM	sample analysis at Mars, the organic detection instrument on NASA's Curiosity rover
SP	surface platform, the element of the ExoMars 2018 DM reaching the surface. Following rover egress, the SP becomes a science station
SPDS	sample processing and distribution system
TGO	Trace Gas Orbiter, part of ExoMars 2016
TV	thermal volatilization, sometimes also called pyrolysis, refers to the release of volatile organic molecules (fragments) by heating a sample
TMAH	25 wt% tetramethylammonium hydroxide in methanol, one of the MOMA derivatization agents
UCZ	ultra clean zone, the part of the ALD enclosing the sample path
UHF	ultrahigh frequency, the radiofrequency band used at present for rover-to-orbiter communications, sometimes called "proximity link"
VIS + IR	visible (red, green, and blue) plus infrared
VS	vertical survey
WAC	wide-angle cameras in PanCam
WISDOM	water, ice, and subsurface deposit observations on Mars, a GPR

12.1 OVERVIEW

The second ExoMars mission will be launched in 2020 to target an ancient landing site interpreted to possess a strong potential for preserving the physical and chemical biosignatures of fossil martian microorganisms, if they existed there. The mission will deliver a lander with instruments designed for atmospheric and geophysical investigations and a rover tasked with searching for signs of extinct life. The ExoMars rover will be equipped with a drill to collect material from outcrops and at depth (between 0 and 2 m). This subsurface sampling capability, coupled with novel analytic instruments, will provide the best chance yet to gain access to and characterize molecular biomarkers.

Starting with a brief discussion of the ExoMars program, this chapter concentrates on the ExoMars rover. We describe its scientific underpinnings, the rover configuration, its Pasteur payload, its drill, and its sample processing systems and present the reference surface mission. We conclude by addressing desirable scientific attributes of the landing-site region.

Large sections of this chapter were published previously in the work by Vago et al. (2017).

12.2 WHAT IS ExoMars?

ExoMars is a partnership program between the European Space Agency (ESA) and the Russian space agency, Roscosmos, for implementing two Mars missions.

The first mission was launched on 14 March 2016 from the Baikonur Cosmodrome, in Kazakhstan, on a Roscosmos-provided Proton rocket and arrived at Mars on 19 October 2016. ExoMars 2016 consists of two elements, the Trace Gas Orbiter (TGO) and the Schiaparelli entry, descent and landing demonstrator module (EDM), both contributed by ESA. The TGO carries European and Russian scientific instruments to conduct a detailed analysis of atmospheric gases, including methane (CH_4) and other minor constituents (Allen et al., 2006; Sherwood Lollar et al., 2006; Yung et al., 2010; Yung and Chen, 2015), and study the surface signatures of possible active processes (e.g., hydrothermal); TGO will also serve as a communication relay for future surface missions. The EDM was a European capsule to test technologies for controlled landing and to perform measurements during descent and while on the martian surface. Unfortunately, the last phase of the landing sequence did not operate as planned, and the lander was lost, but sufficient data were transmitted back to determine the cause of the accident.

The second mission will be launched on July 2020 and arrive on March 2021 to land an ESA rover using a descent module (DM) developed by Roscosmos (with some ESA-provided subsystems). The DM will travel to Mars on an ESA carrier module (CM). Roscosmos will launch the spacecraft composite on a Proton rocket. The rover will accommodate a European and Russian suite of instruments; it will also be equipped with a drill for subsurface sampling and a sample preparation and distribution system (SPDS), supporting the geology and life-seeking experiments in the rover's analytic laboratory. The Russian surface platform (SP) will contain an additional suite of Russian and European instruments, mainly concentrating on environmental and geophysical investigations.

NASA also provides important contributions to ExoMars, such as the Electra ultra-high frequency (UHF) radio package for TGO-to-Mars-surface proximity communications; engineering support for the DM; Deep Space Network services to complement ESA and Russian station coverage; and a major part of Mars organic molecule analyzer (MOMA), the organic molecule characterization instrument on the rover.

12.3 POSSIBLE LIFE ON MARS: WHEN AND WHERE?

As was the case on Earth (Genda, 2016), it is likely that early Mars ($\sim$4.45 Ga ago) also formed a primordial atmosphere through outgassing after accretion. This H_2O-rich atmosphere would have condensed, forming a global warm ocean (or at least large bodies of water) enveloped in a dense carbon dioxide (CO_2)-rich atmosphere (Elkins-Tanton, 2011). If we assume that the first stage of crustal-atmospheric evolution on Mars proceeded more or less as it did on Earth, most CO_2 would have been rapidly fixed in the planet's interior through carbonation and subduction. In a relatively short time, much of the greenhouse contribution from CO_2 would have disappeared. However, as was probably the case on Earth, we can expect that, driven by internal heat, active hydrothermal processes in the crust would have released CH_4 and other gases, helping to raise the planet's surface temperature (Pavlov et al., 2000; Oze and Sharma, 2005; Schulte et al., 2006). Mars' distance from the faint young Sun would have required relatively high greenhouse forcing to keep the surface above freezing conditions. Assuming an increase of 70 K—a plausible upper limit for greenhouse warming on the early Earth due to CO_2 and to the addition of gases liberated by widespread suboceanic serpentinization coupled with Fischer-Tropsch-type synthesis (Vago et al., 2017) (this value is 33 K for present-day Earth)—the average Mars temperature could have hovered around water's freezing point. Hence, we can conclude that for a good part of its early history, Mars could have looked like a colder version of present-day Iceland. Nevertheless, the likelihood of a mostly cold surface scenario does not constitute a serious impediment for the possible appearance of life since abundant subglacial and submerged volcanic/hydrothermal activity would have resulted in numerous liquid-water-rich settings (Warner and Farmer, 2010; Cousins and Crawford, 2011). The right mixture of ingredients, temperature and chemical gradients, organic molecule transport, concentration, and fixation processes could have been found just as well in a plethora of terrestrial submarine vents as in a multitude of vents under possibly top-frozen martian bodies of water (Westall et al., 2013; Russell et al., 2014).

Life seems to have appeared on Earth as soon as the environment allowed it—sometime between 4.4 and 3.8 Ga ago. Sustained by a young planet's internal heat, a similar process could have occurred early in the history of Mars. However, the availability of transport paths between liquid-water-rich environments proceeded differently on the two planets. Sometime during the late Noachian martian, surface habitable regions gradually became more isolated; their lateral connectivity started to dwindle and would eventually disappear (Westall et al., 2013, 2015). This process could be described as "punctuated" habitability. As surface conditions deteriorated, potential microbes could have found refuge in subterranean habitats (Michalski et al., 2013a). Occasionally, impact-formed hydrothermal systems would have resulted in transient liquid water becoming available close to the surface, even if the martian climate was cold (Rathbun and Squyres, 2002). But it does not necessarily follow that these habitats would have been colonized (Cockell et al., 2012). To maximize the chances of finding signs of past life, we must target the "sweet spot" in Mars' geologic history, the one with highest lateral water connectivity—the early Noachian (Poulet et al., 2005)—and look for large areas preserving evidence of prolonged, low-energy, water-rich environments, the type of environment that would have been able to receive, host, and propagate microorganisms.

The absence of a global plate tectonics on Mars (van Thienen et al., 2004) increases the probability that rapidly buried, ancient sedimentary rocks (possibly hosting microorganisms) may have been spared thermal alteration and been shielded from ionizing radiation damage until uncovered by eolian erosion relatively recently (Malin and Edgett, 2000), thus granting us access to rocks of an age no longer existing on Earth.

12.4 BIOSIGNATURES: WHICH AND HOW RELIABLE?

The main challenge for any life-seeking mission consists in determining whether a candidate observation (or better yet a collection of observations) can be uniquely attributed to the action of biology (Cady and Noffke, 2009).

12.4.1 Morphological Biosignatures

The primordial types of microorganisms that could have existed on early Mars would have been too small to distinguish. However, as on Earth, their permineralized or compressed microbial colonies and biofilms would be much larger. Traces of these features may be preserved on martian rocks as mineral-replaced structures and/or as carbonaceous remains trapped in sediments encased in mineral cement. Rover cameras and, especially, high-resolution close-up imagers would be able to investigate candidate microbialites similar to terrestrial thrombolites, stromatolites, layered biofilms, and abiotic/biotic organic particles and laminae (Westall, 2008; Westall et al., 2015; Ruff and Farmer, 2016).

12.4.2 Chemical Biosignatures

Most of the Earth's biological matter exists in the form of carbonaceous macromolecules stored within layered sedimentary rocks. If life existed on ancient Mars, organic material may have been preserved in extensive, organic-rich sedimentary deposits.

When considering molecular biosignatures, the first obvious set of targets is the ensemble of primary biomolecules associated with active microorganisms, such as amino acids, proteins, nucleic acids, carbohydrates, some pigments, and intermediary metabolites. Detecting the presence of these compounds in high abundance would be diagnostic of extant life, but unfortunately, they degrade quickly once microbes die. Lipids and other structural biopolymers, on the other hand, are biologically essential components (e.g., of cell membranes) known to be stable for billions of years when buried (Brocks, 1999; Georgiou and Deamer, 2014). The recalcitrant hydrocarbon backbone is responsible for the high-preservation potential of lipid-derived biomarkers relative to that of other biomolecules (Eigenbrode, 2008).

Along the path from primary compound to molecular fossil, all biological materials undergo in situ chemical reactions dictated by the circumstances of the source organisms' transport, deposition, entombment, and postdepositional conditions. The end product of diagenesis is macromolecular organic matter, which, through the loss of superficial hydrophilic functional groups, slowly degrades into the solvent-insoluble form of fossil carbonaceous matter called kerogen, but not all information is lost. The heterogeneous chemical structure of the kerogen matrix can preserve patterns and distribution diagnostic of biosynthetic pathways. Kerogen also possesses molecular sieve properties allowing it to retain diagenetically altered biomolecules (Tissot and Welte, 1984).

In addition to the direct recognition of biomolecules and/or their degradation products, other characteristics of bioorganic compounds include the following (Summons et al., 2008, 2011):

Isomerism selectivity: Some important biologically produced molecules occur preferentially as one stereoisomer (e.g., left-handed amino acids).

Molecular-weight fingerprints: This category includes the uneven distribution of clusters of structurally related organic compounds, the repetition of constitutional subunits, and the systematic isotopic ordering. These signals can be detected by analyzing chromatograms and mass spectra.

Bulk isotopic fractionation: The isotopic fractionation of stable elements like C, H, O, N, S, and Fe can be used as signature to recognize the action of biological pathways. However, bulk isotopic fractionation is not considered a robust biosignature when applied to locations or epochs for which we have scant knowledge of sources and sinks. In the specific case of carbon, $^{13}C/^{12}C$ ratios may serve as reliable biosignatures for past or present life only if the key components of the C-cycling system (applicable at the time of deposition and since then) are well constrained (Summons et al., 2011). This is certainly not the case for Mars, and one can also wonder to what extent we are sure about our own past carbon dynamics when analyzing ancient samples. We are willing to include bulk isotopic fractionation in this list but only with the caveat that it should be used in association with other, less indirect, biosignatures.

12.4.3 Importance of Geological Context for Boosting Biosignature Confidence

Demonstrating that a sample has been obtained from a geologic setting possessing long-standing aqueous attributes that could have allowed hosting and propagation of microorganisms would help to increase the confidence of any potential biosignature claim.

12.4.4 ExoMars Biosignature Targeting

The ExoMars rover can search for two broad classes of biosignatures: (1) morphological, textural information preserved on outcrops, rocks, and collected samples, and (2) biochemical, in the form of bioorganic compounds and their degradation products. The rover is also capable of exploring the landing site and establishing the geologic environment at the time of deposition and its subsequent evolution.

The biosignatures that cannot be addressed by the Pasteur payload are (1) visual recognition of individual organism microfossils, which is only achievable on Earth with high-magnification instruments, for example, electron microscopy conducted on thin-section, acid-etched samples, and (2) bulk isotope excursions, which we claim are not as robust a diagnostic as others.

Within the available resource envelope, the science team tried to implement the techniques that we believed could, when used in a combined fashion, give us the best chance to achieve a (potential) positive detection.

Please note that in the previous discussion, we have not considered morphological changes with time, movement, or experiments designed to elicit active metabolic responses. As demonstrated by the Viking lander biologic experiments, these "more dynamic" expressions of possible extant life would not be easy to verify (Navarro-Gonzalez et al., 2006; Biemann, 2007; Biemann and Bada, 2011; Navarro-González and McKay, 2011).

12.5 THE NEED FOR SUBSURFACE EXPLORATION

12.5.1 Results From Previous Missions

In the late 1970s, twin Viking landers conducted the first in situ measurements on the martian surface. Their biology package contained three experiments, all designed to test for signs of metabolism in soil samples (Klein et al., 1976). One of them, the Labeled Release Experiment, produced provocative results (Levin and Straat, 2016). If other information had not been also obtained, these data would have been interpreted as proof of biological activity. However, theoretical modeling of the martian atmosphere and regolith chemistry hinted at the existence of powerful oxidants that could, more or less, account for the results of the three biology package experiments (Klein, 1999). The biggest blow was the failure of the gas chromatograph–mass spectrometer (GCMS) to find evidence of organic molecules at the parts-per-billion level. With few exceptions, the majority of the scientific community concluded that the Viking findings did not demonstrate the presence of extant life (Klein, 1998, 1999).

The successful 1996 Mars Global Surveyor and 2003 Mars Exploration Rovers (MER), which were conceived as robotic geologists, have demonstrated the past existence of wet environments (Malin and Edgett, 2000; Squyres et al., 2004a,b, 2012). But perhaps, it has been Mars Express 2003 and Mars Reconnaissance Orbiter 2005 that have most drawn our attention to ancient Mars, revealing many instances of finely layered deposits containing phyllosilicate minerals that could only have formed in the presence of liquid water, which reinforced the hypothesis that early Mars was wetter than today (Poulet et al., 2005; Bibring et al., 2006; Loizeau et al., 2010, 2012; Ehlmann et al., 2011; Bishop et al., 2013; Michalski et al., 2013b).

The next incremental step in our chemical understanding of the martian surface was entirely unexpected. It came as a result of measurements conducted by the 2007 Phoenix lander in the northern subpolar plains. Phoenix included, for the first time, a wet chemistry analysis instrument that detected the presence of the perchlorate (ClO_4^-) anion in soil samples collected by the robotic arm (Hecht et al., 2009; Kounaves et al., 2010, 2014). Perchlorate salts are chemically inert at room temperature, but when heated beyond a few hundred degrees, the four oxygen atoms are released and become reactive oxidation vectors. Soon thereafter, investigators recalled that Viking had relied on thermal volatilization (TV; in other words, heat) to release organics from soil samples (Navarro-González et al., 2010, 2011; Biemann and Bada, 2011; Navarro-González and McKay, 2011). If perchlorate had been present in the soil at the two Viking lander locations, perhaps heating could explain the negative organic carbon results obtained? In fact, some simple chlorinated organic molecules (chloromethane and dichloromethane) had been detected by the Viking experiments (Biemann et al., 1977), but these compounds were interpreted to have resulted from a reaction between adsorbed residual methanol (a cleaning agent used to prepare the spacecraft) and HCl. Today, there is a general suspicion that the results from the Viking biological experiments were influenced by heat-activated perchlorate dissociation and reaction with indigenous organic compounds (Steininger et al., 2012; Glavin et al., 2013; Quinn et al., 2013; Sephton et al., 2014; Goetz et al., 2016; Lasne et al., 2016).

More recently, Curiosity's SAM experiment team detected oxygen (O_2) released by the thermal decomposition of oxychlorine species [i.e., perchlorates and/or chlorates (Archer et al., 2016)] and chlorine-bearing hydrocarbons attributable to the reaction of oxychlorine species with organic compounds, when they analyzed modern sand deposits and when they drilled into much older rocks (Glavin et al., 2013; Freissinet et al., 2015). The inferred presence of perchlorate in the two different types of material (granular, recently transported, and consolidated, ancient) cannot be explained by cross contamination between samples. The ExoMars biosignature identification strategy needs to work also when the material to be analyzed contains perchlorate. We will see that this is indeed the case.

12.5.2 Degradation of Organic Matter

Effective chemical identification of biosignatures requires access to well-preserved organic molecules. Because the martian atmosphere is more tenuous than Earth's, three important physical agents reach the surface of Mars with adverse effects for the long-term preservation of biomarkers: (1) The UV radiation dose is higher than on our planet and will quickly damage exposed organisms or biomolecules. (2) UV-induced photochemistry is responsible for the production of reactive oxidant species that, when activated, can also destroy chemical biosignatures. The diffusion of oxidants into the subsurface is not well characterized and constitutes an important measurement that the mission must perform. Finally, (3) ionizing radiation penetrates into the uppermost meters of the planet's subsurface. This causes a slow degradation process that, operating over many millions of years, can alter organic molecules beyond the detection sensitivity of analytic instruments. Radiation effects are depth-dependent: the material closer to the surface is exposed to higher doses than that buried deeper (Hassler et al., 2014).

12.5.3 Access to Molecular Biosignatures

The molecular record of ancient martian life, if it ever existed, is likely to have escaped radiation and chemical damage only if trapped in the subsurface for long periods. Studies suggest that a subsurface penetration in the range of 2 m is necessary to recover well-preserved organics from the early history of Mars (Kminek and Bada, 2006)—assuming there has been some help from additional, recently eroded overburden (Dartnell et al., 2007, 2012; Parnell et al., 2007; Pavlov et al., 2012; Hassler et al., 2014). It is also essential to avoid loose dust deposits distributed by eolian transport. In the course of being driven by the wind, this material has been processed by UV radiation, ionizing radiation, and potential oxidants in the atmosphere and on the surface of Mars. Any organic biosignatures would be highly degraded in these samples. For all the above reasons, it was decided early on that the ExoMars rover must be able to penetrate and obtain samples from well-consolidated (i.e., hard) formations, such as sedimentary rocks and evaporitic deposits, at various depths from 0 down to 2 m.

The major difficulty with the investigation of biogenic material lies not in the recognition of fossil biosignatures, but in the ability to obtain the correct sample to study. As justified previously, it is water-lain sedimentary deposits from Mars' early history that we are interested in. But not any old, wet location is suitable. We require ancient sites that have been uncovered by the action of wind only recently for molecular biosignature preservation against the ravages of long-term ionizing radiation. In the absence of a deep drill, the rover would need to drive close to a receding scarp to gain access to shallow material having experienced a lower radiation dose (Farley et al., 2014). The samples would need to be not only of the right age, of the right aqueous environment, of the right type of deposit, and with the right exhumation history, but also have been collected at the foot of a scarp? How likely would that be? The ExoMars science team realized early on that having a subsurface drill greatly increases the probability to collect well-preserved material for analysis. It also provides the added benefit of being able to study how the geochemical environment changes with depth.

12.6 THE ExoMars ROVER AND ITS PASTEUR PAYLOAD

12.6.1 From Panoramic to Molecular Scale Through Nested Investigations

The mission strategy to achieve the ExoMars rover's scientific objectives is as follows:
- To land on an ancient location possessing high exobiological interest for past life signatures, that is, access the appropriate geologic environment.
- To collect well-preserved samples (free from radiation and oxidation damage) at different sites using a rover equipped with a drill capable of reaching well into the ground and surface rocks.
- To conduct an integral set of measurements at multiple scales to achieve a coherent understanding of the geologic context and, thus, inform the search for biosignatures. Beginning with a panoramic assessment of the geologic environment, the rover must progress to smaller-scale investigations of surface rock textures and culminate with the collection of well-selected samples to be studied in its analytic laboratory.

The ExoMars rover will have a nominal lifetime of 218 sols (~7 Earth months). During this period, it will ensure a regional mobility of several kilometers relying on solar array electric power. Figs. 12.1 and 12.2, respectively, present front and rear views of the rover with some general size information. Its mass is ~310 kg, with an instrument payload of 26 kg (excluding payload servicing equipment such as the drill and sample processing mechanisms).

The rover's kinematic configuration is based on a six-wheel, triple-bogie concept (Fig. 12.2 bottom left) with locomotion formula $6 \times 6 \times 6 + 6$, denoting six supporting

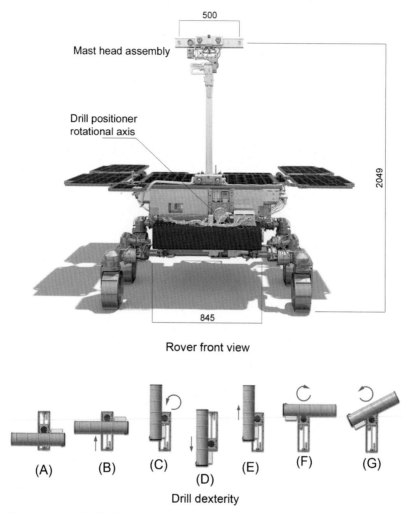

Fig. 12.1 Front view of the ExoMars rover with general dimensions. The drill can acquire samples at depths ranging between 0 and 2 m. The drill box lies horizontally across the rover's front face when traveling (A). It is raised (B), rotated counterclockwise (C), and lowered vertically to commence drilling operations (D). Once a sample has been acquired, the drill is elevated (E), turned clockwise (F), and further inclined to deliver the sample (G). The inlet port to the analytic laboratory can be seen on the rover's front, above the drill box, to the left.

wheels, six driven wheels, and six steered wheels, plus six articulated (deployment) knee drives. This system enables the rover to passively adapt to rough terrains, providing inherent platform stability without the need for a central differential. The rover can perform drive and turn-on-spot maneuvers, double-Ackermann steering, and diagonal crabbing motions; the latter can be useful for moving sideways across an outcrop for imaging.

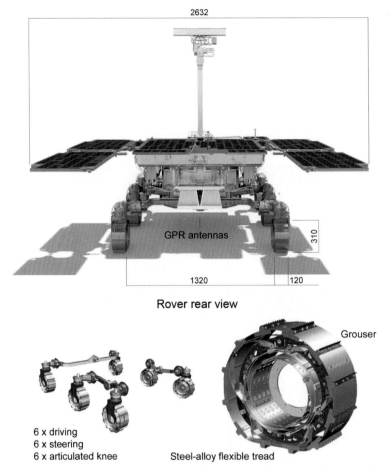

2632

GPR antennas   310

1320   120

Rover rear view

Grouser

6 x driving
6 x steering
6 x articulated knee   Steel-alloy flexible tread

Fig. 12.2 Rear perspective of the ExoMars rover with general dimensions. The rover's locomotion configuration is based on a triple-bogie concept and has flexible wheels to improve tractive performance.

Lander accommodation constraints have imposed the use of relatively small wheels (28.5 cm diameter without grousers and 12.0 cm width). To reduce the traction performance disadvantages of small wheels, flexible wheels have been adopted (Fig. 12.2 bottom right) (Favaedi et al., 2011); their high deformation enlarges the size of the wheel/soil contact patch, reduces ground pressure (to ~10 kPa average), and offers a substantial impact load-absorption capability (Poulakis et al., 2016). For comparison, the average wheel ground pressure of the MER (25 cm wheel diameter without grousers and 16 cm width) and MSL (48 cm wheel diameter without grousers and 40 cm width) rovers is 5.75 kPa (Heverly et al., 2013). So the ExoMars rover exceeds the wheel ground pressure prescription of NASA rovers. This is a concern since, even with less wheel ground pressure, Opportunity experienced serious difficulties with unconsolidated terrain at

Purgatory Ripple (Maimone et al., 2007) and the same happened to MSL when attempting to traverse wind-blown, megaripple deposits (Arvidson et al., 2016). To mitigate this risk, the ExoMars team is investigating the possibility to (re)enable wheel walking (Patel et al., 2010); a coordinated rototranslational wheel gait that our tests have demonstrated can greatly improve dynamic stability during rover egress, provide better traction for negotiating loose soils (in case the rover gets stuck during normal driving), and increase slope gradeability (Azkarate et al., 2015).

12.6.2 Pasteur Payload Instruments

The rover's Pasteur payload will produce comprehensive sets of measurements capable of providing reliable evidence for or against the existence of a range of biosignatures at each search location. The Pasteur payload contains panoramic instruments (cameras, an infrared (IR) spectrometer, a ground-penetrating radar, and a neutron detector); contact instruments for studying rocks and collected samples (a close-up imager and an IR spectrometer in the drill head); a subsurface drill capable of reaching a depth of 2m and obtaining specimens from bedrock; a sample preparation and distribution system (SPDS); and the analytic laboratory, the latter including a visual + IR imaging spectrometer, a Raman spectrometer, and a laser desorption, TV GCMS (with the possibility to use three different derivatization agents)—see Fig. 12.3 and Table 12.1.

If any bioorganic compounds are detected on Mars, it will be important to show that they were not brought from Earth. Great care is being devoted during the assembly, testing, and integration of instruments and rover components. Strict organic cleanliness requirements apply to all parts that come into contact with the sample and to the rover assembly process. Once completed, the analytic laboratory drawer (ALD) will be sealed and kept at positive pressure throughout transport, final integration, launch, cruise, and landing on Mars. The ExoMars rover will carry a blank in each drill tip (nominal and backup) to reliably demonstrate that the entire sample chain from acquisition through handling, processing, and analysis is free from contaminants. An additional six, individually encapsulated blanks will be stored in a dedicated dispenser. When deemed necessary, they can be used to evaluate the organic cleanliness of the sample handling and analysis chain. Upon landing, one of the first science actions will be for the drill to pass a blank sample to the analytic laboratory. After performing a full investigation, the results should indicate "no life" and "no organics."

Below is a brief summary of the Pasteur payload capabilities.

12.6.2.1 Panoramic Camera System

Panoramic camera (PanCam) (Coates et al., 2012, 2017; Cousins et al., 2012; Yuen et al., 2013) is designed to perform digital terrain mapping for the ExoMars rover mission. A suite that consists of a fixed-focus, wide-angle, stereoscopic, color camera pair (WAC) complemented by a focusable, high-resolution, color camera (HRC), PanCam, will enable the science team to characterize the geologic environment at the sites the

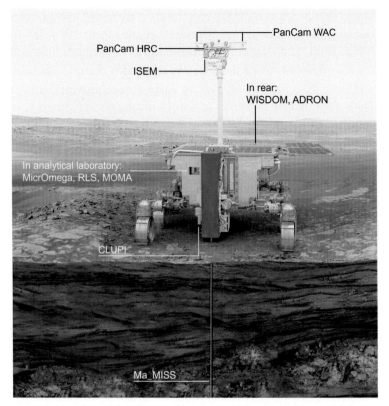

Fig. 12.3 Sketch of ExoMars rover showing the location of the drill and the nine Pasteur payload instruments.

rover will visit—from panoramic (tens of meters) to millimeter scale. It will be used to examine outcrops, rocks, and soils in detail and to image samples collected by the drill before they are delivered to the analytic laboratory for analysis. PanCam will also be a valuable asset for conducting atmospheric studies:

> *PanCam WAC*: 1024 × 1024-pixel, multispectral, stereoscopic images with 32.28° (horizontal/vertical) field of view (FOV).
> *PanCam HRC*: 1024 × 1024-pixel, color, monoscopic images with "telephoto" 4.88° (horizontal/vertical) FOV.

12.6.2.2 IR Spectrometer

Infrared spectrometer for ExoMars (ISEM) (Korablev et al., 2015, 2017) is a pencil-beam infrared spectrometer mounted on the rover mast that is coaligned with the PanCam HRC. ISEM will record IR spectra of solar light reflected off surface targets, such as rocks

Table 12.1 The ExoMars rover's Pasteur payload can perform a detailed mineralogical and chemical characterization of surface and subsurface material collected with the drill

Instrument	Scientific rationale
Panoramic instruments	To characterize the rover's geologic context, both at the surface and the subsurface. Typical scales span from panoramic (100 m) to 1 m, with a spatial resolution in the order of 1 cm for close targets
Panoramic camera system	PanCam: two wide-angle stereo cameras and one high-resolution camera to investigate the rover's environment and landing-site geology. Also important for target selection and for rock textural studies
IR spectrometer	ISEM: for bulk mineralogy characterization, for remote identification of water-related minerals, and for aiding PanCam with target selection
GPR	WISDOM: to establish subsurface stratigraphy down to 3 m depth and help plan the drilling strategy
Neutron detector	ADRON: to determine the level of subsurface hydration and the possible presence of an ice fraction to 1 m depth
Contact instruments	To investigate outcrops, rocks, and soils. Among the scientific interests at this scale are macroscopic textures, structure, and layering. This information will be fundamental to understand the local depositional environment and to search for morphological biosignatures on rocks
Close-up imager	CLUPI: to study rock targets at close range (50 cm) with submillimeter resolution. This instrument will also investigate the fines produced during drilling operations and image samples collected by the drill. The close-up imager has variable focusing and can obtain high-resolution images also at longer distances. Certain morphological biosignatures, such as biolamination, if present, can be identified by CLUPI
IR spectrometer in drill	Ma_MISS: for conducting mineralogical studies in the drill borehole's walls
Support subsystems	These essential devices are devoted to the acquisition and preparation of samples for detailed studies in the analytic laboratory. The mission's ability to break new scientific ground, particularly for "signs of life" investigations, depends on these two subsystems
Subsurface drill	Capable of obtaining samples from 0 to 2 m depth, where organic molecules can be well preserved from radiation damage. Includes a blank sample, temperature sensors, and an IR spectrometer (Ma_MISS)
Sample preparation and distribution system	Receives a sample from the drill system, produces particulate material preserving the organic and water fractions, and presents it to all analytic laboratory instruments. Includes a dispenser with additional blank samples

Continued

Table 12.1 The ExoMars rover's Pasteur payload can perform a detailed mineralogical and chemical characterization of surface and subsurface material collected with the drill—cont'd

Instrument	Scientific rationale
Analytic laboratory	To perform a detailed, coordinated analysis of each collected sample. After sample crushing, the initial step is a visual and spectroscopic investigation. Thereafter follows a first search for organic molecules. In case interesting results are found, the instruments are able to perform more in-depth analyses
VIS + IR imaging spectrometer	MicrOmega: will examine the crushed sample material to characterize structure and composition at grain-size level. These measurements will be used to help point the laser-based instruments (RLS and MOMA)
Raman laser spectrometer	RLS: to identify mineral phases at grain scale in the crushed sample material, determine their composition, and establish the presence of carbon (inorganic/organic)
Mars organic molecule analyzer	MOMA (LD + Der-TV GCMS): MOMA is the rover's largest instrument. Its goal is to conduct a broad-range, high-sensitivity search for organic molecules in the collected sample. It includes two different ways of extracting organics: (1) LD and (2) TV, with or without derivatization (Der) agents, followed by separation using four GC columns. The identification of the evolved organic molecules is achieved with an ion-trap MS

ADRON, active detector for gamma rays and neutrons; *CLUPI*, close-up imager; *GC*, gas chromatograph; *GPR*, ground-penetrating radar; *IR*, infrared; *ISEM*, infrared spectrometer for ExoMars; *LD*, laser desorption; *Ma_MISS*, Mars multispectral imager for subsurface studies; *MOMA*, Mars organic molecule analyzer; *MS*, mass spectrometer; *RLS*, Raman laser spectrometer; *SPDS*, sample processing and distribution system; *TV*, thermal volatilization; *WISDOM*, water, ice, and subsurface deposit observations on Mars.

and soils, to determine their bulk mineralogical composition. ISEM will be a useful tool to discriminate between various classes of minerals at a distance. This information can be employed to decide which target to approach for further studies. ISEM can also be used for atmospheric studies:

ISEM: 1.1–3.3 μm spectral range and $20\,cm^{-1}$ spectral sampling, with 1° FOV.

12.6.2.3 Shallow Ground-Penetrating Radar

The water, ice, and subsurface deposit observations on Mars (WISDOM) radar (Ciarletti et al., 2011, 2017) will characterize stratigraphy to a depth of 3–5 m with vertical resolution of the order of a few centimeters (depending on subsurface electromagnetic properties). WISDOM will allow the team to construct two- and three-dimensional subsurface maps to improve our understanding of the deposition environment. Most

importantly, WISDOM will identify layering and help select interesting buried formations from which to collect samples for analysis. Targets of particular interest for the
ExoMars mission are well-compacted, sedimentary deposits that could have been associated with past water-rich environments. This ability is fundamental to achieve the
rover's scientific objectives, as deep subsurface drilling is a resource-demanding operation
that can require several sols:

> *WISDOM*: Broadband UHF GPR (0.5–3.0 GHz), step frequency, bistatic and polar-
> imetric (XX-XY-YX-YY) measurements, penetration depth ~3 m, and vertical
> resolution of a few centimeters.

12.6.2.4 Subsurface Neutron Detector

Active detector for gamma rays and neutrons (ADRON) (Mitrofanov et al., 2017) will
count the number of thermal and epithermal neutrons scattered in the martian subsurface
to determine hydrogen content (present as grain adsorbed water, water ice, or hydrated
minerals) in the top 1 m. This information will complement the subsurface characterization performed by WISDOM:

> *ADRON*: Detects neutrons in the broad range from 0.01 eV to ~100 keV.

12.6.2.5 Close-Up Imager

Close-up imager (CLUPI) will obtain high-resolution, color images to study the depositional environment (Josset et al., 2017). By observing textures in detail, CLUPI can
recognize potential morphological biosignatures, such as biolamination, preserved on
surface rocks. This is a key function that complements the possibilities of PanCam when
observing close targets at high magnification. CLUPI will be accommodated on the drill
box and has several viewing modes, allowing the study of outcrops, rocks, soils, the fines
produced during drilling, and also imaging collected samples in high resolution before
delivering them to the analytic laboratory:

> *CLUPI*: 2652×1768-pixel, color, z-stacked images and $11.9° \times 8.0°$ FOV, imaging
> resolution varies with distance to target; for example, it is 8 μm/pixel at 11.5 cm dis-
> tance with view area 2.0×1.4 cm, 39 μm/pixel at 50 cm distance with view area
> 10×7 cm, and 79 μm/pixel at 100 cm distance with view area 21×14 cm.

12.6.2.6 Drill IR Spectrometer

Mars multispectral imager for subsurface studies (Ma_MISS) (De Angelis et al., 2013;
De Sanctis et al., 2017) is a miniaturized IR spectrometer integrated in the drill tool for
imaging the borehole wall as the drill is operated. Ma_MISS provides the capability to
study stratigraphy and geochemistry in situ. This is important since deep samples may

be altered following their extraction from their cold, subsurface conditions, typically of the order of −50°C at midlatitudes (Grott et al., 2007). The analysis of unexposed material by Ma_MISS, coupled with other data obtained with spectrometers located inside the rover, will be crucial for the unambiguous interpretation of rock formation conditions:

> *Ma_MISS*: 0.4–2.2 μm spectral range and 20 nm spectral sampling, with spatial resolution of 120 μm (corresponding to one rotational step of the drill tool).

12.6.2.7 Subsurface Drill

ExoMars employs a rotary drill (with no percussion capability) to acquire (~3 cm long by 1 cm diameter) samples (solid, fragments, or powder) at depths ranging between 0 and 2 m. The drill box lies horizontally across the rover's front face when traveling (Fig. 12.1A); it assumes a vertical stance for drilling (Fig. 12.1D) and is raised and rotated for delivering a sample to the analytic laboratory's inlet port (Fig. 12.1G). The drill box's dexterity is also used for orienting CLUPI observations (CLUPI is not shown in sketches A–G).

The drill is composed of the following elements: (1) a drill tool ~70 cm long, equipped with the sample acquisition device (including a shutter, movable piston, and position and temperature sensors) and the Ma_MISS front elements (sapphire window, IR lamp, reflector, and optical fiber); (2) a set of three extension rods, 50 cm each, to achieve the required penetration depth; they contain optical and electric contacts for the transmission of Ma_MISS signals to the spectrometer in the upper part of the drill unit; (3) a backup drill tool without spectrometer; and (4) the rotation-translation group, comprising sliding carriage motors, guides, and sensors.

Preserving the sample's organic and volatile content is of paramount scientific importance. The drill has thermocouples close to the tip to monitor temperature variations in the sample collection region. We have conducted numerous tests in Mars chambers using different geologically representative, simulated stratigraphic columns, including ice lenses varying from 0% to 35% water content.

Temperature increases because of drilling are ephemeral and modest, in the order of 20°C, when we proceed in a continuous manner, and can be reduced to ≤5°C if we implement a variable cutting law ("cut a little, wait a little" to allow dissipating thermal energy) just before collecting the sample; however, this means more time. The low atmospheric pressure on Mars leads to the rapid sublimation of any ice particles directly in contact with the drill tip, resulting in an upward traveling gas jet that can be helpful for evacuating drill fines from the borehole. Considering the typical temperature of subsurface materials on Mars (which at midlatitudes can experience oscillations between −30°C and −80°C at 0.5 m depth and have an average value of about −50°C deeper), we can adapt our strategy to ensure that samples remain sufficiently cold throughout the drilling process.

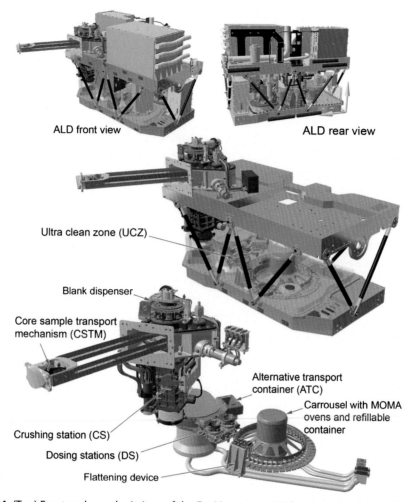

ALD front view

ALD rear view

Ultra clean zone (UCZ)

Blank dispenser

Core sample transport
mechanism (CSTM)

Alternative transport
container (ATC)

Carousel with MOMA
ovens and refillable
container

Crushing station (CS)

Dosing stations (DS)

Flattening device

Fig. 12.4 (Top) Front and rear depictions of the ExoMars rover ALD housing MicrOmega, RLS, MOMA, and SPDS. (Middle) The UCZ envelops the entire sample-handling path and is sealed at positive pressure until open on Mars. (Bottom) SPDS mechanisms: the sample is deposited in the CSTM and, after being imaged with CLUPI and PanCam, is retracted into the ALD. The rock CS crushes the sample and discharges the resulting particulate matter into a DS. The DS pays out the necessary amount of sample material onto the refillable container or into a MOMA oven, as necessary. *ALD*, analytic laboratory drawer; *CLUPI*, close-up imager; *CS*, crushing station; *CSTM*, core sample transport mechanism; *DS*, dosing station; *MOMA*, Mars organic molecule analyzer; *SPDS*, sample preparation and distribution system; *UCZ*, ultra clean zone.

12.6.2.8 Sample Preparation and Distribution System

The entire ALD sample path is enclosed in a so-called ultra clean zone (UCZ), which is shown as a transparent volume in Fig. 12.4 (middle). The SPDS groups the ensemble of ALD mechanisms used for manipulating sample material (Fig. 12.4 bottom). The SPDS receives a sample from the drill by extending its core sample transport mechanism

(CSTM), a sort of *hand* that comes out through a door in the rover's front panel (shown in Fig. 12.3). Once deposited in the CSTM, typically at the end of a sol's work, PanCam HRC and CLUPI can image the sample during a narrow time window of a few minutes; this duration is based on the results of a sample contamination analysis from possible external rover system sources.

After the imaging exercise has been completed, the CSTM retracts, moving the sample into the analytic laboratory. A rock crusher is used to produce particulate matter having a more or less Gaussian size distribution ranging from a few to ~500 μm, with 250 μm as the median value. This is done early in the morning, when the temperature in the ALD is at its lowest, to preserve as much as possible the organic and volatile fractions in the sample. The temperature of the crushing station (CS) is monitored prior to and throughout the crushing process. The SPDS includes a blank dispenser with the capability to provide individual blank samples to the rock crusher. The pulverized sample material drops into one of two, redundant dosing stations (DSs). Their function is to distribute the right amount of sample either to a refillable container—a flat tray where mineral grains can be observed by ALD instruments—or into individual, single-use ovens. A rotating carrousel accommodates the refillable container and ovens under the DS. Both DSs are piezovibrated to improve the flow of granular material. The refillable container is further served by two other mechanisms: The first flattens the crushed sample material at the correct height to present it to the ALD instruments, and the second is utilized to empty the refillable container so that it can be used again.

A number of emergency devices have been implemented to deal with potential offnominal situations. To prevent the CS from becoming blocked, a spring-actuated hammer can impart a strong shock to the fixed jaw, where material may stick. In case of jamming, a special actuator can open the CS jaws to evacuate the entire sample. If both DSs were to fail, they can be bypassed. An alternative transport container allows dropping the entire crushed sample material at once, without control for the quantity provided, either onto the refillable container or into an oven.

12.6.2.9 MicrOmega

Micro observatoire pour la mineralogie, l'eau, les glaces et l'activité (MicrOmega) (Pilorget and Bibring, 2013; Bibring et al., 2017) will be the first instrument to image the crushed sample material. MicrOmega is a near-IR hyperspectral camera that will study mineral grain assemblages in detail to try to unravel their geologic origin, structure, and composition. Its FOV covers a sample area of $5 \times 5 \, mm^2$. These data will be vital for interpreting past and present geologic processes and environments on Mars. The rover computer can analyze a MicrOmega hyperspectral cube's absorption bands at each pixel to identify particularly interesting minerals and assign them as objectives for Raman and MOMA laser-desorption mass spectrometry (LDMS) observations:

MicrOmega: 250×256-pixel $\times$ 320-spectral-step VIS + IR image cubes, 0.95–3.65 μm spectral range, and $20\,cm^{-1}$ spectral sampling, with imaging resolution of 20 μm/pixel.

12.6.2.10 Raman Laser Spectrometer

Raman laser spectrometer (RLS) (Edwards et al., 2013; Foucher et al., 2013; Lopez-Reyes, 2015; Rull et al., 2017) provides geologic and mineralogical information on igneous, metamorphic, and sedimentary processes, especially regarding water-related interactions (chemical weathering, chemical precipitation from brines, etc.). In addition, it also permits the detection of a wide variety of organic functional groups. Raman can contribute to the tactical aspects of exploration by providing a quick assessment of organic content prior to the analysis with MOMA:

RLS: Continuous excitation, 532 nm (green laser) with a 50 μm-size spot on the target, covering an $\sim$150–3800 cm^{-1} spectral shift with Raman resolution $\sim$6 cm^{-1} in the fingerprint spectral region below 2000 cm^{-1} and with slightly degraded spectral resolution beyond this value.

12.6.2.11 Mars Organic Molecule Analyzer

MOMA is the largest instrument in the rover and the one that directly targets chemical biosignatures. MOMA is able to identify a broad range of organic molecules with high analytic specificity, even if present at low concentrations (Arevalo et al., 2015; Goetz et al., 2016; Goesmann et al., 2017).

MOMA has two basic operational modes: LDMS, to study large macromolecules and inorganic minerals (Busch, 1995; Bounichou, 2010), and GCMS, for the analysis of volatile organic molecules. In MOMA-LDMS, crushed drill sample material is deposited in a refillable container. A high-power, pulsed UV laser fires on the sample, and the resulting molecular ions are guided into the mass spectrometer for analysis. In MOMA-GCMS, sample powder is deposited into 1 of 32 single-use ovens. The oven is sealed and heated up stepwise to a high temperature (for some ovens, in the presence of a derivatization agent). The ensuing gases are separated by gas chromatography and delivered to the shared mass spectrometer for analysis. This combined process of derivatization, chromatographic separation, and mass spectrometric identification is useful for small organic molecules, such as amino acids.

The MOMA instrument implements innovative techniques for the extraction and robotic characterization of organic molecules, including the derivatization of refractory molecules such as carboxylic acids and amino acids. For the elucidation of the chirality of martian analytes, the MOMA gas chromatograph employs one chiral stationary phase that is able to resolve and quantify enantiomers of many different families of organic molecules. Furthermore, the MOMA-LDMS mode of operation does not seem to be affected by the presence of perchlorate in the sample (Li et al., 2015); the laser energy deposition pulse is too fast for perchlorates to dissociate and trigger oxidative reactions but effective enough to desorb organic molecules.

An early MOMA-GCMS prototype was tested in the field during the AMASE11 field Campaign in Svalbard, Norway (Siljeström et al., 2014):

MOMA Pyr GCMS: 20 pyrolysis ovens, each ~150 mm^3 sample volume heated to any desired temperature <800°C, and four different GC columns (including one enantioselective). For volatile organics (e.g., alkanes, amines, alcohols, and carboxylic acids), detection mass range of 50–500 Da, and detection limit ≤ nmol analyte (signal-to-noise ratio (SNR) ≥10).

MOMA derivatization (Der) GCMS and thermochemolysis (Pyr ± Der) GCMS: four ovens for each of three derivatization agents: (1) MTBSTFA/DMF (for carboxylic and amino acids, nucleobases, amines, and alcohols), (2) DMF-DMA (for amino acids, fatty acids, and primary amines with chiral centers; this agent preserves the asymmetrical center C* so will be used together with the enantioselective GC column for chiral separation), and (3) TMAH (for lipids, for fatty acids, and—when driven to higher temperatures, e.g., 700°C—for more refractory compounds such as PAHs and kerogen); each ~150 mm^3 sample volume heated to some moderately high temperature from 100°C to 300°C, detection mass range of 50–500 Da, and detection limit ≤ nmol analyte (SNR ≥10).

MOMA-LDMS: λ=266 nm UV laser, ~1.3 ns, ≤135 μJ pulses in bursts ≤100 Hz (average 2 Hz), with a spot size of 400 × 600 μm and a depth of 10 nm/shot. For non-volatile organics (e.g., macromolecular carbonaceous compounds, proteins, and inorganic species), detection mass range of 100–1000 Da, and detection limit ≤ pmol/mm^2 analyte (SNR ≥3).

12.7 THE REFERENCE SURFACE MISSION

The reference surface mission (RSM) defines a rover exploration scenario that has two *raisons d'être*. The first is scientific. The RSM specifies a logical sequence of science steps, proceeding from large- to small-scale studies, concluding with the collection and analysis of samples from both surface and subsurface targets. The RSM is ambitious and thus affords sufficient rover resources and operational scope to carry out something completely different in case the landing site would not match our expectations. For instance, the rover could travel a distance far longer than originally planned to reach a suitable prime science location, at the expense, of course, of time for investigations. The second purpose of the RSM is formal. When ESA placed a contract with European industry to procure the ExoMars rover, it specified that it must be capable of executing the RSM within the nominal mission duration. The demonstration—

through simulation and tests—that the rover can complete the RSM is an agreed requirement.

The RSM begins with the rover deployment and egress sequence in either of two possible directions. Thereafter, a short functional commissioning phase is performed in the vicinity of the surface platform (duration 10 sols). The rover then moves (~60 m) away from the descent engine blast contamination zone. Once sufficiently far, the drill and the ALD may be opened. The first ALD science operation is the drill blank analysis run to perform a full calibration and assess organic cleanliness. Only after this has been completed can the search for biosignatures begin (duration 5 sols).

The exploration part of the RSM includes six experiment cycles (ECs) and two vertical surveys (VSs). During the course of an EC, the rover exercises all Pasteur instruments and analyzes two samples, one surface and one subsurface—the latter specified to be obtained at 1.5 m depth. The distance between EC locations is arbitrarily assumed to increase in 100 m steps (100 m to the first spot, an additional 200 m to the second, and so on) for a total surface travel of approximately 1.5 km. Twice during the nominal mission, most likely in case something particularly interesting is found, a VS can be executed. During a VS, samples are collected (and analyzed) at 0, 50, 100, 150, and 200 cm depth from the same place. The objective of a VS is to understand how organic compound preservation and overall geochemistry vary with depth.

At least for the first few months, rover operations will be conducted on "Mars time." We assume two communication sessions per sol with Earth through TGO; this is the nominal condition. Typically, ground control instructs the rover what to do during a morning pass, and the rover reports back the results of its travails in the evening pass. This means that all critical data required to define the next sol's activities must reach the Rover Operations and Control Center (ROCC) with the evening pass. However, since TGO's orbit is not Sun-synchronous, the local time of communication sessions drifts in a complex but (fortunately) predictable manner—they take place ~30 min earlier each sol for three consecutive sols and then jump forward ~2 h to start the cycle again. Moreover, the duration and data volume capacity of TGO overflights are not constant because of the varying geometry between TGO and the landed asset. These two conditions introduce an additional constraint on strategic operation planning: it may not always be possible to complete the required tasks and provide the data "just in time" for the next TGO pass. Under such circumstances, we would need to tailor rover activities to the available time (as part of the daily tactical planning exercise) or accept to skip a communication session.

The secret to successful deep drilling is to proceed slowly. The RSM assumes a conservative vertical progress of 50 cm per sol, mostly through loose regolith, to reach the target sedimentary deposit at a prescribed depth. What can actually be achieved on any given sol will depend on the rover resources, the nature of the terrain being drilled, and the time available. Progress will be less when the terrain is harder and/or the drill goes deeper. We have demonstrated that the tool can sample formations of up to 150 MPa unconfined compressive strength (this covers most sedimentary rocks and weathered basalt, but not hard basalt or chert) and collect cores, fragments, or unconsolidated material (Magnani et al., 2011).

The desired baseline approach for drill operations is to be able to "park" (a part of) the drill string in the subsurface overnight, minimizing "dead" string assembly/disassembly periods to afford Ma_MISS sufficient science time. Operationally, however, this will be achieved in a stepwise manner. In the beginning, the rover will disassemble and store all segments at the end of each sol. Thereafter, part of the tip will be left in the borehole to evaluate the torque necessary to get it to move again in the morning. As confidence builds up, progressively longer drill sections will be allowed in the borehole overnight.

Tests performed in a Mars atmospheric chamber during which the drill was exercised to its full penetration length through different Mars-representative stratigraphic sequences permeated with 10%–35% water-content ice lenses, stopped, and left in the (simulated) subsurface overnight at −110°C showed that it was possible to restart the drill in the morning and extract it safely out of the borehole. Nonetheless, if the drill were to get stuck, it is feasible to command its counter rotation to disengage the string at the last blocked element, recovering the top portion. Further drilling would need to be performed with the backup drill tool and any remaining extension rods. The positioning system is equipped with an emergency ejection unit to be used as a last resort in case the drill becomes permanently immobilized in the terrain. However, without the drill, it would no longer be possible to provide samples to the ALD.

Summarizing, the RSM provides a step-by-step model exploration scenario that indicates how the mission objectives could be fulfilled. Its scope secures a level of resources affording a good degree of operational flexibility. Nevertheless, the real mission is likely to be different. The rover and instrument teams will adapt science operations as necessary to perform the best possible mission with the available resources.

12.8 A SUITABLE LANDING SITE

Barring the minor issues of landing and egressing safely, it is the scientific characteristics of the landing-site region that will have the greatest effect on what the ExoMars rover will be able to discover. Attributes like (1) age; (2) nature, duration, and connectivity of aqueous environments; (3) sediment deposition, burial, and diagenesis; and (4) exhumation history are decisive for the successful (or otherwise) trapping and preservation of possible chemical biosignatures. Other aspects related to how we may gain access to good samples are also important. For example, how many prime targets can we identify from orbit? What is their relative spacing and distribution in the landing ellipse? Do obstacles exist for rover locomotion? How extensive?

During 2013, ESA and Roscosmos appointed a Landing Site Selection Working Group (LSSWG) for the second ExoMars mission. The LSSWG includes the necessary scientific and engineering expertise to evaluate the suitability of candidate landing sites to meet science, engineering, and planetary protection constraints (Vago et al., 2015). Combining scientific and engineering competence in one body was considered paramount to the success of the landing-site selection process. Two separate bodies, one scientific and another engineering, would have likely resulted in incompatible recommendations. In this manner, the successful combination of science interest and landing safety must be achieved within the LSSWG.

12.8.1 Scientific Constraints

The ExoMars rover mission must target a geologically diverse, ancient site interpreted to possess strong potential for past habitability and for preserving physical and chemical biosignatures (as well as abiotic/prebiotic organics):

1. *Age*: The site must be older than 3.6 Ga, from Mars' early habitable period: pre-Noachian to late Noachian (Phyllosian).
2. *Preservation*: Regarding the search for molecular biosignatures, the site must provide easy access to locations with reduced radiation accumulation in the subsurface. The presence of fine-grained sediments in units of recent exposure age would be desirable (on Earth, organic molecules are better preserved in fine-grained sediments—which are more resistant to the penetration of biologically damaging agents, such as oxidants—than they are in porous, coarse materials). Young craters can provide the means to access deeper sediments, and studies on Earth suggest that fossil biomarkers can survive moderate impact heating (Parnell and Lindgren, 2006). Additionally, impact-related hydrothermal fractures might have contributed to creating habitats for microbial life in the past. However, for landing safety reasons, it is better not to have many craters in the ellipse, so sites recently exposed by high erosion rates would be preferable.
3. *Aqueous history*: The site must show abundant morphological and mineralogical evidence for long-duration (preferred) or frequently reoccurring (acceptable), low-energy transport, aqueous activity. We seek a geologic setting with a water-rich/hydrothermal history consistent with conditions favorable to life (e.g., evidence of slow-circulating or ponded water).
4. *Outcrop distribution*: The site must include numerous sedimentary rock outcrops. The outcrops must be well distributed over the landing ellipse to ensure that the rover can get to some of them.
5. *Little dust and drift sand*: It is essential to avoid loose dust deposits and drift sand distributed by eolian transport. Scientifically, there are two reasons for this: (1) Dust and mobile sand are not an interesting target for the rover. (2) The usefulness of the drill

will be nullified if the landing site has a dust/sand layer thicker than the drill's maximum penetration depth. Additionally, dunes constitute a serious risk for the rover's locomotion system.

12.8.2 Engineering Constraints

Engineering constraints are criteria that, in case they are not satisfied, can result in a landing site being judged unfeasible for the mission and therefore rejected:

1. *Altitude*: The terrain elevation in the landing ellipse must be less than or equal to −2 km MOLA.
2. *Landing ellipse size*: Including margin to account for offtrack radar operations (i.e., while oscillating under the parachute), the initial landing ellipse for site selection has been assumed to be 104×19 km, although it may vary according to the selected site's location and other dynamic constraints on trajectory imposed by the launch period.
3. *Terrain relief*: Surface features and slopes are entry, descent, and landing performance drivers since they can impact radar measurements and affect the stability of the landing platform. They can also constitute trafficability obstacles for the rover.
4. *Rock distribution*: The landing platform is designed with a clearance between nozzles and terrain of 0.50 m as the legs touch down and 0.35 m following deformation of the legs' shock absorbers. The site must have an areal fraction occupied by surface rocks (commonly referred to as rock abundance) $\leq 7\%$.
5. *Latitude*: The ExoMars rover can operate in the latitude range between 5 and 25°N. These engineering constraints and others, including thermal inertia, eolian deposit cover, radar reflectivity, and wind speed limits, are more precisely described in the work of Vago et al. (2015).

12.8.3 Planetary Protection Constraints

The ExoMars mission is not compatible with landing or operating in a Mars special region (Kminek and Rummel, 2015; Kminek et al., 2016). For the mission to be able to access a location where Earth microorganisms could multiply, the complete lander plus rover combination would need to be sterilized to satisfy Category 4c bioburden levels (as was done for the Viking landers). This will not be the case. Instead, the ExoMars mission has been classified as Category 4b. It is a mission including analytic instruments that can detect signatures of extinct and extant life; hence, all parts of the spacecraft that can come into contact with samples (i.e., the drill, the SPDS, and all mechanisms and volumes) have to be isolated, organically clean, and sterile throughout the mission to avoid potential false-positive detections (as per Category 4b rules). The rest of the rover (and indeed the lander) will comply with Category 4a prescriptions, those used for the MSL and MER rovers. A Category 4b classification allows exploring for signs of (extinct) life outside Mars special regions. Since ExoMars focuses on the search for ancient life bio-signatures and landing-site selection is tailored accordingly, this is the right approach.

The work to ensure that a candidate landing site does not include surface features that must be treated as special region (evaluated on a case-by-case basis) or experience environmental conditions that would meet the threshold levels of the parameters defined for special regions is based on a detailed analysis of orbital data, laboratory-based experiments, and modeling. A team appointed by the European Science Foundation will perform an independent review of the mission team's results.

12.8.4 Possible Locations for Landing

Two candidate landing sites have been identified: Oxia Planum and Mawrth Vallis (Bridges et al., 2016). Both will need to be verified in detail before the final landing location can be selected.

12.8.4.1 Oxia Planum (18.159°N, 335.666°E; −3 km MOLA)

The Oxia Planum area is situated at the eastern margin of the Chryse Basin, along the martian dichotomy border, and at the outlet of the Coogoon Valles system. The ellipse lies in the lower part of a wide basin where extensive exposures of Fe/Mg-rich phyllosilicates (>80% of the ellipse surface area) have been detected with both OMEGA and CRISM hyperspectral and multispectral data (Quantin et al., 2016). Smectite clays (Fe-Mg-rich saponite) and smectite/mica (e.g., vermiculite) are the dominant minerals within the ellipse. Hydrated silica, possibly opal, and Al-rich phyllosilicates may be present to the east of the ellipse (Carter et al., 2016). The Fe-/Mg-rich clay detections are associated with early/middle- to late-Noachian layered rocks (with layering thickness ranging from a few meters to <1 m for several tens of meters). They may represent the southwestern expansion (lowest member) of the Mawrth Vallis clay-rich deposits, pointing to a geographically extended aqueous alteration process.

The large Fe/Mg phyllosilicate-bearing unit overlaps the preexisting topography; is cut by valleys and inverted channels; and is overlain by younger, presumably Hesperian, alluvial and deltaic sediments to the east of the ellipse. A 10 km-wide, 80 km-long, low-thermal-inertia feature interpreted as a potential delta and bearing hydrated silica signatures in its stratum is observed at the outlet of Coogoon Valles. The putative delta waterline suggests the presence of a standing body of water after the formation of the clay-rich unit over the entire landing ellipse area (Quantin et al., 2016). A 20 m-thick, dark, capping unit covers both the layered formation and the fluvial morphologies and is interpreted to be Amazonian lava material. Crater counts yield ages of 4.0 Ga for the clay-rich unit and 2.6 Ga for the capping unit. The region has undergone extensive eolian erosion, as attested by anomalies in crater density, forming geologic windows to fresh exposures (<100 Ma) where material has been recently removed (Quantin et al., 2016).

12.8.4.2 Mawrth Vallis (22.160°N, 342.050°E; −2 km MOLA)

The Mawrth Vallis area is located at the boundary between the cratered Noachian terrains and the northern lowlands and represents one of the largest exposures of phyllosilicates detected on Mars (Poulet et al., 2005; Bibring et al., 2006; Loizeau et al., 2007). The proposed ellipse lies in early/middle- to late-Noachian clay-rich terrains southwest of the Mawrth Vallis channel (Gross et al., 2016).

The phyllosilicates are arranged in light-toned, finely layered deposits (~1 m thickness) of unknown origin, but their extent—covering thousands of square kilometers—is suggestive of a large, stable aqueous system. Outcrops in Mawrth Vallis are compositionally diverse, with a >300 m-thick sequence of various Al-rich phyllosilicates overlying Fe-/Mg-rich smectites, including local outcrops of sulfates (alunite, jarosite, and bassanite) and hydrated silica (Poulet et al., 2014).

These rocks show the highest degree of mineralogical diversity identified so far on Mars, which suggests a rich geologic history that may have included multiple aqueous environments. The deposition and aqueous alteration of the smectites are ancient (dated at 4.0 Ga) and have most likely been followed by episodes of acid leaching (as evidenced by the detection of kaolinite, alunite, and ferrous clays) and the deposition of an anhydrous dark capping unit of volcanic/pyroclastic origin during the early Hesperian (3.7 Ga).

Possible formation mechanisms for the phyllosilicate-rich deposits are the alteration of volcanic ash layers, eolian, or fluvial sediments in a wet environment, either due to top-down leaching in a pedogenesis context or through concurrent weathering and sedimentation (Gross et al., 2016). Given that the dark capping layer is relatively resistant to erosion, it is expected that the main target outcrops will be well preserved.

12.9 CONCLUSIONS

In this work, we have tried to show that microorganisms could have appeared and flourished on early Mars, as they did on our planet. Finding signs of their possible existence would be an important discovery, although ultimately we would want to understand to what extent their biochemical nature was similar to ours: Did Mars life have an independent genesis, or do we share a common ancestor (McKay, 2010)?

1. The ExoMars rover's design, payload, and exploration strategy focus on the search for extinct life; however, the mission also has the potential to recognize chemical indicators of extant life. Only if we were to detect abundant, nondegraded, primary biomolecules—as one would expect to find in association with living (or recently deceased) microorganisms—could we postulate the possible presence of extant life in the samples we have analyzed. Considering the harsh near-surface conditions on Mars and the fact that we are targeting low-latitude, relatively

water-poor landing sites, we do not believe we have high chances of encountering active life. We mention this payload capability because the possibility, albeit small, exists.

The rover will be equipped with a drill to collect material from outcrops and at depth, down to 2 m. This subsurface sampling capability is quite unique and provides the best chance yet to gain access to well-preserved chemical biosignatures.

2. Using the Pasteur instruments, the ExoMars science team will conduct a holistic search for biosignatures (morphological and chemical) and seek corroborating geologic context information.

   Although SAM's means to characterize indigenous organics have been hindered by their reaction with oxychlorine species in the martian soil, we have learned much from Curiosity to help us prepare future investigations. In fact, the ExoMars organic detection instrument, MOMA, is a joint undertaking of the SAM and COSAC (Rosetta lander) teams. Our work shows that the laser-desorption extraction method implemented in MOMA is not affected by perchlorates. In other words, we are able to detect (relatively large) organic molecules quite effectively even when oxychlorine species are present in the sample.

   Another powerful capability of ExoMars is that it can investigate the same mineral grains with LDMS, VIS-IR, and Raman, allowing us the opportunity to observe a target with all three techniques, although the MOMA-LDMS footprint is larger than that of the RLS and MicrOmega spectrometers.

3. Targeting an early Noachian location will grant us access to deposits of an age no longer available for study on our own planet. The absence of plate tectonics on Mars increases the probability that rapidly buried ancient sedimentary rocks (possibly hosting microorganisms) may have been spared thermal alteration and been shielded from ionizing radiation damage until denuded relatively recently. The scientific quality of the landing site in terms of suitable age; nature, duration, and connectivity of aqueous environments; sediment deposition, burial, and diagenesis; and exhumation history will play a determinant role in shaping the mission's outcome.

The ExoMars rover is well suited to search for signs of life. Nevertheless, the ultimate confirmation of a collection of potential biosignature detections may require more thorough analyses than can be performed with our present robotic means. Even a tentative finding would constitute a powerful catalyst for a Mars sample return (MSR) mission. Because of the ExoMars rover's special ability to explore the third dimension—depth—its discoveries will contribute immensely to determining what types of samples we should return to Earth.

ACKNOWLEDGMENTS

The rover mission would not be possible without the work and unwavering dedication of the ExoMars project team and industry. We would also like to recognize the help and support of ESA, Roscosmos, the European states and agencies participating in the ExoMars program, and NASA. We are doing this together for the benefit of all.

REFERENCES

Allen, M., Sherwood Lollar, B., Runnegar, B., Oehler, D.Z., Lyons, J.R., Manning, C.E., Summers, M.E., 2006. Is Mars alive? EOS Trans. Am. Geophys. Union 87 (41), 433. https://doi.org/10.1029/2006EO410001.

Archer, P.D., Ming, D.W., Sutter, B., Morris, R.V., Clark, B.C., Mahaffy, P.H., Wray, J.J., Fairen, A.G., Gellert, R., Yen, A.S., Blake, D.F., Vaniman, D.T., Glavin, D.P., Eigenbrode, J.L., Trainer, M.G., Navarro-González, R., McKay, C.P., Freissinet, C., 2016. In: Oxychlorine species on Mars: implications from Gale Crater Samples. Abstract 2947, 47th Lunar and Planetary Science Conference (2016).

Arevalo, R., Brinckerhoff, W., van Amerom, F., Danell, R., Pinnick, V., Li, X., Getty, S., Hovmand, L., Grubisic, A., Mahaffy, P., Goesmann, F., Steininger, H., 2015. In: Design and demonstration of the Mars organic molecule analyzer (MOMA) on the ExoMars 2018 rover. 2015 IEEE Aerospace Conference. IEEE Conference Publications, pp. 1–11. https://doi.org/10.1109/AERO.2015.7119073.

Arvidson, R.E., Iagnemma, K.D., Maimone, M., Fraeman, A.A., Zhou, F., Heverly, M.C., Bellutta, P., Rubin, D., Stein, N.T., Grotzinger, J.P., Vasavada, A.R., 2016. Mars Science Laboratory Curiosity rover megaripple crossings up to Sol 710 in Gale crater. J. Field Rob. 7 (PART 1), 1–24. https://doi.org/10.1002/rob.21647.

Azkarate, M., Zwick, M., Hidalgo-Carrio, J., Nelen, R., Wiese, T., Poulakis, P., Joudrier, L., Visentin, G., 2015. In: First experimental investigations on wheel-walking for improving triple-bogie rover locomotion performances. Proceedings Advanced Space Technologies for Robotics and Automation (ASTRA). European Space Agency, Noordwijk.

Bibring, J.-P., Langevin, Y., Mustard, J.F., Poulet, F., Arvidson, R., Gendrin, A., Gondet, B., Mangold, N., Pinet, P., Forget, F., Berthe, M., Bibring, J.-P., Gendrin, A., Gomez, C., Gondet, B., Jouglet, D., Poulet, F., Soufflot, A., Vincendon, M., Combes, M., Drossart, P., Encrenaz, T., Fouchet, T., Merchiorri, R., Belluci, G., Altieri, F., Formisano, V., Capaccioni, F., Cerroni, P., Coradini, A., Fonti, S., Korablev, O., Kottsov, V., Ignatiev, N., Moroz, V., Titov, D., Zasova, L., Loiseau, D., Mangold, N., Pinet, P., Doute, S., Schmitt, B., Sotin, C., Hauber, E., Hoffmann, H., Jaumann, R., Keller, U., Arvidson, R., Mustard, J.F., Duxbury, T., Forget, F., Neukum, G., 2006. Global mineralogical and aqueous Mars history derived from OMEGA/Mars express data. Science 312 (5772), 400–404. https://doi.org/10.1126/science.1122659.

Bibring, J.-P., Hamm, V., Pilorget, C., Vago, J.L., The MicrOmega Team, 2017. The MicrOmega investigation onboard ExoMars. Astrobiology 17 (6–7), 621–626. https://doi.org/10.1089/ast.2016.1642.

Biemann, K., 2007. On the ability of the Viking gas chromatograph-mass spectrometer to detect organic matter. Proc. Natl. Acad. Sci. 104 (25), 10310–10313. https://doi.org/10.1073/pnas.0703732104.

Biemann, K., Bada, J.L., 2011. Comment on "Reanalysis of the Viking results suggests perchlorate and organics at midlatitudes on Mars" by Rafael Navarro-González et al. J. Geophys. Res. 116 (E12), E12001. https://doi.org/10.1029/2011JE003869.

Biemann, K., Oro, J., Toulmin, P., Orgel, L.E., Nier, A.O., Anderson, D.M., Simmonds, P.G., Flory, D., Diaz, A.V., Rushneck, D.R., Biller, J.E., Lafleur, A.L., 1977. The search for organic substances and inorganic volatile compounds in the surface of Mars. J. Geophys. Res. 82 (28), 4641–4658. https://doi.org/10.1029/JS082i028p04641.

Bishop, J.L., Loizeau, D., McKeown, N.K., Saper, L., Dyar, M.D., Des Marais, D.J., Parente, M., Murchie, S.L., 2013. What the ancient phyllosilicates at Mawrth Vallis can tell us about possible habitability on early Mars. Planet. Space Sci. 86, 130–149. https://doi.org/10.1016/j.pss.2013.05.006.

Bounichou, M., 2010. LA METHODE DIAMS: Desorption/Ionization on Self-Assembled Monolayer Surface. Une Nouvelle Technique de désorption Ionisation Laser Sans Matrice Pour la Spectrométrie de Masse. Université d'Angers, Angers.

Bridges, J.C., Hensoin, R.A., Vago, J.L., Loizeau, D., Williams, R.M.E., Hauber, E., Sefton-Nash, E., 2016. In: ExoMars landing site characterization and selection. Abstract 2170, 47th Lunar and Planetary Science Conference. Lunar and Planetary Institute, Houston, TX.

Brocks, J.J., 1999. Archean molecular fossils and the early rise of eukaryotes. Science 285 (5430), 1033–1036. https://doi.org/10.1126/science.285.5430.1033.

Busch, K.L., 1995. Special feature: tutorial. Desorption ionization mass spectrometry. J. Mass Spectrom. 30 (2), 233–240. https://doi.org/10.1002/jms.1190300202.

Cady, S., Noffke, N., 2009. Geobiology: evidence for early life on Earth and the search for life on other planets. GSA Today 19 (11), 4–10. https://doi.org/10.1130/GSATG62A.1.

Carter, J., Quantin, C., Thollot, P., Loizeau, D., Ody, A., Lozach, L., 2016. In: Oxia planum, a clay-laden landing site proposed for the ExoMars rover mission: aqueous mineralogy and alteration scenarios. Abstract 2064, 47th Lunar and Planetary Science Conference. Lunar and Planetary Institute, Houston, TX.

Ciarletti, V., Corbel, C., Plettemeier, D., Cais, P., Clifford, S.M., Hamran, S., 2011. WISDOM GPR designed for shallow and high-resolution sounding of the Martian subsurface. Proc. IEEE 99 (5), 824–836. https://doi.org/10.1109/JPROC.2010.2100790.

Ciarletti, V., Clifford, S., Plettemeier, D., Le Gall, A., Hervé, Y., Dorizon, S., Quantin-Nataf, C., Benedix, W.-S., Schwenzer, S., Pettinelli, E., Heggy, E., Herique, A., Berthelier, J.-J., Kofman, W., Vago, J.L., Hamran, S.-E., the WISDOM Team, 2017. The WISDOM radar: unveiling the subsurface beneath the ExoMars rover and identifying the best locations for drilling. Astrobiology 17 (6–7), 565–584. https://doi.org/10.1089/ast.2016.1532.

Coates, A.J., Griffiths, A.D., Leff, C.E., Schmitz, N., Barnes, D.P., Josset, J.-L., Hancock, B.K., Cousins, C.R., Jaumann, R., Crawford, I.A., Paar, G., Bauer, A., 2012. Lunar PanCam: adapting Exo-Mars PanCam for the ESA lunar lander. Planet. Space Sci. 74 (1), 247–253. https://doi.org/10.1016/j.pss.2012.07.017.

Coates, A.J., Jaumann, R., Griffiths, A.D., Leff, C.E., Schmitz, N., Josset, J.-L., Paar, G., Gunn, M., Hauber, E., Cousins, C.R., Cross, R.E., Grindrod, P.M., Bridges, J.C., Balme, M., Gupta, S., Crawford, I.A., Irwin, P., Stabbins, R., Tirsch, D., Vago, J.L., Theodorou, T., Caballo-Perucha, M., Osinski, G.R., The PanCam Team, 2017. The PanCam instrument for the ExoMars rover. Astrobiology 17 (6–7), 511–541. https://doi.org/10.1089/ast.2016.1548.

Cockell, C.S., Balme, M., Bridges, J.C., Davila, A., Schwenzer, S.P., 2012. Uninhabited habitats on Mars. Icarus 217 (1), 184–193. https://doi.org/10.1016/j.icarus.2011.10.025.

Cousins, C.R., Crawford, I.A., 2011. Volcano-ice interaction as a microbial habitat on earth and Mars. Astrobiology 11 (7), 695–710. https://doi.org/10.1089/ast.2010.0550.

Cousins, C.R., Gunn, M., Prosser, B.J., Barnes, D.P., Crawford, I.A., Griffiths, A.D., Davis, L.E., Coates, A.J., 2012. Selecting the geology filter wavelengths for the ExoMars panoramic camera instrument. Planet. Space Sci. 71 (1), 80–100. https://doi.org/10.1016/j.pss.2012.07.009.

Dartnell, L.R., Desorgher, L., Ward, J.M., Coates, A.J., 2007. Modelling the surface and subsurface Martian radiation environment: Implications for astrobiology. Geophys. Res. Lett. 34 (2), L02207. https://doi.org/10.1029/2006GL027494.

Dartnell, L.R., Page, K., Jorge-Villar, S.E., Wright, G., Munshi, T., Scowen, I.J., Ward, J.M., Edwards, H.G.M., 2012. Destruction of Raman biosignatures by ionising radiation and the implications for life detection on Mars. Anal. Bioanal. Chem. 403 (1), 131–144. https://doi.org/10.1007/s00216-012-5829-6.

De Angelis, S., De Sanctis, M.C., Ammannito, E., Di Iorio, T., Carli, C., Frigeri, A., Capria, M.T., Federico, C., Boccaccini, A., Capaccioni, F., Giardino, M., Cerroni, P., Palomba, E., Piccioni, G., 2013. VNIR spectroscopy of Mars analogues with the ExoMars-Ma_Miss instrument. Mem. Soc. Astronom. Ital. Suppl. 26, 121–127.

De Sanctis, M.C., Altieri, F., Ammannito, E., Biondi, D., De Angelis, S., Meini, M., Mondello, G., Novi, S., Paolinetti, R., Soldani, M., Mugnuolo, R., Pirrotta, S., Vago, J.L., the Ma_MISS team,

2017. Ma_MISS on ExoMars: mineralogical characterization of the martian subsurface. Astrobiology 17 (6–7), 612–620. https://doi.org/10.1089/ast.2016.1541.

Edwards, H.G.M., Hutchinson, I.B., Ingley, R., Parnell, J., Vítek, P., Jehlička, J., 2013. Raman spectroscopic analysis of geological and biogeological specimens of relevance to the ExoMars mission. Astrobiology 13 (6), 543–549. https://doi.org/10.1089/ast.2012.0872.

Ehlmann, B.L., Mustard, J.F., Murchie, S.L., Bibring, J.-P., Meunier, A., Fraeman, A.A., Langevin, Y., 2011. Subsurface water and clay mineral formation during the early history of Mars. Nature 479 (7371), 53–60. https://doi.org/10.1038/nature10582.

Eigenbrode, J.L., 2008. Fossil lipids for life-detection: a case study from the early earth record. Space Sci. Rev. 135 (1–4), 161–185. https://doi.org/10.1007/s11214-007-9252-9.

Elkins-Tanton, L.T., 2011. Formation of early water oceans on rocky planets. Astrophys. Space Sci. 332 (2), 359–364. https://doi.org/10.1007/s10509-010-0535-3.

Farley, K.A., Malespin, C., Mahaffy, P., Grotzinger, J.P., Vasconcelos, P.M., Milliken, R.E., Malin, M., Edgett, K.S., Pavlov, A.A., Hurowitz, J.A., Grant, J.A., Miller, H.B., Arvidson, R., Beegle, L., Calef, F., Conrad, P.G., Dietrich, W.E., Eigenbrode, J., Gellert, R., Gupta, S., Hamilton, V., Hassler, D.M., Lewis, K.W., McLennan, S.M., Ming, D., Navarro-Gonzalez, R., Schwenzer, S.P., Steele, A., Stolper, E.M., Sumner, D.Y., Vaniman, D., Vasavada, A., Williford, K., Wimmer-Schweingruber, R.F., Blake, D.F., Bristow, T., DesMarais, D., Edwards, L., Haberle, R., Hoehler, T., Hollingsworth, J., Kahre, M., Keely, L., McKay, C., Wilhelm, M.B., Bleacher, L., Brinckerhoff, W., Choi, D., Dworkin, J.P., Floyd, M., Freissinet, C., Garvin, J., Glavin, D., Harpold, D., Martin, D.K., McAdam, A., Raaen, E., Smith, M.D., Stern, J., Tan, F., Trainer, M., Meyer, M., Posner, A., Voytek, M., Anderson, R.C., Aubrey, A., Behar, A., Blaney, D., Brinza, D., Christensen, L., Crisp, J.A., DeFlores, L., Feldman, J., Feldman, S., Flesch, G., Hurowitz, J., Jun, I., Keymeulen, D., Maki, J., Mischna, M., Morookian, J.M., Parker, T., Pavri, B., Schoppers, M., Sengstacken, A., Simmonds, J.J., Spanovich, N., de la Torre Juarez, M., Webster, C.R., Yen, A., Archer, P.D., Cucinotta, F., Jones, J.H., Morris, R.V., Niles, P., Rampe, E., Nolan, T., Fisk, M., Radziemski, L., Barraclough, B., Bender, S., Berman, D., Dobrea, E.N., Tokar, R., Williams, R.M.E., Yingst, A., Leshin, L., Cleghorn, T., Huntress, W., Manhes, G., Hudgins, J., Olson, T., Stewart, N., Sarrazin, P., Vicenzi, E., Wilson, S.A., Bullock, M., Ehresmann, B., Peterson, J., Rafkin, S., Zeitlin, C., Fedosov, F., Golovin, D., Karpushkina, N., Kozyrev, A., Litvak, M., Malakhov, A., Mitrofanov, I., Mokrousov, M., Nikiforov, S., Prokhorov, V., Sanin, A., Tretyakov, V., Varenikov, A., Vostrukhin, A., Kuzmin, R., Clark, B., Wolff, M., Botta, O., Drake, D., Bean, K., Lemmon, M., Anderson, R.B., Herkenhoff, K., Lee, E.M., Sucharski, R., de Pablo Hernández, M.A., Avalos, J.J.B., Ramos, M., Kim, M.-H., Plante, I., Muller, J.-P., Ewing, R., Boynton, W., Downs, R., Fitzgibbon, M., Harshman, K., Morrison, S., Kortmann, O., Palucis, M., Williams, A., Lugmair, G., Wilson, M.A., Rubin, D., Jakosky, B., Balic-Zunic, T., Frydenvang, J., Jensen, J.K., Kinch, K., Koefoed, A., Madsen, M.B., Stipp, S.L.S., Boyd, N., Campbell, J.L., Perrett, G., Pradler, I., VanBommel, S., Jacob, S., Owen, T., Rowland, S., Savijarvi, H., Boehm, E., Bottcher, S., Burmeister, S., Guo, J., Kohler, J., Garcia, C.M., Mueller-Mellin, R., Bridges, J.C., McConnochie, T., Benna, M., Franz, H., Bower, H., Brunner, A., Blau, H., Boucher, T., Carmosino, M., Atreya, S., Elliott, H., Halleaux, D., Renno, N., Wong, M., Pepin, R., Elliott, B., Spray, J., Thompson, L., Gordon, S., Newsom, H., Ollila, A., Williams, J., Bentz, J., Nealson, K., Popa, R., Kah, L.C., Moersch, J., Tate, C., Day, M., Kocurek, G., Hallet, B., Sletten, R., Francis, R., McCullough, E., Cloutis, E., ten Kate, I.L., Kuzmin, R., Fraeman, A., Scholes, D., Slavney, S., Stein, T., Ward, J., Berger, J., Moores, J.E., 2014. In situ radiometric and exposure age dating of the martian surface. Science 343 (6169), 1247166. https://doi.org/10.1126/science.1247166.

Favaedi, Y., Pechev, A., Scharringhausen, M., Richter, L., 2011. Prediction of tractive response for flexible wheels with application to planetary rovers. J. Terrramech. 48 (3), 199–213. https://doi.org/10.1016/j.jterra.2011.02.003.

Foucher, F., Lopez-Reyes, G., Bost, N., Rull-Perez, F., Rüßmann, P., Westall, F., 2013. Effect of grain size distribution on Raman analyses and the consequences for in situ planetary missions. J. Raman Spectrosc. 44 (6), 916–925. https://doi.org/10.1002/jrs.4307.

Freissinet, C., Glavin, D.P., Mahaffy, P.R., Miller, K.E., Eigenbrode, J.L., Summons, R.E., Brunner, A.E., Buch, A., Szopa, C., Archer, P.D., Franz, H.B., Atreya, S.K., Brinckerhoff, W.B., Cabane, M., Coll, P., Conrad, P.G., Des Marais, D.J., Dworkin, J.P., Fairén, A.G., François, P., Grotzinger, J.P., Kashyap, S., ten Kate, I.L., Leshin, L.A., Martin, M.G., Martin-Torres, F.J., McAdam, A.C., Ming, D.W., Navarro-González, R., Pavlov, A.A., Prats, B.D., Squyres, S.W., Steele, A., Stern, J.C., Sumner, D.Y., Sutter, B., Zorzano, M.-P., 2015. Organic molecules in the Sheepbed Mudstone, Gale Crater, Mars. J. Geophys. Res. Planets 120 (3), 495–514. https://doi.org/10.1002/2014JE004737.

Genda, H., 2016. Origin of Earth's oceans: an assessment of the total amount, history and supply of water. Geochem. J. 50 (1), 27–42. https://doi.org/10.2343/geochemj.2.0398.

Georgiou, C.D., Deamer, D.W., 2014. Lipids as universal biomarkers of extraterrestrial life. Astrobiology 14 (6), 541–549. https://doi.org/10.1089/ast.2013.1134.

Glavin, D.P., Freissinet, C., Miller, K.E., Eigenbrode, J.L., Brunner, A.E., Buch, A., Sutter, B., Archer, P.D., Atreya, S.K., Brinckerhoff, W.B., Cabane, M., Coll, P., Conrad, P.G., Coscia, D., Dworkin, J.P., Franz, H.B., Grotzinger, J.P., Leshin, L.A., Martin, M.G., McKay, C., Ming, D.W., Navarro-González, R., Pavlov, A., Steele, A., Summons, R.E., Szopa, C., Teinturier, S., Mahaffy, P.R., 2013. Evidence for perchlorates and the origin of chlorinated hydrocarbons detected by SAM at the Rocknest aeolian deposit in Gale Crater. J. Geophys. Res. Planets 118 (10), 1955–1973. https://doi.org/10.1002/jgre.20144.

Goesmann, F., Brinckerhoff, W.B., Raulin, F., Goetz, W., Danell, R.M., Getty, S.A., Siljeström, S., Mißbach, H., Steininger, H., Arevalo, R.D., Buch, A., Freissinet, C., Grubisic, A., Meierhenrich, U.J., Pinnick, V.T., Stalport, F., Szopa, C., Vago, J.L., Lindner, R., Schulte, M.D., Brucato, J.R., Glavin, D.P., Grand, N., Li, X., van Amerom, F.H.W., The MOMA Science Team, 2017. The Mars Organic Molecule Analyzer (MOMA) instrument: characterization of organic material in martian sediments. Astrobiology 17 (6–7), 655–685. https://doi.org/10.1089/ast.2016.1551.

Goetz, W., Brinckerhoff, W.B., Arevalo, R., Freissinet, C., Getty, S., Glavin, D.P., Siljeström, S., Buch, A., Stalport, F., Grubisic, A., Li, X., Pinnick, V., Danell, R., van Amerom, F.H.W., Goesmann, F., Steininger, H., Grand, N., Raulin, F., Szopa, C., Meierhenrich, U., Brucato, J.R., 2016. MOMA: the challenge to search for organics and biosignatures on Mars. Int. J. Astrobiol. 15 (3), 239–250. https://doi.org/10.1017/S1473550416000227.

Gross, C., Poulet, F., Michalski, J., Horgan, B., Bishop, J.L., 2016. In: Mawrth Vallis—proposed landing site for ExoMars 2018/2020. Abstract 1421, 47th Lunar and Planetary Science Conference Abstracts. Lunar and Planetary Institute, Houston, TX.

Grott, M., Helbert, J., Nadalini, R., 2007. Thermal structure of Martian soil and the measurability of the planetary heat flow. J. Geophys. Res. 112 (E9), E09004. https://doi.org/10.1029/2007JE002905.

Hassler, D.M., Zeitlin, C., Wimmer-Schweingruber, R.F., Ehresmann, B., Rafkin, S., Eigenbrode, J.L., Brinza, D.E., Weigle, G., Bottcher, S., Bohm, E., Burmeister, S., Guo, J., Kohler, J., Martin, C., Reitz, G., Cucinotta, F.A., Kim, M.-H., Grinspoon, D., Bullock, M.A., Posner, A., Gomez-Elvira, J., Vasavada, A., Grotzinger, J.P., Team, M.S., Kemppinen, O., Cremers, D., Bell, J.F., Edgar, L., Farmer, J., Godber, A., Wadhwa, M., Wellington, D., McEwan, I., Newman, C., Richardson, M., Charpentier, A., Peret, L., King, P., Blank, J., Schmidt, M., Li, S., Milliken, R., Robertson, K., Sun, V., Baker, M., Edwards, C., Ehlmann, B., Farley, K., Griffes, J., Miller, H., Newcombe, M., Pilorget, C., Rice, M., Siebach, K., Stack, K., Stolper, E., Brunet, C., Hipkin, V., Leveille, R., Marchand, G., Sanchez, P.S., Favot, L., Cody, G., Steele, A., Fluckiger, L., Lees, D., Nefian, A., Martin, M., Gailhanou, M., Westall, F., Israel, G., Agard, C., Baroukh, J., Donny, C., Gaboriaud, A., Guillemot, P., Lafaille, V., Lorigny, E., Paillet, A., Perez, R., Saccoccio, M., Yana, C., Armiens-Aparicio, C., Rodriguez, J.C., Blazquez, I.C., Gomez, F.G., Hettrich, S., Malvitte, A.L., Jimenez, M.M., Martinez-Frias, J., Martin-Soler, J., Martin-Torres, F.J., Jurado, A.M., Mora-Sotomayor, L., Caro, G.M., Lopez, S.N., Peinado-Gonzalez, V., Pla-Garcia, J., Manfredi, J.A.R., Romeral-Planello, J.J., Fuentes, S.A.S., Martinez, E.S., Redondo, J.T., Urqui-O'Callaghan, R., Mier, M.-P.Z., Chipera, S., Lacour, J.-L., Mauchien, P., Sirven, J.-B., Manning, H., Fairen, A., Hayes, A., Joseph, J., Squyres, S., Sullivan, R., Thomas, P., Dupont, A., Lundberg, A., Melikechi, N., Mezzacappa, A., Berger, T., Matthia, D., Prats, B., Atlaskin, E.,

Genzer, M., Harri, A.-M., Haukka, H., Kahanpaa, H., Kauhanen, J., Kemppinen, O., Paton, M., Polkko, J., Schmidt, W., Siili, T., Fabre, C., Wray, J., Wilhelm, M.B., Poitrasson, F., Patel, K., Gorevan, S., Indyk, S., Paulsen, G., Gupta, S., Bish, D., Schieber, J., Gondet, B., Langevin, Y., Geffroy, C., Baratoux, D., Berger, G., Cros, A., D'Uston, C., Forni, O., Gasnault, O., Lasue, J., Lee, Q.-M., Maurice, S., Meslin, P.-Y., Pallier, E., Parot, Y., Pinet, P., Schroder, S., Toplis, M., Lewin, E., Brunner, W., Heydari, E., Achilles, C., Oehler, D., Sutter, B., Cabane, M., Coscia, D., Israel, G., Szopa, C., Dromart, G., Robert, F., Sautter, V., Le Mouelic, S., Mangold, N., Nachon, M., Buch, A., Stalport, F., Coll, P., Francois, P., Raulin, F., Teinturier, S., Cameron, J., Clegg, S., Cousin, A., DeLapp, D., Dingler, R., Jackson, R.S., Johnstone, S., Lanza, N., Little, C., Nelson, T., Wiens, R.C., Williams, R.B., Jones, A., Kirkland, L., Treiman, A., Baker, B., Cantor, B., Caplinger, M., Davis, S., Duston, B., Edgett, K., Fay, D., Hardgrove, C., Harker, D., Herrera, P., Jensen, E., Kennedy, M.R., Krezoski, G., Krysak, D., Lipkaman, L., Malin, M., McCartney, E., McNair, S., Nixon, B., Posiolova, L., Ravine, M., Salamon, A., Saper, L., Stoiber, K., Supulver, K., Van Beek, J., Van Beek, T., Zimdar, R., French, K.L., Iagnemma, K., Miller, K., Summons, R., Goesmann, F., Goetz, W., Hviid, S., Johnson, M., Lefavor, M., Lyness, E., Breves, E., Dyar, M.D., Fassett, C., Blake, D.F., Bristow, T., DesMarais, D., Edwards, L., Haberle, R., Hoehler, T., Hollingsworth, J., Kahre, M., Keely, L., McKay, C., Wilhelm, M.B., Bleacher, L., Brinckerhoff, W., Choi, D., Conrad, P., Dworkin, J.P., Floyd, M., Freissinet, C., Garvin, J., Glavin, D., Harpold, D., Jones, A., Mahaffy, P., Martin, D.K., McAdam, A., Pavlov, A., Raaen, E., Smith, M.D., Stern, J., Tan, F., Trainer, M., Meyer, M., Voytek, M., Anderson, R.C., Aubrey, A., Beegle, L.W., Behar, A., Blaney, D., Calef, F., Christensen, L., Crisp, J.A., DeFlores, L., Ehlmann, B., Feldman, J., Feldman, S., Flesch, G., Hurowitz, J., Jun, I., Keymeulen, D., Maki, J., Mischna, M., Morookian, J.M., Parker, T., Pavri, B., Schoppers, M., Sengstacken, A., Simmonds, J.J., Spanovich, N., de la Torre Juarez, M., Webster, C.R., Yen, A., Archer, P.D., Jones, J.H., Ming, D., Morris, R.V., Niles, P., Rampe, E., Nolan, T., Fisk, M., Radziemski, L., Barraclough, B., Bender, S., Berman, D., Dobrea, E.N., Tokar, R., Vaniman, D., Williams, R.M.E., Yingst, A., Lewis, K., Leshin, L., Cleghorn, T., Huntress, W., Manhes, G., Hudgins, J., Olson, T., Stewart, N., Sarrazin, P., Grant, J., Vicenzi, E., Wilson, S.A., Hamilton, V., Peterson, J., Fedosov, F., Golovin, D., Karpushkina, N., Kozyrev, A., Litvak, M., Malakhov, A., Mitrofanov, I., Mokrousov, M., Nikiforov, S., Prokhorov, V., Sanin, A., Tretyakov, V., Varenikov, A., Vostrukhin, A., Kuzmin, R., Clark, B., Wolff, M., McLennan, S., Botta, O., Drake, D., Bean, K., Lemmon, M., Schwenzer, S.P., Anderson, R.B., Herkenhoff, K., Lee, E.M., Sucharski, R., de Pablo Hernández, M.A., Avalos, J.J.B., Ramos, M., Malespin, C., Plante, I., Muller, J.-P., Navarro-Gonzalez, R., Ewing, R., Boynton, W., Downs, R., Fitzgibbon, M., Harshman, K., Morrison, S., Dietrich, W., Kortmann, O., Palucis, M., Sumner, D.Y., Williams, A., Lugmair, G., Wilson, M.A., Rubin, D., Jakosky, B., Balic-Zunic, T., Frydenvang, J., Jensen, J.K., Kinch, K., Koefoed, A., Madsen, M.B., Stipp, S.L.S., Boyd, N., Campbell, J.L., Gellert, R., Perrett, G., Pradler, I., VanBommel, S., Jacob, S., Owen, T., Rowland, S., Atlaskin, E., Savijarvi, H., Garcia, C.M., Mueller-Mellin, R., Bridges, J.C., McConnochie, T., Benna, M., Franz, H., Bower, H., Brunner, A., Blau, H., Boucher, T., Carmosino, M., Atreya, S., Elliott, H., Halleaux, D., Renno, N., Wong, M., Pepin, R., Elliott, B., Spray, J., Thompson, L., Gordon, S., Newsom, H., Ollila, A., Williams, J., Vasconcelos, P., Bentz, J., Nealson, K., Popa, R., Kah, L.C., Moersch, J., Tate, C., Day, M., Kocurek, G., Hallet, B., Sletten, R., Francis, R., McCullough, E., Cloutis, E., ten Kate, I.L., Kuzmin, R., Arvidson, R., Fraeman, A., Scholes, D., Slavney, S., Stein, T., Ward, J., Berger, J., Moores, J.E., 2014. Mars' surface radiation environment measured with the Mars Science Laboratory's Curiosity Rover. Science 343 (6169), 1244797. https://doi.org/10.1126/science.1244797.

Hecht, M.H., Kounaves, S.P., Quinn, R.C., West, S.J., Young, S.M.M., Ming, D.W., Catling, D.C., Clark, B.C., Boynton, W.V., Hoffman, J., Deflores, L.P., Gospodinova, K., Kapit, J., Smith, P.H., 2009. Detection of perchlorate and the soluble chemistry of martian soil at the Phoenix lander site. Science 325, 64–67. https://doi.org/10.1126/science.1172466.

Heverly, M., Matthews, J., Lin, J., Fuller, D., Maimone, M., Biesiadecki, J., Leichty, J., 2013. Traverse performance characterization for the Mars Science Laboratory Rover. J. Field Rob. 30 (6), 835–846. https://doi.org/10.1002/rob.21481.

Josset, J.-L., Westall, F., Hofmann, B.A., Spray, J., Cockell, C., Kempe, S., Griffiths, A.D., De Sanctis, M.C., Colangeli, L., Koschny, D., Föllmi, K., Verrecchia, E., Diamond, L., Josset, M., Javaux, E.J., Esposito, F., Gunn, M., Souchon-Leitner, A.L., Bontognali, T.R.R., Korablev, O., Erkman, S., Paar, G., Ulamec, S., Foucher, F., Martin, P., Verhaeghe, A., Tanevski, M., Vago, J.L., 2017. The close-up imager onboard the ESA ExoMars rover: objectives, description, operations, and science validation activities. Astrobiology 17 (6–7), 595–611. https://doi.org/10.1089/ast.2016.1546.

Klein, H.P., 1998. The search for life on Mars: what we learned from Viking. J. Geophys. Res. 103 (E12), 28463–28466. https://doi.org/10.1029/98JE01722.

Klein, H.P., 1999. Did viking discover life on Mars? Orig. Life Evol. Biosph. 29 (6), 625–631. https://doi.org/10.1023/A:1006514327249.

Klein, H.P., Lederberg, J., Rich, A., Horowitz, N.H., Oyama, V.I., Levin, G.V., 1976. The Viking mission search for life on Mars. Nature 262 (5563), 24–27.

Kminek, G., Bada, J., 2006. The effect of ionizing radiation on the preservation of amino acids on Mars. Earth Planet. Sci. Lett. 245 (1–2), 1–5. https://doi.org/10.1016/j.epsl.2006.03.008.

Kminek, G., Rummel, J.D., 2015. COSPAR planetary protection policy. Space Res. Today 193, 7–19. https://doi.org/10.1016/j.srt.2015.06.008.

Kminek, G., Hipkin, V.J., Anesio, A.M., Barengoltz, J., Boston, P.J., Clark, B.A., Conley, C.A., Coustenis, A., Detsis, E., Doran, P., Grasset, O., Hand, K., Hajime, Y., Hauber, E., Kolmasová, I., Lindberg, R.E., Meyer, M., Raulin, F., Reitz, G., Rennó, N.O., Rettberg, P., Rummel, J.D., Saunders, M.P., Schwehm, G., Sherwood, B., Smith, D.H., Stabekis, P.E., Vago, J., 2016. COSPAR panel on planetary protection colloquium, Bern, Switzerland, September 2015. Space Res. Today 195, 42–67. https://doi.org/10.1016/j.srt.2016.03.013.

Korablev, O., Ivanov, A., Fedorova, A., Kalinnikov, Y.K., Shapkin, A., Mantsevich, S., Viazovetsky, N., Evdokimova, N., Kiselev, A.V., Region, M., Measurements, R., Region, M., State, M., Gory, L., 2015. In: Development of a mast or robotic arm-mounted infrared AOTF spectrometer for surface moon and Mars probes.Proc. of SPIE 9608, Infrared Remote Sensing and Instrumentation XXIII, 960807. vol. 9608, pp. 1–10. https://doi.org/10.1117/12.2190450.

Korablev, O.I., Dobrolensky, Y., Evdokimova, N., Fedorova, A.A., Kuzmin, R.O., Mantsevich, S.N., Cloutis, E.A., Carter, J., Poulet, F., Flahaut, J., Griffiths, A., Gunn, M., Schmitz, N., Martín-Torres, J., Zorzano, M.-P., Rodionov, D.S., Vago, J.L., Stepanov, A.V., Titov, A.Y., Vyazovetsky, N.A., Trokhimovskiy, A.Y., Sapgir, A.G., Kalinnikov, Y.K., Ivanov, Y.S., Shapkin, A.A., Ivanov, A.Y., 2017. Infrared spectrometer for ExoMars: a mast-mounted instrument for the rover. Astrobiology 17 (6–7), 542–564. https://doi.org/10.1089/ast.2016.1543.

Kounaves, S.P., Hecht, M.H., Kapit, J., Gospodinova, K., DeFlores, L., Quinn, R.C., Boynton, W.V., Clark, B.C., Catling, D.C., Hredzak, P., Ming, D.W., Moore, Q., Shusterman, J., Stroble, S., West, S.J., Young, S.M.M., 2010. Wet Chemistry experiments on the 2007 Phoenix Mars Scout Lander mission: data analysis and results. J. Geophys. Res. 115 (E1), E00E10. https://doi.org/10.1029/2009JE003424.

Kounaves, S.P., Chaniotakis, N.A., Chevrier, V.F., Carrier, B.L., Folds, K.E., Hansen, V.M., McElhoney, K.M., O'Neil, G.D., Weber, A.W., 2014. Identification of the perchlorate parent salts at the Phoenix Mars landing site and possible implications. Icarus 232, 226–231. https://doi.org/10.1016/j.icarus.2014.01.016.

Lasne, J., Noblet, A., Szopa, C., Navarro-González, R., Cabane, M., Poch, O., Stalport, F., François, P., Atreya, S.K., Coll, P., 2016. Oxidants at the surface of Mars: a review in light of recent exploration results. Astrobiology 16 (12), 977–996. https://doi.org/10.1089/ast.2016.1502.

Levin, G.V., Straat, P.A., 2016. The case for extant life on Mars and its possible detection by the viking labeled release experiment. Astrobiology 16(10). https://doi.org/10.1089/ast.2015.1464.

Li, X., Danell, R.M., Brinckerhoff, W.B., Pinnick, V.T., van Amerom, F., Arevalo, R.D., Getty, S.A., Mahaffy, P.R., Steininger, H., Goesmann, F., 2015. Detection of trace organics in Mars analog samples containing perchlorate by laser desorption/ionization mass spectrometry. Astrobiology 15 (2), 104–110. https://doi.org/10.1089/ast.2014.1203.

Loizeau, D., Mangold, N., Poulet, F., Bibring, J.-P., Gendrin, A., Ansan, V., Gomez, C., Gondet, B., Langevin, Y., Masson, P., Neukum, G., 2007. Phyllosilicates in the Mawrth Vallis region of Mars. J. Geophys. Res. 112 (E8), E08S08. https://doi.org/10.1029/2006JE002877.

Loizeau, D., Mangold, N., Poulet, F., Ansan, V., Hauber, E., Bibring, J.-P., Gondet, B., Langevin, Y., Masson, P., Neukum, G., 2010. Stratigraphy in the Mawrth Vallis region through OMEGA, HRSC color imagery and DTM. Icarus 205 (2), 396–418. https://doi.org/10.1016/j.icarus.2009.04.018.

Loizeau, D., Werner, S.C., Mangold, N., Bibring, J.-P., Vago, J.L., 2012. Chronology of deposition and alteration in the Mawrth Vallis region, Mars. Planet. Space Sci. 72 (1), 31–43. https://doi.org/10.1016/j.pss.2012.06.023.

Lopez-Reyes, G., 2015. Development of Algorithms and Methodological Analyses for the Definition of the Operation Mode of the Raman Laser Spectrometer Instrument. Universidad de Valladolid, Valladolid.

Magnani, P., Re, E., Fumagalli, A., Senese, S., Ori, G.G., Gily, A., Baglioni, P., 2011. In: Testing of Exo-Mars EM drill tool in Mars analogous materials. Proceedings Advanced Space Technologies for Robotics and Automation (ASTRA). European Space Agency, Noordwijk.

Maimone, M., Cheng, Y., Matthies, L., 2007. Two years of visual odometry on the Mars Exploration Rovers. J. Field Rob. 24 (3), 169–186. https://doi.org/10.1002/rob.20184.

Malin, M.C., Edgett, K.S., 2000. Sedimentary rocks of early Mars. Science 290 (5498), 1927–1937. https://doi.org/10.1126/science.290.5498.1927.

McKay, C.P., 2010. An origin of life on Mars. Cold Spring Harb. Perspect. Biol. 2 (4), a003509. https://doi.org/10.1101/cshperspect.a003509.

Michalski, J.R., Cuadros, J., Niles, P.B., Parnell, J., Deanne Rogers, A., Wright, S.P., 2013a. Groundwater activity on Mars and implications for a deep biosphere. Nat. Geosci. 6 (1), 1–6. https://doi.org/10.1038/ngeo1706.

Michalski, J.R., Niles, P.B., Cuadros, J., Baldridge, A.M., 2013b. Multiple working hypotheses for the formation of compositional stratigraphy on Mars: insights from the Mawrth Vallis region. Icarus 226 (1), 816–840. https://doi.org/10.1016/j.icarus.2013.05.024.

Mitrofanov, I.G., Litvak, M.L., Nikiforov, S.Y., Jun, I., Bobrovnitsky, Y.I., Golovin, D.V., Grebennikov, A.S., Fedosov, F.S., Kozyrev, A.S., Lisov, D.I., Malakhov, A.V., Mokrousov, M.I., Sanin, A.B., Shvetsov, V.N., Timoshenko, G.N., Tomilina, T.M., Tret'yakov, V.I., Vostrukhin, A.A., 2017. The ADRON-RM instrument onboard the ExoMars rover. Astrobiology 17 (6–7), 585–594. https://doi.org/10.1089/ast.2016.1566.

Navarro-González, R., McKay, C.P., 2011. Reply to comment by Biemann and Bada on "Reanalysis of the Viking results suggests perchlorate and organics at midlatitudes on Mars". J. Geophys. Res. 116 (E12), E12002. https://doi.org/10.1029/2011JE003880.

Navarro-Gonzalez, R., Navarro, K.F., de la Rosa, J., Iniguez, E., Molina, P., Miranda, L.D., Morales, P., Cienfuegos, E., Coll, P., Raulin, F., Amils, R., McKay, C.P., 2006. The limitations on organic detection in Mars-like soils by thermal volatilization-gas chromatography-MS and their implications for the Viking results. Proc. Natl. Acad. Sci. 103 (44), 16089–16094. https://doi.org/10.1073/pnas.0604210103.

Navarro-González, R., Vargas, E., de la Rosa, J., Raga, A.C., McKay, C.P., 2010. Reanalysis of the Viking results suggests perchlorate and organics at midlatitudes on Mars. J. Geophys. Res. 115 (E12), E12010. https://doi.org/10.1029/2010JE003599.

Navarro-González, R., Vargas, E., de la Rosa, J., Raga, A.C., McKay, C.P., 2011. Correction to "Reanalysis of the Viking results suggests perchlorate and organics at midlatitudes on Mars". J. Geophys. Res. 116 (E8), E08011. https://doi.org/10.1029/2011JE003854.

Oze, C., Sharma, M., 2005. Have olivine, will gas: serpentinization and the abiogenic production of methane on Mars. Geophys. Res. Lett. 32 (10), L10203. https://doi.org/10.1029/2005GL022691.

Parnell, J., Lindgren, P., 2006. In: The processing of organic matter in impact craters: implications for the exploration for life. 40th ESLAB Proceedings. European Space Agency, Noordwijk, pp. 147–152.

Parnell, J., Cullen, D., Sims, M.R., Bowden, S., Cockell, C.S., Court, R., Ehrenfreund, P., Gaubert, F., Grant, W., Parro, V., Rohmer, M., Sephton, M., Stan-Lotter, H., Steele, A., Toporski, J., Vago, J., 2007. Searching for life on Mars: selection of molecular targets for ESA's aurora ExoMars mission. Astrobiology 7 (4), 578–604. https://doi.org/10.1089/ast.2006.0110.

Patel, N., Slade, R., Clemmet, J., 2010. The ExoMars rover locomotion subsystem. J. Terrramech. 47 (4), 227–242. https://doi.org/10.1016/j.jterra.2010.02.004.

Pavlov, A.A., Kasting, J.F., Brown, L.L., Rages, K.A., Freedman, R., 2000. Greenhouse warming by CH 4 in the atmosphere of early Earth. J. Geophys. Res. Planets 105 (E5), 11981–11990. https://doi.org/10.1029/1999JE001134.

Pavlov, A.A., Vasilyev, G., Ostryakov, V.M., Pavlov, A.K., Mahaffy, P., 2012. Degradation of the organic molecules in the shallow subsurface of Mars due to irradiation by cosmic rays. Geophys. Res. Lett. 39 (13), L13202. https://doi.org/10.1029/2012GL052166.

Pilorget, C., Bibring, J.-P., 2013. NIR reflectance hyperspectral microscopy for planetary science: Application to the MicrOmega instrument. Planet. Space Sci. 76 (2013), 42–52. https://doi.org/10.1016/j.pss.2012.11.004.

Poulakis, P., Vago, J.L., Loizeau, D., Vicente-Arevalo, C., Hutton, A., McCoubrey, R., Arnedo-Rodriguez, J., Smith, J., Boyes, B., Jessen, S., Otero-Rubio, A., Durrant, S., Gould, G., Joudrier, L., Yushtein, Y., Alary, C., Zekri, E., Baglioni, P., Cernusco, A., Maggioni, F., Yague, R., Ravera, F., 2016. In: Overview and development status of the ExoMars rover mobility system. Proceedings—ASTRA 2015. European Space Agency, p. 8. Available at http://robotics.estec.esa.int/ASTRA/Astra2015/Papers/Session 1A/96038_Poulakis.pdf.

Poulet, F., Bibring, J.-P., Mustard, J.F., Gendrin, A., Mangold, N., Langevin, Y., Arvidson, R.E., Gondet, B., Gomez, C., Berthé, M., Bibring, J.-P., Langevin, Y., Erard, S., Forni, O., Gendrin, A., Gondet, B., Manaud, N., Poulet, F., Poulleau, G., Soufflot, A., Combes, M., Drossart, P., Encrenaz, T., Fouchet, T., Melchiorri, R., Bellucci, G., Altieri, F., Formisano, V., Fonti, S., Capaccioni, F., Cerroni, P., Coradini, A., Korablev, O., Kottsov, V., Ignatiev, N., Titov, D., Zasova, L., Mangold, N., Pinet, P., Schmitt, B., Sotin, C., Hauber, E., Hoffmann, H., Jaumann, R., Keller, U., Arvidson, R., Mustard, J., Forget, F., 2005. Phyllosilicates on Mars and implications for early martian climate. Nature 438 (7068), 623–627. https://doi.org/10.1038/nature04274.

Poulet, F., Carter, J., Bishop, J.L., Loizeau, D., Murchie, S.M., 2014. Mineral abundances at the final four curiosity study sites and implications for their formation. Icarus 231, 65–76. https://doi.org/10.1016/j.icarus.2013.11.023.

Quantin, C., Carter, J., Thollot, P., Broyer, J., Lozach, L., Davis, J., Grindrod, P., Pajola, M., Baratti, E., Rossato, S., Allemand, P., Bultel, B., Leyrat, C., Fernando, J., Ody, A., 2016. In: Oxia Planum, the Landing Site for ExoMars 2018. Abstract 2863, 47th Lunar and Planetary Science Conference. Lunar and Planetary Institute, Houston, TX.

Quinn, R.C., Martucci, H.F.H., Miller, S.R., Bryson, C.E., Grunthaner, F.J., Grunthaner, P.J., 2013. Perchlorate radiolysis on Mars and the origin of Martian soil reactivity. Astrobiology 13 (6), 515–520. https://doi.org/10.1089/ast.2013.0999.

Rathbun, J.A., Squyres, S.W., 2002. Hydrothermal systems associated with Martian impact craters. Icarus 157 (2), 362–372. https://doi.org/10.1006/icar.2002.6838.

Ruff, S.W., Farmer, J.D., 2016. Silica deposits on Mars with features resembling hot spring biosignatures at El Tatio in Chile. Nat. Commun. 7, 13554. https://doi.org/10.1038/ncomms13554.

Rull, F., Maurice, S., Hutchinson, I., Moral, A., Perez, C., Diaz, C., Colombo, M., Belenguer, T., Lopez-Reyes, G., Sansano, A., Forni, O., Parot, Y., Striebig, N., Woodward, S., Howe, C., Tarcea, N., Rodriguez, P., Seoane, L., Santiago, A., Rodriguez-Prieto, J.A., Medina, J., Gallego, P., Canchal, R., Santamaría, P., Ramos, G., Vago, J.L., on behalf of the RLS Team, 2017. The Raman laser spectrometer for the ExoMars rover mission to Mars. Astrobiology 17 (6–7), 627–654. https://doi.org/10.1089/ast.2016.1567.

Russell, M.J., Barge, L.M., Bhartia, R., Bocanegra, D., Bracher, P.J., Branscomb, E., Kidd, R., McGlynn, S., Meier, D.H., Nitschke, W., Shibuya, T., Vance, S., White, L., Kanik, I., 2014. The drive to life on wet and icy worlds. Astrobiology 14 (4), 308–343. https://doi.org/10.1089/ast.2013.1110.

Schulte, M., Blake, D., Hoehler, T., McCollom, T., 2006. Serpentinization and its implications for life on the early earth and Mars. Astrobiology 6 (2), 364–376. https://doi.org/10.1089/ast.2006.6.364.

Sephton, M.A., Lewis, J.M.T., Watson, J.S., Montgomery, W., Garnier, C., 2014. Perchlorate-induced combustion of organic matter with variable molecular weights: implications for Mars missions. Geophys. Res. Lett. 41 (21), 7453–7460. https://doi.org/10.1002/2014GL062109.

Sherwood Lollar, B., Lacrampe-Couloume, G., Slater, G.F., Ward, J., Moser, D.P., Gihring, T.M., Lin, L.-H., Onstott, T.C., 2006. Unravelling abiogenic and biogenic sources of methane in the Earth's deep subsurface. Chem. Geol. 226 (3–4), 328–339. https://doi.org/10.1016/j.chemgeo.2005.09.027.

Siljeström, S., Freissinet, C., Goesmann, F., Steininger, H., Goetz, W., Steele, A., Amundsen, H., 2014. Comparison of prototype and laboratory experiments on MOMA GCMS: results from the AMASE11 campaign. Astrobiology 14 (9), 780–797. https://doi.org/10.1089/ast.2014.1197.

Squyres, S.W., Arvidson, R.E., Bell, J.F., Brückner, J., Cabrol, N.A., Calvin, W., Carr, M.H., Christensen, P.R., Clark, B.C., Crumpler, L., Des Marais, D.J., D'Uston, C., Economou, T., Farmer, J., Farrand, W., Folkner, W., Golombek, M., Gorevan, S., Grant, J.A., Greeley, R., Grotzinger, J., Haskin, L., Herkenhoff, K.E., Hviid, S., Johnson, J., Klingelhöfer, G., Knoll, A., Landis, G., Lemmon, M., Li, R., Madsen, M.B., Malin, M.C., McLennan, S.M., McSween, H.Y., Ming, D.W., Moersch, J., Morris, R.V., Parker, T., Rice, J.W., Richter, L., Rieder, R., Sims, M., Smith, M., Smith, P., Soderblom, L.A., Sullivan, R., Wänke, H., Wdowiak, T., Wolff, M., Yen, A., 2004a. The Spirit Rover's Athena science investigation at Gusev Crater, Mars. Science 305 (5685), 794–799. https://doi.org/10.1126/science.1100194.

Squyres, S.W., Arvidson, R.E., Bell, J.F., Brückner, J., Cabrol, N.A., Calvin, W., Carr, M.H., Christensen, P.R., Clark, B.C., Crumpler, L., Des Marais, D.J., D'Uston, C., Economou, T., Farmer, J., Farrand, W., Folkner, W., Golombek, M., Gorevan, S., Grant, J.A., Greeley, R., Grotzinger, J., Haskin, L., Herkenhoff, K.E., Hviid, S., Johnson, J., Klingelhöfer, G., Knoll, A.H., Landis, G., Lemmon, M., Li, R., Madsen, M.B., Malin, M.C., McLennan, S.M., McSween, H.Y., Ming, D.W., Moersch, J., Morris, R.V., Parker, T., Rice, J.W., Richter, L., Rieder, R., Sims, M., Smith, M., Smith, P., Soderblom, L.A., Sullivan, R., Wänke, H., Wdowiak, T., Wolff, M., Yen, A., 2004b. The Opportunity Rover's Athena science investigation at Meridiani Planum, Mars. Science 306 (5702), 1698–1703. https://doi.org/10.1126/science.1106171.

Squyres, S.W., Arvidson, R.E., Bell, J.F., Calef, F., Clark, B.C., Cohen, B.A., Crumpler, L.A., de Souza, P.A., Farrand, W.H., Gellert, R., Grant, J., Herkenhoff, K.E., Hurowitz, J.A., Johnson, J.R., Jolliff, B.L., Knoll, A.H., Li, R., McLennan, S.M., Ming, D.W., Mittlefehldt, D.W., Parker, T.J., Paulsen, G., Rice, M.S., Ruff, S.W., Schroder, C., Yen, A.S., Zacny, K., 2012. Ancient impact and aqueous processes at Endeavour crater, Mars. Science 336 (6081), 570–576. https://doi.org/10.1126/science.1220476.

Steininger, H., Goesmann, F., Goetz, W., 2012. Influence of magnesium perchlorate on the pyrolysis of organic compounds in Mars analogue soils. Planet. Space Sci. 71 (1), 9–17. https://doi.org/10.1016/j.pss.2012.06.015.

Summons, R.E., Albrecht, P., McDonald, G., Moldowan, J.M., 2008. Molecular biosignatures. Space Sci. Rev. 135 (1–4), 133–159. https://doi.org/10.1007/s11214-007-9256-5.

Summons, R.E., Amend, J.P., Bish, D., Buick, R., Cody, G.D., Des Marais, D.J., Dromart, G., Eigenbrode, J.L., Knoll, A.H., Sumner, D.Y., 2011. Preservation of Martian organic and environmental records: final report of the Mars Biosignature Working Group. Astrobiology 11 (2), 157–181. https://doi.org/10.1089/ast.2010.0506.

Tissot, B.P., Welte, D.H., 1984. Petroleum Formation and Occurrence. Springer Berlin Heidelberg, Berlin, Heidelberg, p. 702. https://doi.org/10.1007/978-3-642-87813-8.

Vago, J.L., Lorenzoni, L., Calantropio, F., Zashchirinskiy, A.M., 2015. Selecting a landing site for the ExoMars 2018 mission. Sol. Syst. Res. 49 (7), 538–542. https://doi.org/10.1134/S0038094615070205.

Vago, J.L., Westall, F., Coates, A.J., Jaumann, R., Korablev, O., Ciarletti, V., Mitrofanov, I., Josset, J.-L., De Sanctis, M.C., Bibring, J.-P., Rull, F., Goesmann, F., Steininger, H., Goetz, W., Brinckerhoff, W., Szopa, C., Raulin, F., Westall, F., Edwards, H.G.M., Whyte, L.G., Fairén, A.G., Bibring, J.-P., Bridges, J., Hauber, E., Ori, G.G., Werner, S., Loizeau, D., Kuzmin, R.O., Williams, R.M.E., Flahaut, J., Forget, F., Vago, J.L., Rodionov, D., Korablev, O., Svedhem, H., Sefton-Nash, E., Kminek, G., Lorenzoni, L., Joudrier, L., Mikhailov, V., Zashchirinskiy, A., Alexashkin, S., Calantropio, F., Merlo, A., Poulakis, P., Witasse, O., Bayle, O., Bayón, S., Meierhenrich, U., Carter, J., García-Ruiz, J.M., Baglioni, P., Haldemann, A., Ball, A.J., Debus, A., Lindner, R., Haessig, F., Monteiro, D., Trautner, R., Voland, C., Rebeyre, P., Goulty, D., Didot, F., Durrant, S., Zekri, E., Koschny, D., Toni, A., Visentin, G., Zwick, M., van Winnendael, M., Azkarate, M., Carreau, C., Team, the E. P., 2017. Habitability on early Mars and the search for biosignatures with the ExoMars rover. Astrobiology 17 (6–7), 471–510. https://doi.org/10.1089/ast.2016.1533.

van Thienen, P., Vlaar, N., van den Berg, A., 2004. Plate tectonics on the terrestrial planets. Phys. Earth Planet. Inter. 142 (1–2), 61–74. https://doi.org/10.1016/j.pepi.2003.12.008.

Warner, N.H., Farmer, J.D., 2010. Subglacial hydrothermal alteration minerals in Jökulhlaup deposits of southern Iceland, with implications for detecting past or present habitable environments on Mars. Astrobiology 10 (5), 523–547. https://doi.org/10.1089/ast.2009.0425.

Westall, F., 2008. Morphological biosignatures in early terrestrial and extraterrestrial materials. Space Sci. Rev. 135 (1–4), 95–114. https://doi.org/10.1007/s11214-008-9354-z.

Westall, F., Loizeau, D., Foucher, F., Bost, N., Betrand, M., Vago, J., Kminek, G., 2013. Habitability on Mars from a microbial point of view. Astrobiology 13 (9), 887–897. https://doi.org/10.1089/ast.2013.1000.

Westall, F., Foucher, F., Bost, N., Bertrand, M., Loizeau, D., Vago, J.L., Kminek, G., Gaboyer, F., Campbell, K.A., Bréhéret, J.-G., Gautret, P., Cockell, C.S., 2015. Biosignatures on Mars: what, where, and how? Implications for the search for Martian life. Astrobiology 15 (11), 998–1029. https://doi.org/10.1089/ast.2015.1374.

Yuen, P., Gao, Y., Griffiths, A., Coates, A., Muller, J.-P., Smith, A., Walton, D., Leff, C., Hancock, B., Shin, D., 2013. ExoMars rover PanCam: autonomous and computational intelligence. IEEE Comput. Intell. Mag. 8 (4), 52–61. https://doi.org/10.1109/MCI.2013.2279561.

Yung, Y.L., Chen, P., 2015. Methane on Mars. J. Astrobiol. Outreach 3 (1), 3–5. https://doi.org/10.4172/2332-2519.1000125.

Yung, Y.L., Russell, M.J., Parkinson, C.D., 2010. The search for life on Mars. J. Cosmol. 5, 1121–1130.

CHAPTER 13

Concluding Remarks: Bridging Strategic Knowledge Gaps in the Search for Biosignatures on Mars—A Blueprint ☆

Nathalie A. Cabrol, Edmond A. Grin, Janice L. Bishop, Sherry L. Cady, Nancy W. Hinman, Jeffrey Moersch, Nora Noffke, Cynthia Phillips, Pablo Sobron, David Summers, Kimberley Warren-Rhodes, David S. Wettergreen

Contents

13.1 OVERVIEW: THE CURRENT CHALLENGE

Searching for biosignatures on Mars is changing the focus of exploration from the characterization of habitability to the investigation of a potential coevolution, that is, the spatiotemporal interactions of life with its environment. Yet, the intellectual framework underpinning the preparation of Mars 2020 and ExoMars along with future life-seeking missions is essentially the same as the one that has guided the exploration of Mars for the past 15 years (e.g., Mustard et al., 2013; Hays et al., 2017; Cabrol, 2018); see also Chapters 11 and 12. This framework is articulated around the terrestrial analogy principle of habitability. While this principle is helpful in characterizing Mars habitability potential over time, it is limiting and potentially misleading for the exploration of biosignatures as its focus is primarily on the spatiotemporal dynamics of environmental factors.

☆ Parts of this chapter's content were submitted as a white paper to the NASA Astrobiology Program on 8 January 2018 in preparation for and as an input to the next decadal surveys in astronomy, astrophysics, and planetary sciences by the National Academies of Sciences, Engineering, and Medicine.

From Habitability to Life on Mars
https://doi.org/10.1016/B978-0-12-809935-3.00014-1

On the other hand, coevolution synergistically considers both life and environment, and how they modify each other. In that, it is a more effective, systemic, and dynamic approach than habitability alone for understanding how to detect, identify, and characterize past and present microbial habitats and biosignatures. As a result, new paths of investigations are necessary to advance our understanding of plausible coevolution models on early Mars and to support biosignature exploration. They include (1) revisiting intellectual frameworks, theories, hypotheses, and science questions from a coevolutionary perspective; (2) injecting an ecosystem view at all levels of biosignature exploration (Cabrol, 2018), that is, spatiotemporal scales, spectral resolution, orbit-to-ground detection and identification thresholds (Phillips et al., 2017), landing site selection, and exploration strategies; and (3) designing and deploying new mission concepts to gain a high-resolution view of environmental variability at scales that are relevant to past and present martian microbial habitats.

13.2 HISTORIC PERSPECTIVE ON COEVOLUTION

The idea that life and environment are interdependent is a modern approach to evolution. Before Darwin, uniformitarianism (e.g., Hutton, 1788, 1795; Lyell, 1830–1833) and catastrophism (e.g., Cuvier, 1818) offered different and opposite evolutionary views. For historic perspectives, see Hooykaas (1970) and Baker (1998). The former proposed that species were adapted to slow environmental changes over time, while the latter supported the idea that the environment was transformed through sudden and violent events and species killed off and replaced. Darwin (1859) understood that change was permanent and life is either adapted or disappeared. The concept whereby biology, geology, and chemistry constantly and mutually transformed each other, which is foundational to astrobiology, to biogeology, and to the notion of biosphere, can be traced to Vernadsky (1998). At its extreme with the Gaia hypothesis, coevolution is viewed as a self-regulating system involving biosphere, atmosphere, hydrosphere, and pedosphere tightly coupled as an evolving system (Lovelock and Margulis, 1974), where homeostatic balance is actively pursued to maintain optimal conditions for life. Regarded by some as a scientific theory that passed predictive tests (e.g., Lovelock, 2007), the Gaia hypothesis remains controversial (e.g., Kirchner, 2002).

Today, coevolution is a pivotal theme in astrobiology's three primary questions (e.g., Des Marais et al., 2008; Achenbach et al., 2015):

With *how does life begin and evolve?* It introduces the notions of habitable zone, habitable environments, and habitats and how primary physicochemical conditions on a planet drive the range of plausible biochemistries and may allow prebiotic chemistry to transition to biology (e.g., Pross, 2012). Earth is the only planet where coevolution has been demonstrated so far, and it shows that from the time life first developed, environmental changes have accompanied its evolution as either causes or effects (Knoll, 2009).

With *is there life beyond Earth, and if so, how can we detect it?* Our planet serves as a guide to inform the potential for other models of coevolution within and outside the solar system through the study of planetary analogs and modeling.

With *what is the future of life on Earth and in the universe?* It compels us to fully grasp its systemic implications at a time when human activity has upset the delicate balance of environmental conditions that led to its emergence, which globally threatens environment and biodiversity. It also drives us to reflect upon the possible generational aspect of life in the universe, one that could be linked to elements produced by the successive generations of stars since the Big Bang. Ultimately, it questions what type of planetary environments and biochemistries could emerge in the distant future (and where), as well as the role and possible duration of the biological process in the evolution of the universe.

Envisioned broadly, this synergy may range from sterile worlds where no life ever developed and environmental processes exclusively dominate, to others that are "bio-obvious" like the Earth, where environment and life have shaped each other almost in equal terms over time. On Earth, the signatures of this coevolution are visible everywhere, including from space in the modification of the landscape, the spectral signatures of biomediated minerals, or the technological footprint of the human species. As life evolves and biomass increases, this biofootprint becomes more visible and diversified. These are end-members.

The exploration of the solar system suggests that many worlds may have experienced extreme environmental conditions, potentially constraining life as we know it for most of their history. Such conditions may control life's ability to develop and limit its expansion and complexity and its environmental impact to specific sheltered niches that may only have very subtle to no surface expression. Finding those niches requires an in-depth knowledge of the evolution of each world to understand where to start searching and at what scale and resolution exploration should be taking place.

In the past decades, synergies between Earth, space and life sciences, and planetary exploration have given us a deeper understanding of the spatiotemporal scales associated with this coevolution and the various types of changes that may be driven, for instance, by fluctuations in solar activity (e.g., Beer et al., 2006; Fröhlich, 2009); the astronomical characteristics of planets (obliquity, eccentricity, and precession), which determine climate cycles over tens of thousands to millions of years (e.g., Laskar et al., 2002); the presence, size, and stability of moons; geologic and atmospheric changes; catastrophic events (e.g., cosmic, solar, and environmental); and other stochastic events of biological or environmental nature (e.g., gene mutations, asteroid impacts, volcanic eruptions, and solar and cosmic events). Regardless of scale, changes reverberate throughout the entire system with transient or lasting impact and with subtle or deeply transformative effects.

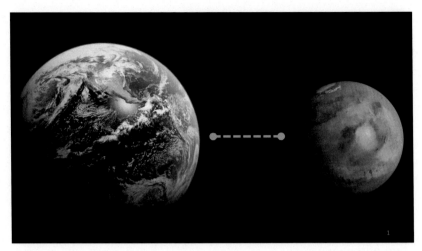

Fig. 13.1 While the Earth is "bio-obvious" even from space, Mars does not show any signs that biology ever had a significant impact on surface processes. If it did, it is at a resolution yet to be reached or in ways we do not understand yet, and this alone is a critical piece of information for biosignature exploration.

In that context, Mars is important for comparing the effects of these various spatiotemporal changes on an earthlike, albeit different, planet of the solar system. In the case of Mars, the early climate change would have led to a strong, complex, and broadscale physical control of habitats, which may have critically impacted all aspects of the ecosystem and biological evolution (Cabrol, 2018), biodiversity, biosignature formation, and the preservation of a biological record (Fig. 13.1). Here, the study of terrestrial analogs in extreme environments is essential to envision the types of signatures that may be present on Mars—and identify where to search for them, guide the development of adapted exploration strategies, and mitigate current data gaps to better support upcoming missions. They also suggest a necessary change in intellectual framework adapted to the search for biosignatures.

13.3 REVISITING THE INTELLECTUAL FRAMEWORK: COEVOLUTION AS A GUIDING EXPLORATION PRINCIPLE

Biological processes on Mars, if any, would have taken place within the distinct context of an irreversible early collapse of the magnetosphere and atmosphere (Ehlmann et al., 2016; Wordsworth, 2016; Jakosky et al., 2017); greater climate variability and gradients; and specific geographic, planetary, and astronomical characteristics. These comprised the unique factors of a coevolution that would have separated a martian biosphere from that of Earth very early. To evaluate their full effect on biosignatures, these factors should be envisioned within an intellectual framework that includes life as an interactive agent of transformation of its environment and a piece of a dynamic system of polyextreme environmental conditions with complex loops and feedback mechanisms (i.e., an ecosystem).

The concept of habitability currently driving exploration defines the environmental range (e.g., astronomical and planetary) within which life as we know it could survive. In this definition, life is regarded as a passive actor in an environment that provides (or not) water, energy, nutrients, carbon sources, and shelter for prebiotic and biological processes. In itself, the definition of habitability does not imply life; it simply considers environmental conditions for its emergence and sustainability.

The habitability of early Mars has now been demonstrated by 20 years of orbital and landed missions (e.g., Knoll et al., 2005; Des Marais, 2010; Summons et al., 2011; Grotzinger et al., 2015). Organic molecules have been detected (e.g., Chastain and Chevrier, 2007; Mumma et al., 2009; Blamey et al., 2015; Freissinet et al., 2015; Webster et al., 2015). The upcoming missions will test the hypothesis that life has developed on Mars and left evidence of its presence. Testing this hypothesis requires to search for traces left by two dynamic agents (life and environment) that possibly modified each other for over 4 billion years (e.g., Knoll, 2009). As Earth shows, coevolution affects physicochemical, geochemical, and biological processes at all scales, including the biological architecture, metabolic activity, morphology, the mineralogy and texture of soils and sediments, topography, the atmosphere, microbial habitats, biological dispersal, biomass production and repositories, and biosignature preservation. Coevolution is, therefore, a concept essential to biosignature exploration, one that allows core hypotheses and science questions to be reframed on the basis of plausible spatiotemporal synergies between life and environment and importantly to refine our understanding of the instruments, methods, and spatial scales and spectral resolution necessary to identify biosignatures. Examples include the following:

Hypothesis A   Prebiotic Chemistry and Life as We Know It
Example questions: (1) What role did environmental differences between Earth and Mars play in an early evolution of life on Mars? (2) What was the impact of unique physical features (e.g., global dichotomy, high obliquity, lost magnetosphere and atmosphere, and volcanic and tectonic characteristics) on the formation and spatiotemporal evolution of environmental pathways for biological dispersal and biomass/biosignature repositories? (3) What does the comparison between the timing of early-life evolution on Earth and the current environmental models for early Mars suggest about ancient habitable environments, habitat development potential, biological dispersal, biosignature preservation, and detection thresholds? (4) What does the lack of obvious biosignatures at current resolution suggest about (a) the extent and duration of subaerial habitats, biomass accumulation, and preservation potential and (b) the detection and identification thresholds of integrated instrument payloads required from orbit to the ground?

Hypothesis B   Second Genesis
Example questions: (1) What distinct biological traits (e.g., metabolism, structure, size, and biogeochemical cycles) could have evolved from the unique terms of a martian coevolution (astronomical, planetary, environmental, geographic, climatic, and others), and (2)

what distinct traces of coevolution could they have left in the geologic or spectral records? For instance, how can existing datasets be searched for unique geochemical, mineralogical, textural, and biochemical markers that could have stemmed from life's adaptation to the martian polyextreme environment?

Hypothesis C   Coevolution Both From Planetary Transfer (Earth) and In Situ Abiogenesis
Example questions are comparable with those of hypotheses A and B.

Hypothesis D   No Coevolution
Example questions: What are the critical exploratory steps to complete at the surface, subsurface, and deep underground (and where), before such a conclusion can be reached?

13.4 UNDERSTANDING COEVOLUTION IN A POLYEXTREME ENVIRONMENT

A martian coevolution would have been imprinted early by the development of a polyextreme environment (Cabrol, 2018; Cabrol et al., 2007a,b). While the current approach to biosignature exploration considers multiple extreme factors, it often analyzes their impact individually, with limited attempts at a systemic approach, that is, the characterization of these interactions and their effects (e.g., Jakosky and Phillips, 2001; Kreslavsky and Head, 2005; Atri et al., 2013), which leads to two major knowledge gaps: unknown interaction of multiple environmental extremes and unknown metabolic pathways and their adaptation.

We currently lack an understanding of the spatiotemporal interplay of environmental polyextremes, their resulting interactions with biological processes, and the resulting biogeosignatures. Terrestrial analogs of such environments demonstrate that interactions between multiple extreme environmental factors (e.g., UV radiation, thin atmosphere, and aridity) generate complex loops and feedback mechanisms at various scales through combinations that may alternatively either magnify, decrease, and/or cancel their individual effects and often override global (planetary) trends at the scale of microbial habitats (e.g., Cabrol et al., 2007a,b; Cabrol, 2018).

At global to regional scale, the unique complexity of Mars—including in its early geologic history—resided in the relative dominance of these polyextreme factors over space and time. Some parameters declined with time (magnetosphere, atmosphere, and energy), while others had distinct spatiotemporal effects depending on obliquity (e.g., water and ice distribution). For example, while the loss of the atmosphere was ultimately linked to the loss of the magnetic field, weak fields play a lesser role in surface radiation doses than the loss of the atmospheric depth (e.g., Atri et al., 2013). Changes in atmospheric shielding were therefore not only a factor of time but also a factor of obliquity (e.g., Jakosky and Phillips, 2001; Kreslavsky and Head, 2005), and this unpredictability in the radiation environment was only one of many variables (e.g., changes in temperature, desiccation, geochemistry and sediment texture, and acidity) living organisms would have had to contend with. Understanding how this variability affected prebiotic and biological processes, as well as the development and footprint of microbial habitats, is critical for evaluating plausible biomass production, potential biosignature formation and preservation, and appropriate detection levels for instruments.

At local (habitat) scale, the footprint and sustainability of microbial habitats in terrestrial analogs in extreme environments depend on microclimates generated by synergies between microbial (metabolic) activity and local environmental factors, which trigger unique loops and feedback mechanisms. Changing environmental conditions would have thus affected habitats in a systemic way, with the modifications and/or loss in connectivity networks, formation and isolation of microniches, and production of very localized and specific sets of ecosystem conditions.

We also do not have a clear understanding of the plausible metabolic pathways and responses to variable polyextreme environmental factors that would have been necessary to martian microbial organisms to adapt in changing subaerial habitats over time, their spatiotemporal distribution, and biosignature production and preservation potential.

13.5 EXPLORING COEVOLUTION

Assuming that a coevolution was initiated on Mars, the upcoming missions will have only limited (contextual) support at this point to search for biosignatures (Fig. 13.2).

Data at relevant spatial scales and spectral resolution are only available at the three rover landing sites, and unless a mission returns to one of them (Ruff and Farmer, 2016), knowledge acquired at these sites will be only partially transferable to exploration of a new site, that is, only if sets of environmental conditions are repeated at a habitat-relevant scale, for example, sediment mineralogy, geochemistry, texture, structure, insolation, slope, and moisture (Cabrol, 2018).

Current knowledge gaps will not be filled by the time the Mars 2020 and ExoMars missions launch. However, significant advances can be made in the coming decade and support provided to upcoming and future missions through data analysis, theoretical modeling, machine learning, lab experiments, and fieldwork. Examples of what could be accomplished include the following:

13.5.1 Investigating Loops and Feedback Mechanisms in Polyextreme Environments

Mars' ability to preserve subaerial habitats, ecotones, connectivity networks, and microbial dispersal pathways over time would have depended on fluctuating interactions between multiple environmental extremes and their relative dominance at any given

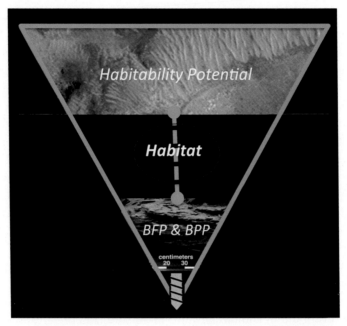

Fig. 13.2 Current orbital spatial scales and spectral resolution are adapted to the characterization of broad planetary environmental habitability, not that of a coevolution and the detection of habitats. Critical intermediate detection and identification thresholds are needed to access geosignature suites associated with potential habitats (and possibly ancient biomediated mineral deposits) and understand environmental parameters that matter at the *habitat level*, which can change sediment origin, texture, mineralogy, and biosignature preservation potential (see also Phillips et al., 2017; Cabrol, 2018). Acronyms: *BFP*, biosignature formation potential; *BPP*, biosignature preservation potential.

time (Cabrol, 2018). This relative dominance would have impacted the interactions between life and environment and the spatiotemporal nature (e.g., distribution, type, biochemistry, geochemistry, and mineralogy) of biosignatures.

Relative dominance must be thus characterized over geologic timescales and with changing obliquities, including along a depth gradient. This can be accomplished by (a) conducting lab experiments and fieldwork in extreme environments that combine multiple extreme factors relevant to Mars. Emphasizing the characterization of their interactions and their effects on prebiotic, biological processes, and microbial habitats should be prioritized; (b) developing libraries of biogeosignatures resulting from these interactions (e.g., spectral, morphological, metabolic, and genomic). These libraries should be generated at integrated scales from orbit to ground to lab; (c) characterizing biosignature formation through the lens of polyextreme environmental factors and their role on local scale microclimates, including the characterization of microbial habitats (e.g., geology, morphology, mineralogy, sediment texture, structure, and composition) and their preservation potential; and (d) developing theoretical modeling using datasets from past and present missions to support the quantitative and qualitative characterization of the spatiotemporal evolution of polyextreme interactions on Mars, including through episodic changes in obliquity. This characterization should include present-day Mars, which reflects 3.5 billion years of environmental history.

13.5.2 Crossing the Uncertainty Threshold

Crossing the uncertainty threshold, that is, how do we go from suspected biosignature to confirmed biosignature, requires (a) the development of knowledge on how coevolution could have shaped a martian biological architecture (e.g., chemical structure, morphology, size, genetic makeup, and metabolism) and its interactions with and response to a polyextreme environment, (b) the prioritization of observations, and (c) the understanding of when a suite of observations constitutes an unambiguous and definitive confirmation of the presence of life.

Filling the current knowledge gaps in the origin and nature of life, biological architecture, and biosignatures demands the analysis of vast amounts of data from many scientific domains and envisions countless probabilistic occurrences. This is an area where artificial intelligence (AI), specifically machine learning, can provide a critical support for standard lab, field, and theoretical approaches and significantly speed up breakthrough discoveries. For instance, computational modeling could determine the systemic environmental envelop in which to test scenarios for an origin of life (Section 13.5.3) and the spatiotemporal evolution of habitats. Coevolution models for life as we know it can be generated by exploring datasets relative to prebiotic and biotic processes known from early Earth, which can be run through the environmental models. AI and machine learning could accelerate the identification of unique (biogeo)signatures across past and present mission data (orbital, landed, ground-based, and space observations) and foster the discovery of patterns of interactions from biological processes unique to Mars (life as may not know it) with the environment.

13.5.3 Moving Biosignature Exploration Forward: An Ecosystem Approach to Landing Site Selection and Surface Operations

A coevolution approach calls for novel integrated investigation methods and techniques at specific spatial scales, spectral resolution, and detection/identification thresholds relevant to past and present microbial ecosystems. Support may involve to (a) engage microbiologists, geneticists, environmental, extreme environments, and AI specialists early and at all stages including programmatic level, missions (concept and instrument payload design and science teams), and surface operations (exploration templates); (b) develop an integrated suite of missions and instruments that allow the identification of biogeosignatures from the orbit to the ground. This requires a quantum leap in instrument capabilities and the development of novel analytic tools (e.g., Thompson et al., 2011; Tao and Muller, 2016; Candela et al., 2017). This is critical because mission simulations in extreme environments show that current orbital resolution is of limited support for Mars-relevant biosignature detection and because finding evidence of potentially limited and scattered biomass may prove difficult from the ground alone (Cabrol, 2018); and (c) integrate survey techniques developed in microbial ecology into surface operation templates (e.g., Cabrol et al., 2007b; Hock et al., 2007; Warren-Rhodes et al., 2007).

13.6 BEYOND MARS

The study of coevolution and that of the existing strategic knowledge gaps identify promising key research areas, science questions, and technology challenges in the field of astrobiology. While they are presented here in the context of the exploration of Mars, coevolution—along with the questions, hypotheses, and approaches suggested here—may be considered primary guiding principles for the search for life in the solar system and beyond.

ACKNOWLEDGMENT

The concepts and research directions presented here are being developed within the research program of the SETI Institute NAI team, which is supported by the National Aeronautics and Space Administration Astrobiology Institute's Grant No. NNX15BB01A, under the project entitled *Changing Planetary Environment and the Fingerprints of Life*.

REFERENCES

Achenbach, L. et al., 2015. NASA Astrobiology Strategy, L. Hays (ed.), 235 p.

Atri, D., Hariharan, B., Grießmeier, J.M., 2013. Galactic cosmic ray-induced radiation dose on terrestrial exoplanets. Astrobiology 13, 910–919.

Baker, V.R., 1998. Catastrophism and uniformitarianism: logical roots and current relevance in geology. In: Blundell, D.J., Scott, A.C. (Eds.), Lyell: The Past is Key to the Present. vol. 143. Geological Society, London, pp. 171–182. Special Publication.

Beer, J., Vonmoos, M., Muscheler, R., 2006. Solar variability over the past several millennia. Space Sci. Rev. 125, 67–79.

Blamey, N.J.F., Parnell, J., McMahon, S., Mark, D.F., Tomkinson, T., Lee, M., Shivak, J., Izawa, M.R.M., Banerjee, N.R., Flemming, R.L., 2015. Evidence for methane in martian meteorites. Nat. Commun. 6. https://doi.org/10.1038/ncomms8399.

Cabrol, N.A., 2018. The coevolution of life and environment on Mars: an ecosystem perspective on the robotic exploration of biosignatures. Astrobiology. 18(1). https://doi.org/10.1089/ast.2017.1756.

Cabrol, N.A., Grin, E.A., Hock, A.N., 2007a. Mitigation of environmental extremes as a possible indicator of extended habitat sustainability for lakes on early Mars. Proc SPIE vol. 6694. https://doi.org/10.1117/12.731506.

Cabrol, N.A., et al., 2007b. Life in the Atacama: searching for life with rovers (science overview). J. Geophys. Res. Biogeosci. 112. https://doi.org/10.1029/2006JG000298.

Candela, A., Thompson, D., Noe Dobrea, E., Wettergreen, D., 2017. In: Planetary robotic exploration driven by science hypotheses for geologic mapping. IEEE/RSJ International Conference on Intelligent Robots and Systems, Vancouver, BC, Canada, September, 2017.

Chastain, B.K., Chevrier, V., 2007. Methane clathrate hydrates as a potential source for martian atmospheric methane. Planet. Space Sci. 55, 1246–1256.

Cuvier, G., 1818. Essay on the Theory of the Earth. Kirk & Mercein, New York.

Darwin, C., 1859. On the Origin of Species, first ed. John Murray, London.

Des Marais, D.J., 2010. Exploring Mars for evidence of habitable environments and life. Proc. Am. Philos. Soc. 154, 402–421.

Des Marais, D., et al., 2008. The NASA astrobiology roadmap. Astrobiology 8 (4), 715–730.

Ehlmann, B.L., et al., 2016. The sustainability of habitability on terrestrial planets: insights, questions, and needed measurements from Mars for understanding the evolution of Earth-like worlds. J. Geophys. Res. Planets. 121. https://doi.org/10.1002/2016JE005134.

Freissinet, C., et al., 2015. Organic molecules in the Sheepbed Mudstone, Gale Crater, Mars. J. Geophys. Res. Planets 120, 495–514.

Fröhlich, C., 2009. Evidence of a long-term trend in total solar irradiance. Astron. Astrophys. 501, L27–L30.

Grotzinger, J.P., et al., 2015. Deposition, exhumation, and paleoclimate of an ancient lake deposit, Gale Crater, Mars. Science. 350. https://doi.org/10.1126/science.aac7575.

Hays, L.E., Graham, H.V., Des Marais, D.J., Hausrath, E.M., Horgan, B., McCollom, T.M., Parenteau, M.N., Potter-McIntyre, S.L., Williams, A.J., Lynch, K.L., 2017. Biosignature preservation and detection in Mars analog environments. Astrobiology 17, 363–400.

Hock, A.N., et al., 2007. Life in the Atacama: a scoring system for habitability and the robotic exploration for life. J. Geophys. Res. Biogeosci. 112. https://doi.org/10.1029/2006JG000321.

Hooykaas, R., 1970. Catastrophism in geology, its scientific character in relation to actualism and uniformitarianism. Koninklijke Nederlandse kademie van Loefenschappen, afd. Letterkunde, Med. (n.r.) 33 (7), 271–316.

Hutton, J., 1788. *Theory of the Earth*; or an investigation of the laws observable in the composition, dissolution, and restoration of land upon the globe. Trans. R. Soc. Edinb. 1, 209–304.

Hutton, J., 1795. Theory of the Earth with Proofs and Illustrations. Creech, Edinburgh.

Jakosky, B.M., Phillips, R.J., 2001. Mars' volatile and climate history. Nature 412, 237–244.

Jakosky, B.M., Slipski, M., Benna, M., Mahaffy, P., Elrod, M., Yelle, R., Stone, S., Alsaeed, N., 2017. Mars' atmospheric history derived from upper-atmosphere measurements of $^{38}Ar/^{36}Ar$. Science 355, 1408–1410.

Kirchner, J.W., 2002. The Gaia hypothesis: fact, theory, and wishful thinking. Climate Change 52 (4), 391–408.

Knoll, A.H., 2009. The coevolution of life and environments. Rend. Lincei Sci. Fis. Nat. 20, 301–306.

Knoll, A.H., Carr, M., Clark, B., Des Marais, D.J., Farmer, J.D., Fischer, W.W., Grotzinger, J.P., McLennan, S.M., Malin, M., Schröder, C., Squyres, S., Tosca, N.J., Wdowiak, T., 2005. An astrobiological perspective on Meridiani Planum. Earth Planet. Sci. Lett. 240, 179–189.

Kreslavsky, M.A., Head, J.W., 2005. Mars at very low obliquity: atmospheric collapse and the fate of volatiles. Geophys. Res. Lett. 32. https://doi.org/10.1029/2005GL022645.

Laskar, J., Levrard, B., Mustard, J.F., 2002. Orbital forcing of the martian polar layered deposits. Nature 419, 375–377.

Lovelock, J. 2007. The Revenge of Gaia, Basic Books Publisher, ISBN-10: 0465041698, 208 pp.

Lovelock, J.E., Margulis, L., 1974. Atmospheric homeostasis by and for the biosphere. Tellus Ser. A 26 (1–2), 2–10. Stockholm: International Meteorological Institute.

Lyell, C., 1830–1833. Principles of Geology. 3 vols. London, Murray.

Mumma, M.J., Villanueva, G.L., Novak, R.E., Hewagama, T., Bonev, B.P., DiSanti, M.A., Mandell, A.M., Smith, M.D., 2009. Strong release of methane on Mars in northern summer 2003. Science 323, 1041–1045.

Mustard, J., et al., 2013. Appendix to the Report of the Mars 2020 Science Definition Team. Mars Exploration Program Analysis Group (MEPAG), .pp. 155–205. Available from:http://mepag.jpl.nasa.gov/reports/MEP/Mars_2020_SDT_Report_Appendix.pdf.

Phillips, M.S., Moersch, J.E., Cabrol, N.A., 2017. In: Thresholds of detectability for habitable environments in the Altiplano of Chile with implications for Mars exploration [abstract 3373]. Astrobiology Science Conference. Lunar and Planetary Institute, Houston.

Pross, A., 2012. What Is Life? How Chemistry Becomes Biology. Oxford University Press, Oxford, 200 pp.

Ruff, S.W., Farmer, J.D., 2016. Silica deposits on Mars with features resembling hot spring biosignatures at El Tatio in Chile. Nat. Commun. 7, https://doi.org/10.1038/ncomms13554.

Summons, R.E., Amend, J.P., Bish, D., Buick, R., Cody, G.D., Des Marais, D.J., Dromart, G., Eigenbrode, J.L., Knoll, A.H., Sumner, D.Y., 2011. Preservation of martian organic and environmental records: final report of the Mars Biosignature Working Group. Astrobiology 11 (2), 157–181. https://doi.org/10.1089/ast.2010.0506.

Tao, Y., Muller, J.P., 2016. A novel method for surface exploration: super-resolution restoration of Mars repeat-pass orbital imagery. Planet. Space Sci. 121, 104–113.

Thompson, D.R., Wettergreen, D.S., Calderóon Peralta, F.J., 2011. Autonomous science during large-scale robotic survey. J. Field Rob. 28, 542–564.

Vernadsky, V. I. 1998. The Biosphere, Copernicus, 1998th ed., M. A. S. McMenamin (Editor), ISBN-10: 038798268X. Available at: https://publications.copernicus.org.

Warren-Rhodes, K., et al., 2007. Robotic ecological mapping: habitats and the search for life in the Atacama Desert. J. Geophys. Res. Biogeosci. 112. https://doi.org/10.1029/2006JG000301.

Webster, C.R., et al., 2015. Mars methane detection and variability at Gale Crater. Science 347 (6220), 415–417.

Wordsworth, R.D., 2016. The climate of early Mars. Annu. Rev. Earth Planet. Sci. 44. https://doi.org/10.1146/annurev-earth-060115-012355.

INDEX

Note: Page numbers followed by *f* indicate figures, and *t* indicate tables.

Printed in the United States
By Bookmasters